ABHANDLUNGEN
DER AKADEMIE DER WISSENSCHAFTEN DER DDR
Abteilung Mathematik — Naturwissenschaften — Technik

Jahrgang 1977 · Nr. 1 N

Theory of Nonlinear Operators

Constructive Aspects

Proceedings of an international summer school
held at Berlin, GDR
from September 22 to 26, 1975

Herausgegeben von
Professor Dr. sc. Reinhard Kluge

bearbeitet von
Dr. habil. Wolfdietrich Müller

Akademie der Wissenschaften der DDR
Zentralinstitut für Mathematik und Mechanik

Mit 16 Abbildungen

AKADEMIE-VERLAG · BERLIN · 1977

Herausgegeben im Auftrage des Präsidenten
der Akademie der Wissenschaften der DDR
von Vizepräsident Prof. Dr. Heinrich Scheel

Erschienen im Akademie-Verlag, 108 Berlin, Leipziger Str. 3—4

Lizenznummer: 202 · 100/429/77
Gesamtherstellung: VEB Druckhaus „Maxim Gorki", 74 Altenburg
Bestellnummer: 762 300 2 (2001/77/1 N) · LSV 1065
Printed in GDR
DDR 48,— M

Inhaltsverzeichnis

Lectures

Short communications

Preface

From September 22 to 26, 1975, the fourth international summer school "Theory of Nonlinear Operators", with special emphasis on "Constructive Aspects", was held in Berlin at the Central Institute of Mathematics and Mechanics of the Academy of Sciences of the GDR. It continued the summer schools devoted to nonlinear problems organized alternately by the Czechoslovak Academy of Sciences and by the Central Institute in Berlin.

The first summer school took place at Babylon (CSSR) in 1971, the second at Neuendorf/Hiddensee (GDR) in 1972, the third at Stara Lesná (CSSR) in 1974.

The 82 participants of the fourth summer school came from Czechoslovakia, the Federal Republic of Germany, France, Italy, Poland, Switzerland, the USSR and the German Democratic Republic.

In the 22 lectures and 22 short communications special attention was given to qualitative aspects and, in particular, to approximation methods for the solution of various problems.

It is hoped that the 36 papers contained in this volume provide an informative survey of the scientific programme of this summer school and arouse the readers' interest in the summer schools of this kind to come.

Berlin, November 1975

Prof. Dr. sc. Reinhard Kluge
Chairman of the summer school

Lectures

A theory of finite elastic deformations under second and third boundary conditions for general structures

Herbert Beckert, Leipzig

Prof. Dr. E. Hölder to his 75th birthday

In this paper I deal with the second and third boundary value problem of nonlinear elasticity. In accordance with earlier papers of the author concerning the first boundary value problem [1, 2] the starting point of our theory is the second variation of the elastic potential in a generalized sense of E. Trefftz [8] with suitable transition rules for the stresses. We are able to describe elastic deformations in the large rigorously under general conditions whatever. To overcome typical difficulties in case of the second boundary value problem, an $H_{1,2} \ominus V_6$ theory is developed.

Let G be an elastic body with boundary S of class $C^{2+\mu}$, $0 < \mu < 1$, consisting of r closed surfaces, on a part of which, say, $S^1, S^2, \ldots, S^{r'}$, $S'' = S^1 + S^2 + \cdots + S^{r'}$, the displacement vector $\varphi(\sigma) \subset C^{2+\mu}$, $\sigma \in S''$, and along the other surfaces $S' = S - S''$ boundary forces $T(\sigma) \subset C^{1+\mu}$, $\sigma \in S'$ are prescribed. The determination of the corresponding elastic deformation of $G + S$ under additional volume forces $X \subset C^\mu$ to $G^e + S^e$ in the sense that the given data are realized over $G^e + S^e$ leads to the third boundary value problem of nonlinear elasticity and to the second boundary value problem if S'' is empty. We make the convenient assumption that volume and boundary forces, acting on the volume element dx and boundary element $d\sigma$, respectively, do not change by the deformation of these elements.

It is well known that in the theory of nonlinear elasticity mechanical, thermodynamical and geometrical principles alone do not lead rigorously to an adequate intrinsic expression of the elastic potential as a function of displacements and strains as is the case in the linear theory of elasticity. In literature, therefore, we find a lot of different proposals for the elastic potential more or less sufficient for practical purposes. In general it is problematic to work with an universal expression of the elastic potential, when considering far distant states. One has to take notice of the influence of regions of plasticity and regions of fading memory of material even in states of reversibility.

What do we actually know about the increase of the elastic potential

$$U = \int_G F\left(x_j, u, \frac{\partial u_i}{\partial x_j}\right) dx \tag{1}$$

in respect to the undeformed state over $G + S$, when deforming $G + S$ in a stable state over the area $G^\lambda + S^\lambda$

$$x_j^\lambda = x_j + u_j(x_i), \quad x \in G + S,$$

with displacements $u_j(x)$?

First: If the state over $G + S$ is stable, we have by Trefftz stability criterion, following easily from energy considerations:

$$\varDelta U > \varDelta A. \tag{2}$$

Here $\varDelta U$ is the increase of U under arbitrary admissible displacements of the points of $G + S$ and $\varDelta A$ the corresponding amount of work done by the given forces. The linear terms on both sides of (2) must be equal, so we have:

A stable elastic state is described by the solution of the regular variational problem

$$U - A \to \mathrm{Min} \tag{3}$$

under the prescribed conditions. The second variation Q of U is positive because of (2), which proves regularity of (3).

Second: We can deduce the expression of the first variation of (3) in respect to the deformed state over $G^\lambda + S^\lambda$ from mechanical principles:

$$\begin{aligned} V(w) &= \frac{1}{2}\int_{G^\lambda} \sigma^\lambda_{\mu\nu}(x^\lambda)\,\gamma_{\mu\nu}\,dx^\lambda = \int_{G^\lambda} \lambda X_\nu w_\nu\,dx^\lambda - \int_{S^\lambda} \lambda T_\nu w_\nu\,d\sigma^\lambda; \\ \gamma_{\mu\nu} &= \frac{\partial w_\mu}{\partial x_\nu{}^\lambda} + \frac{\partial w_\nu}{\partial x_\mu{}^\lambda}. \end{aligned} \tag{4}$$

$\sigma^\lambda_{\mu\nu}(x)$ defines the stress tensor over $G^\lambda + S^\lambda$, and $w(x^\lambda)$ are arbitrary admissible displacements.

Third: In his paper [8] E. Trefftz derived the explicit expression

$$Q(w, w) = \frac{1}{2}\int_{G^\lambda} \sigma^\lambda_{\mu\nu}(x^\lambda)\,\frac{\partial w_h}{\partial x_\mu{}^\lambda}\,\frac{\partial w_h}{\partial x_\nu{}^\lambda} + 2a\,dx^\lambda \tag{5}$$

for the second variation of (3) over G^λ, under the assumption that the stress distribution $k_{\mu\nu}(x^{\lambda'})$ over $G^{\lambda'}$ after:

$$G^\lambda \to G^{\lambda'}: \qquad x^\lambda + w(x^\lambda) = x^{\lambda'}$$

may be split into the sum:

$$k_{\mu\nu}(x^{\lambda'}) = \sigma^\lambda_{\mu\nu}(x^\lambda) + \tau_{\mu\nu}(x^\lambda), \tag{6}$$

at which the additional stresses $\tau_{\mu\nu}$ are to be calculated within the framework of the linear theory:

$$\tau_{\mu\nu} = \frac{\partial a}{\partial \gamma_{\mu\nu}}, \tag{7}$$

$a(w, w)$ being the well known expression of the elastic potential in the linear elastic theory. For further generalizations of (5), (6) see (17), (18) or [1].

In [1] I introduced a new method for describing finite deformations in nonlinear elasticity, dealing with the first boundary value problem. Here we generalize this method to the second and third boundary value problem, for further details see [5]. We neither postulate an explicitly known analytical expression of the elastic potential U, globally valid, nor do we try to solve the variational problem (3) directly. Our method consists in a rigorous mathematical description of the deformation process itself, only using (4), (5), (6), or (17), (18). Introducing a parameter λ, we follow the deformation of $G + S$ for simplicity along the straight path $\lambda\varphi \times \lambda T \times \lambda X$ in the Banach space $C^{2+\mu} \times C^{1+\mu} \times C^{\mu}$ of the data. To get a suitable approximation scheme, after dividing the interval I_1: $0 \leqq \lambda \leqq 1$ by $0 < \lambda_1 < \lambda_2 < \cdots \lambda_n = 1$, the following steps of deformation $G \to G^1 \to G^2 \cdots \to G^n$ are defined.

Third boundary value problem

For $G \to G^1$: $x + v^0(x) = x^1$ the displacements $v^0(x)$ are to be calculated from the mixed free variational problem

$$J_0(v^0, v^0) = \int_G a(v^0, v^0)\,dx - \lambda_1(X, v^0)_0 - \lambda_1 \int_S T_k(\sigma)\, v_k{}^0(\sigma)\, d\sigma \to \text{Min}, \qquad (8)$$

$$\lambda_1\varphi_j(\sigma) = v_j{}^0(\sigma), \qquad \sigma \in S'',$$

by which the stress distribution

$$\sigma^1_{\mu\nu}(x^1) = \tau_{\mu\nu}(x), \quad \tau_{\mu\nu} = \frac{\partial a(v^0, v^0)}{\partial \gamma_{\mu\nu}}$$

over G^1 is defined. $G \to G^1 \to \cdots \to G^l$ being constructed, we then define inductively the transition $G^l \to G^{l+1}$:

$$x^{l+1} = x^l + v^l(x^l) \qquad (9)$$

by the solution of the variational problem

$$J_l(v^l, v^l) - Q_l(v^l, v^l) - \Delta\lambda(X, v^l)_0 - \Delta\lambda \int_{S'} T_k(\sigma^l)\, v_k{}^l(\sigma^l)\, d\sigma^l \to \text{Min}, \qquad (10)$$

$$\Delta\lambda\varphi^l(\sigma^l) = v^l(\sigma^l), \quad \sigma^l \in S'', \quad v^l(x^l) \in \overline{H}_{1,2}(G^l).$$

We calculate the additional stresses $\tau^l_{\mu\nu}(x^l)$ according to (7), receiving the stress distribution

$$\sigma^{l+1}_{\mu\nu}(x^{l+1}) = \sigma^l_{\mu\nu}(x^l) + \tau^l_{\mu\nu}(x^l) \qquad (11)$$

over G^{l+1}.

Some remarks about the solutions of (8), (10). As to (8), we note the generalized Korn's inequality, see [4] or (24):

$$\int_G a(w, w)\,dx \geqq d_1\|w\|^2_{1,2}, \quad d_1 > 0, \quad w \in \overline{H}_{1,2}(G), \qquad (12)$$

$\overline{H}_{1,2}(G)$ being the closure of all regular zero boundary functions along S'' in the $H_{1,2}(G)$-

norm. From this and (5) follows

$$Q(w, w) \geqq k_1 \|w\|^2_{1,2}, \qquad w \in \overline{H}_{1,2}, \tag{13}$$

with a suitable small constant $k_1 > 0$, within a ball

$$E_\gamma: \ |\sigma_{\mu\nu}| < \delta \tag{14}$$

in the Banach space $C(\sigma_{\mu\nu})$ of stresses.

The region E in $C(\sigma_{\mu\nu})$, over which the inequality (13) holds with a positive constant k_1, defines the states of strong stability of our mechanical system.

Under (13) it is well known, how to solve the homogeneous problems of (8), (10) over $\overline{H}_{1,2}$, we reduce the inhomogeneous problems (8), (10), to it, by setting $v^l(x^l) = w^l(x^l) - \Delta\lambda\varphi^l(x^l)$. $\varphi^l(x^l)$ arises from $\varphi(x)$ by transplantation along (9). If

$$S^r \subset C^{2+\mu}, \qquad \sigma^r_{\mu\nu} \subset C^{1+\mu} \cap E, \qquad T_k \subset S^{1+\mu}, \qquad \varphi(\sigma^r) \subset C^{2+\mu}$$

holds, the classical theory of regularity for (10) tells us:

$$v^r(x^r) \subset C^{2+\mu}, \qquad \sigma^{r+1}_{\mu\nu}(x^{r+1}) \subset C^{1+\mu}, \qquad S^{r+1} \subset C^{2+\mu}.$$

So we have no loss of regularity during our procedure. We surely arrive at the extreme point $\varphi \times T \times X$ of the data, if our deformation does not leave E and the boundary does not intersect.

The fundamental Schauder estimates

$$\begin{aligned} &\|v^l(x^l)\|_{2+\mu} \leqq \Delta\lambda L(H_{1+\mu}, c, k_1 R_{2+\mu}) \, (A + \|v^l\|_{0,2}), \\ &A = (\|X\|_\mu + \|T\|_{1+\mu} + \|\varphi\|_{2+\mu}) \end{aligned} \tag{15}$$

are valid for the solutions of (10). This follows from the general Schauder estimates for the solutions of strongly elliptic systems under complementing boundary conditions in [1]. $H_{1+\mu}$ is a bound for the $C^{1+\mu}$-norms of the stresses, c the ellipticity constant and $R_{2+\mu}$ a bound of the $C^{2+\mu}$-norms of the boundary representation. The norm $\|v^l\|_{0,2}$ can easily be eliminated on the right of (15) in our case, [4, 5]. (15) is fundamental for our convergence proof if $\Delta\lambda \to 0$. Summing equations (9), gives:

$$G \to G^r: x_k^r = f_k^r(x), \qquad x \subset G + S, \qquad k = 1, 2, 3. \tag{16}$$

We interpolate the functions $f_k^r(x)$, $D^1 f_k^r(x)$, $D^2 f_k^r(x)$ linear between $\lambda_r \leqq \lambda \leqq \lambda_{r+1}$ and prove their equi-boundedness and equi-continuity along a suitably chosen fixed interval $I_s: 0 \leqq r\Delta\lambda \leqq s$ in the same manner as we did in [1]. Along suitable subsequences $f_k^r(x)$ converge uniformly in C^2 to functions $x_k^\lambda = f_k(\lambda, x) \subset C^{2+\mu}$, $0 \leqq \lambda \leqq s$ according to Ascoli-Arzela's theorem. In a similar way we receive the corresponding stresses $\sigma_{\mu\nu}(\lambda, x)$.

All remains valid if Trefftz's expression (5) for the second variation of the elastic potential is replaced by an arbitrary positive definite quadratic functional

$$Q(w, w) = \frac{1}{2} \int_G A_{i,j,k,l}(\sigma^\lambda_{\mu\nu}, \ldots) \frac{\partial w_i}{\partial x_k^\lambda} \frac{\partial w_j}{\partial x_l^\lambda} \, dx^\lambda \tag{17}$$

with general transition rules

$$\begin{aligned} \sigma_{\mu\nu}^{l+1}(x^{l+1}) &= \sigma_{\mu\nu}^{l}(x^{l}) + \varphi_{\mu\nu}\left(\sigma_{\mu\nu}^{l}, \frac{\partial v^{l}}{\partial x^{l}}, \ldots\right), \\ \varphi_{\mu\nu}(\sigma_{\mu\nu}^{l}, 0, \ldots) &= 0, \end{aligned} \tag{18}$$

moreover, all can be generalized to fading memory models. Transition rules of type (18) must be considered for long range deformations, where, clearly, the formula (6) of Trefftz becomes inadequate. In spite of this, Trefftz's expression (5) of the second variation of the elastic potential has general validity. One has only to modify the elastic constants in the quadratic expression of $a(v, v)$ to depend on the present or more general past stress distributions to obtain an adequate description by means of the corresponding transitions rules (7) for long range deformations. By the same arguments as in [1] we prove the theorem:

Theorem I. *The functions* $f_k(\lambda, x) = x^\lambda$ *with the corresponding stresses describe the deformation of the elastic body* $G + S$ *in states of equilibrium under the given volume-boundary forces and boundary displacements* λX, λT, $\lambda\varphi$ *along the interval* I_s: $0 \leqq \lambda \leqq s$, *under the condition, that* (5) *or* (17) *is the second variation of the elastic potential with transition rules* (11) *or* (18).

We are allowed to put forward our deformation if

a) *the border of stability or ellipticity is not reached,*

b) *the* $C^{1+\mu}$*-norms of the stresses do not tend to infinity,*

c) *the surfaces of our elastic body do not tend to intersect each other.*

The second boundary value problem

In dealing with the second boundary value problem we see from (4), by choosing variations $w(x)$ from V_6 (V_6: $a_i + c_{ij}x_j = w_i(x)$, with constants a_i, $c_{ij} = -c_{ji}$, being the six dimensional linear space of "infinitesimal displacements" in the linear theory) that a state of equilibrium demands the given forces to fulfil the momentum conditions

$$\int_{G^\lambda} (\mathfrak{x} \times X)\, dx^\lambda + \int_{S^\lambda} (\mathfrak{x} \times T)\, d\sigma^\lambda = 0, \tag{19}$$

and further

$$\int_{G^\lambda} X_i\, dx^\lambda + \int_{S^\lambda} T_i\, d\sigma^\lambda = 0, \qquad i = 1, 2, 3. \tag{20}$$

Therefore we must provide (19), (20) in the initial state over $G + S$. According to our assumptions the conditions (20) will hold during the deformation, on the contrary, the total momentum of the given forces will change with the deformation of the body. This causes some difficulties, when generalizing our method to the second boundary value problem.

Now, in the first step of our approximation procedure the free variational problem (8), S'' being empty, must be solved, which can be done by the use of Korn's second inequality

$$\int_G a(w, w)\, dx \geqq d_2 \|w\|^2_{1,2}, \qquad d_2 > 0$$

$$w \in H_{1,2} \ominus V_6 \tag{21}$$

and classical lower continuity arguments. In reality the solutions $v^0(x)$ over $H_{1,2} \ominus V_6$ are solutions over $H_{1,2}$ because of (19), (20). The undeformed state over $G + S$ represents a boundary point of the region of strong stability E, above defined in the Banach space $C(\sigma_{\mu\nu})$ of stresses in case of the second boundary value problem, as the left side of (21) vanishes for $w \in V_6$. Our deformation $G \to G^1 : x + v^0(x) = x^1$, therefore, may firstly lead into E or secondly into the region of instability. In the first case (22) is valid, for λ_1, sufficiently small,

$$Q_1(w, w) \geqq k_1' \|w\|^2_{1,2}, \quad k_1' > 0, \quad k_1' \to 0 \quad \text{for} \quad \lambda_1 \to 0, \quad w \subset H'_{1,2}(G^1). \tag{22}$$

Here $H'_{1,2}(G^1)$ denotes the subspace of $H_{1,2}(G^1)$, consisting of all vector functions orthogonal to the constant vector. It is essential to point out that it is sufficient for (22) to have

$$Q_1(w, w) > 0, \text{ for } \lambda_1, \text{ sufficiently small, only, } w \neq 0. \tag{23}$$

This follows from Korn's inequality in the form

$$\int_{G^1} a(w, w)\, dx + \|w\|^2_{0,2} \geqq d_3 \|w\|^2_{1,2}, \qquad d_3 > 0$$

$$w \in H_{1,2}(G^1), \tag{24}$$

[3, 6], which points out that $Q_1(w, w)$ fulfils a Garding inequality for small λ_1 over $H'_{1,2}(G^1)$. Applying a result of M. R. Hestenes, $Q_1(w, w)$ must be positive definite in $H'_{1,2}(G^1)$ because of (23). On the other hand the stresses, resulting from the deformation $G \to G^1$, may cause the second variation $Q_1(w, w)$ to become indefinite. According to (2) the corresponding states are unstable. In the stable case (22) we are allowed to approximate the deformation under the conditions of the second boundary value problem in E in the same manner as we did in the case of the third boundary value problem. Because of (20) the solution $v^l(x^l)$ of (10), S'' being empty, in $H'_{1,2}(G^l)$ is the solution over $H_{1,2}(G^l)$ in reality. There are no additional difficulties, when extending our convergence proof in [1] to this approximation of the deformation in E. The same is true for the proof, that the so constructed deformation in E leads under the given forces over states of equilibrium. By choosing variations out of V_6, it is seen from (4) that the momentum conditions are fulfilled along the deformation. When studying elastic deformations in regions of instability under second boundary conditions, we take notice of the linear Neumann-eigenvalue problem of the selfadjoint Lagrangean operator corresponding to $Q_1(w, w)$ and make the assumption (A): the constant vector is the only eigenfunction of the eigenvalue zero. We can then solve the inhomogeneous Neumann boundary value problem, corresponding to (10), by Rieß-Schauder theory, the necessary and sufficient conditions of solvability (20) being fulfilled. After normalizing this solution orthogonal to the constant vector, the $H_{0,2}$-norm on the right side of the Schauder estimates (15)

can be eliminated, using the boundedness of the pseudoresolvent operator. Using the methods above, we can now prove the same existence results concerning elastic deformations in regions of instability as in the stable case, [5].

Theorem II. *An arbitrary small deformation of $G + S$ over states of equilibrium into $G' + S'$ under given boundary and volume forces, fulfilling the conditions* (19), (20) *over $G + S$, may lie in the domain E of stability or in the domain of instability under* (A). *The functions $f_k(\lambda, x) = x^\lambda$, constructed in our existence proof, just mentioned, describe the continuation of the elastic deformation of $G' + S'$ over states of equilibrium under the given second boundary conditions as stated in Theorem* I. *The momentum conditions are fulfilled during this deformation.*

We have the same limit conditions of continuation.

$H_{1,2} \ominus V_6$ Theory

To overcome the complications in the theory of the second boundary value problem of nonlinear elasticity, arising from the fact, that the undeformed state represents a boundary point of the domains of stability and instability in $C(\sigma_{\mu\nu})$, we introduce a $H_{1,2} \ominus V_6$ Theory, [5].

According to Korn's second inequality there exists a ball E_δ of type (14), in which

$$Q(w, w) \geqq k_2 \|w\|^2_{1,2}, \qquad k_2 > 0, \qquad w \in H_{1,2} \ominus V_6 \tag{25}$$

is valid. Being in the same position as in the case of the first and third boundary value problem, we start our approximation over $H_{1,2} \ominus V_6$ by solving the free variational problems (8), (10), S'' being empty, over $H_{1,2} \ominus V_6$, inductively defining by (9) the transitions $G^l \to G^{l+1}$. For our convergence proof the Schauder estimates (15) are necessary, which we receive for solutions of (8), (10) over $H_{1,2}$. We, therefore, replace (10) by the variational problem

$$J_l(v^l, v^l) + \int_{S^l} v_j{}^l(\sigma^l) \cos(n, x_k{}^l)\, c^l_{j,k}\, d\sigma^l \to \text{Min} \tag{26}$$

and try to determine the constants $c^l_{j,k}$, $c^l_{j,k} = -c^l_{j,k}$, in such a manner that the solutions $v^l(x^l)$ of (10), just constructed over $H_{1,2} \ominus V_6$, are the desired solutions of (26) over $H_{1,2}(G^l)$. Indeed, this is possible, as we have

$$\begin{gathered} \int_{S^l} \eta_j(\sigma^l) \cos(n, x_k{}^l)\, c^l_{j,k}\, d\sigma^l = 0, \\ \eta_j(x^l) \in H_{1,2} \ominus V_6 . \end{gathered} \tag{27}$$

Thus, $v^l(x^l)$ is a solution of (26) in the subspace $H_{1,2} \ominus V_6$ of $H_{1,2}$ as constructed. The demand the first variation of (26) for variations out of V_6 to become zero, too, leads us to a system of linear equations, from which the desired constants $c^l_{j,k}$ can be uniquely calculated, [5]. Moreover, we get the estimates

$$\begin{gathered} \text{Max}\, |c^l_{j,k}| \leqq \Delta\lambda k_3(\|X\|_\mu + \|T\|_\mu), \\ \|v^l\|_{0,2} \leqq \Delta\lambda k_4(\|X\|_\mu + \|T\|_\mu) \end{gathered} \tag{28}$$

with fixed constants k_3, k_4. Inserting these bounds into the Schauder estimates for the solutions $v^l(x^l)$ of (26), we receive the desired estimate, the starting point for our existence proof, which can be carried out in the same manner as in the cases already mentioned, [5]. Clearly, we have to assume that our deformation has not yet left the region of ellipticity.

Having now constructed the deformation $x^\lambda = f(\lambda, x) \subset C^{2+\mu}$ with corresponding stress distributions $\sigma_{\mu\nu}(\lambda, x) = \sigma^\lambda_{\mu\nu}(x^\lambda)$, it can be proved that the constants $c^i_{r,k}/\Delta\lambda$ fulfil the estimate

$$\left|\frac{c^i_{rk}}{\Delta\lambda} - \frac{c^j_{rk}}{\Delta\lambda}\right| = O(|\Delta\lambda'|), \qquad \lambda_j - \lambda_i = \Delta\lambda'. \tag{29}$$

$O(|\Delta\lambda'|)$ is the well known Landau symbol not dependend upon the specially chosen subdivision of J_s. It follows from this

$$\lim_{\Delta\lambda\to 0} \frac{c^i_{rk}}{\Delta\lambda} = \varphi_{rk}(\lambda) \subset C^0(J_s) \tag{30}$$

Setting $\int\limits_0^\lambda \varphi_{rk}(\lambda')\, d\lambda' = \psi_{rk}(\lambda)$, we prove the following theorem.

Theorem III. *The functions $f(\lambda, x) = x^\lambda$, just constructed, with stresses $\sigma_{\mu\nu}(\lambda, x)$, describe the deformation of our elastic body $G + S$ along the interval I_s: $0 \leqq \lambda \leqq s$ over states of equilibrium under the given volume forces λX and the modified boundary forces $\lambda T - \psi_{r,k}(\lambda) \cos(n, x_k)$ along S into a domain of stresses in $C(\sigma_{\mu\nu})$ of strong stability or instability with condition* (A). *The momentum condition holds with this forces during the deformation. We have similar limit conditions of continuation as in the other Theorems.*

As of course the additional forces

$$\psi_{r,k}(\lambda) \cos(n, x_k), \qquad r = 1, 2, 3, \tag{$*$}$$

depend upon the given forces $\lambda' T$ along the interval J_λ: $0 \leqq \lambda' \leqq \lambda$ only, it is important to note that we are allowed to make the boundary forces (*) retrograde again, if we have arrived at the domain of stability or instability with condition (A), applying Theorem II, in order to reach the prescribed forces at the end of the deformation. The forces (*), working on an arbitrary small interval J_s, are used to stabilize the deformation from the undeformed state under second boundary conditions only.

References

[1] Beckert, H., Zur ersten Randwertaufgabe in der nichtlinearen Elastizitätstheorie. Z. Angew. Math. Mech. **55**, 47–58 (1975).

[2] Beckert, H., Über eine neue Existenztheorie in der nichtlinearen Elastizitätstheorie. Beiträge zur Analysis **8**, 49–57 (1976).

[3] Beckert, H., Über die klassischen Randwertaufgaben in der Theorie der Wärmespannungen in stückweise stetigen, anisotropen Körpern unter Kopplungsbedingungen. Z. Angew. Math. Mech. **52**, 111–122 (1972).

[4] Beckert, H., Die mathematische Beschreibung endlicher Deformationen bei der dritten Randwertaufgabe in der nichtlinearen Elastizitätstheorie. Wiss. Z. Karl-Marx-Univ. Leipzig, Math.-Natur. Reihe **25**, 1—17 (1976).
[5] Beckert, H., Zur zweiten und dritten Randwertaufgabe in der nichtlinearen Elastizitätstheorie. Z. angew. Math. Mech. **57**, 153—164 (1977).
[6] Fichera, G., Linear elliptic differential systems and eigenvalue problems. Lecture Notes in Math. 8. Springer, Berlin—New York 1965.
[7] Hestenes, M. R., Applications of the theory of quadratic forms in Hilbert space to the calculus of variations, Pacific J. Math. **1**, 524—581 (1951).
[8] Trefftz, E., Zur Theorie der Stabilität des elastischen Gleichgewichts. Z. Angew. Math. Mech. **13**, 160—165 (1933).
[9] Trefftz, E., Über die Ableitung der Stabilitätskriterien des elastischen Gleichgewichts. Verhandl. III. Intern. Kongr. für Techn. Mech., Stockholm, 1930, **3**, 44—48.
[10] Truesdell, C., Noll, W., The nonlinear field theories of mechanics. Handb. d. Physik Bd. III/3, Springer, Berlin 1965.
[11] Agmon, S., Douglis, A., Nirenberg, L., Estimate near the boundary for solutions of elliptic partial differential equations satisfying general boundary conditions I. II. Comm. Pure Appl. Math. **12**, 623—727 (1959); **17**, 35—92 (1964).

Address of the author:

Karl-Marx-Universität Leipzig
Sektion Mathematik
701 Leipzig
Karl-Marx-Platz 1
DDR

Successive approximations in finite linear viscoelasticity and the existence of the solution

Jozef Brilla, Bratislava

1. Introduction

Many different stress-strain relations called constitutive equations have been proposed for characterizing the mechanical behaviour of nonlinear, inelastic solids. The most general mathematical theory of constitutive equations has been initiated by Green and Rivlin in their paper [1] on the theory of nonlinear materials with memory. This theory has been developed by Coleman, Noll, Truesdell and others [2]. The mathematical theories are probably general enough to take into account almost all types of observed nonlinearities. However, they can be quite impractical, when dealing with concrete problems.

For practical purposes it is necessary to consider different approximations of constitutive equations. Coleman and Noll [3] derived a special nonlinear theory known as finite linear viscoelasticity. This theory considers the finite deformation, but it restricts the deformation to have been slowly changing in the recent past. We shall deal with the solution of boundary value problems of finite linear viscoelasticity for materials of the differential type by the method of successive approximation.

2. Constitutive equations

The motion of the body $\mathfrak{B}$ may be described by a vector function $x(t) = \chi(X, t)$, where $x(t)$ is the position of a material point at time t which has the position X in a fixed reference configuration $\mathfrak{R}$. We assume that $\chi(X, t)$ is always smoothly invertible in X. The motion of the body $\mathfrak{B}$ defines its deformation from the reference configuration.

In the case of isothermal deformations the constitutive equations may be expressed in the form

$$T(x, t) = \mathop{\mathscr{F}}_{t'=-\infty,\, X'\in\mathfrak{B}}^{t'=t} [\chi(X', t'), X, t], \tag{2.1}$$

where $T(x, t)$ is a stress tensor and $\mathscr{F}$ is a symmetric second order tensor operator restricted only by some physical principles. From a mathematical point of view a body acts as an operator $\mathscr{F}$ which assigns to each continuous vector valued function $\chi(X, t)$ a tensor valued function $T(x, t)$.

Restricting the dependence of T only on the first deformation gradient

$$F = \nabla\chi(X, t), \quad F_{\alpha}{}^{k} = \frac{\partial x^k}{\partial X^\alpha} = \chi^k_{;\alpha} \tag{2.2}$$

we obtain the constitutive equations of simple materials

$$T(x, t) = \mathop{\mathscr{G}}_{t'=-\infty}^{t} [F(X, t'), X, t]. \tag{2.3}$$

For slow deformations the general constitutive equation of simple materials with fading memory can be approximated [2, 4] by a constitutive equation of the form

$$R^T T R = f(C) + \sum_{j_1,\ldots,j_k} l_{j_1,\ldots,j_k}(C) \left[A^*_{j_1}, \ldots, A^*_{j_k}\right], \tag{2.4}$$

where each $l_{j_1,\ldots,j_k}(C)$ [.], for each choice of C, is a multilinear tensor function of k tensor variables and the summation is to be extended over all sets of indices $j_1, \ldots, j_k$, $k = 1, \ldots, n$, that satisfy inequalities

$$1 \leqq j_1 \leqq \cdots \leqq j_k \leqq n, \qquad j_1 + \cdots + j_k \leqq n. \tag{2.5}$$

Here R is to be determined from the polar decomposition of F

$$F = RU = VR \tag{2.6}$$

in which R is orthogonal and U and V are symmetric and positive definite. Further

$$C = U^2 = F^T F \tag{2.7}$$

is the right Cauchy-Green tensor and

$$A_n{}^* = R^T A_n R, \qquad A_n = \overset{-1}{F^T} \overset{(n)}{C} \overset{-1}{F}, \tag{2.8}$$

where A_n is the n-th Rivlin-Ericksen tensor. A material obeying (2.4) is called [2] a material of the differential type of grade n.

For the sake of simplicity we restrict our analysis to materials of the grade 1. Then the constitutive equation assumes the form

$$R^T T R = f(C) + V(C) [R^T A_1 R]. \tag{2.9}$$

Materials obeying this equation are called linearly viscous materials.

An alternative form of the constitutive equation for linearly viscous materials is

$$\overline{T} = s(C) + P(C) [F^T D F], \tag{2.10}$$

where

$$\overline{T} = F^T T F \tag{2.11}$$

is the convected stress tensor and

$$D = \frac{1}{2} A_1. \tag{2.12}$$

When dealing with traction boundary value problems it is convenient to make use of Piola-Kirchhoff stress tensors. By using the second Piola-Kirchhoff stress tensor

$$\tilde{T} = JF^{-1}T(F^{-1})^T, \tag{2.13}$$

where

$$J = |\det F| \tag{2.14}$$

we may write the constitutive equation in the form

$$\tilde{T} = t(C) + S(C)\,[F^T D F]. \tag{2.15}$$

Here we have used the relation

$$\tilde{T} = JF^{-1}(F^{-1})^T\,\overline{T}F^{-1}(F^{-1})^T = JC^{-1}\overline{T}C^{-1} \tag{2.16}$$

and

$$t(C) = JC^{-1}s(C)\,C^{-1}, \tag{2.17}$$

$$S(C)\,[F^T D F] = JC^{-1}P(C)\,[F^T D F]\,C^{-1}. \tag{2.18}$$

3. Approximate solution of traction boundary value problems

Quasistatic traction boundary value problems for linearly viscous bodies are defined by

i. the equilibrium equation

$$\mathrm{Div}\, T + \varrho b = 0, \tag{3.1}$$

where b is the specific body force and ϱ is the mass density at time t,

ii. the constitutive equation

$$T = J^{-1}F\{t(C) + S(C)\,[F^T D F]\}\,F^T, \tag{3.2}$$

iii. boundary conditions

$$t = Tn, \tag{3.3}$$

where t is the surface traction acting on the deformed boundary and n is the outward unit vector normal to the boundary at time t, and

iv. initial conditions

$$F(X, 0) = F_0(X). \tag{3.4}$$

When dealing with boundary value problems we assume that the reference configuration is a homogeneous natural state, i.e., that the constitutive equation is a function of F only, independent of X, and that the body is free of stresses when in reference configuration. We have also to take into account the fact that the domain of definition and its boundary are changing with time and to transform all values to one frame of reference, for example to the reference configuration X.

Similarly as it was done by Signorini [5] for finite elasticity problems, we suppose that the magnitudes of the external forces are proportional to the parameter ε.

Thus we have

$$t_{\mathfrak{R}} = \varepsilon t_{\mathfrak{R}_1}, \qquad b = \varepsilon b_1. \tag{3.5}$$

The subscript $\mathfrak{R}$ denotes use of the reference configuration.

Further we assume that there exists a solution, expressed in terms of the displacement

$$u(X, t) = x(X, t) - X, \tag{3.6}$$

of the boundary value problem defined by the loads (3.5) and the initial condition

$$F(X, 0) = \varepsilon F_1(X), \tag{3.7}$$

which depends analytically on ε and vanishes when $\varepsilon = 0$. Thus we assume that

$$u = \sum_{n=1}^{\infty} \varepsilon^n u_n. \tag{3.8}$$

Then the displacement gradient

$$\nabla u = H = F - I \tag{3.9}$$

has also a series expansion

$$H = \sum_{n=1}^{\infty} \varepsilon^n \nabla u_n = \sum_{n=1}^{\infty} \varepsilon^n H_n. \tag{3.10}$$

Now, it is necessary to insert the expansion (3.10) into (3.1—4) and to find equations for the determination of u_n.

For this purpose we rewrite the constitutive equation (3.2) in the form

$$\tilde{T} = t(2E + I) + S(2E + I)[\dot{E}], \tag{3.11}$$

where

$$E = \frac{1}{2}(C - I) = \frac{1}{2}(H + H^T + H^T H) \tag{3.12}$$

is the Green-St. Venant strain tensor and

$$\dot{E} = \frac{D}{Dt} E = F^T D F. \tag{3.13}$$

Thus the constitutive equation (3.11) may be rewritten in the form

$$\tilde{T} = l(E) + K(E)[\dot{E}]. \tag{3.14}$$

According to the assumption of the free equilibrium state in the reference configuration it is necessary that

$$l(0) + K(0)[0] = 0. \tag{3.15}$$

Thus the constitutive equation can be expanded in the form of a series

$$\tilde{T}^{\alpha\beta} = L_1^{\alpha\beta\gamma\delta}E_{\gamma\delta} + L_2^{\alpha\beta\gamma\delta\mu\nu}E_{\gamma\delta}E_{\mu\nu} + \cdots + K_1^{\alpha\beta\gamma\delta}\dot{E}_{\gamma\delta} + K_2^{\alpha\beta\gamma\delta\mu\nu}E_{\gamma\delta}\dot{E}_{\mu\nu} + \cdots \tag{3.16}$$

Making use of (3.16) E and $\dot{E}$ can be expanded in

$$2E = \varepsilon(H_1 + H_1^T) + \cdots + \varepsilon^n\left(H_n + H_n^T + \sum_{k=1}^{n-1} H_k^T H_{n-k}\right) + \cdots \tag{3.17}$$

and

$$2\dot{E} = \varepsilon(\dot{H}_1 + \dot{H}_1^T) + \cdots + \varepsilon^n\left[\dot{H}_n + \dot{H}_n^T + \sum_{k=1}^{n-1}(\dot{H}_k^T H_{n-k} + H_k^T\dot{H}_{n-k})\right] + \cdots \tag{3.18}$$

Substitution of (3.17) and (3.18) into (3.16) shows that $\tilde{T}$ has also a series expansion

$$\tilde{T} = \sum_{n=1}^{\infty} \varepsilon^n \tilde{T}_n. \tag{3.19}$$

Then collecting the terms proportional to ε^n we find that $\tilde{T}_n$ has to satisfy the equation

$$\tilde{T}_n = \Omega_1(H_n) + \Pi_n(H_1, \ldots, H_{n-1}), \tag{3.20}$$

where

$$\begin{aligned} &\Omega_1(H_n) = \frac{1}{2}\left[L_1(H_n + H_n^T) + K_1(\dot{H}_n + \dot{H}_n^T)\right] = L_1E_n + K_1\dot{E}_n, \\ &\Pi_1 = 0, \\ &\Pi_n(H_1, \ldots, H_{n-1}) = l_n(H_1, \ldots, H_{n-1}) + k_n(H_1, \ldots, H_{n-1}, \dot{H}_1, \ldots, \dot{H}_{n-1}), \\ &n \geqq 2. \end{aligned} \tag{3.21}$$

and l_n, k_n are polynomial tensor functions of $(n-1)$ and $2(n-1)$ tensor variables, respectively, that are uniquely determined by the constitutive equation (3.16). E_n is the infinitesimal strain tensor and $\dot{E}_n$ is the rate of the infinitesimal strain tensor.

The equilibrium equation (3.1) assumes in terms of the second Piola-Kirchhoff stress tensor the form

$$(x^k_{;\alpha}\tilde{T}^{\alpha\beta})_{;\beta} + \varrho_R b^k = 0. \tag{3.22}$$

From

$$x^k = g_\alpha{}^k(X^\alpha + u^\alpha), \tag{3.23}$$

where $g_\alpha{}^k$ are components of contravariant shift vector, we arrive at

$$x^k_{;\alpha} = g_\gamma{}^k(\delta_\alpha{}^\gamma + u^\gamma_{;\alpha}). \tag{3.24}$$

Inserting (3.24) into (3.22) by some transformation we obtain

$$\tilde{T}^{\alpha\beta}_{;\beta} + (u^\alpha_{;\gamma}\tilde{T}^{\gamma\beta})_{;\beta} + \varrho_{\Re} b^\alpha = 0. \tag{3.25}$$

Making use of the series expansion of $\tilde{T}$ and u and collecting the terms proportional to ε^n yields

$$\varepsilon(\tilde{T}^{\alpha\beta}_{(1);\beta} + \varrho_{\Re} b^{\alpha}_{(1)}) + \sum_{n=2}^{\infty} \varepsilon^n \left[\tilde{T}^{\alpha\beta}_{(n);\beta} + \sum_{k=1}^{n-1} (u^{\alpha}_{(k);\gamma} \tilde{T}^{\gamma\beta}_{(n-k)})_{;\beta} \right] = 0 . \tag{3.26}$$

Finally it is necessary to express the boundary condition (3.3) in a similar form. In the reference configuration we have

$$t_{\Re} = T_{\Re} n_{\Re} , \tag{3.27}$$

where

$$T_{\Re} = JT(F^{-1})^T = F\tilde{T} \tag{3.28}$$

is the first Piola-Kirchhoff stress tensor. Then in terms of the second Piola-Kirchhoff stress tensor we have

$$t_{\Re} = F\tilde{T} n_{\Re} \tag{3.29}$$

or in the component form

$$t^k{}_{\Re} = x^k_{;\alpha} \tilde{T}^{\alpha\beta} n_{\Re\beta} . \tag{3.30}$$

Inserting (3.23) for $x^k_{;\alpha}$ and shifting the components we arrive at

$$t^{\alpha}{}_{\Re} = \tilde{T}^{\alpha\beta} n_{\Re\beta} + u^{\alpha}_{;\gamma} \tilde{T}^{\gamma\beta} n_{\Re\beta} . \tag{3.31}$$

Finally using series expansions for $\tilde{T}$ and u and collecting the terms proportional to ε^n yields

$$\varepsilon t^{\alpha}_{(1)\Re} = \varepsilon \tilde{T}^{\alpha\beta}_{(1)} n_{\Re\beta} + \sum_{n=2}^{\infty} \varepsilon^n \left[\tilde{T}^{\alpha\beta}_{(n)} + \sum_{k=1}^{n-1} u^{\alpha}_{(k);\gamma} \tilde{T}^{\gamma\beta}_{(n-k)} \right] n_{\Re\beta} . \tag{3.32}$$

Thus we have arrived at series expansions of the constitutive equation (3.20), the equilibrium equation (3.26) and of the boundary condition (3.32). We assume that there exists such $\bar{\varepsilon}$ that the problem under consideration has the solution for any ε which satisfies the inequality

$$0 \leqq \varepsilon < \bar{\varepsilon} . \tag{3.33}$$

Then equating like powers of ε yields the following successive systems of equations

$$\begin{aligned} &\tilde{T}^{\alpha\beta}_{(n);\beta} + \varrho_{\Re} b^{\alpha}_{(n)} = 0 , \\ &\tilde{T}^{\alpha\beta}_{(n)} = \Omega_1{}^{\alpha\beta}(E_n) + \Pi_n{}^{\alpha\beta}(H_1, \ldots, H_{n-1}) , \\ &\tilde{T}^{\alpha\beta}_{(n)} n_{\Re\beta} = t^{\alpha}_{(n)\Re} , \\ &u_n(x, 0) = u_n(x) , \end{aligned} \tag{3.34}$$

where b_1, $t_{1\Re}$ and $u_1(x)$ are prescribed external forces and the initial displacement, respectively, and

$$\varrho_{\Re} b^{\alpha}_{(n)} = \sum_{k=1}^{n-1} (u^{\alpha}_{(k);\gamma} \tilde{T}^{\gamma\beta}_{(n-k)})_{;\beta},$$

$$t^{\alpha}_{(n)\Re} = -\sum_{k=1}^{n-1} u^{\alpha}_{(k);\gamma} \tilde{T}^{\gamma\beta}_{(n-k)} n_{\Re\beta}, \tag{3.35}$$

$$u_n(x) = 0, \qquad n \geqq 2.$$

Substituting the values of the stress components according to the constitutive equation in the equilibrium equation and the boundary condition gives

$$\Omega^{\alpha\beta}_{(1)}(E_n)_{;\beta} + \varrho_{\Re} b^{*\alpha}_{(n)} = 0,$$

$$\Omega^{\alpha\beta}_{(1)}(E_n)\, n_{\Re\beta} = t^{*\alpha}_{(n)\Re}, \tag{3.36}$$

$$u_n(x, 0) = u_n(x),$$

where

$$\varrho_{\Re} b^{*\alpha}_{(1)} = \varrho_{\Re} b^{\alpha}_{(1)}, \qquad t^{*\alpha}_{(1)\Re} = t^{\alpha}_{(1)\Re}, \qquad u_1(x, 0) = u_1(x),$$

$$\varrho_{\Re} b^{*\alpha}_{(n)} = \sum_{k=1}^{n-1} (u^{\alpha}_{(k);\gamma} \tilde{T}^{\gamma\beta}_{(n-k)})_{;\beta} + \Pi^{\alpha\beta}_{(n)}(H_1, \dots, H_{n-1})_{;\beta},$$

$$t^{*\alpha}_{(n)\Re} = -\left[\sum_{k=1}^{n-1} u^{\alpha}_{(k);\gamma} \tilde{T}^{\gamma\beta}_{(n-k)} + \Pi^{\alpha\beta}_{(n)}(H_1, \dots, H_{n-1})\right] n_{\Re\beta} \tag{3.37}$$

$$u_n(x, 0) = 0, \qquad n \geqq 2.$$

Thus we have arrived at a successive system of traction boundary value problems of the classical infinitesimal viscoelasticity for the same material and the same boundary. For $n = 1$ we have to consider the given external forces and the initial condition and for $n \geqq 2$ fictitious external forces which depend in an explicitly known way upon previously determined solutions $u_1, \dots, u_{n-1}$ and thus through $T_1, \dots, T_{n-1}$ in (3.37) also upon material coefficients of higher orders. Linear viscoelastic problems can be effectively solved by variational methods [6—8].

4. Conditions of existence of the solution

The solution of a static boundary value problem must be such that the prescribed external forces are in equilibrium in the deformed state for each value of t. Hence the resultant force and the resultant moment acting upon the body in the configuration of equilibrium vanish:

$$\int_{\partial\mathfrak{B}} t\, ds + \int_{\mathfrak{B}} \varrho b\, dv = 0, \tag{4.1}$$

$$\int_{\partial\mathfrak{B}} p \times t\, ds + \int_{\mathfrak{B}} p \times \varrho b\, dv = 0, \tag{4.2}$$

where p is the position vector of x from a fixed origin. With respect to the reference configuration we have the equivalent conditions

$$\int_{\partial\mathfrak{B}} t_{\mathfrak{R}}\, ds_{\mathfrak{R}} + \int_{\mathfrak{B}} \varrho_{\mathfrak{R}} b\, dv_{\mathfrak{R}} = 0, \tag{4.3}$$

$$\int_{\partial\mathfrak{B}} p \times t_{\mathfrak{R}}\, ds_{\mathfrak{R}} + \int_{\mathfrak{B}} p \times \varrho_{\mathfrak{R}} b\, dv_{\mathfrak{R}} = 0, \tag{4.4}$$

where p is the position vector of x but now is expressed as a function of X.

According to Da Silva's theorem we can always find such a change of reference configuration by a rigid rotation that

$$\int_{\partial\mathfrak{B}} p_{\mathfrak{R}} \times t_{\mathfrak{R}}\, ds_{\mathfrak{R}} + \int_{\mathfrak{B}} p_{\mathfrak{R}} \times \varrho_{\mathfrak{R}} b\, dv_{\mathfrak{R}} = 0, \tag{4.5}$$

where $p_{\mathfrak{R}}$ is the position vector in the reference configuration.

In order that the system (3.36) have a solution u_n, it is necessary that the prescribed and fictitious external forces satisfy the equilibrium conditions

$$\begin{aligned} &\int_{\partial\mathfrak{B}} t^*_{n\mathfrak{R}}\, ds_{\mathfrak{R}} + \int_{\mathfrak{B}} \varrho_{\mathfrak{R}} b_n{}^*\, dv_{\mathfrak{R}} = 0, \\ &\int_{\partial\mathfrak{B}} p_{\mathfrak{R}} \times t^*_{n\mathfrak{R}}\, ds_{\mathfrak{R}} + \int_{\mathfrak{B}} p_{\mathfrak{R}} \times \varrho_{\mathfrak{R}} b_n{}^*\, dv_{\mathfrak{R}} = 0. \end{aligned} \tag{4.6}$$

Applying the divergence theorem and the equilibrium equation $(3.36)_1$ we find easily that $(4.6)_1$ is automatically fulfilled. Similarly we find out that $(4.6)_2$ is satisfied if

$$\int_{\mathfrak{B}} G_\alpha \times G_\beta \left[\sum_{k=1}^{n-1} u^\alpha_{(k);\gamma} \tilde{T}^{\gamma\beta}_{(n-k)} + \Pi^{\alpha\beta}_{(n)}(H_1, \ldots, H_{n-1})\right] dv_{\mathfrak{R}} = 0, \tag{4.7}$$

where G_α, G_β are the base vectors at X. This condition is satisfied if and only if

$$\int_{\mathfrak{B}} \left[\sum_{k=1}^{n-1} u^\alpha_{(k);\gamma} \tilde{T}^{\gamma\beta}_{(n-k)} + \Pi^{\alpha\beta}_{(n)}(H_1, \ldots, H_{n-1})\right] dv_{\mathfrak{R}} \tag{4.8}$$

is a symmetric tensor.

Since the second Piola-Kirchhoff stress tensor is symmetric, it follows from the constitutive equation (3.20) that $\Pi^{\alpha\beta}_{(n)}(H_1, \ldots, H_{n-1})$ is a symmetric tensor, too. Thus it remains to analyse the conditions under which the tensor

$$S^{\alpha\beta} = \int_{\mathfrak{B}} \sum_{k=1}^{n-1} u^\alpha_{(k);\gamma} \tilde{T}^{\gamma\beta}_{(n-k)}\, dv_{\mathfrak{R}} \tag{4.9}$$

is symmetric.

Making use of Green's theorem we arrive at

$$S^{\alpha\beta} = \int_{\mathfrak{B}} \sum_{k=1}^{n-1} u^\alpha_{(k)} \tilde{T}^{\gamma\beta}_{(n-k);\gamma}\, dv_{\mathfrak{R}} + \int_{\partial\mathfrak{B}} \sum_{k=1}^{n-1} u^\alpha_{(k)} \tilde{T}^{\gamma\beta}_{(n-k)} n_{\mathfrak{R}\beta}\, ds_{\mathfrak{R}}. \tag{4.10}$$

Now, in view of (3.34) and (3.35), we find that (4.10) may be written in the form

$$S^{\alpha\beta} = \int_{\mathfrak{B}} u^{\alpha}_{(n-1)}\varrho_{\mathfrak{R}} b^{\beta}_{(1)}\, dv_{\mathfrak{R}} + \int_{\partial\mathfrak{B}} u^{\alpha}_{(n-1)} t_{(1)\mathfrak{R}}\, ds_{\mathfrak{R}}$$
$$+ \int_{\mathfrak{B}} \sum_{k=1}^{n-2} \sum_{q=1}^{n-k-1} u^{\alpha}_{(q);\gamma} u^{\beta}_{(n-k-1);\delta} \overset{*}{T}{}^{\gamma\delta}_{(k)}\, dv_{\mathfrak{R}}. \tag{4.11}$$

As the terms

$$\sum_{q=1}^{n-k-1} u^{\alpha}_{(q);\gamma} u^{\beta}_{(n-k-1);\delta} \overset{*}{T}{}^{\gamma\delta}_{(k)} \tag{4.12}$$

are symmetric, the condition of existence for u_n assumes the form

$$\int_{\mathfrak{B}} (u_{\alpha}{}^{(n-1)}\varrho_{\mathfrak{R}} b^{\beta}_{(1)} - u_{\beta}{}^{(n-1)}\varrho_{\mathfrak{R}} b^{\alpha}_{(1)})\, dv_{\mathfrak{R}}$$
$$+ \int_{\partial\mathfrak{B}} (u_{\alpha}{}^{(n-1)} t^{\beta}_{(1)\mathfrak{R}} - u_{\beta}{}^{(n-1)} t^{\alpha}_{(1)\mathfrak{R}})\, ds_{\mathfrak{R}} = 0, \qquad n = 2, 3, \ldots, \tag{4.13}$$

which may be expressed in the vectorial form

$$\int_{\mathfrak{B}} u_{n-1} \times \varrho_{\mathfrak{R}} b_1\, dv_{\mathfrak{R}} + \int_{\partial\mathfrak{B}} u_{n-1} \times t_{1\mathfrak{R}}\, ds_{\mathfrak{R}} = 0, \qquad n = 2, 3, \ldots \tag{4.14}$$

This equation represents necessary conditions for the existence of solutions of the successive system (3.36). In fact, the approximate solution has to satisfy the equilibrium conditions in deformed states at all times $t \geq 0$, what is in agreement with (4.1). We have:

Existence Theorem. *The solutions of the successive system of approximations exist only if the external forces satisfy the equilibrium conditions* (4.14).

The sufficient conditions of the existence of the solution can be analyzed similarly as in infinitesimal viscoelasticity.

In order to fulfil the equilibrium conditions (4.14) we have to proceed in the following way. The solutions of the system (3.36) are solutions of the traction boundary value problems of infinitesimal viscoelasticity and thus are determined at most to within a uniform infinitesimal rotation. This rotation can be uniquely determined from (4.14), applying the following procedure.

We suppose that the solutions $u_1, \ldots, u_{k-1}$ of (3.36) satisfying both equilibrium conditions have already been found. Let $u_k{}^*$ be any solution of (3.36) for $n = k$ in general not fulfilling the moment equilibrium condition. Then

$$u_k = u_k{}^* + \omega_k \times p_{\mathfrak{R}}, \tag{4.15}$$

where ω_k is any vector depending only on t, is also a solution of (3.36). This solution will satisfy the equilibrium condition (4.14) if

$$\int_{\mathfrak{B}} (\omega_k \times p_{\mathfrak{R}}) \times \varrho_{\mathfrak{R}} b_1\, dv_{\mathfrak{R}} + \int_{\partial\mathfrak{B}} (\omega_k \times p_{\mathfrak{R}}) \times t_{1\mathfrak{R}}\, ds_{\mathfrak{R}}$$
$$= -\left[\int_{\mathfrak{B}} u_k \times \varrho_{\mathfrak{R}} b_1\, dv_{\mathfrak{R}} + \int_{\partial\mathfrak{B}} u_k \times t_{1\mathfrak{R}}\, ds_{\mathfrak{R}}\right] = -r_k. \tag{4.16}$$

If we denote the components of the astatic load by

$$A^{\alpha}_{(1)\beta} = \int_{\mathfrak{B}} X^{\alpha} \varrho_{\mathfrak{R}} b_{\beta}{}^{(1)} \, dv_{\mathfrak{R}} + \int_{\partial \mathfrak{B}} X^{\alpha} t^{(1)}_{\mathfrak{R}\beta} \, ds_{\mathfrak{R}} \tag{4.17}$$

then the equation (4.16) may be expressed in the form

$$(A_{\beta}{}^{\alpha} - \delta_{\beta}{}^{\alpha} A_{\gamma}{}^{\gamma}) \, \omega_{(k)\alpha} = r_{(k)\beta}. \tag{4.18}$$

If

$$A = \|A_{\beta}{}^{\alpha} - \delta_{\beta}{}^{\alpha} A_{\gamma}{}^{\gamma}\| \neq 0 \tag{4.19}$$

the vector ω is uniquely determined by the equation (4.18).

Grioli [9] has obtained similar results only independent of time in the case of finite elasticity. He has not considered terms corresponding to $\Pi^{\alpha\beta}_{(n)}(H_1, \ldots, H_{n-1})$. However these terms fulfil equilibrium conditions automatically.

When $\Delta = 0$, the external forces have an axis of equilibrium and the vector ω is partially or totally indeterminate. Then in order to ensure the unique solution, it is necessary to consider some displacement constraints.

Finally it is clear that this method is applicable only to those problems of finite deformation that reduce smoothly to problems of infinitesimal viscoelasticity with vanishing external forces.

References

[1] Green, A. E., Rivlin, R. S., The mechanics of non-linear materials with memory, Part I. Arch. Rat. Mech. Anal. 1, 1—21 (1957).

[2] Truesdell, C., Noll, W., The non-linear field theories of mechanics. Handbuch der Physik Bd. III/3, Springer-Verlag, Berlin 1965.

[3] Coleman, B. D., Noll, W., Foundations of linear viscoelasticity. Rev. Modern Phys. 33, 239 to 249 (1961).

[4] Coleman, B. D., Noll, W., An approximation theorem for functionals with applications in continuum mechanics. Arch. Rat. Mech. Anal. 6, 355—370 (1960).

[5] Signorini, A., Trasformazioni termoelastiche finite. Memoria 2a. Ann. Mat. Pura Appl. Ser. IV, 30, 1—72 (1949).

[6] Brilla, J., Variational methods in mathematical theory of viscoelasticity. In: Proceedings of the Czechoslovak Conference on Differential Equations and their Applications, 211—216. Univerzita J. E. Purkyně, Brno 1973.

[7] Brilla, J., Finite element method in linear viscoelasticity. Z. Angew. Math. Mech. 54, T 47 bis T 48 (1974).

[8] Brilla, J., Generalized variational methods in linear viscoelasticity. *In:* IUTAM Symposium on the Mechanics of Visco-Elastic Media and Bodies, Gothenburg 1974. Springer-Verlag, Berlin—Göttingen—Heidelberg 1975.

[9] Grioli, G., Mathematical theory of elastic equilibrium. Springer-Verlag, Berlin—Göttingen—Heidelberg 1962.

Address of the author:

Institute of Applied Mathematics and Computing Techniques
Mlynská dolina
816 31 Bratislava
Czechoslovakia

Optimal control of stochastic processes

Ф. Л. Черноусько, Москва

Abstract. The paper gives a survey of some results in the field of optimal control for dynamical systems subject to random disturbances. The problem of the optimal feedback control of stochastic processes is reduced to the solution of a certain non-linear parabolic partial differential equation of the second order. Some methods for obtaining the solution of this equation are described. The first approach is based on the expansion of the solution in a series with respect to a small parameter presenting an intensity of the random disturbances. The explicit general formulae for the approximate solution of the equation mentioned above and for the approximate optimal control are obtained. The error estimates for this method are given. The second approach consists in obtaining classes of invariant group solutions which depend on a smaller number of independent variables than the general solution does. The third approach is a numerical solution. The last two approaches are demonstrated by solving the stochastic optimal control problems with integral constraints. The solutions are obtained for the cases of both continuous and impulse control functions.

1. Introduction

This paper presents a survey of some results in the field of the optimal control for dynamical systems subject to random disturbances. It is well known that the problem of optimal feedback control for a stochastic dynamical system is reduced to the solution of a nonlinear partial differential parabolic equation of the second order (Bellman's equation). Exact solutions of this equation are known only in few cases, namely in the case of minimization of an integral quadratic functional for a linear controlled system subject to a white noise. In general the problem is very complicated both for theoretical investigation and practical solution because of its nonlinearity and the great number of phase coordinates arising in practical problems. Therefore it seems useful to develop constructive approaches for obtaining an optimal control of stochastic processes.

In this paper several approaches are presented which can be used for constructing an optimal feedback control for dynamical systems governed by differential equations with random disturbances.

The first approach consists in approximate solving of Bellman's equation in the case of a small intensity of the random disturbances. In this case the partial differential equation contains a small parameter multiplied by the second derivatives, and the solution is obtained as an expansion with respect to this parameter. The optimal feedback control for the stochastic system in the first approximation can be obtained constructively in terms of the solution of the corresponding deterministic problem. The error estimates for this method are given. Some other possibilities using perturbation technique are discussed. An example is presented.

Another approach is based on the methods of the theory of groups. The classes of invariant group solutions are found which depend on a smaller number of independent variables than the general solution does. Some new exact analytical solutions are obtained.

The third approach consists in numerical solving of the partial differential equation. Two further approaches are demonstrated by applying them to problems of optimal stochastic control with integral constraints. In the case of a quadratic integral constraint (energy constraint) the optimal feedback control is a bounded function, while in the case of a modulus integral constraint (fuel consumption constraint) the control has impulses. The problem of impulse control with and without restrictions on the number of impulses are considered. In the impulse control case the parabolic equation degenerates in some region where it becomes a partial differential equation of the first order. A boundary value problem with unknown boundary arises here. The results of numerical solutions are presented.

This paper is based on the results of the references [1—5], where more details can be found. The main attention is given to the constructive aspects rather than to the questions of existence and uniqueness of solution. The problems of optimal stochastic control for small noise intensities were considered by a different approach in the paper [6]. The application of the quasi-variational inequalities to the numerical solution of stochastic optimal control problems is described in the paper [7].

2. Small parameter technique

We consider a problem of optimal control for a system

$$\frac{dx}{dt} = A(t)\,x(t) + b(t, u) + C(t)\,\xi(t), \qquad x(t_0) = x^0. \tag{2.1}$$

Here t is time, x is the n-dimensional vector of phase coordinates, u is the m-dimensional vector of control functions, $\xi(t)$ is the s-dimensional vector function of random disturbances, t_0 and x^0 are initial data. The matrices $A(t)$, $C(t)$ and the vector $b(t, u)$ are given functions. The distribution of the random vector $\xi(t)$ at each time moment is normal with zero mathematical expectation and with covariance matrix $\varepsilon G(t)$; there is no time correlation. Therefore $\xi(t)$ is a Gaussian white noise; its intensity is assumed to be small: ε is a small parameter ($\varepsilon \ll 1$), $G(t)$ is a bounded matrix. The vector of control functions is subject to the following constraint

$$u \in U \tag{2.2}$$

for all $t \in [t_0, T]$. Here U is a closed bounded set. It is assumed that at each time moment $t \in [t_0, T]$ all coordinates of the system $x(t)$ can be measured exactly and the feedback control in the form $u = u(t, x)$ can be realized.

The problem is to find the optimal control $u(t, x)$ as a function of t, x which satisfies the condition (2.2) and minimizes the mathematical expectation of a given function $F(x(T))$ at the terminal time moment T, i.e.

$$J = \mathrm{M}F(x(T)) \to \min. \tag{2.3}$$

We introduce the Bellman function $S(t, x)$ which is equal to the minimal value of the functional J for the process starting at the time moment t from the point x. The problem of optimal control formulated above is reduced under some conditions to the solution of the following nonlinear parabolic equation

$$S_t - H(t, x, S_x) + \frac{1}{2}\varepsilon \operatorname{Sp}(LS_{xx}) = 0 \tag{2.4}$$

with the initial condition

$$S(T, x) = F(x). \tag{2.5}$$

Here S_t is the partial derivative with respect to time, S_x is the vector of the first partial derivatives, S_{xx} is the matrix of the second partial derivatives, Sp denotes the trace of a matrix. The following notations are used

$$\begin{aligned} H(t, x, S_x) &= -\min_{u \in U} \left(S_x, Ax + b(t, u)\right), \\ L(t) &= CGC'. \end{aligned} \tag{2.6}$$

The symbol (,) denotes the scalar product of two vectors.

The problem of optimal control in the case of a small intensity of random disturbances is reduced to the initial-value problem (2.4), (2.5) for a nonlinear parabolic equation with a small parameter. The optimal control $u(t, x)$ can be determined by means of the minimum condition (2.6) after the function $S(t, x)$ has been found.

We begin with a formal construction of the approximate solution. We assume that the functions H, L, F, S are sufficiently smooth and try to find the solution of the initial-value problem (2.4), (2.5) in the form of a regular expansion

$$S(t, x) = S^0(t, x) + \varepsilon S^1(t, x) + \cdots \tag{2.7}$$

The function H can be presented in the form

$$H(t, x, S_x) = H(t, x, S_x^0) + \varepsilon\left(\nabla H(t, x, S_x^0), S_x^1(t, x)\right) + \cdots \tag{2.8}$$

Here ∇H is the gradient of the function H with respect to the vector S_x. Inserting the expansions (2.7), (2.8) into (2.4), (2.5), we obtain the following initial-value problems for the zeroth approximation

$$S_t^0 = H(t, x, S_x^0), \qquad S^0(T, x) = F(x), \tag{2.9}$$

and for the first approximation

$$S_t^1 - \left(\nabla H(t, x, S_x^0), S_x^1\right) + \frac{1}{2} \operatorname{Sp}(LS_{xx}^0) = 0, \qquad S^1(T, x) = 0. \tag{2.10}$$

Similar equations hold for the higher approximations. Equations (2.6) and (2.9) show that H is a Hamiltonian and S^0 is Bellman's function for a deterministic problem of optimal control ($\varepsilon = 0$). We assume that the solution of the deterministic problem is known. It consists of the optimal feedback control $u^0(t, x)$, Bellman's function $S^0(t, x)$ and the field of optimal trajectories

$$x = \varphi(t, a). \tag{2.11}$$

Here $\varphi(t, a)$ is a vector function, a is an n-dimensional vector of arbitrary constants. We assume that the equality (2.11) can be solved with respect to a in some domain, i.e.

$$a = \varphi^{-1}(t, x). \tag{2.12}$$

The equation (2.10) for the first approximation is a linear partial differential equation of the first order. Its characteristics satisfy the following system of ordinary differential equations

$$\frac{dx}{dt} = -\nabla H(t, x, S_x{}^0), \quad \frac{dS^1}{dt} = -\frac{1}{2} \operatorname{Sp} (L S^0_{xx}). \tag{2.13}$$

The first equation (2.13) describes according to L. S. Pontrjagin's maximum principle the optimal trajectories of the deterministic ($\varepsilon = 0$) system.

The general solution of this equation and its first integrals are given by the equalities (2.11), (2.12) respectively. Then from the equalities (2.11)—(2.13) we obtain the solution of the initial-value problem (2.10) in the form

$$\begin{gathered} S^1(t, x) = \frac{1}{2} \int_t^T \operatorname{Sp} [L(\tau) S^0_{yy}(\tau, y)] \, d\tau, \\ y = \varphi\big(\tau, \varphi^{-1}(t, x)\big). \end{gathered} \tag{2.14}$$

The optimal feedback control in the zeroth and first approximations $u^0(t, x)$, $u^1(t, x)$ are obtained from the conditions

$$\begin{gathered} \big(S_x{}^0, b(t, u^0)\big) = \min_{u \in U} \big(S_x{}^0, b(t, u)\big), \\ \big(S_x{}^0 + \varepsilon S_x{}^1, b(t, u^1)\big) = \min_{u \in U} \big(S_x{}^0 + \varepsilon S_x{}^1, b(t, u)\big). \end{gathered} \tag{2.15}$$

Thus the approximate solution of the stochastic optimal control problem is formally built up. It is expressed explicitly by the formulae (2.14), (2.15) in terms of the solution of the corresponding deterministic problem.

Some error estimates for this method are given in the paper [3]. Let the following conditions be satisfied: function b, A, L are bounded, L is a positive definite matrix for all $t \in [t_0, T]$, the functions $F(x)$, $S^0(t, x)$, $S(t, x)$ have continuous partial derivatives of the third order with respect to t, x, the function H from (2.6) has continuous partial derivatives with respect to S_x in some domain. Then the following estimates are true

$$\begin{gathered} |S(t, x) - S^0(t, x)| \leqq C_1 \varepsilon, \\ |S(t, x) - S^0(t, x) - \varepsilon S^1(t, x)| \leqq C_2 \varepsilon^2. \end{gathered} \tag{2.16}$$

Let $Z^0(t, x)$, $Z^1(t, x)$ denote the values of the functional J (2.3) for control processes starting at a time moment t from a point x if the controls u^0, u^1 of the zeroth and the first approximations (2.15) respectively are applied. Then the following estimates analogous to (2.16) are true

$$|S(t, x) - Z^0(t, x)| \leqq C_3\varepsilon,$$
$$|S(t, x) - Z^1(t, x)| \leqq C_4\varepsilon^2. \tag{2.17}$$

The estimates (2.16), (2.17) are proved to be true for all $x, t \leqq T$ under the conditions mentioned above. These conditions are very hard and often are not fulfilled in specific problems. For instance, the functions $F(x)$, $S^0(t, x)$ or their derivatives often have jumps. In these cases the solution of the problem can be found in the form

$$S = S^0 + W^0 + \varepsilon(S^1 + W^1) + \cdots \tag{2.18}$$

Here $S^0, S^1, \ldots$ are (as before) terms of regular expansion with respect to ε while $W^0, W^1, \ldots$ are functions of the boundary layer type which vanish rapidly at the distance of the order of ε from the surfaces of discontinuities. Such technique for the equation (2.4) is considered in the paper [8].

As an example we consider the following one-dimensional controlled system

$$\frac{dx}{dt} = u + \xi, \qquad |u| \leqq k, \qquad k > 0. \tag{2.19}$$

Here k is a constant, ξ is a white noise of a small intensity ε. The problem is to find the control u which satisfies the constraint (2.19) and minimizes the functional $J = Mx^2(T)/2$, where T is the terminal time moment. The equation (2.4) and the initial condition (2.5) for this problem can be written as follows

$$S_t - k|S_x| + \frac{\varepsilon}{2} S_{xx} = 0, \qquad S(T, x) = x^2/2. \tag{2.20}$$

The optimal control here is $u = -\operatorname{sgn} x$. Applying the technique (2.7) — (2.14) we obtain in the first approximation

$$S^0 + \varepsilon S^1 = \begin{cases} \frac{1}{2}\,[|x| + k(t - T)]^2 + \frac{\varepsilon}{2}\,(T - t), & |x| > k(T - t), \\ 0, & |x| < k(T - t). \end{cases} \tag{2.21}$$

The approximate solution (2.21) has a jump at the lines $|x| = k(T - t)$. If the initial point is in the region $|x| < k(T - t)$, then the determistic system ($\varepsilon = 0$) can attain the point $x = 0$ at the moment T, therefore here $S^0 = S^1 = 0$. The conditions under which the estimates (2.16), (2.17) were proved, are not satisfied for this example. In order to obtain a smooth and uniformly valid approximation for all t, x, one has to apply the expansions of the boundary layer type. However, the comparison of the approximate solution (2.21) with an exact solution which can be constructed for this example shows [3] that the difference between the two solutions has the order $O(\varepsilon^2)$

everywhere except in a small vicinity of the lines $|x| = k(T - t)$. In this vicinity this difference is of the order $O(\varepsilon)$. This situation is typical for the technique described above. More complicated examples solved by this technique see in the papers [3, 8].

3. Stochastic optimal control problems with integral constraints

We now consider a motion of a material point upon which control and disturbance forces act. The motion is described by the following equation

$$\frac{d^2x}{dt^2} = a(t)\, u + b(t)\, \xi, \qquad x(t_0) = x_0, \qquad \left.\frac{dx}{dt}\right|_{t_0} = \dot{x}_0. \tag{3.1}$$

Here x is the n-dimensional vector of Cartesian coordinates of the point, u is the n-dimensional vector of control functions, ξ is an n-dimensional vector of Gaussian independent white noises of unit intensity, t_0, x_0 and $\dot{x}_0$ are initial data. The scalar functions $a(t) \geqq 0$ and $b(t) \geqq 0$ denote respectively the control efficiency and the noise intensity. The problem is to find the optimal feedback control satisfying the integral constraint

$$\int_{t_0}^{T} |u|^m\, dt \leqq q_0, \qquad |u| = \left(\sum_{i=1}^{n} u_i^2\right)^{1/2}, \qquad m \geqq 1, \qquad q_0 \geqq 0, \tag{3.2}$$

which minimizes the mathematical expectation of the following function of the terminal coordinates

$$J = \mathrm{M}F\big(r(T)\big), \qquad r(T) = \left[\sum_{i=1}^{n} x_i^2(T)\right]^{1/2}. \tag{3.3}$$

The function F is assumed to be a non-negative and strictly monotone function of r. The constraint (3.2) has the meaning of a resource, fuel expenditure or energy constraint. We introduce a new variable q which is equal to the amount of control rèsource available at the moment t, by equations

$$dq(t)/dt = -|u|^m, \qquad q(t_0) = q_0. \tag{3.4}$$

The feedback control u is to be expressed as a function of arbitrary t, x, $\dot{x}$, q ($t \leqq T$, $q \geqq 0$). Introducing a new variable

$$y = x + (T - t)\, \dot{x} \tag{3.5}$$

we transform the system (3.2) to the form

$$\begin{aligned} &dy/dt = a_1(t)\, u + b_1(t)\, \xi, \qquad && y(t_0) = x_0 + (T - t_0)\, \dot{x}_0, \\ &a_1(t) = (T - t)\, a(t), && b_1(t) = (T - t)\, b(t). \end{aligned} \tag{3.6}$$

According to the equation (3.5) we have $y = x$ for $t = T$, therefore the formulae (3.3) can be rewritten as follows

$$J = \mathrm{M}F\big(r(T)\big), \qquad r(T) = \left[\sum_{i=1}^{n} y_i^2(t)\right]^{1/2}. \tag{3.7}$$

We denote by S the minimal value of the functional (3.3) or (3.7) which can be attained for given f, x, $\dot{x}$, q. It follows from the equalities (3.4), (3.6), (3.7) that the Bellman function S depends only on t, y, q.

Introducing one more variable

$$\tau = \int_t^T b_1^2(t_1)\, dt_1 \tag{3.8}$$

instead of t, we can write the parabolic equation and the initial condition for the case $m > 1$ as follows

$$S_\tau = \frac{1}{2} \Delta S + b_1^{-2}(t) \min_u \left[a_1(t)\,(u, S_y) - S_q|u|^m\right], \qquad S|_{\tau=0} = F(r). \tag{3.9}$$

Here t is a function of τ according to the equality (3.8), Δ is a Laplacian with respect to y, the minimum in (3.9) is to be found among all u. The minimized functional (3.7) cannot increase when the control resource q grows, therefore $S_q \leqq 0$. The minimum in (3.9) is attained for

$$u = -(S_y/|S_y|)\,[a_1(t)\,|S_y|/(-\,mS_q)]^{1/(m-1)}. \tag{3.10}$$

Inserting (3.10) into (3.9), we obtain a nonlinear partial differential equation which is invariant with respect to arbitrary rotations in the space of vectors y. The initial condition (3.9) is also invariant with respect to these rotations. Therefore the function S can be written in the form $S = S(\tau, r, q)$. Substituting (3.10) into (3.9) we obtain

$$S_\tau = \frac{1}{2}\left(S_{rr} + \frac{n-1}{r} S_r\right) - \frac{m-1}{m^{\frac{m}{m-1}}} |\alpha^m(\tau)\, S_r^m\, S_q^{-1}|^{1/(m-1)} \tag{3.11}$$

The initial condition and the boundary condition at $r = 0$ are

$$S(0, r, q) = F(r), \qquad S_r(\tau, 0, q) = 0. \tag{3.12}$$

The following notation is used

$$\alpha(\tau) = a_1(t)\,[b_1(t)]^{2(1-m)/m}, \tag{3.13}$$

where t is a function of τ according to (3.8). Instead of $2n + 2$ independent variables t, x, $\dot{x}$, q in the initial formulation of the problem we have now only 3 independent variables in the equation (3.11). These variables τ, r, q belong to the domain $D = \{\tau \geqq 0, r \geqq 0, q \geqq 0\}$.

The solution of the problem (3.11), (3.12) has a singularity at $\tau \to 0$ because we have $S_q = 0$ at $\tau = 0$ according to the initial condition (3.12). The asymptotic solution of the problem (3.11), (3.12) for $\tau \to 0$ can be found in the form [1]

$$S(\tau, r, q) = F(r) - q^p F_1(r)\, f_1(\tau) + \cdots \tag{3.14}$$

Inserting (3.14) into (3.11) and making equal the principal terms in the equation (3.11) for small τ, we obtain

$$p = m^{-1}, \qquad F_1(r) = F'(r),$$
$$f_1'(\tau) = \frac{m-1}{m} \alpha^{\frac{m}{m-1}}(\tau) f_1^{\frac{1}{1-m}}(\tau). \tag{3.15}$$

Integrating the last equation (3.15) with initial condition $f_1(0) = 0$, we have

$$f_1(\tau) = \left[\int_0^\tau \alpha^{\frac{m}{m-1}}(\tau')\, d\tau'\right]^{\frac{m-1}{m}}. \tag{3.16}$$

The formulae (3.14)—(3.16) define the asymptotic behaviour of the function S for $\tau \to 0$ and make it possible to step away from the plane $\tau = 0$ during numerical solution of the equation (3.11).

Because of the symmetry of the problem the optimal control vector u is always opposite to the vector y for all $m \geqq 1$. For $m > 1$ it is given by

$$u = -\frac{y}{r}\left[\frac{a_1(t)\, S_r}{-mS_q}\right]^{1/(m-1)} \tag{3.17}$$

4. Impulse control of stochastic systems

The important case $m = 1$ needs a special consideration. This case arises in the problems with fuel expenditure constraints. The Bellman function $S(\tau, r, q)$ in the case $m = 1$ is to be defined in the domain $D = \{\tau \geqq 0, r \geqq 0, q \geqq 0\}$. It can be shown [1] that everywhere in this domain $Q \leqq 0$, where Q is introduced by the equality

$$Q = \alpha(\tau)\, S_r + S_q \quad \text{in} \quad D_1. \tag{4.1}$$

Here $\alpha(t)$ is defined by the formula (3.13) for $m = 1$. The type of equation for S depends on the value of Q from (4.1), see [1]. In the region D_1 defined by the condition $Q < 0$ the function S satisfies the linear parabolic equation

$$S_\tau = \frac{1}{2}\left(S_{rr} + \frac{n-1}{r} S_r\right) \quad \text{in} \quad D_1. \tag{4.2}$$

In the region $D_2 = D \setminus D_1$ we have $Q = 0$, therefore we obtain from (4.1)

$$S(\tau, r, q) = R\big(r - \alpha(\tau)\, q, \tau\big) \qquad \text{in} \quad D_2. \tag{4.3}$$

Here R is an unknown function of two variables.

The meaning of the regions D_1, D_2 is the following. D_1 is a region of uncontrolled motion where only random disturbances act on the system. The control is equal to zero in D_1. The region D_2 corresponds to impulse control. In this region the phase point τ, r, q moves along the characteristics of the equation $Q = 0$ until it reaches either the boundary Γ between D_1 and D_2 or the plane $q = 0$ (in this case all the fuel is over). This condition defines the magnitude of control impulses.

The function S satisfies the initial and symmetry conditions (3.12) and the following smoothness condition at the boundary Γ

$$Q|_{\Gamma} = 0, \quad \left.\frac{\partial Q}{\partial r}\right|_{\Gamma} = 0. \tag{4.4}$$

The problem of optimal control in the case $m = 1$ is reduced to determination of the boundary Γ by means of solving the equation (4.2) in the region D_1 under the conditions (3.12), (4.3), (4.4).

The control in this problem can have impulses, and the number of impulses is not restricted. In particular, a special sliding regime along the boundary Γ with an infinite number of infinitesimal impulses can take place.

Another formulation of the problem implies constraints for the number of impulses. We shall consider the optimal control problem under the same assumptions as above for the case $m = 1$, but under one additional constraint: the control consists of a single impulse. The Bellman function for this one-impulse problem we denote $S^q(\tau, r)$. This function is defined in the domain D. In the unknown region of uncontrolled motion $\delta_1 \subset D$ this function satisfies the equation

$$S_\tau^q = \frac{1}{2}\left(S_{rr}^q + \frac{n-1}{r} S_r^q\right). \tag{4.5}$$

The following conditions at the boundaries of this region are deduced in the paper [5]

$$\begin{aligned} &S^q(0, r) = F(r), \ S_r^q(\tau, 0) = 0, \\ &S^q(\tau, r)|_\gamma = S^0(\tau, r - \alpha(\tau)\, q), \\ &S_r^q(\tau, r)|_\gamma = S_r^0(\tau, r - \alpha(\tau)\, q). \end{aligned} \tag{4.6}$$

Here γ is an unknown boundary between the region δ_1 and the region $\delta_2 = D \setminus \delta_1$ where the single impulse of control is to be made. The magnitude of the impulse is equal to q. By S^0 in the formulae (4.6) we denote the solution of the linear problem

$$S_\tau^0 = \frac{1}{2}\left(S_{rr}^0 + \frac{n-1}{r} S_r^0\right), \qquad S^0(0, r) = F(r), S_r^0(\tau, 0) = 0 \tag{4.7}$$

corresponding to the uncontrolled motion with $q = 0$ in the whole domain D.

Thus the determination of the impulse control in both cases (with or without constraints for the number of impulses) is reduced to some boundary value problems with unknown boundaries for parabolic equations.

5. Invariant group solutions

Let the following equalities be true for $m \geqq 1$

$$F(r) = r^l, \qquad \alpha(\tau) = A\tau^k, \tag{5.1}$$

where l, A, k are positive constants. In particular, if the functions $a(t)$ and $b(t)$ in the equation (3.1) are equal to

$$a(t) = A_1(T - t)^\mu, \qquad b(t) = B_1(T - t)^\nu \tag{5.2}$$

with constant A_1, B_1, μ, ν, then according to the equalities (3.6), (3.8), (3.13) the function $\alpha(\tau)$ has the form (5.1) with

$$k = \frac{m(\mu + 1) - 2(m - 1)(\nu + 1)}{m(2\nu + 3)}. \tag{5.3}$$

If the equalities (5.1) are true, then the equation (3.11) for $m > 1$, the boundary conditions (3.12) and the equations (4.2), (4.4)—(4.7) for $m = 1$ are invariant with respect to the following one-parameter group of transformations

$$r \to Cr, \ \tau \to C^2\tau, \ q \to C^{2-m-km}q, \ S \to C^l S \tag{5.4}$$

with parameter C. Therefore the boundary value problems from the sections 3 and 4 have solutions which are invariant with respect to the group (5.4). These solutions can be represented in terms of the invariants of the group (5.4), for example, in the following form

$$S = \tau^{l/2} W(y, z), \ y = Aq^{1/m}\tau^{k+1/2-1/m}, \ z = r\tau^{-1/2}. \tag{5.5}$$

Inserting the equalities (5.1), (5.5) into the equation and the boundary conditions (3.11), (3.12) we obtain

$$lW + yW_y(2k + 1 - 2m^{-1}) - zW_z = W_{zz} + (n - 1)z^{-1} W_z - 2(m - 1)\, m^{-1}\, y|W_y|^{1/(1-m)}$$
$$\times |W_z|^{m/(m-1)} \ (m \geqq 1), \quad W_z(y, 0) = 0, \quad W_z^{-l} \to 1 \quad (z \to \infty). \tag{5.6}$$

The boundary value problem (5.6) has only two independent variables. The solutions of the problems using similar independent variables (5.5) we shall consider for the cases of $m = 2$ and $m = 1$.

6. Optimal control for integral quadratic constraint

We assume that $m = 2$ what corresponds to the quadratic integral constraint (3.2), or energy constraint. We assume also that the minimized functional is quadratic too, i.e., $l = 2$, and that $k = 0$ in the equality (5.1). The case $k = 0$ takes place, for instance, if $m = 2$ and $\mu = \nu$ in the equalities (5.2), (5.3).

Thus we put $k = 0$, $m = l = 2$ in the formulae (5.5), (5.6). In this case we have $y = Aq^{1/2}$, and from $S_q \leqq 0$ it follows that $S_y \leqq 0$. The boundary value problem (5.6) takes the form

$$\begin{gathered} W_y[2W - zW_z - W_{zz} - (n - 1)\, z^{-1} W_z] = yW_z^2, \\ W_z(y, 0) = 0, \quad Wz^{-2} \to 1 \quad \text{for} \quad z \to \infty. \end{gathered} \tag{6.1}$$

The solution of the boundary value problem (6.1) in the domain $y \geqq 0$, $z \geqq 0$ can be constructed numerically by steps in y, if this solution is already known for some $y > 0$. The solution of the problem (6.1) has a singularity for $y = 0$. The asymptotic solution for small y can be built up in the form

$$W(y, z) = W_0(z) + y\, W_1(z) + \cdots \tag{6.2}$$

Here $W_0(z)$ is a solution of the problem (6.1) for $y = 0$. It is easily verified that the function

$$W_0(z) = z^2 + n \tag{6.3}$$

satisfies (6.1) for $y = 0$. The way to determine the next term of the expansion (6.2) and the technique of the numerical solution for the problem (6.1) is given in [4]. The partial derivative W_y is replaced by a finite difference, and for each discrete y we obtain from (6.1) the non-linear boundary value problem for the ordinary differential equation where z is an independent variable. This problem was solved numerically in the interval $0 \leqq z < \infty$ for each discrete y. The results of computer calculations are presented in the paper [4].

7. Optimal impulse controls

We now consider the invariant group solutions for the impulse control ($m = 1$) in the case $l = 2$. After the change of variables (5.5) the boundary value problem (4.2), (4.4), (3.12) in the region D_1 is transformed into the problem for the function $W(y, z)$ in the region D_1^0 with an unknown boundary Γ^0

$$\begin{gathered} W_{zz} + \left(z + \frac{n-1}{z}\right) W_z - 2W = (2k-1)\, yW_y, \\ W_z(y, 0) = 0, \qquad Wz^{-2} \to 1 \qquad (z \to \infty), \\ (W_z + W_y)|_{\Gamma^0} = (W_{zz} + W_{yz})|_{\Gamma^0} = 0. \end{gathered} \tag{7.1}$$

In the region D_2^0 which is the image of the region D_2, the function W satisfies the equation

$$W_z + W_y = 0. \tag{7.2}$$

The boundary value problem (4.5), (4.6) for the one-impulse control can be transformed in a similar way. We note that for our case ($l = 2$) we have $F = r^2$, and the solution of the problem (4.7) has the explicit form

$$S^0(\tau, r) = r^2 + n\tau. \tag{7.3}$$

Taking into consideration the expression (7.3) we transform the boundary value problem (4.5), (4.6) into the following problem in the region δ_1^0

$$\begin{gathered} W_{zz}^* + \left(z + \frac{n-1}{z}\right) W_z^* - 2W^* = (2k-1)\, yW_y^*, \\ W_z^*(y, 0) = 0, \qquad W^*z^{-2} \to 1 \qquad (z \to \infty), \\ W^*|_{\gamma^0} = (z-y)^2 + n, \qquad W_z^*|_{\gamma^0} = 2(z-y). \end{gathered} \tag{7.4}$$

Here W^* is analogous to W, γ^0 is the boundary between the regions δ_1^0, δ_2^0 which are the images of the appropriate regions D_1^0, D_2^0. The function W^* in the region δ_2^0 is given by the explicit formula, following from (7.3),

$$W^*(y, z) = (z - y)^2 + n. \tag{7.5}$$

The boundary value problems (7.1) and (7.4) have exact analytical solutions if $k = 1/2$, see papers [1, 5]. These exact solutions are

$$W(y, z) = \begin{cases} (z^2 + n) \exp(-yn^{-1/2}), & 0 \leqq z \leqq n^{1/2}, \\ 2n \exp[(z - y) n^{-1/2} - 1], & n^{1/2} \leqq z \leqq y + n^{1/2}, \\ (z - y)^2 + n, & y + n^{1/2} \leqq z \end{cases} \tag{7.5}$$

for the problem (7.1) and

$$W^*(y, z) = \begin{cases} (z^2 + n) [(y^2 + 4n)^{1/2} - y]^2/(4n), & 0 \leqq z \leqq g(y), \\ (z - y)^2 + n, \quad z \geqq g(y), \end{cases} \tag{7.7}$$

$$g(y) = [(y^2 + 4n)^{1/2} + y]/2$$

for the problem of one-impulse control (7.4). The boundary Γ^0 between the regions D_1^0, D_2^0 for the problem (7.1) is given by the equality $z = n^{1/2}$. In the region D_1^0 defined by the inequality $z < n^{1/2}$, the uncontrolled stochastic motion takes place.

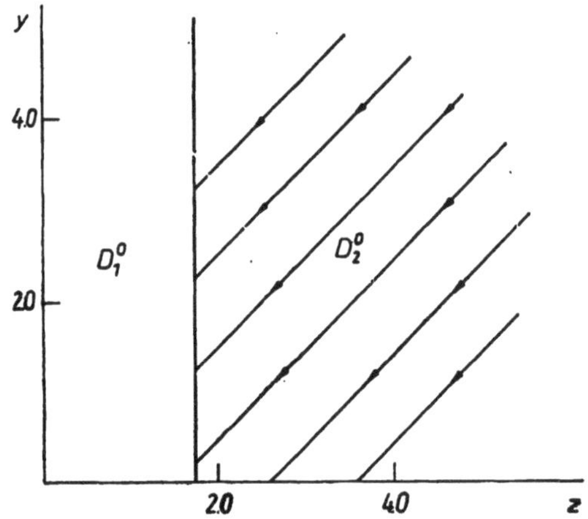

Fig. 1. Optimal impulse feedback control for $k = 1/2$, $n = 3$ (exact solution)

The region D_2^0 ($z > n^{1/2}$) corresponds to the impulse control and consists of two parts. If $z > y + n^{1/2}$ then the whole control resource is spent in one impulse, and after this impulse the system reaches the set $y = q = 0$. On the other hand, if $n^{1/2} < z < y + n^{1/2}$ then the impulse control tranfers the system to the boundary $\Gamma^0 : z = n^{1/2}$. The sliding impulse regime takes place along the boundary Γ^0. Fig. 1 shows the boundary Γ^0 between the regions D_1^0, D_2^0 for $k = 1/2$ and $n = 3$. The lines with arrows denote the trajectories $y - z = \text{const}$, corresponding to the impulse motion. The similar picture for the one-impulse control is shown at Fig. 2. Here the boundary γ^0 between the regions δ_1^0, δ_2^0 is given by the equation $z = g(y)$ where $g(y)$ is defined by (7.7). The region δ_1^0 is a region of uncontrolled stochastic motion, while the region δ_2^0 corresponds to the impulse control with a single impulse using the whole resource.

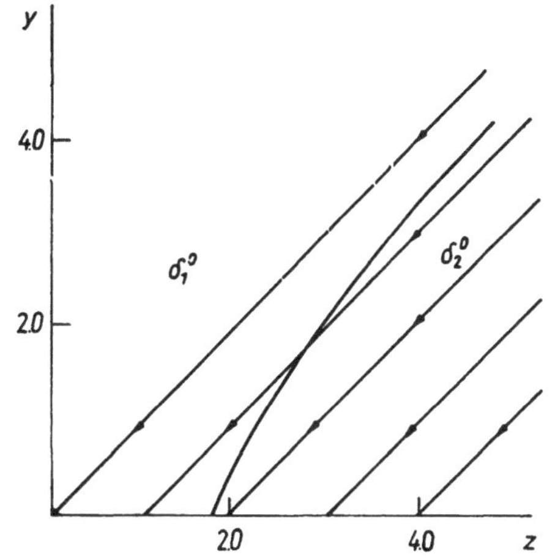

Fig. 2. Optimal one-impulse control for $k = 1/2$, $n = 3$ (exact solution)

The boundary value problems (7.1), (7.4) for $k \neq 1/2$ were solved numerically in the papers [4, 5]. The procedure of solution is the following. For small y the solution of the problem (7.1) and the boundary Γ^0 are represented in the form

$$\begin{aligned} W(y, z) &= W_0(z) + yW_1(z) + y^2W_2(z) + \cdots, \\ z &= \Gamma^0(y) = \Gamma_0 + \Gamma_1 y + \Gamma_2 y^2. \end{aligned} \tag{7.8}$$

The unknown functions W_0, W_1, W_2 and constants Γ_0, Γ_1, Γ_2 we find inserting the expansions (7.8) into the equations (7.1) and equating the coefficients at equal powers of y. Thus we obtain the expression (6.3) for W_0. Having calculated three terms of the asymptotic expansions (7.8) we obtained the function $W(y, z)$ for $y = 10^{-4}$. Further the implicit finite difference scheme was used with steps $\Delta y = \Delta z = 0.005$. At each step for y we solved numerically the corresponding system of finite difference equations and boundary conditions and determined the appropriate point of the unknow boundary Γ^0. The problem of one-impulse control (7.4) was treated similarly.

Some results obtained by computer calculations are presented at Figs. 3, 4. The unbroken curves correspond to the case of the impulse control without constraints for a number of impulses while the dotted lines belong to the case of one-impulse control. Fig. 3 shows the boundaries Γ^0, γ^0 for $k = 2$. The regions D_1^0, δ_2^0 lie to the left and the regions D_2^0, δ_2^0 are to the right of the corresponding curves Γ^0, γ^0. The region δ_1^0 always contains D_1^0, and for $y = 0$ these two regions coincide; this fact can be deduced analytically, see reference [5]. Fig. 4 shows the functions $W(1, z)$ and $W^*(1, z)$ for $k = 1$ and for different n. The following qualitative results are true. It is obvious that $W \leqq W^*$, but the difference between W and W^* is small. Thus the one-impulse control is only slightly worse than the control with an infinite number of impulses. The integer n is the dimension of the physical space (the number of coordinates), therefore the growth of n makes the control problem more difficult. That is why the values of the functionals W and W^* increase (see Fig. 4) and the regions of the uncontrolled motion D_1^0 and δ_1^0 widen (see Fig. 3) when n increases.

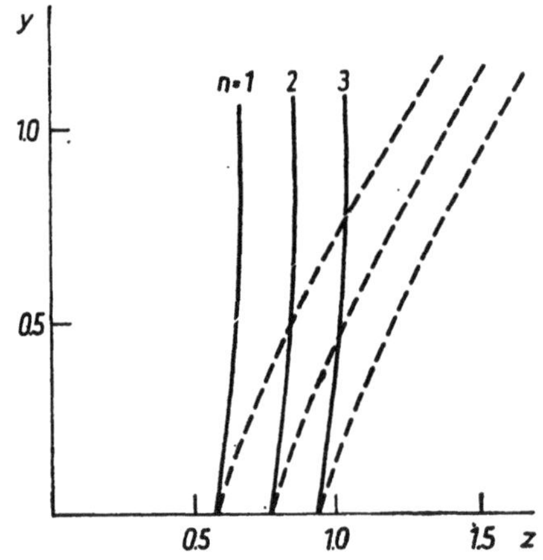

Fig. 3. Optimal multi-impulse and one-impulse (dotted lines) control for $k = 2$ (numerical solutions)

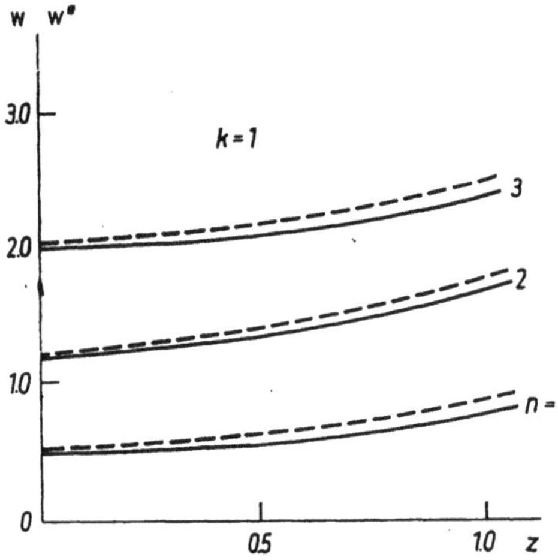

Fig. 4. Functions $W(1, z)$, $W^*(1, z)$ for $k = 1$ (numerical solutions)

Therefore the desired solutions of the optimal control problems were obtained by means of a combination of invariant group, asymptotic and numerical methods. The optimal feedback control for a multidimensional stochastic system is expressed here through a single curve (Γ^0 or γ^0) in a plane of variables y, z.

References

[1] Черноусько, Ф. Л., Автомодельные решения уравнения Беллмана для задач оптимальной коррекции случайных возмущений. Приклад. Мат. Мех. **35**, 333–342 (1971); English translation: J. Appl. Math. Mech. **35**, 291–300 (1971).

[2] Колмановский, В. Б., Черноусько, Ф. Л., Задачи оптимального управления при

неполной информации. Труды Четвертой Зимней школы по математическому программированию и смежным вопросам, выпуск 1, 5—120, Москва 1971.

[3] Соляник, А. И., Черноусько, Ф. Л., Приближенный метод синтеза оптимального управления системой, подверженной случайным возмущениям. Приклад. Мат. Мех. **36**, 829—839 (1972); English translation: J. Appl. Math. Mech. **36**, 780—790 (1972).

[4] Братусь, А. С., Черноусько, Ф. Л., Численное решение задач оптимальной коррекции при случайных возмущениях. Ж. вычисл. Мат. и мат. Физ. **14**, 68—78 (1974).

[5] Бородовский, М. Ю., Братусь, А. С., Черноусько, Ф. Л., Оптимальная импульсная коррекция при случайных возмущениях. Приклад. Мат. Мех. **39**, 797—805 (1975); English translation: J. Appl. Math. Mech. **39**, 767—775 (1975).

[6] Fleming, W. H., Stochastic control for small noise intensities. SIAM J. Control **9**, 473—517 (1971).

[7] Lions, J. L., On the numerical approximation of problems of impulse controls. *In:* Lecture Notes in Computer Science, vol. 27, Optimization Techniques IFIP Technical Conference Novosibirsk 1974, 232—251. Springer-Verlag, Berlin—Heidelberg—New York 1975.

[8] Братусь, А. С., Приближенное решение уравнения Беллмана для одного класса задач оптимального управления конечным состоянием. Приклад. Мат. Мех. **37**, 414—425 (1973); English translation: J. Appl. Math. Mech. **37**, 396—406 (1973).

Address of the author:

Институт проблем механики АН СССР
проспект Вернагского, 101
117526 Москва В-526
СССР

Eine gleichmäßige asymptotische Fehlerabschätzung zur Methode der finiten Elemente bei quasilinearen elliptischen Randwertproblemen

Jens Frehse, Bonn

Es sei Ω eine beschränkte offene Punktmenge des euklidischen $\boldsymbol{R}^2$ und F bzw. F_i, $i = 0, 1, 2$, vorgegebene stetige Funktionen auf $\boldsymbol{R}^2 \times \boldsymbol{R}^3$. Wir betrachten das Variationsproblem

$$\textit{Minimiere} \qquad \int_\Omega F(., u, \nabla u)\, dx, \qquad u \in H_0^1(\Omega) \tag{1}$$

bzw. das quasilineare Randwertproblem

$$\textit{Gesucht ist} \quad u \in H_0^1(\Omega) \quad \textit{mit}$$

$$\sum_{i=0}^{2} \big(F_i(., u, \nabla u), D_i v\big) = 0 \quad \textit{für alle} \quad v \in H_0^1(\Omega) \tag{2}$$

Hierbei ist $H_0^1 = H_0^1(\Omega)$ die Abschließung des Raumes $C_0^\infty(\Omega)$ der Testfunktionen im Sobolevraum $H^1(\Omega)$ der L^2-Funktionen mit verallgemeinerten Ableitungen in $L^2(\Omega)$. Es wurde $(w, z) = \int wz\, dx$, $\int = \int_{\boldsymbol{R}^n}$, gesetzt. (Beachte, daß $v = 0$ auf $\boldsymbol{R}^n - \Omega$).

Das Problem (1) bzw. (2) soll mit Hilfe der einfachsten Methode der finiten Elemente approximiert werden:

$$\textit{Minimiere} \quad \int_\Omega F(., v, \nabla v) dx, \qquad v \in V_h \tag{3}$$

bzw.

$$\textit{Gesucht ist} \quad u_h \in V_h \ \textit{mit}$$

$$\sum_{i=0}^{2} \big(F_i(., u_h, \nabla u_h), D_i v_h\big) = 0 \quad \textit{für alle} \quad v_h \in V_h. \tag{4}$$

V_h ist hierbei ein endlichdimensionaler Teilraum von stetigen, bez. einer Triangulation $\mathscr{T}_h$ von Ω *stückweise linearen* Funktionen mit Nullrandbedingungen. Die Triangulation $\mathscr{T}_h$ von Ω wird folgendermaßen definiert: Es sei $\Omega_h \subset \Omega$ ein polygonales Gebiet des $\boldsymbol{R}^2$, dessen Abschließung endliche Vereinigung einer Menge $\mathscr{E}_h$ von nichtentarteten Dreiecken ist, deren Inneres paarweise disjunkt ist. Die maximale Kantenlänge der Dreiecke aus $\mathscr{E}_h$ sei gleich h. Außerdem gelte für jedes Dreieck aus $\mathscr{E}_h$, daß jede seiner Seiten entweder eine Dreiecksseite eines anderen Dreiecks aus $\mathscr{E}_h$ ist oder eine Seite des Polygonzuges $\partial\Omega_h$ ist. Wir bezeichnen die Menge der Eckpunkte der Dreiecke aus $\mathscr{E}_h$ mit $\bar{Q}_h$, die Menge derjenigen Punkte aus $\bar{Q}_h$, die im Inneren von Ω_h enthalten sind, mit Q_h und setzen $\partial Q_h = \bar{Q}_h - Q_h$.

Wir nehmen nun an, daß zu einer Nullfolge von Zahlen h Triangulationen $\mathscr{E}_h$ gegeben sind, die überdies für $h \to 0$ die unten genannten Regularitätseigenschaften besitzen sollen.

Das Anliegen der Arbeit besteht darin, unter geeigneten Regularitäts- und Elliptizitätseigenschaften (siehe weiter unten) für die Koeffizienten F_i die *asymptotische Fehlerabschätzung*

$$\|u - u_h\|_{L^\infty(\Omega)} = O(h^2 \ln h) \tag{5}$$

zu beweisen. u ist die Lösung von (1) oder (2), u_h von (3) bzw. (4). Bisher verfügte man über folgende Ergebnisse, die z. B. unter den Voraussetzungen des Satzes dieser Arbeit gelten:

$$\|u - u_h\|_{L^2} = O(h), \qquad \|\nabla u - \nabla u_h\|_{L^2} = O(h), \tag{6}$$

$$\|u - u_h\|_{L^p} = O(h^2), \qquad 1 \leqq p < 2, \tag{7}$$

$$[u - u_h]_\alpha = O(h), \qquad \|u - u_h\|_{L^\infty} = O(h^{1+\delta}) \tag{8}$$

mit Konstanten α und $\delta \in (0, 1)$, die von den Daten abhängen, sowie $\|u - u_h\|_{L^\infty} = O(h^{3/2} \ln h^{1/2})$ (H. D. Mittelmann [11]). (6) folgt aus den Standardtechniken zur Theorie der finiten Elemente, siehe etwa [9, 10, 1]; (7) stammt von Johnson und Thomeé [12], die den bekannten Nitsche-Trick [4] auf den nichtlinearen Fall übertragen. (8) stammt vom Verfasser [2] — es bedeutet $[z]_\alpha = \sup \{|x - y|^{-\alpha} |z(x) - z(y)| \mid x \neq y, x, y \in \Omega\}$.

Für den Fall von linearen elliptischen Randwertproblemen ist die Abschätzung (5) durch Arbeiten von R. Scott [7], F. Natterer [3], J. Nitsche [5] und R. Rannacher [6] gesichert.

Für den Fall $n > 2$ ist ein Beweis von (5) noch nicht abzusehen.

An die Funktionen F_i sowie deren Ableitungen $F_{ik}(x, \eta) = \frac{\partial}{\partial \eta_k} F_i(x, \eta)$, $F_{ikj}(x, \eta) = \frac{\partial}{\partial \eta_k} \frac{\partial}{\partial \eta_j} F_i(x, \eta)$, $i, k, j = 0, \ldots, 2$, $\eta = (\eta_0, \ldots, \eta_2) \in \mathbf{R}^3$, stellen wir die folgenden Bedingungen:

$$F_i \in C^{1+\gamma}(\mathbf{R}^2 \times \mathbf{R}^3), \qquad i = 0, \ldots, 2, \text{ mit einem } \gamma \in (0, 1), \tag{9}$$

$$F_{ik} \in C^1(\mathbf{R}^2 \times \mathbf{R}^3), \qquad i, k = 0, \ldots, 2. \tag{10}$$

Es existiert eine Konstante K mit

$$|F_{ik}| \leqq K, \qquad i, k = 0, \ldots, 2 \tag{11}$$

$$|F_{ikx}| + |F_{ikj}| \leqq K, \quad i, k, j = 0, \ldots, 2 \tag{12}$$

gleichmäßig auf $\mathbf{R}^2 \times \mathbf{R}^3$.

Für die Funktionen F_{ik} und F_{ikj}, für die ein i, k, j gleich 0 wird, wäre allgemeineres

Wachstum als (11) und (12) zulässig. Wir verzichten jedoch darauf, um die Darstellung nicht zu überladen.

Schließlich benötigen wir die Elliptizitätsbedingung

$$\sum^{2} F_{ik}(x, \eta)\, \xi_i \xi_k \geqq \lambda(\xi_1^2 + \xi_2^2) \tag{13}$$

mit einer Konstanten $\lambda > 0$ gleichmäßig für $x \in \Omega$, $\xi \in \boldsymbol{R}^2$, $\eta \in \boldsymbol{R}^3$ sowie die Monotoniebedingung

$$\sum_{i=0}^{2} \left(F_i(\cdot, v, \nabla v) - F_i(\cdot, w, \nabla w), D_i v - D_i w\right) \geqq c\|v - w\|_{H^1}^2 \tag{14}$$

mit einer von v und $w \in H_0^1$ unabhängigen Konstanten $c > 0$.

Für die Triangulationen $\mathcal{T}_h$ benötigen wir die folgenden Regularitätsbedingungen:

Das Verhältnis zwischen maximaler und minimaler Seitenlänge aller Dreiecke aus $\mathcal{E}_h$ ist für $h \to 0$ gleichmäßig beschränkt. (15)

Die Winkel der Dreiecke aus $\mathcal{E}_h$ sind für $h \to 0$ gleichmäßig durch einen Winkel $\alpha_0 > 0$ nach unten beschränkt. (16)

$$d(\partial\Omega, \partial\Omega_h) = O(h^2). \tag{17}$$

Für den Rand $\partial\Omega$ von Ω setzen wir mit einem $\gamma \in (0, 1)$ voraus:

$$\partial\Omega \in C^{2+\gamma}. \tag{18}$$

Damit können wir den angekündigten Satz formulieren:

Satz. *Es mögen die Voraussetzungen* (9)—(14) *für die Funktionen F_i und die Regularitätsbedingungen* (15)—(17) *für die Triangulationen $\mathcal{T}_h$ ($h \to 0$) sowie Bedingung* (18) *für $\partial\Omega$ gelten. Dann gilt für die Lösung u von* (2) *und die Lösung u_h von* (4) *die Abschätzung*

$$\|u - u_h\|_{L^\infty} = O(h^2 \ln h).$$

Beweis. Die Voraussetzungen sichern die Existenz einer eindeutigen Lösung $u \in C^{2+\delta}$, $\delta \in (0, 1)$. Wegen der Regularitätsvoraussetzungen (15) und (16) haben die Punkte aus Q_h (also die „inneren" Gitterpunkte der Triangulation) einen Abstand von $\partial\Omega_h$, der größer ist als bh mit einer Konstanten $b > 0$. b sei nicht von h abhängig.

Es sei $y \in Q_h$ und $B_{ah}(y)$ ein Kreis um y mit Radius ah, $0 < a < b$. Über a wird noch verfügt. Es gilt $B_{ah}(y) \subset \Omega_h$. Ferner sei $\omega_h = \omega_h^y$ eine symmetrische *Mittelfunktion* (mit Symmetriezentrum y) aus $C_0^\infty(B_{ah}(y))$ mit $\omega_h \geqq 0$ und $\|\omega_h\|_\infty \leqq \frac{K}{h^2}$, wobei K unabhängig von h ist. Wir definieren eine „regularisierte" Greensche Funktion Γ_h des linearen Differentialoperators L, der definiert ist durch $Lz = -\sum_{i,k=1}^{2} D_i\left(F_{ik}(\cdot, u, \nabla u)\, D_k z\right)$; Γ_h sei definiert als Lösung des Dirichletproblems

$$L\Gamma_h = \omega_h, \qquad \Gamma_h \in H_0^1(\Omega).$$

Da $\omega_h \in C_0^\infty$ und die $F_{ik} \in C^1$ sind, gilt $\Gamma_h \in C^2(\Omega)$ und wegen der Randregularität $\Gamma_h \in C(\bar{\Omega})$.

Entsprechend definieren wir die diskrete Approximation $G_h \in V_h$ von Γ_h durch

$$\sum_{i,k=1}^{2} \left(F_{ik}(.,u,\nabla u)\, D_k G_h, D_i\varphi\right) = (\omega_h, \varphi) \quad \text{für alle} \quad \varphi \in V_h.$$

Für Γ_h und G_h gelten die folgenden, fundamentalen Abschätzungen:

$$\|\Gamma_h - G_h\|_{L^1} = O(h^2 \ln h), \tag{19}$$

$$\|\nabla\Gamma_h - \nabla G_h\|_{L^1} = O(h \ln h). \tag{20}$$

Diese Abschätzungen wurden in der Arbeit von Scott [7] für die regularisierten Greenschen Funktionen für das Neumann-Problem bewiesen, sie gelten jedoch — wie in [7] behauptet — auch für das Dirichletproblem elliptischer Operatoren mit variablen Koeffizienten. Ein unabhängiger Beweis zu (19) und (20) erscheint demnächst. Man verwendet bei den hiesigen Überlegungen die regularisierte Greensche Funktion Γ_h anstelle der klassischen Greenschen Funktion, da dies technisch bequemer ist. Nach diesen Vorbereitungen können wir mit dem eigentlichen Beweis beginnen:

Es gilt

$$\int \omega_h(u-u_h)\,dx = \sum_{i,k=1}^{2} \{(F_{ik}D_k u, D_i\Gamma_h) - (F_{ik}D_k u_h, D_i G_h)\}$$
$$= \sum_{i,k=1}^{2} \left\{\left(F_{ik}D_k(u-u_h), D_i G_h\right) - \left(D_i(F_{ik}D_k u), \Gamma_h - G_h\right)\right\}. \tag{21}$$

Hierbei ist $F_{ik} = F_{ik}(.,u,\nabla u)$; $\int$ bedeutet die Integration über Ω. Da $u \in C^{2+\delta}$ ist, folgt mit (19)

$$\left|\int \omega_h(u-u_h)\,dx\right| \leq \left|\sum_{i,k=1}^{2} \left(F_{ik}D_k(u-u_h), D_i G_h\right)\right| + O(h^2 \ln h).$$

Um zu einer Abschätzung des ersten Summanden zu kommen, benutzen wir (2) und (4) und erhalten

$$0 = \sum_{i,k=0}^{2} \left(F_i(.,u,\nabla u) - F_i(.,u_h,\nabla u_h), D_i G_h\right)$$
$$= \sum_{i,k=0}^{2} \left(F_{ik}D_k(u-u_h), D_i G_h\right)$$
$$+ \sum_{i,k,j=0}^{2} \left(\int_0^1 (1-t)\, F_{ikj}(t)\, D_k(u-u_h)\, D_j(u-u_h), D_i G_h\right).$$

Hierbei ist $F_{ik} = F_{ik}(.,u,\nabla u)$, $F_{ikj}(t) = F_{ikj}\left(.,u+t(u-u_h), \nabla u + t\nabla(u-u_h)\right)$, D_0 ist die Identität.

Daraus folgt unter Benutzung der Wachstumsbedingungen für F_{ik} und F_{ikj} sowie F_{xij}

$$\left|\sum_{i,k=1}^{2} \left(F_{ik}D_k(u-u_h), D_i G_h\right)\right|$$
$$\leq K' \int |u-u_h|\,(|\nabla G_h| + |G_h|)\,dx + K' \int |\nabla u - \nabla u_h|^2\, |\nabla G_h|\,dx. \tag{22}$$

Hierbei wurde noch die partielle Integration

$$\left(F_{0k}D_k(u - u_h), G_h\right) = -\left(u - u_h, D_k(F_{0k}G_h)\right)$$

durchgeführt und anschließend abgeschätzt. Wegen der gleichmäßigen Beschränktheit von $\|u_h\|_\infty$ (siehe [2]) können die Terme $|u - u_h|^2$, $|u - u_h| \cdot |\nabla u - \nabla u_h|$ durch $K|u - u_h| + |\nabla u - \nabla u_h|^2$ abgeschätzt werden.

Mit (21) und (22) ergibt sich

$$\left|\int \omega_h(u - u_h)\, dx\right| \leqq O(h^2 \ln h) + K' \int |u - u_h| \left(|\nabla G_h| + |G_h|\right) dx$$
$$+ K' \int |\nabla u - \nabla u_h|^2\, |\nabla G_h|\, dx. \tag{23}$$

Es sei $I_h u$ diejenige stetige Funktion aus V_h, die in den Punkten aus Q_h mit u übereinstimmt, auf den Dreiecken der zugrunde liegenden Triangulation linear ist und auf $\partial\Omega_h$ verschwindet. Da $u \in C^{2+\delta}$ und $\partial\Omega \in C^{2+\delta}$ und außerdem die Bedingung $d(\partial\Omega, \partial\Omega_h) \leqq ch^2$ gilt, haben wir die Beziehung

$$\left|\int \omega_h I_h u\, dx\right| \leqq \left|\int \omega_h u\, dx\right| + O(h^2).$$

Wir erinnern daran, daß ω_h seinen Träger in einer Kugel mit Mittelpunkt $y \in Q_h$ und Radius ah hat, wobei über $a > 0$ noch verfügt werden kann, wenn es nur von h unabhängig ist. Wegen der Regularitätsbedingungen für die Triangulationen kann man a so klein wählen, daß für beliebiges $v \in V_h$ die Beziehung

$$\left|\int \omega_h v\, dx\right| \geqq |v(y)| - \frac{1}{4}\, \|v\|_\infty$$

gilt.

Daraus folgt für $y \in Q_h$

$$|I_h u(y) - u_h(y)| \leqq \frac{1}{4}\, \|u - u_h\|_\infty + \left|\int \omega_h(u - u_h)\, dx\right| + O(h^2). \tag{24}$$

Hierbei haben wir die Beziehung $\|I_h v\|_\infty \leqq \|v\|_\infty$ und $I_h u(y) = u(y)$, $y \in Q_h$, benutzt.

Wir schätzen nun die rechte Seite von (23) ab, und zwar zunächst den Term

$$A = \int |u - u_h| \left(|\nabla G_h| + |G_h|\right) dx.$$

Es gilt

$$A \leqq \int |u - u_h| \left(|\nabla G_h - \nabla \Gamma_h| + |G_h - \Gamma_h|\right) dx$$
$$+ \int |u - u_h| \left(|\nabla \Gamma_h| + |\Gamma_h|\right) dx$$
$$\leqq K_1\, \|u - u_h\|_\infty\, h \ln h + \int |u - u_h| \left(|\nabla \Gamma_h| + |\Gamma_h|\right) dx.$$

K_1 hängt nicht von h oder y ab; es wurde Abschätzung (21) und (22) benutzt. Da Γ_h die Differentialgleichung

$$-\sum_{i,k=1}^{2} D_i(F_{ik} D_k \Gamma_h) = \omega_h, \qquad \Gamma_h = 0 \quad \text{auf} \quad \partial\Omega$$

erfüllt (ω_h war eine Mittelfunktion mit Träger in der Kugel $B_{ah}(y)$), so folgt aus der Gärdingschen Ungleichung in L^p (siehe [8]), daß

$$\|\nabla \Gamma h\|_{L^{1+1/3}} \leqq K_2 \sup \left\{ \sum_{i,k=1}^{2} (F_{ik} D_k \Gamma_h, D_i \varphi) \,\Big|\, \|\nabla \varphi\|_{L^4} \leqq 1 \right\} \leqq K_3$$

gleichmäßig für $y \in Q_h$ und $h \to 0$. (Beachte, daß $n = 2$.) Daraus folgt die Existenz einer von h unabhängigen Zahl $\varrho > 0$, so daß $K' \int_* (|\nabla \Gamma_h| + |\Gamma_h|)\, dx \leqq \frac{1}{4}$, wenn $\int_*$ die Integration über die Kugel $B_\varrho(y)$ mit Radius ϱ und Mittelpunkt y bedeutet. (Γ_h wird außerhalb von Ω durch Null fortgesetzt.) Wir erhalten damit

$$\begin{aligned} K'A &\leqq K_1' \|u - u_h\|_\infty h \ln h + K' \int |u - u_h| (|\nabla \Gamma_h| + |\Gamma_h|)\, dx \\ &\leqq K_1 \|u - u_h\|_\infty h \ln h + K' \int_* |u - u_h| (|\nabla \Gamma_h| + |\Gamma_h|)\, dx \\ &\quad + K' \int_{**} |u - u_h| (|\nabla \Gamma_h| + |\Gamma_h|)\, dx \\ &\leqq \|u - u_h\|_\infty \left(K_1 h \ln h + \frac{1}{4} \right) + K' \int_{**} |u - u_h| (|\nabla \Gamma_h| + |\Gamma_h|)\, dx. \end{aligned} \tag{25}$$

Hierbei ist $\int_{**}$ die Integration über $\Omega - B_\varrho(y)$. Da Γ_h der Differentialgleichung

$$-\sum_{i,k=1}^{2} D_i(F_{ik} D_k \Gamma_h) = \omega_h, \qquad \Gamma_h = 0 \text{ auf } \partial\Omega$$

genügt und ω_h seinen Träger in $B_{ah}(y)$ hat, folgt die gleichmäßige Beschränktheit von $\nabla \Gamma_h$ und Γ_h auf $\Omega - B_\varrho(y)$. (Dies sind Standardüberlegungen aus der Theorie der partiellen Differentialgleichungen. Die $C^{2+\delta}$-Regularität von $\partial\Omega$ wird dabei benutzt.) Wir schätzen daher weiter ab:

$$\int_{**} |u - u_h| (|\nabla \Gamma_h| + |\Gamma_h|)\, dx \leqq K_4 \|u - u_h\|_{L^1}. \tag{26}$$

Aus den Abschätzungen von Johnson-Thomée folgt $\|u - u_h\|_{L^1} = O(h^2)$.

Es ist zu beachten, daß bei der Anwendung der Abschätzungstechnik aus [12] zusätzlich Terme niederer Ordnung auftreten, da $F_i(., u, \nabla u)$ auch von u abhängt. Diese Terme lassen sich jedoch ebenfalls mit Techniken aus der Theorie der partiellen Differentialgleichungen behandeln.

Wir erhalten somit aus (25) und (26)

$$K'A \leqq \|u - u_h\|_\infty \left(K_1 h \ln h + \frac{1}{4} \right) + O(h^2)$$

und für genügend kleines h

$$K'A \leqq \frac{1}{2} \|u - u_h\|_\infty + O(h^2). \tag{27}$$

Mit (23) folgt daraus

$$\left| \int \omega_h (u - u_h)\, dx \right| \leqq O(h^2 \ln h) + \frac{1}{2} \|u - u_h\|_\infty + K' \int |\nabla u - \nabla u_h|^2 |\nabla G_h|\, dx. \tag{28}$$

Wir schätzen nun den Term

$$B = \int |\nabla u - \nabla u_h|^2 \, |\nabla G_h| \, dx$$

ab. Es sei $B_\varrho(y)$ eine Kugel um $y \in Q_h$ mit von h unabhängigem Radius ϱ, über den noch verfügt wird, und es bezeichne $\int_*$ die Integration über $B_\varrho(y)$, $\int_{**}$ diejenige über $\Omega - \Omega \cap B_\varrho(y)$. Es gilt dann

$$B \leqq \int_{**} |\nabla u - \nabla u_h|^2 \, |\nabla G_h| \, dx + \|\nabla u - \nabla u_h\|_\infty \int_* |\nabla u - \nabla u_h| \, |\nabla G_h| \, dx. \tag{29}$$

Zur Abschätzung von $\int_* |\nabla u - \nabla u_h| \, |\nabla G_h| \, dx$ verwenden wir nun unsere Abschätzung der Morreynorm von $\nabla u - \nabla u_h$ aus [2]. Es wurde dort bewiesen, daß ein $\alpha > 0$ existiert, so daß

$$\sup \left\{ R^{-\alpha} \left(\int_{B_R(x)} |\nabla u - \nabla u_h|^2 \, dx \right)^{1/2} \middle| \, x \in \Omega, R > 0 \right\} = O(h).$$

Daraus folgt für alle $\beta \in (0, \alpha)$ — wir wählen ein festes solches —

$$\sup \left\{ \left(\int |x - y|^{-\beta} \, |\nabla u - \nabla u_h|^2 \, dx \right)^{1/2} \middle| \, y \in \Omega \right\} = O(h).$$

Damit gilt für $y \in Q_h$

$$\begin{aligned} \int_* |\nabla u - \nabla u_h| \, |\nabla G_h| \, dx &= \int_* |x - y|^{-\beta/2} \, |\nabla u - \nabla u_h| \cdot |x - y|^{\beta/2} \, |\nabla G_h| \, dx \\ &\leqq Kh \left(\int_* |x - y|^\beta \, |\nabla G_h|^2 \, dx \right)^{1/2} \leqq Kh \left(\int_* |x - y|^\beta \, |\nabla G_h - \nabla \Gamma_h|^2 \, dx \right)^{1/2} \\ &\quad + Kh \left(\int_* |x - y|^\beta \, |\nabla \Gamma_h|^2 \, dx \right)^{1/2}. \end{aligned}$$

Wie im Anhang bewiesen wird, ist $\int |\nabla G_h - \nabla \Gamma_h|^2 \, dx = O(1)$; wählt man ϱ genügend klein, so ist $|x - y|$ für $x \in B_\varrho(y)$ klein, und wir erhalten

$$\int_* |\nabla u - \nabla u_h| \, |\nabla G_h| \, dx \leqq \varepsilon h + Kh \left(\int_* |x - y|^\beta \, |\nabla \Gamma_h|^2 \, dx \right)^{1/2}.$$

Zur Abschätzung des letzten Summanden benutzen wir die Differentialgleichung für Γ_h und erhalten

$$\sum_{i,k=1}^{2} \left(F_{ik} D_k \Gamma_h, D_i(\Gamma_h \, |x - y|^{\beta'}) \right) = (\omega_h, \Gamma_h \, |x - y|^{\beta'}), \qquad \beta > \beta' > 0.$$

Da $\omega_h \in C_0^\infty(B_{ah}(y))$ und $|\omega_h| \leqq \dfrac{K}{h^2}$, folgt $(\omega_h, \Gamma_h \, |x - y|^{\beta'}) \leqq K'$ und mit der Elliptizitäts- und Wachstumsbedingung

$$\int |\nabla \Gamma_h|^2 \, |x - y|^{\beta'} \, dx \leqq K_0 \int \Gamma_h^2 \, |x - y|^{\beta' - 2} \, dx + K''.$$

Da $-\sum_{i,k=1}^{2} D_i(F_{ik} D_k \Gamma_h) = \omega_h$, $\Gamma_h = 0$ auf $\partial\Omega$, und $\|\omega_h\|_{L^1} \leqq K_3$ gleichmäßig, folgt unter Hinzuziehung der Sobolevschen Ungleichung, daß $\|\Gamma_h\|_p \leqq C_p$ gleichmäßig für $h \to 0$ und $y \in Q_h$.

Da die durch $|x - y|^{\beta'-2}$ definierte Funktion in einem $L^{1+\varepsilon}$, $\varepsilon > 0$, liegt, erhält man die gleichmäßige Beschränktheit von $\int \Gamma_h^2 \, |x - y|^{\beta'-2} \, dx$ und damit

$$\int |\Gamma_h|^2 \, |x - y|^{\beta'} \, dx \leqq K_4 \tag{30}$$

gleichmäßig für $y \in Q_h$ und $h \to 0$. Wählt man nun den Radius ϱ von $B_\varrho(y)$ klein genug, so folgt aus (30)

$$K \int_* |\nabla \Gamma_h|^2 \, |x - y|^\beta \, dx \leqq K_4 \varrho^{\beta - \beta'},$$

und wir erhalten

$$\int_* |\nabla u - \nabla u_h| \, |\nabla G_h| \, dx \leqq 2\varepsilon h. \tag{31}$$

Wir schätzen nun den Term $\int_{**} |\nabla u - \nabla u_h|^2 \, |\nabla G_h| \, dx$ ab. Es gilt

$$\int_{**} |\nabla u - \nabla u_h|^2 \, |\nabla G_h| \, dx \leqq \int_{**} |\nabla u - \nabla u_h|^2 \, |\nabla G_h - \nabla \Gamma_h| \, dx + \int_{**} |\nabla u - \nabla u_h|^2 \, |\nabla \Gamma_h| \, dx. \tag{32}$$

Aus dem Randwertproblem für Γ_h folgt, daß $|\nabla \Gamma_h|$ außerhalb von $B_\varrho(y)$ gleichmäßig für $y \in Q_h$ und $h \to 0$ beschränkt ist, daraus folgt

$$\int_{**} |\nabla u - \nabla u_h|^2 \, |\nabla \Gamma_h| \, dx = O(h^2).$$

Zur Abschätzung des anderen Summanden schreiben wir

$$\int_{**} |\nabla u - \nabla u_h|^2 \, |\nabla G_h - \nabla \Gamma_h| \, dx$$
$$\leqq \|\nabla u - \nabla u_h\|_\infty \, \|\nabla u - \nabla u_h\|_{L^2} \left(\int_{**} |\nabla G_h - \nabla \Gamma_h|^2 \, dx\right)^{1/2}.$$

Wie im Anhang (Lemma 2) bewiesen wird, gilt

$$\left(\int_{**} |\nabla G_h - \nabla \Gamma_h|^2 \, dx\right)^{1/2} \leqq K_\varrho \left(\int |G_h - \Gamma_h|^2 \, dx\right)^{1/2} + O(h)$$

und aus der Sobolewschen Ungleichung sowie (20)

$$\left(\int_{**} |\nabla G_h - \nabla \Gamma_h|^2 \, dx\right)^{1/2} \leqq K_\varrho' \left(\int |\nabla G_h - \nabla \Gamma_h| \, dx\right) = O(h \ln h).$$

Daraus ergibt sich zusammengefaßt

$$\int_{**} |\nabla u - \nabla u_h|^2 \, |\nabla G_h| \, dx \leqq O(h^2) + C \, \|\nabla u - \nabla u_h\|_\infty \, h^2 \ln h$$

mit einer Konstanten C. (Hierbei wurde $\|\nabla u - \nabla u_h\|_{L^2} = O(h)$ benutzt.) Wählt man $h_0 > 0$ genügend klein, so ist für alle $h \in (0, h_0)$ mit vorgegebenem $\varepsilon > 0$

$$\int_{**} |\nabla u - \nabla u_h|^2 \, |\nabla G_h| \, dx \leqq O(h^2) + \varepsilon h \, \|\nabla u - \nabla u_h\|_\infty. \tag{33}$$

Faßt man nun die Abschätzungen (28), (29), (31) und (33) zusammen, so erhält man für $0 < h < h_0$

$$\left| \int \omega_h (u - u_h) \, dx \right| \leqq \frac{1}{2} \, \|u - u_h\|_\infty + K\varepsilon h \, \|\nabla u - \nabla u_h\|_\infty + O(h^2 \ln h). \tag{34}$$

Wir erinnern daran, daß $\omega_h = \omega_h^y \in C_0^\infty(B_{ah}(y))$ als eine nichtnegative Mittelfunktion definiert wurde. Über a durfte noch verfügt werden. (Die Konstanten K und h_0 hängen selbstverständlich von a ab.)

Da $I_h u - u_h \in V_h$ ist und wir die Regularitätsbedingungen (15)—(17) für die Triangulation vorausgesetzt haben, gilt

$$\left(1 - \frac{1}{4}\right) \cdot \|I_h u - u_h\|_\infty \leqq \sup \left\{\left|\int \omega_h^y (I_h u - u_h)\, dx\right|; y \in Q_h\right\},$$

sofern wir a genügend klein (jedoch unabhängig von h) wählen. a hängt von den Konstanten der Regularitätsbedingungen (15)—(17) ab. Da $u \in C^2$, gilt $\|I_h u - u\|_\infty = O(h^2)$. Daraus folgt mit (34) und der Gl. $\int \omega_h\, dx = 1$

$$\left(1 - \frac{1}{4}\right) \|u - u_h\|_\infty \leqq \frac{1}{2} \|u - u_h\|_\infty + K\varepsilon h \|\nabla u - \nabla u_h\|_\infty + O(h^2 \ln h)$$

und

$$\|u - u_h\|_\infty \leqq K'\varepsilon h \|\nabla u - \nabla u_h\|_\infty + O(h^2 \ln h). \tag{35}$$

Da $u_h \in V_h$ und $u \in C^2$, folgt

$$\|\nabla u - \nabla u_h\|_\infty \leqq \|\nabla(I_h u - u_h)\|_\infty + O(h) \leqq \frac{C}{h} \|u - u_h\|_\infty + O(h).$$

Setzt man dies in (35) ein, ergibt sich

$$\|u - u_h\|_\infty \leqq C'\varepsilon \|u - u_h\|_\infty + O(h^2 \ln h).$$

Wählt man $\varepsilon = \dfrac{1}{2C}$, ergibt sich

$$\|u - u_h\|_\infty = O(h^2 \ln h),$$

und der Satz ist bewiesen.

Anhang. Abschätzungen für Γ_h und G_h

Lemma 1. *Unter den Voraussetzungen des Satzes gilt*

$$\|\nabla \Gamma_h - \nabla G_h\|_{L^2} = O(1).$$

Beweis. Für $\varphi \in V_h$ gilt

$$\sum_{i,k=1}^{2} \left(F_{ik} D_k(\Gamma_h - G_h), D_i \varphi\right) = 0. \tag{36}$$

Es sei $I_h \Gamma_h \in V_h$ Interpolierende von Γ_h auf V_h bez. $\|.\|_{H^2}$. Es gilt

$$\|\nabla I_h \Gamma_h - \nabla \Gamma_h\|_2 \leqq Kh \|\nabla^2 \Gamma_h\|_2.$$

Aus (36) und der gleichmäßigen Elliptizität folgt mit $\varphi = I_h\Gamma_h - G_h$ in geläufiger Weise ($\Gamma_h' = I_h\Gamma_h$)

$$\|\nabla\Gamma_h - \nabla G_h\|_{L^2} \leqq K\, \|\nabla\Gamma_h - \nabla\Gamma_h'\|_{L^2} \leqq K'h\, \|\nabla^2\Gamma_h\|_{L^2}. \tag{37}$$

Da $F_{ik} \in C^\alpha$ und $\partial\Omega \in C^{2+\gamma}$, gilt

$$\|\nabla^2\Gamma_h\|_{L^2} \leqq C\, \|\omega_h^y\|_{L^2} \leqq \frac{C}{h^2}\left(\int\limits_{B_{\alpha h}} dx\right)^{1/2} \leqq \frac{K}{h}. \tag{38}$$

Aus (37) und (38) folgt die Behauptung des Lemmas.

Lemma 2. *Es sei $\varrho > 0$ und $y \in Q_h$. Mit den bisherigen Voraussetzungen und Bezeichnungen gilt dann*

$$\int_{**} |\nabla\Gamma_h - \nabla G_h|^2\, dx \leqq K_\varrho \int |\Gamma_h - G_h|^2\, dx + O(h^2)$$

mit einer von y und h unabhängigen Konstanten K_ϱ.

$\int_{**}$ bedeutet die Integration über $\Omega - \Omega \cap B_\varrho(y)$.

Beweis. Es gilt für alle $\varphi \in V_h$

$$\sum_{i,k=1}^{2} \big(F_{ik}D_k(\Gamma_h - G_h),\, D_i\varphi\big) = 0. \tag{39}$$

Es sei

$$(I_h\Gamma_h - G_h)(x) = \sum_{p\in Q_h} C_p\omega_p(x).$$

Q_h waren die inneren Knotenpunkte der Triangulation; ω_p sind die zu der Triangulation gehörigen Basisfunktionen, deren Graphen die Gestalt von Pyramiden haben und deren Träger einen Durchmesser $\leqq Kh$ hat. Schließlich sei $\tau \in C_0^\infty\big(\boldsymbol{R}^2 - B_{\varrho/2}(y)\big)$, $0 \leqq \tau \leqq 1$, $\tau = 1$ auf $\Omega - B_\varrho(y)$, und

$$|\nabla\tau| \leqq \frac{K}{\varrho},\quad |\nabla^2\tau| \leqq \frac{K}{\varrho^2}, \tag{40}$$

unabhängig von y.

Wir setzen in (39)

$$\varphi(x) = \sum_{p\in Q_h} \tau^2(p)\, C_p\omega_p(x)$$

und erhalten

$$\sum_{i,k=1}^{2} \big(F_{ik}D_k(\Gamma_h - G_h),\, \tau^2 D_i(I_h\Gamma_h - G_h)\big)$$

$$\sum_{i,k=1}^{2} \left(F_{ik}D_k(\Gamma_h - G_h),\quad \sum_{p\in Q_h} \big(\tau^2(p) - \tau^2(.)\big)\, C_pD_i\omega_p(.)\right)$$

$$= \sum_{p\in Q_h} \sum_{i,k=1}^{2} \int_p F_{ik}D_k(\Gamma_h - G_h)\cdot\big(\tau^2(p) - \tau^2(x)\big)\, C_pD_l\omega_p(.)\, dx = (*).$$

$\int_p$ bedeutet hierbei die Integration über den Träger von ω_p, also über eine Menge von Dreiecken mit Seitenlänge $\leqq h$; die Anzahl der Dreiecke dieser Menge ist wegen der Regularitätsbedingung für die Triangulation unabhängig von h und p beschränkt.

Wir schätzen (*) ab, indem wir benutzen, daß $|\nabla \omega_p| \leqq \frac{K}{h}$ ist und nach dem Mittelwertsatz und (40)

$$|\tau^2(p) - \tau^2(x)| \leqq \frac{Kh}{\varrho} \tau(x^*) \leqq \frac{Kh}{\varrho} \tau(x) + \frac{K'h^2}{\varrho^2}$$

gilt, hierbei ist $x^* \in \operatorname{Trg} \omega_p$. Somit gilt

$$\begin{aligned}(*) &\leqq \frac{\overline{K}}{\varrho} \sum_{p \in Q_h - B_{\varrho/4}} \int_p \left\{|\nabla \Gamma_h - \nabla G_h| \, \tau C_p + \frac{C'h}{\varrho^2} |\nabla \Gamma_h - \nabla G_h| \, C_p\right\} dx \\ &\leqq \varepsilon \, \|\tau \cdot (\nabla \Gamma_h - \nabla G_h)\|_{L^2}^2 + K_\varrho(\varepsilon) \, h^2 \sum_{p \in Q_h - B_{\varrho/4}} C_p^2 \\ &\quad + K_\varrho h \, \|\nabla \Gamma_h - \nabla G_h\|_{L^2} \Big(h^2 \sum_{p \in Q_h - B_{\varrho/4}} C_p^2\Big)^{1/2}.\end{aligned}$$

Hierbei ist $B_{\varrho/4}$ ein Kreis um y mit Radius $\varrho/4$, wir haben benutzt, daß $\tau \in C_0^\infty(\mathbf{R}^2 - B_{\varrho/2})$, und setzen $h < h_0(\varrho)$ voraus. Wegen der Regularitätsbedingung für die Triangulationen gilt

$$h^2 \sum_{p \in Q_h - B_{\varrho/4}} C_p^2 \leqq K \int_0 |I_h \Gamma_h - G_h|^2 \, dx$$

— $\int_0$ bedeutet die Integration über $\Omega - B_{\varrho/8}$ —, außerdem ist wegen Lemma 1

$$\|\nabla \Gamma_h - \nabla G_h\|_{L^2} = O(1).$$

Daraus folgt

$$(*) \leqq \varepsilon \, \|\tau(\nabla \Gamma_h - \nabla G_h)\|_{L^2}^2 + K_\varrho(\varepsilon) \int_0 |I_h \Gamma_h - G_h|^2 \, dx + K_\varrho' h \Big(\int_0 |I_h \Gamma_h - G_h|^2 \, dx\Big)^{1/2}.$$

Aus der mit (*) endenden Gleichungskette folgt unter Benutzung der gleichmäßigen Elliptizität

$$\begin{aligned}\lambda \int \tau^2 (\nabla \Gamma_h - \nabla G_h)^2 \, dx &\leqq K \int \tau^2 (\nabla \Gamma_h - \nabla I_h \Gamma_h)^2 \, dx \\ &\quad + \varepsilon \int \tau^2 (\nabla \Gamma_h - \nabla G_h)^2 \, dx + K_\varrho(\varepsilon) \int_0 |I_h \Gamma_h - G_h|^2 \, dx \\ &\quad + K_\varrho' h \Big(\int_0 |I_h \Gamma_h - G_h|^2 \, dx\Big)^{1/2}.\end{aligned}$$

Da $\tau \in C_0^\infty(\mathbf{R}^2 - B_{\varrho/2})$, gilt

$$\int \tau^2 (\nabla \Gamma_h - \nabla I_h \Gamma_h)^2 \, dx = O(h^2)$$

und aus ähnlichem Grunde

$$\int_0 |I_h \Gamma_h - \Gamma_h| \, dx = O(h^2),$$

da außerhalb eines Kreises $B_{\varrho/2}$ um die (regularisierte) „Singularität“ von Γ_h die

Funktion Γ_h gleichmäßig beschränkte zweite Ableitungen hat. Damit ergibt sich

$$\int \tau^2(\nabla\Gamma_h - \nabla G_h)^2\,dx \leqq O(h^2) + \int |\Gamma_h - G_h|^2\,dx,$$

und das Lemma ist bewiesen.

Der Verfasser dankt an dieser Stelle Herrn H. D. Mittelmann für wichtige Literaturhinweise, dessen Ergebnis $\|u - u_h\|_\infty = O(h^{3/2} \ln h)$ aus [11] zur Beschäftigung mit dem vorliegenden Problem anregte.

Anmerkung bei der Korrektur: Die Abschätzung (19) ist noch offen. Wir kommen jedoch ohne sie alleine mit Abschätzung (20) aus, indem wir den zweiten Term in (21) in der folgenden Weise behandeln

$$\sum_{i,k=1}^{2} \big(D_i(F_{ik}D_k u), \Gamma_h - G_h\big) = \sum_{i,k=1}^{2} (F_{ik}D_k u, D_i\Gamma_h - D_iG_h) = \sum_{i,k=1}^{2} \big(F_{ik}D_k(u - \bar{u}_h), D_i\Gamma_h - D_iG_h\big)$$
$$= O(h^2 \ln h)$$

Hierbei ist $\bar{u}_h \in V_h$ eine Interpolierende von u.

Literatur

[1] Bramble, J. H., Hilbert, S. R., Bounds for a class of linear functionals with applications to Hermite interpolation. Numer. Math. 16, 362—369 (1971).

[2] Frehse, J., Eine a-priori-Abschätzung zur Methode der finiten Elemente in der numerischen Variationsrechnung. Bonner Mathematische Schriften 77, 115—126 (1975).

[3] Natterer, F., Über die punktweise Konvergenz finiter Elemente. Num. Math. 25, 67—77 (1975).

[4] Nitsche, J., Lineare Spline Funktionen und die Methode von Ritz für elliptische Randwertprobleme. Arch. Rat. Mech. Anal. 36, 348—355 (1970).

[5] Nitsche, J., L^∞-convergence of finite element approximation. Second Conference on Finite Elements, Rennes (France) 1975.

[6] Rannacher, R., Zur L^∞-Konvergenz linearer finiter Elemente. Math. Z. 149, 69—77 (1976).

[7] Scott, R., Optimal L^∞-estimates for the finite element method on irregular meshes. Math. Comp. 30, 681—697 (1976).

[8] Simader, C. G., On Dirichlet's boundary value problem. Lecture Notes in Mathematics 268, Springer-Verlag, Berlin—Heidelberg—New York (1972).

[9] Strang, G., und Fix, G. J., An analysis of the finite element method. Prentice Hall, Inc., Englewood Cliffs, N. J. (1973).

[10] Zlámal, M., On the finite element method. Num. Math. 12, 394—409 (1968).

[11] Mittelmann, H. D., On pointwise estimates for a finite element solution of nonlinear boundary value problems. Preprint, Darmstadt.

[12] Johnson, C., and Thomée, V., Error estimates for a finite element approximation of a minimal surface, preprint no. 8, Chalmers Institute of Technology and the University of Göteborg (1974).

Anschrift des Verfassers:

Universität Bonn
Institut für Angewandte Mathematik und Informatik
53 Bonn
Beringstr. 4—6
BRD

Solvability and non-solvability of weakly nonlinear equations

Svatopluk Fučík, Praha

Abstract. The solvability of the equation

$$x - \mu S x^{+} + \nu S x^{-} = f$$

in a semiordered Banach space ($x^{+} = \max(x, 0)$, $x = x^{+} - x^{-}$) in the dependence on real parameters μ, ν is studied and the application to the boundary value problems for differential equations is given.

1. Introduction

Let X be a semiordered Hilbert space with positive cone C and let S be a linear completely continuous and selfadjoint map in X. In this paper we study the solvability and unsolvability of the equation

$$x - \mu S x^{+} + \nu S x^{-} = f, \tag{1}$$

where $f \in X$, μ and ν are real parameters, $x = x^{+} - x^{-}$ and $x^{+}, x^{-} \in C$ are uniquely determined. Such results play an important role if the equation

$$x - \mu S x^{+} + \nu S x^{-} + G x = f \tag{2}$$

is considered, where G is a nonlinear completely continuous operator satisfying some additional assumptions.

In a paper [5], the author has given sufficient conditions for the solvability of boundary value problems of second order ordinary differential equations of the type (1). These results were generalized in many directions by E. N. Dancer [1]. E.g., he has considered also periodic problems for second order ordinary differential equations and boundary value problems for second order partial differential equations. Moreover, he has given an almost complete description of all parameters μ and ν for which the boundary value problem

$$\begin{cases} -u'' - \mu u^{+} + \nu u^{-} = f \\ u(0) = u(\pi) = 0 \end{cases}$$

is solvable for arbitrary right hand side f.

In the present paper we give the abstract theorems and the applications of these results to the higher order differential equations. For the proofs there are used the methods from the papers [1—4]. The results are not complete and there are problems (see the pictures) which, to the best of author's knowledge, have not been solved till now.

The author is very much indebted to E. N. Dancer (University of New England, Armidale, Australia) for the sending of a manuscript of [1]. The good correspondence contacts with him played the main role during the preparation of present paper.

2. Ranges of positively a-homogeneous operators

The purpose of this section is to give the precise formulation of the problem considered and to establish the basic results.

Until further comment, X, Y, Z will denote Banach spaces with zero elements O_X, O_Y, O_Z and with norms $\|x\|_X, \|y\|_Y, \|z\|_Z$, respectively. The convergence of the sequence $\{x_m\} \subset X$ to $x_0 \in X$ in the norm-topology of X will be denoted by $x_m \xrightarrow{X} x_0$. Analogously we shall denote the convergences in Y and Z.

2.1. The spaces X and Z. A subset C of Z is called a cone if it is closed, convex, invariant under multiplication by nonnegative real numbers, and if $C \cap (-C) = \{O_Z\}$. We suppose that a given cone C in Z has the following properties:

(Z 1) *If $z \in Z$ then there exists a uniquely determined couple $z^+, z^- \in C$ such that $z = z^+ - z^-$.*

(Z 2) *The mapping $z \mapsto z^+$ is continuous.*

In the sequel we shall suppose that

(Z 3) *$X \subset Z$ and the identity mapping is continuous* (i.e., $x_n \xrightarrow{X} x_0$ *implies* $x_n \xrightarrow{Z} x_0$).

Let $a > 0$ be fixed.

2.2. The operator J. Let J be a mapping defined on X with values in the space Y, and suppose that the following assumptions are fulfilled:

(J 1) *J is positively a-homogeneous,* i.e.

$$x \in X, \ t > 0 \Rightarrow J(tx) = t^a Jx;$$

(J 2) *J is a homeomorphism of X onto Y* (denote by J^{-1} the inverse of J);

(J 3) *J is odd,* i.e., $x \in X \Rightarrow J(-x) = -Jx$.

2.3. The operator S. Let S be a mapping from Z into Y and suppose:

(S 1) *S is positively a-homogeneous,* i.e., $z \in Z, t > 0 \Rightarrow S(tz) = t^a Sz$;

(S 2) *S is continuous*;

(S 3) *The mappings $x \mapsto Sx^+$, $x \mapsto Sx^-$ are completely continuous operators from X into Y.*

In the sequel we shall suppose for simplicity that the assumptions (*Zi*), (*Ji*), (*Si*) ($i = 1, 2, 3$) are fulfilled. (Some of the following assertions do not require all of these assumptions.)

2.4. The Leray-Schauder degree. Let $K_Y(\varrho) = \{y \in Y; \|y\|_Y < \varrho\}$ and let $F: \overline{K_Y(\varrho)} \subset Y \to Y$ (the bar denotes the closure) be a completely continuous mapping. Denote by I the identity mapping in Y. Let $y - Fy \neq O_Y$ for each $y \in Y$ with $\|y\|_Y = \varrho$. Then it is possible to define the Leray-Schauder degree $d[I - F; K_Y(\varrho), O_Y]$ of the mapping $I - F$ with respect to $K_Y(\varrho)$ and the point O_Y. For the basic properties of the Leray-Schauder degree which will be used in the sequel see e.g. [6, 7], etc.

2.5. Notation, formulation of problem. Let $G: X \to Y$ be a completely continuous mapping and let μ, ν be real numbers. We shall investigate the set

$$\mathscr{R}_{[\mu,\nu]}(J, S, G) = \{f \in Y; \exists x_0 \in X: Jx_0 - \mu S x_0^+ + \nu S x_0^- + G x_0 = f\}.$$

If G is a zero operator we shall write only $\mathscr{R}_{[\mu,\nu]}(J, S)$. Let

$$\begin{aligned}
A_{-1} &= \{[\mu, \nu] \in R^2;\ \ \exists x_0 \neq O_X: Jx_0 - \mu S x_0^+ + \nu S x_0^- = O_Y\},\\
A_0 &= \{[\mu, \nu] \in R^2;\ \ Jx_0 - \mu S x_0^+ + \nu S x_0^- = O_Y \Rightarrow x_0 = O_X\},\\
A_1 &= \{[\mu, \nu] \in A_0;\ \ d[y - \mu S(J^{-1}y)^+ + \nu S(J^{-1}y)^-;\ \ K_Y(1), O_Y] \neq 0\},\\
A_2 &= \{[\mu, \nu] \in A_0;\ \ \mathscr{R}_{[\mu,\nu]}(J, S) \neq Y\},\\
A_3 &= \{[\mu, \nu] \in R^2;\ \ \mathscr{R}_{[\mu,\nu]}(J, S) = Y\}.
\end{aligned}$$

The following two lemmas are an immediate consequence of definitions.

2.6. Lemma. *For $i = -1, 0, 1, 2, 3$ it is $[\mu, \nu] \in A_i$ if and only if $[\nu, \mu] \in A_i$.*

2.7. Lemma. *Let $[\alpha, \beta] \in R^2$ and let*

$$c_1(\alpha, \beta) = \sup_{\|x\|_X = 1} \|\alpha S x^+ - \beta S x^-\|_Y .$$

Then

$$c_1(\alpha, \beta) \leqq (|\alpha| + |\beta|)\, s,$$

where $s = \max\left(\sup\limits_{\|x\|_X=1} \|Sx^+\|_Y, \sup\limits_{\|x\|_X=1} \|Sx^-\|_Y\right)$.

Especially:

(i) $c_1(\alpha, \beta) < \infty$;

(ii) $\lim\limits_{[\alpha,\beta] \to [0,0]} c_1(\alpha, \beta) = 0$;

(iii) $x \in X \Rightarrow \|\alpha S x^+ - \beta S x^-\|_Y \leqq c_1(\alpha, \beta)\, \|x\|_X^a$.

2.8. The properties of the set A_0. *Let $[\mu, \nu] \in A_0$. Then:*

(i) $c_2(\mu, \nu) = \inf_{\|x\|_X = 1} \|Jx - \mu Sx^+ + \nu Sx^-\|_Y > 0$;

(ii) $x \in X \Rightarrow \|Jx - \mu Sx^+ + \nu Sx^-\|_Y \geqq c_2(\mu, \nu) \|x\|_X^a$;

(iii) *$\mathscr{R}_{[\mu,\nu]}(J, S)$ is a closed subset of Y;*

(iv) *A_0 is an open subset of R^2. Moreover, if $[\alpha, \beta] \in R^2$ satisfies*

$$\frac{c_2(\mu, \nu)}{s} > |\alpha| + |\beta|,$$

then $[\mu + \alpha, \nu + \beta] \in A_0$.

Sketches of proofs. The assertion (i) can be proved via contrary, (ii) is an immediate consequence of (i) and the a-homogeneity of the operators J, S.

(iii) Let $Jx_n - \mu Sx_n^+ + \nu Sx_n^- = f_n$ and let $f_n \xrightarrow{Y} f$. Then $\{x_n\}$ contains a subsequence converging to the solution of the considered equation with a right hand side f. Thus $f \in \mathscr{R}_{[\mu,\nu]}(J, S)$.

(iv) Let $[\alpha, \beta] \in R^2$ be such that $c_1(\alpha, \beta) < c_2(\mu, \nu)$. Then $\|Jx - (\mu + \alpha) Sx^+ + (\nu + \beta) Sx^-\|_Y \geqq [c_2(\mu, \nu) - c_1(\alpha, \beta)] \|x\|_X^a$.

2.9. The properties of the set A_1.

(i) *$A_1 \subset A_3$;*

(ii) *A_1 is an open subset of R^2;*

(iii) *A_1 is a union of some components of A_0;*

(iv) *Let T be a component of A_0 containing a point $[\lambda, \lambda]$. Then $T \subset A_1$. Especially,*

$$[\lambda, \lambda] \in A_0, [\alpha, \beta] \in R^2, |\alpha| + |\beta| < \frac{c_2(\lambda, \lambda)}{s} \Rightarrow [\lambda + \alpha, \lambda + \beta] \in A_1.$$

(v) *Let $[\mu, \nu] \in A_1$. Suppose that*

$$\limsup_{\|x\|_X \to \infty} \frac{\|Gx\|_Y}{\|x\|_X^a} < c_2(\mu, \nu).$$

Then $\mathscr{R}_{[\mu,\nu]}(J, S, G) = Y$.

Sketches of proofs. The assertion (i) is a formal generalization of [2, Theorem 1] (see also [6, Chap. II, Theorem 1.1]). The assertions (ii) and (iii) follow from the homotopy invariance of the degree.

(iv) Since $[\lambda, \lambda] \in A_0$ and since the mapping $y \mapsto \lambda [S(J^{-1}y)^+ - S(J^{-1}y)^-]$ is odd the Borsuk theorem implies that $d[y - \lambda[S(J^{-1}y)^+ - S(J^{-1}y)^-]; K_Y(1), O_Y]$ is an odd number (i.e., different from zero) and thus $[\lambda, \lambda] \in A_1$. Hence $T \subset A_1$ by (iii). The addendum follows from the previous consideration and from 2.8 (iv).

The proof of (v) is analogous to (i).

2.10. The properties of the set A_2.

(i) *For a given $[\mu, \nu] \in A_2$ there exists $c_3(\mu, \nu) > 0$ such that if*

$$\limsup_{\|x\|_X \to \infty} \frac{\|Gx\|_Y}{\|x\|_X^a} \leqq c_3(\mu, \nu)$$

then $\mathscr{R}_{[\mu,\nu]}(J, S, G) \neq Y$.

(ii) *A_2 is an open subset of R^2. Especially:*

$$[\mu, \nu] \in A_2, \ [\alpha, \beta] \in R^2, \ |\alpha| + |\beta| < \frac{c_3(\mu, \nu)}{s} \Rightarrow [\mu + \alpha, \nu + \beta] \in A_2 .$$

(The assertion (i) is essentially Theorem 2 in [1] only formally rewritten for a-homogeneous operators instead of 1-homogeneous ones.)

3. Linear operators

In this section we shall investigate the set $\mathscr{R}_{[\mu,\nu]}(J, S)$ under some additional assumptions. First of all we shall suppose $X = Y$ and X is a real Hilbert space with the inner product (x, y). Further, let J be the identity operator and let S be linear. According to the linearity of S and the assumption (S 3) the mapping S is completely continuous from X into X. For simplicity suppose that $S: X \to X$ is selfadjoint.

Notice that if $[\lambda, \lambda] \in A_0$ then

$$c_3(\lambda, \lambda) = \|(J - \lambda S)^{-1}\|^{-1} \geqq d(\lambda),$$

where $d(\lambda)$ means the distance of λ from the spectrum of the operator S. Thus according to the assertions 2.9 (iv) and (i) we have $[\lambda + \alpha, \lambda + \beta] \in A_3$ provided $|\alpha| + |\beta| < \frac{d(\lambda)}{s}$ (see Fig. 2).

Now we shall study the case $[\lambda, \lambda] \in A_1$. Denote $L_\lambda = J - \lambda S$, $N(L_\lambda)$ the null-space of L_λ and $\text{Im}\,(L_\lambda)$ the range of L_λ. Let P be the orthogonal projection of X onto $N(L_\lambda)$ and let $Q = I - P$, i.e., Q is the orthogonal projection of X onto $\text{Im}\,(L_\lambda)$. Under our assumptions there exists a linear continuous map (the so-called right inverse) M_λ: $\text{Im}\,(L_\lambda) \to \text{Im}\,(L_\lambda)$ satisfying

$$M_\lambda L_\lambda x = Qx \qquad (x \in X),$$

$$L_\lambda M_\lambda y = y \qquad \big(y \in \text{Im}\,(L_\lambda)\big).$$

3.1. Theorem. *Let* $[\alpha, \beta] \in R^2$ *and* $\dim N(L_\lambda) \geqq 1$. *Suppose that there exists* $\delta > 0$ *satisfying the following condition:*

$$\inf_{\substack{h \in N(L_\lambda) \\ \|h\|_X = 1 \\ v \in \operatorname{Im}(L_\lambda), \|v\|_X \leqq \delta}} \left(\alpha S(h+v)^+ - \beta S(h+v)^-, h\right) > 0 . \tag{3}$$

Then $[\lambda + \alpha, \lambda + \beta] \in A_3$ *provided*

$$c_1(\alpha, \beta) < \frac{\delta}{\delta + 1} \|M_\lambda\|^{-1} .$$

The proof follows immediately from the more general Theorem 1 from [3].

The assumption (3) can be verified by the following criterion the proof of which one obtains immediately by way of contradiction.

3.2. Lemma. *Let* $[\lambda, \lambda] \in A_1$. *Suppose that*

$$\inf_{\substack{h \in N(L_\lambda) \\ \|h\|_X = 1}} (\alpha S h^+ - \beta S h^-, h) > 0 . \tag{4}$$

Then

$$\varphi(\delta) = \inf_{\|v\|_X \leqq \delta} \inf_{\substack{h \in N(L_\lambda) \\ \|h\|_X = 1}} \left(\alpha S(h+v)^+ - \beta S(h+v)^-, h\right) \tag{5}$$

is decreasing and

$$\delta_0(\alpha, \beta) = \sup \{\delta \geqq 0;\ \varphi(\delta) > 0\} > 0 .$$

Now we summarize the results in the following theorem.

3.3. Theorem. *Let* $[\lambda, \lambda] \in A_1$ *and let* $[\alpha, \beta] \in R^2$. *Suppose* (4) and

$$c_1(\alpha, \beta) < \frac{\delta_0(\alpha, \beta)}{1 + \delta_0(\alpha, \beta)} \|M_\lambda\|^{-1} .$$

Then $[\lambda + \alpha, \lambda + \beta] \in A_3$.

3.4. Remark. Analogously it is possible to prove: *Let* $[\lambda, \lambda] \in A_1$ *and let* $[\alpha, \beta] \in R^2$. *Suppose*

$$\sup_{\substack{h \in N(L_\lambda) \\ \|h\|_X = 1}} (\alpha S h^+ - \beta S h^-, h) < 0 \tag{4'}$$

and denote

$$\delta_0(\alpha, \beta) = \sup \left\{\delta \geqq 0;\ \sup_{\substack{v \in X \\ \|v\|_X \leqq \delta}} \sup_{\substack{h \in N(L_\lambda) \\ \|h\|_X = 1}} \left(\alpha S(h+v)^+ - \beta S(h+v)^-, h\right) < 0\right\} . \tag{5'}$$

Then $\delta_0(\alpha, \beta) > 0$ *and* $[\lambda + \alpha, \lambda + \beta] \in A_3$ *provided*

$$c_1(\alpha, \beta) < \frac{\delta_0(\alpha, \beta)}{1 + \delta_0(\alpha, \beta)} \|M_\lambda\|^{-1}.$$

3.5. Corollary. *Let* $\dim N(L_\lambda) = 1$ *and let*

$$N(L_\lambda) = \{th; t \in R^1\}, \tag{6}$$

where $\|h\|_X = 1$. *Suppose*

$$(Sh^+, h) = -(Sh^-, h) \neq 0.$$

Let $\alpha \neq -\beta$. *Then* $\delta_0(\alpha, \beta) = \delta_0(1, 0)$ *and* $[\lambda + \alpha, \lambda + \beta] \in A_3$ *provided*

$$c_1(\alpha, \beta) < \frac{\delta_0(1, 0)}{1 + \delta_0(1, 0)} \|M_\lambda\|^{-1}.$$

Using [3, Theorem 1] once more one can derive immediately

3.6. Theorem. *Let the assumptions of Theorem* 3.3 *be fulfilled. Let*

$$\limsup_{t\to\infty} \sup_{\substack{h\in N(L_\lambda)\\ \|h\|_X=1\\ \|v\|_X\leq\delta}} \frac{\big(G(t(h + v)), h\big)}{t} < \varphi(\delta)$$

(for $\varphi(\delta)$ *see* (5)*). Then* $\mathscr{R}_{[\lambda+\alpha,\lambda+\beta]}(J, S, G) = X$.

Now we shall solve our problem in the case the condition (4) is not valid. Slightly modifying the proof of [4, Lemma 1] we obtain the following result.

3.7. Theorem. *Let*

(Z 2′) *the mapping* $z \mapsto z^+$ *be lipschitzian, i.e., there exists* $\varrho > 0$ *such that* $\|z_1^+ - z_2^+\|_Z \leqq \varrho \|z_1 - z_2\|_Z$ *for each* $z_1, z_2 \in Z$.

Denote by η *the norm of the identity mapping from* X *into* Z. *Suppose* (6) *and let there exist* $c_4 > 0$ *such that*

$$\|Gx_1 - Gx_2\|_X \leqq c_4 \|x_1 - x_2\|_X$$

for each $x_1, x_2 \in X$. *Let*

$$k = \|M_\lambda\| \big(\varrho \|S\| \eta (|\alpha| + |\beta|) + c_4\big) < 1.$$

Then for arbitrary $t \in R^1$ *and* $f \in X$ *there exists exactly one* $v_{t,f} \in \operatorname{Im}(L_\lambda)$ *satisfying*

$$-M_\lambda Q\big(\alpha S(th + v_{t,f})^+ - \beta S(th + v_{t,f})^- + G(th + v_{t,f}) - f\big) = v_{t,f}.$$

Moreover,

$$\|v_{t_1,f_1} - v_{t_2,f_2}\|_X \leqq \frac{k}{1-k}|t_1 - t_2| + \frac{\|M_\lambda\|}{1-k}\|f_1 - f_2\|_X$$

for all $t_1, t_2 \in R^1$ *and* $f_1, f_2 \in X$. *The equation*

$$L_\lambda x - \alpha S x^+ + \beta S x^- + Gx = f$$

has a solution $x_0 \in X$ *if and only if there exists* $t_0 \in R^1$ *such that*

$$P\big(-\alpha S(t_0 h + v_{t_0,f})^+ + \beta S(t_0 h + v_{t_0,f})^- + G(t_0 h + v_{t_0,f})\big) = Pf. \tag{7}$$

In the case just considered we can also prove an analogous criterion as in Lemma 3.2.

3.8. Lemma. *Let the assumptions of Theorem* 3.7 *be fulfilled. Suppose*

$$(\alpha S h^+ - \beta S h^-, h) < 0, \qquad (\beta S h^+ - \alpha S h^-, h) > 0. \tag{8}$$

Then

$$\delta_0(\alpha, \beta) = \sup\Big\{\delta \geqq 0;\ \inf_{\|v\|_X \leqq \delta}\big(\alpha S(h+v)^+ - \beta S(h+v)^-, h\big) > 0 \quad \textit{and}$$

$$\sup_{\|v\|_X \leqq \delta}\big(\beta S(h+v)^+ - \alpha S(h+v)^-, h\big) < 0\Big\} > 0.$$

3.9. Theorem. *Let the assumptions of Lemma* 3.8 *be fulfilled. Let*

$$[\alpha, \beta] \in R^2, \frac{k}{1-k} < \delta_0(\alpha, \beta) \textit{ and let } \sup_{x \in X}\|Gx\|_X = c_5 \in R^1.$$

Then $[\lambda + \alpha, \lambda + \beta] \notin A_3$.

Proof. Denote $\Phi(t) = -\alpha\big(S(th + v_{t,f})^+, h\big) + \beta\big(S(th + v_{t,f})^-, h\big) + \big(G(th + v_{t,f}), h\big)$. The function $\Phi(t)$ is continuous with limit $+\infty$ in $\pm\infty$. Thus it is bounded from below and now (7) implies the assertion.

3.10. Remark. The assumption (8) can be replaced by

$$(\alpha S h^+ - \beta S h^-, h) > 0, \qquad (\beta S h^+ - \alpha S h^-, h) < 0 \tag{8'}$$

and the assertion of Theorem 3.9 remains valid.

3.11. Corollary. *Let* $(Sh^+, h)^2 \neq (Sh^-, h)^2$. *Then there exists* $[\alpha, \beta] \in R^2$ *such that* $[\lambda + \alpha, \lambda + \beta] \notin A_3$.

4. Applications

4.1. Nonlinear Sturm-Liouville equation of the second order. Let $p \geqq 2$ be an even integer. Put $X = W_0^{1,p}(0, \pi)$ the Sobolev space of all absolutely continuous functions on $\langle 0, \pi\rangle$ with zero values at 0 and π and whose derivatives are L_p-functions. Further put $Z = X$, $Y = X^*$ (the adjoint of X) and

$$C = \{x \in X;\ x(\xi) \geqq 0 \quad \text{for every} \quad \xi \in \langle 0, \pi\rangle\}.$$

Let $a(\xi)$, $b(\xi)$, $c(\xi)$ be three nonnegative functions on $\langle 0, \pi\rangle$ and A_0, A_1, B_0, B_1 nonnegative numbers. Suppose: $a(\xi) > 0$, $a(\xi)$ is continuously differentiable on $\langle 0, \pi\rangle$, $c(\xi) > 0$, $b(\xi)$, $c(\xi)$ are continuous on $\langle 0, \pi\rangle$.

For $u \in X$ put

$$j(u) = \frac{1}{p}\left(\int_0^\pi [a(\xi)\,|u'(\xi)|^p + b(\xi)\,|u(\xi)|^p]\,d\xi\right) + \frac{1}{p}\,A_0\,|u(0)|^p + \frac{1}{p}\,A_1\,|u(\pi)|^p$$

and

$$s(u) = \frac{1}{p}\int_0^\pi c(\xi)\,|u(\xi)|^p\,d\xi + \frac{1}{p}\,B_0\,|u(0)|^p + \frac{1}{p}\,B_1\,|u(\pi)|^p.$$

The Fréchet derivatives of the functionals j and s at the point $u \in X$ denote by $J(u)$ and $S(u)$, respectively.

It is possible to verify that all assumptions from 2.1—2.3 (with $a = p - 1$) are fulfilled. Moreover, the nonlinear operator S satisfies the equation $Su^+ - Su^- = Su$ for each $u \in X$. Under the assumptions mentioned above there exists the smallest $\lambda_1 \in R^1$ for which $Ju - \lambda_1 Su = O_Y$ has a nontrivial solution. J. Nečas [8] (see also [6, Appen-

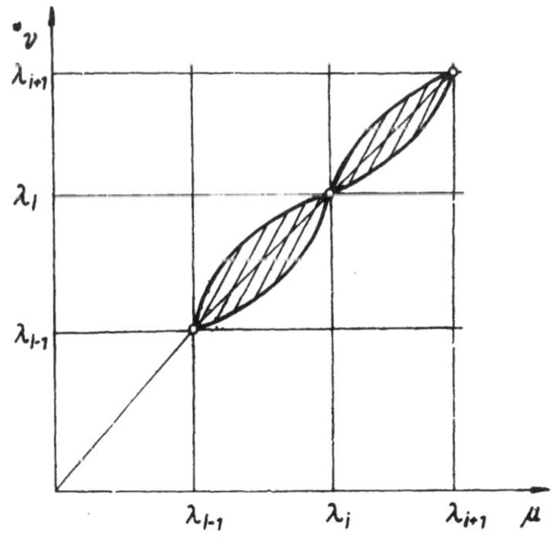

Fig. 1

dix V]) proved: If $\lambda_1 c(\xi) - b(\xi) > 0$ then all eigenvalues of $Ju - \lambda Su$ form a countable set $0 < \lambda_1 < \lambda_2 < \cdots$ such that $\lim_{n\to\infty} \lambda_n = \infty$. Using now 2.9 it is possible to say that the range of the operator $Ju - \mu Su^+ + \nu Su^-$ is whole $X^* = Y$ provided $[\mu, \nu]$ lies in the hatched parts of Fig. 1.

4.2. Nonlinear Sturm-Liouville equation of the fourth order. Let $p \geqq 2$ be an even integer. Put $X = W_0^{2,p}(0, \pi)$, $Z = L_p(0, \pi)$, C the set of nonnegative functions from Z, $Y = X^*$,

$$j(u) = \frac{1}{p} \int_0^\pi a(\xi) \, |u''(\xi)|^p \, d\xi + \frac{1}{p} \int_0^\pi b(\xi) \, |u(\xi)|^p \, d\xi ,$$

$$s(u) = \frac{1}{p} \int_0^\pi c(\xi) \, |u(\xi)|^p \, d\xi ,$$

where a, b, c are the functions introduced in 4.1. The Fréchet derivatives of j and s denote by J and S, respectively. The same result as in 4.1 is true.

4.3. Boundary value problems for higher order differential equations. Let k be an integer. Let Ω be a bounded domain in $R^N (N \geqq 1)$ with boundary $\partial\Omega$ which is lipschitzian in case of $N > 1$. Let $a_{ij}(\xi) \in L_\infty(\Omega)$, $a_{ij} = a_{ji}$. Suppose that there exists $c > 0$ such that

$$\sum_{|i|=|j|=k} a_{ij}(\xi) \, \eta_i \eta_j \geqq c \sum_{|i|=k} \eta_i^2$$

for all $\eta_i \in R^1$ ($|i| = k$) and almost all $\xi \in \Omega$. Put $X = W_0^{k,2}(\Omega)$, $Y = X$, $Z = L_2(\Omega)$. Let C be the set of all almost everywhere nonnegative functions in Z. Define the inner product (u, v) in X by

$$(u, v) = \sum_{|i|=|j|=k} \int_\Omega a_{ij}(\xi) \, D^i v(\xi) \, D^j u(\xi) \, d\xi$$

and now define the mapping S by

$$(Sz, v) = \int_\Omega z(\xi) \, v(\xi) \, d\xi$$

for all $v \in X$ and $z \in Z$.

Since the assumptions from 2.1—2.3 are fulfilled the mapping $Ju - \mu Su^+ + \nu Su^-$ is onto X provided $[\mu, \nu]$ lies in the hatched parallelograms (see Fig. 2). It is also possible to verify the conditions in Section 3.

4.4. Boundary value problem for second order ordinary differential equations. Almost complete E. N. Dancer's result about solvability of

$$\begin{cases} -u'' - \mu u^+ + \nu u^- = f \\ u(0) = u(\pi) = 0 \end{cases}$$

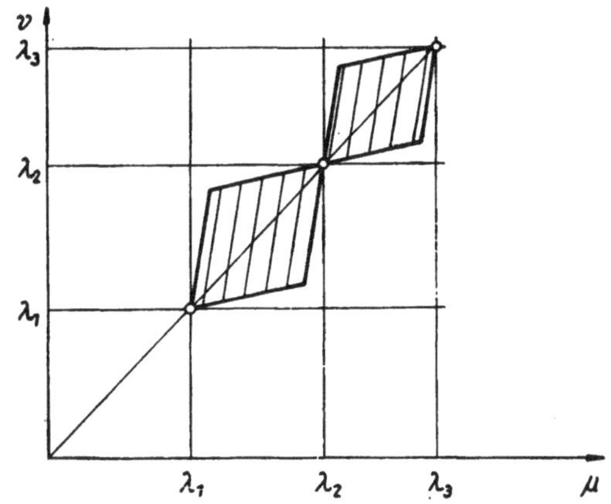

Fig. 2

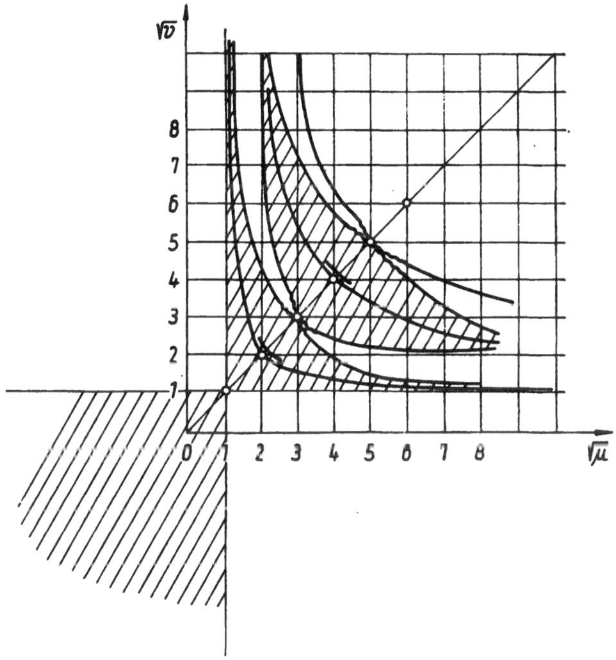

Fig. 3

may be best illustrated by Fig. 3, where the hatched parts mean the set A_1, white parts mean the set A_2 and the black lines form the set A_{-1}. It is also possible to use 3.5 and 3.9: we get that the double lines are subsets of A_3 and the wave lines are not subsets of A_3. (All constants in Section 3 may be easily enumerated.)

References

[1] Dancer, E. N., On the Dirichlet problem for weakly nonlinear elliptic partial differential equations (to appear).
[2] Fučík, S., Note on the Fredholm alternative for nonlinear operators. Comment. Math. Univ. Carolinae **12**, 213–226 (1971).
[3] Fučík, S., Surjectivity of operators involving linear noninvertible part and nonlinear compact perturbation. Funkc. Ekvacioj **17**, 73–83 (1974).
[4] Fučík, S., Remark on a result by A. Ambrosetti and G. Prodi. Boll. Un. Mat. Ital. **11**, 259 to 267 (1975).
[5] Fučík, S., Boundary value problems with jumping nonlinearities. Čas. Pěstování Mat. **101**, No 1, 69–87 (1976).
[6] Fučík, S., Nečas, J., Souček, J., and Souček, V., Spectral analysis of nonlinear operators. Lecture Notes in Mathematics vol. **346**, Springer-Verlag, Berlin 1973.
[7] Krasnoselskij, M. A., Topological methods in the theory of nonlinear integral equations. Pergamon Press, London 1964.
[8] Nečas, J., On the discreteness of the spectrum of nonlinear Sturm-Liouville equation (in Russian). Dokl. Akad. Nauk SSSR **201**, 1045–1048 (1971). English translation: Soviet Math. Dokl. **12**, 1779–1783 (1971).

Address of the author:

Dept. of Math. Analysis
Faculty of Mathematics and Physics of Charles University
Sokolovská 83
186 00 Praha 8
Czechoslovakia

On conjugate evolution equations and a posteriori error estimates

Herbert Gajewski, Berlin

Since the classical works of Trefftz [17], Friedrichs [5], Diaz-Weinstein [4] and Prager-Synge [16] many authors have considered conjugate problems of given original elliptic variational problems in order to obtain a posteriori error estimates. (See e.g. [1, 2, 3, 11, 13, 14, 15] and papers cited therein.) In the present paper we introduce conjugate problems for a class of nonlinear evolution equations and also use them to establish a posteriori error estimates. Our formulation is related to that of conjugate problems as defined by Aubin-Burchard [1] for elliptic variational problems and by Hlaváček [12] for some linear parabolic equations. A more general model to construct conjugate problems of a given original problem has been considered by Gajewski and Gröger [6].

The paper consists of five sections. In the first section we formulate the original problem. In the second section we introduce the conjugate problem. A posteriori error estimates are established in the third section. In the fourth section we state convergence theorems for some approximation procedures assuring that the given a posteriori errors tends to zero. In the last section we consider two simple examples.

1. Notation. The original problem

Throughout this paper let $S = [0, T]$ be a finite interval of the real axis. For an arbitrary Hilbert space X we denote by $L^2(S; X)$ and $C(S; X)$ the usual spaces of the quadratically Bochner integrable and the continuous functions with values in X, respectively. Let $\|.\|$ and $(.,.)$ be the norm and the scalar product in X. Then we write $\|.\|_S$, $\|.\|_C$ and $(.,.)_S$ for the norms in $L^2(S; X)$, $C(S; X)$ and the scalar product in $L^2(S; X)$, respectively. Finally we denote by $H^1(S; X)$ the space of functions u belonging to $L^2(S; X)$ together with their derivatives u' taken in the sense of distributions on $]0, T[$ with values in X.

Definition 1.1. We say that an operator $A \in \big(L^2(S; X) \to L^2(S; X)\big)$ is a *Volterra operator* if for arbitrary $u, v \in L^2(S; X)$ from $u(s) = v(s)$ for almost all $s \in [0, t]$, $t \in S$, it follows that $(Au)(s) = (Av)(s)$.

Let V be a Hilbert space lying densely in another Hilbert space H. Moreover, let the imbedding of V into H be continuous such that between the norms $\|.\|$ and $|.|$ in V and H the following relation holds

$$|x| \leq \gamma\|x\| \qquad \forall x \in V, \qquad \gamma = \text{const}. \tag{1.1}$$

Identifying H with its dual space H^*, we obtain $V \subset H \subset V^*$, where V^* denotes the dual space of V. The pairing between V^* and V as well as the scalar product in H will be denoted by $(.,.)$. Finally, $\|.\|_*$ is the norm in V^*.

Let Y be a further Hilbert space being identified with its dual space Y^* and let the symbols $|||.|||$ and $((.,.))$ denote the norm and the scalar product in Y, respectively.

We consider the following initial value problem

$$v' + L^*ALv = 0, \qquad v(0) = b, \qquad v \in Z, \tag{1.2}$$

where

(A1) $\begin{cases} L \in (V \to Y) \text{ is a continuous linear operator with} \\ |||Lx||| = \|x\| \qquad \forall x \in V; \end{cases}$

(A2) $\begin{cases} A \in \big(L^2(S; Y) \to L^2(S; Y)\big) \text{ is an in general nonlinear Volterra operator with} \\ \big((Ay_1 - Ay_2, y_1 - y_2)\big)_S \geqq m|||y_1 - y_2|||_S^2, \\ |||Ay_1 - Ay_2|||_S \leqq M|||y_1 - y_2|||_S \qquad \forall y_1, \forall y_2 \in L^2(S; Y); \end{cases}$

(A3) $b \in H$ is a given initial value;

Z is the space defined by

$Z = \{v \in L^2(S; V) \mid v' \in L^2(S; V^*)\}, \qquad \|v\|_Z^2 = \|v\|_S^2 + \|v'\|_{*S}^2.$

For the sake of simplicity we suppose throughout this paper that the basic assumptions (A1)—(A3) are fulfilled although some of our results are valid under weaker hypotheses.

Remark 1.1. The space Z is continuously imbedded into the space $C(S; H)$ (see e.g. [11]). Therefore the initial condition $v(0) = b \in H$ makes sense.

Remark 1.2. Taking into account the relation

$$(L^*y, x) = \big((y, Lx)\big) \qquad \forall y \in Y, \qquad \forall x \in V,$$

we obtain by assumption (A1)

$$\|L^*y\|_* \leqq |||y||| \qquad \forall y \in Y. \tag{1.3}$$

The starting-point of our investigations is the following well known result on evolution equations (see e.g. [11]).

Theorem 1.1. *The problem* (1.2) *has a unique solution.*

2. The conjugate problem

In order to formulate the conjugate problem of (1.2) we have to make some preparations now.

Lemma 2.1. *The inverse operator* $A^{-1} \in \big(L^2(S; Y) \to L^2(S; Y)\big)$ *of* A *exists, is Volterra operator and fulfils the conditions*

$$\big((A^{-1}y_1 - A^{-1}y_2, y_1 - y_2)\big)_S \geqq m_* |||y_1 - y_2|||_S^2$$
$$|||A^{-1}y_1 - A^{-1}y_2|||_S \leqq M_* |||y_1 - y_2|||_S \qquad \forall y_1, \forall y_2 \in L^2(S; Y).$$

Here the constants m_* *and* M_* *satisfy*

$$m_* \geqq \frac{m}{M^2} \qquad \text{and} \qquad M_* \leqq \frac{1}{m}. \tag{2.1}$$

If A *is a potential operator then* $m_* \geqq \dfrac{1}{M}$ *holds.*

Proof. The existence, strong monotonicity and Lipschitz continuity of A^{-1} and the estimates (2.1) for the constants m_* and M_* are proved e.g. in [11]. The estimate $m^* \geqq \dfrac{1}{M}$ for a potential operator A we obtain as follows. By [11], Kap. III, Folgerung 4.5, the operator $U = I - \dfrac{2}{M} A$ is nonexpansive. Thus we have

$$|||y_1 - y_2 - \frac{2}{M}(Ay_1 - Ay_2)|||_S \leqq |||y_1 - y_2|||_S \qquad \forall y_1, \forall y_2 \in L^2(S; Y)$$

and therefore

$$|||y_1 - y_2|||_S^2 - \frac{4}{M}\big((Ay_1 - Ay_2, y_1 - y_2)\big)_S + \frac{4}{M^2}|||Ay_1 - Ay_2|||_S^2 \leqq |||y_1 - y_2|||_S^2.$$

Hence it follows

$$\big((Ay_1 - Ay_2, y_1 - y_2)\big)_S \geqq \frac{1}{M}|||Ay_1 - Ay_2||_S^2.$$

Now we show that A^{-1} is a Volterra operator. Let for that y_1, y_2 be arbitrary elements from $L^2(S; Y)$ with $y_1(s) = y_2(s)$ for almost all $s \in [0, t]$, $t \in S$. Setting

$$z_i(s) = \begin{cases} (A^{-1}y_i)(s) & \text{for} \quad s \leqq t, \\ 0 & \text{for} \quad t < s \leqq T, \end{cases}$$

$i = 1, 2$, we obtain

$$m \cdot |||z_1 - z_2|||_S^2 \leqq \int_S \big((Az_1 - Az_2, z_1 - z_2)\big)\,ds = \int_0^t \big((Az_1 - Az_2, z_1 - z_2)\big)\,ds$$
$$= \int_0^t \big((AA^{-1}y_1 - AA^{-1}y_2, z_1 - z_2)\big)\,ds$$
$$= \int_S \big((AA^{-1}y_1 - AA^{-1}y_2, z_1 - z_2)\big)\,ds$$
$$= \int_S \big((y_1 - y_2, z_1 - z_2)\big)\,ds = 0.$$

From this it follows

$$(A^{-1}y_1)(s) = z_1(s) = z_2(s) = (A^{-1}y_2)(s) \qquad \text{a.e. in } [0, t].$$

Lemma 2.1 is proved.

To formulate the conjugate problem of (1.2) we need the following spaces:

$$U = \{y \in Y \mid L^*y \in V\}, \qquad \|y\|_U = |||y + LL^*y|||,$$

$$G = \{y \in Y \mid L^*y \in H\}, \qquad |y|_G^2 = |||y|||^2 + |L^*y|^2,$$

$$W = \{u \in L^2(S; U) \mid u' \in L^2(S; Y)\}, \qquad \|u\|_W^2 = \|u\|^2_{L^2(S;U)} + |||u'|||_S^2.$$

Lemma 2.2. *The space W is continuously imbedded into the space $C(S; G)$.*

Proof. From the corresponding properties of the spaces V and H it follows that the space U is imbedded into G continuously and densely. We identify G with G^*. Then we have obviously

$$U \subset G \subset U^* = Y.$$

Using the result mentioned in Remark 1.1, we obtain the Lemma.

The problem

$$A^{-1}u' + LL^*u = 0, \qquad u(0) = a \in G, \qquad L^*a = -b,\ u \in W \tag{2.2}$$

will be called the conjugate problem of the original problem (1.2)

Remark 2.1. In (2.2) $a \in G$ is an arbitrary element with $L^*a = -b$. The existence of such an element we get as follows. The duality mapping J_V from V onto V^* has the representation $J_V = L^*L$. Thus the equation $L^*Lx = -b$ has a unique solution x and we can set $a = Lx$, for instance. In many special cases an element a with $L^*a = -b$ can be found easier (compare e.g. our first example in section 5).

Remark 2.2. Because of Lemma 2.2 the initial condition $u(0) = a$ makes sense.

In the following theorem we state the relation between solutions of the original and the corresponding conjugate problem.

Theorem 2.1. *Let v be the solution of* (1.2). *Then*

$$u(t) = \int_0^t (ALv)(s)\, ds + a \tag{2.3}$$

is a solution of (2.2). *Conversely, let u be a solution of* (2.2). *Then*

$$v = -L^*u \tag{2.4}$$

is the solution of (1.2).

Proof (→). Let v be solution of (1.2). We set $u(t) = \int_0^t ALv\,ds + a$ and obtain

$$u(0) = a, \qquad u' = ALv \in L^2(S; Y). \tag{2.5}$$

Using (1.2), we find

$$L^*u = L^* \int_0^t ALv\,ds + L^*a = \int_0^t L^*ALv\,ds - b = -v \in L^2(S; V). \tag{2.6}$$

From (2.5) and (2.6) it follows $u \in W$ and $A^{-1}u' + LL^*u = Lv - Lv = 0$.

(←). Now let u be a solution of (2.2). We set $v = -L^*u$. By definition of a it holds $v(0) = b$. Using (1.3) and $u \in W$, we obtain $v \in Z$. Finally, we find from (2.2) $Lv = A^{-1}u'$ and therefore $v' + L^*ALv = -L^*u' + L^*u' = 0$. Theorem 2.1 is proved.

Theorem 2.2. *The conjugate problem* (2.2) *has a unique solution.*

Proof. The existence of a solution of (2.2) follows immediately from Theorem 1.1 and 2.1. Let u_1 and u_2 be solutions of (2.2). Using the strong monotonicity of A^{-1} we find

$$0 = \big((A^{-1}u_1' - A^{-1}u_2' + LL^*u_1 - LL^*u_2, u_1 - u_2)\big)_S$$

$$\geqq m_* |||u_1' - u_2'|||_S^2 + \frac{1}{2} \big|L^*\big(u_1(T) - u_2(T)\big)\big|^2.$$

Thus from $u_1(0) = u_2(0)$ the assertion follows. Theorem 2.2 is proved.

For the sake of simplicity we restrict ourselves to homogeneous evolution equations. The following remark shows that this restriction is not essential.

Remark 2.4. The inhomogeneous initial value problem

$$v' + L^*ALv = g, \qquad v(0) = b,$$

with $g \in L^2(S; V^*)$ can be reduced to the homogeneous problem

$$v' + L^*BLv = 0, \qquad v(0) = b, \qquad Bx = Ax - f$$

by introducing a function $f \in L^2(S; Y)$ with $L^*f = g$. Obviously, then we have $B^{-1}y \equiv A^{-1}(y + f)$.

In the case $g \in L^2(S; V)$ we can also substitute $v = w + \int_0^t g\,ds$. Then we obtain for w the homogeneous problem

$$w' + L^*CLw = 0, \qquad w(0) = b, \qquad Cx \equiv A\left(x + \int_0^t Lg\,ds\right).$$

We note that the operator C^{-1} has the representation

$$C^{-1}y \equiv A^{-1}y - \int_0^t Lg\,ds.$$

3. A posteriori error estimates

In what follows we denote by v_* and u_* the solutions of the original problem (1.2) and the conjugate problem (2.2) and by e and e_* the (explicitly calculable) norms of the imbedding operators from Z into $C(S; H)$ and from W into $C(S; G)$, respectively.

Theorem 3.1. *There are constants $c_1 = c_1(m, M)$ and $c_2 = c_2(M, e)$ such that the a posteriori error estimates*

$$c_1\|v - v_*\|_Z \leqq |v(0) - b| + \|v' + L^*ALv\|_{*S} \leqq c_2 \|v - v_*\|_Z \tag{3.1}$$

hold for an arbitrary approximation $v \in Z$ of v_*.

Proof. Using the strong monotonicity and the Volterra property of A, we find for each $t \in S$

$$\frac{1}{2} |v(t) - v_*(t)|^2 - \frac{1}{2} |v(0) - b|^2 + m \int_0^t \|v - v_*\|^2 \, ds$$

$$\leqq \int_0^t (v' - v_*' + L^*ALv - L^*ALv_*, v - v_*) \, ds = \int_0^t (v' + L^*ALv, v - v_*) \, ds$$

$$\leqq \frac{1}{2m} \int_0^t (\|v' + L^*ALv\|_*^2 + m^2 \|v - v_*\|^2) \, ds$$

or

$$\|v - v_*\|_S^2 \leqq \frac{1}{m} \left(|v(0) - b|^2 + \frac{1}{m} \|v' + L^*ALv\|_{*S}^2\right). \tag{3.2}$$

Using (1.3) and the Lipschitz continuity of A we obtain

$$\|v' - v_*'\|_S = \|v' + L^*ALv - L^*ALv + L^*ALv_* - L^*ALv_* - v_*'\|_{*S}$$

$$\leqq \|v' + L^*ALv\|_{*S} + M \|v - v_*\|_S. \tag{3.3}$$

Finally, we have

$$|v(0) - b| + \|v' + L^*ALv\|_{*S} = |v(0) - b| + \|v' - v_*' + v_*' + L^*ALv_* - L^*ALv_*$$

$$+ L^*ALv\|_{*S} \leqq |v(0) - b| + \|v' - v_*'\|_{*S} + M \|v - v_*\|_S \leqq c \|v - v_*\|_Z. \tag{3.4}$$

From (3.2)—(3.4) we conclude easily the desired estimates (3.1).

Remark 3.1. For the constants c_1 and c_2 we find immediately the bounds

$$c_1 \geqq 1/\max\left(\frac{1}{\sqrt{m}} (1 + M), 1 + \frac{1 + M}{m}\right), \qquad c_2 \leqq \sqrt{2} \max(1, M) + e.$$

Similarly as (3.2) we get also the estimate

$$|v - v_*|_C^2 \leqq |v(0) - b|^2 + \frac{1}{2m} \|v' + L^*ALv\|_{*S}^2. \tag{3.5}$$

Theorem 3.2. *There are constants* $c_1^* = c_1^*(m_*, M_*, \gamma, T)$ *and* $c_2^* = c_2^*(M_*, e_*)$ *such that the a posteriori error estimates*

$$c_1^*\|u - u_*\|_W \leqq |u(0) - a|_G + |||A^{-1}u' + LL^*u|||_S \leqq c_2^*\|u - u_*\|_W \tag{3.6}$$

hold for an arbitrary approximation $u \in W$ *of* u_*.

Proof. Using the strong monotonicity and the Volterra property of A^{-1}, we find for each $t \in S$

$$m_* \int_0^t |||u' - u_*'|||^2\, ds + \frac{1}{2}\, |L^*u(t) - L^*u_*(t)|^2 - \frac{1}{2}\, |L^*u(0) - L^*a|^2$$

$$\leqq \int_0^t \left((A^{-1}u' - A^{-1}u_*' + LL^*(u - u_*), u' - u_*'\right) ds$$

$$= \int_0^t \left((A^{-1}u' + LL^*u, u' - u_*')\right) ds \leqq \int_0^t \left(\frac{1}{2m_*}\, |||A^{-1}u' + LL^*u|||^2 + \frac{m_*}{2}\, |||u' - u_*'|||^2\right) ds$$

or

$$|||u' - u_*'|||_S^2 \leqq \frac{1}{m_*} \left(\left|L^*(u(0) - a)\right|^2 + \frac{1}{m_*}\, |||A^{-1}u' + LL^*u|||_S^2\right). \tag{3.7}$$

Using the Lipschitz continuity of A^{-1}, we obtain

$$|||LL^*u - LL^*u_*|||_S = |||LL^*u + A^{-1}u' - A^{-1}u' + A^{-1}u_*' - A^{-1}u_*' - LL^*u_*|||_S$$

$$\leqq |||LL^*u + A^{-1}u'|||_S + M_*|||u' - u_*'|||_*. \tag{3.8}$$

Finally, we have

$$|u(0) - a|_G + |||A^{-1}u' + LL^*u|||_S$$

$$= |u(0) - a|_G + |||A^{-1}u' - A^{-1}u_*' + A^{-1}u_*' + LL^*u_* - LL^*u_* + LL^*u|||_S$$

$$\leqq |u(0) - a|_G + M_*\, |||u' - u_*'|||_S + |||LL^*u - LL^*u_*|||_S \leqq c_*\|u - u_*\|_W.$$

Since $|||u - u_*|||_S \leqq \sqrt{2T}\, |||u(0) - a||| + T|||u' - u_*'|||_S$ and $|L^*(u - u_*)|_S \leqq \gamma|||LL^*(u - u_*)|||_S$, the desired estimates (3.6) can be concluded from (3.7)—(3.9).

Remark 3.2. By elementary arguments we find the following bounds for the constants c_1^* and c_2^*:

$$c_1^* \geqq 1 \Big/ \max\left(\frac{1}{m_*}\left(1 + T + \left(1 + \sqrt{2}\gamma\right) M^*\right) + 1 + \sqrt{2}\gamma,\right.$$

$$\left.\max\left(\frac{\sqrt{2}}{\sqrt{m_*}}\left(1 + T + \left(1 + \sqrt{2}\gamma\right)M_*\right), 2\sqrt{T}\right)\right), \quad c_2^* \leqq \sqrt{2}\max(1, M_*) + e_*.$$

Similarly as (3.7) we get also the estimate

$$|L^*(u - u_*)|_C^2 \leqq |L^*u(0) + b|^2 + \frac{1}{2m_*} |||A^{-1}u' + LL^*u|||_S^2. \tag{3.1}$$

Remark 3.3. The estimates (3.7) and (3.10) lead immediately to the following error estimates with respect to v_*:

$$|||A^{-1}u' - Lv^*|||_S^2 \leqq \frac{M_*^2}{m_*}\left(|L^*u(0) + b|^2 + \frac{1}{m_*} |||A^{-1}u' + LL^*u|||_S^2\right),$$

$$|L^*u + v_*|_C^2 \leqq |L^*u(0) + b|^2 + \frac{1}{2m_*} |||A^{-1}u' + LL^*u|||_S^2.$$

Remark 3.4. Each of the error estimates (3.1) and (3.6) has a special drawback. So (3.1) involves the defect $\|v' + L^*ALv\|_{*S}$ which can be calculated in many concrete cases (see e.g. our examples given in section 5) only with difficulty. On the other hand (3.6) involves the defect $|||A^{-1}u' + LL^*u|||_S$. But for instance for approximations $u = u_n$ found by projection methods we cannot prove that this defect tends to zero for $n \to \infty$ unless a special basis (compare Definition 4.2) is available.

Let us now suppose that we have approximations v for v_* as well as u for u_*. Then the drawbacks just mentioned can be overcome by the following theorems.

Theorem 3.3. *Suppose $v_*' \in L^2(S; H)$. Then there are constants $k_1 = k_1(m, M, \gamma)$ and $k_2 = k_2(M, e)$ such that the a posteriori error estimates*

$$k_1 \|v - v_*\|_Z \leqq |v(0) - b| + |v' + L^*u'|_S + |||u' - ALv|||_S$$
$$\leqq k_2(\|v - v_*\|_S + |v' - v_*'|_S + |u' - u_*'|_{GS}) \tag{3.11}$$

hold for arbitrary approximations $v \in Z$ of v_* and $u \in W$ of u_* provided $v' \in L^2(S; H)$ and $u' \in L^2(S; G)$.

Proof. We get from (3.1)

$$c_1\|v - v_*\|_Z \leqq |v(0) - b| + \|v' + L^*ALv\|_{*S} = |v(0) - b| + \|v' + L^*u' - L^*u'$$
$$+ L^*ALv\|_{*S} \leqq |v(0) - b| + \gamma\,|v' + L^*u'|_S + |||u' - ALv|||_S$$
$$\leqq |v(0) - b| + \gamma\left(|v' - v_*'|_S + |L^*(u' - u_*')|_S\right) + |||u' - u_*'|||_S + M\|v - v_*\|_S$$
$$\leqq k(\|v - v_*\|_S + |v' - v_*'|_S + |u' - u_*'|_{GS}).$$

Hence the assertion follows.

Remark 3.5. Obviously, the constants k_1, k_2 have the bounds

$$k_1 \geqq 1 \bigg/ \left(\max(1, \gamma) \cdot \max\left(\frac{1}{\sqrt{m}}(1 + M), 1 + \frac{1 + M}{m}\right)\right),$$

$$k_2 \leqq \max\left(\sqrt{2}, M + e\right).$$

In the same way as (3.11) we obtain from (3.2) and (3.5)

$$\|v - v_*\|_S^2 \leqq \frac{1}{m}\left(|v(0) - b|^2 + \frac{2}{m}\left(\gamma^2 |v' + L^*u'|_S^2 + |||u' - ALv|||_S^2\right)\right),$$

$$|v - v_*|_S^2 \leqq |v(0) - b|^2 + \frac{1}{m}\left(\gamma^2 |v' + L^*u'|_S^2 + |||u' - ALv|||_S^2\right).$$

In order to complete our results we formulate both the following theorems.

Theorem 3.4. Let v, u *be approximations as in Theorem* 3.3. *Then*

$$\max\left(|v - v_*|_C^2, \frac{4m}{5}\|v - v_*\|_S^2\right) \leqq 2|v(0) - b|^2 + 4T|v' + L^*u'|_S^2 + \frac{1}{m}|||u' - ALv|||_S^2.$$

Theorem 3.5. *Let* v, u *be approximations as in Theorem* 3.3. *Then*

$$\max\left(|L^*(u - u_*)|_C^2, \frac{6m_*}{7}|||u' - u_*'|||_S^2\right) \leqq 4\left(|L^*u(0) + b| + \frac{1}{2}|v(0) + L^*u(0)|^2\right.$$

$$\left. + \frac{1}{4m_*}|||A^{-1}u' - Lv|||_S^2 + 2|v + L^*u|_C^2 + 2T|v' + L^*u'|_S^2\right).$$

We omit here the simple proofs of these theorems.

Finally let us give an a posteriori error estimate which is suitable for potential operators.

Theorem 3.6. *Let* v, u *be approximations as in Theorem* 3.3. *Suppose that* A *(and thus* A^{-1}*) is potential operator. Let* F *and* F^* *be the potentials of* A *and* A^{-1}*, respectively. Then*

$$\frac{m}{2}\|v - v_*\|_S^2 + \frac{1}{M}|||u' - u_*'|||_S^2 + |v(T) - v_*(T)|^2 \leqq |v(0) - b|^2$$

$$+ \frac{2\gamma^2}{m}|L^*u' + v'|_S^2 + 2\left(F(Lv) + F^*(u') - ((u', Lv))_S\right).$$

Proof. Using known results on conjugate functionals (see e.g. [11]), we obtain

$$\frac{m}{2}\|v - v_*\|_S^2 + \frac{1}{2M}|||u' - u_*'|||_S^2$$

$$\leqq (v_*', v - v_*)_S + \big((LL^*u_*, u' - u_*')\big)_S + F(Lv) - F(Lv_*) + F^*(u') - F^*(u_*')$$

$$= (v_*', v - v_*)_S - \big((Lv_*, u' - ALv_*)\big)_S + F(Lv) + F^*(u') - F(Lv_*) - F^*(ALv_*)$$

$$= (v_*', v - v_*)_S - \big((Lv_*, u')\big)_S + F(Lv) + F^*(u')$$

$$= (L^*u' + v_*', v - v_*)_S + F(Lv) + F^*(u') - \big((u', Lv)\big)_S$$

$$= -(v' - v_*', v - v_*)_S + (L^*u' + v', v - v_*)_S + F(Lv) + F^*(u') - \big((u', Lv)\big)_S.$$

Hence the desired estimation follows immediately.

Remark 3.6. The term $F(Lv) + F^*(u') - \big((u', Lv)\big)_S$ appearing in Theorem 3.6 can be bounded as follows (compare [11], Kap. III, 5)

$$0 \leqq F(Lv) + F^*(u') - \big((u', Lv)\big)_S = F^*(u') - F^*(ALv) - \big((Lv, u' - ALv)\big)_S$$

$$= \int_0^1 \Big(\Big(A^{-1}\big(ALv + s(u' - ALv)\big) - A^{-1}ALv, u' - ALv\Big)\Big)_S \, ds$$

$$\leqq \int_0^1 \frac{s}{m} |||u' - ALv|||_S^2 \, ds = \frac{1}{2m} |||u' - ALv|||_S^2.$$

4. Approximation methods

To apply successfully the a posteriori error estimates proved in the preceding section we need sequences (v_n) and (u_n) of approximations of v^* and u^* for which the errors tend to zero. In what follows we formulate some convergence results of this type. Their proofs may be found in the papers [7—10]. (Indeed it was not exactly the problem (2.2) which we studied in [9, 10]. There we considered problems of the kind $A^{-1}u' + J_U u = 0$, $u(0) = a$, where $J_U = I + LL^*$ is the duality mapping from U onto $U^* = Y$. However, the proofs given in [9, 10] can be modified easily to get the results needed here.) We consider two cases. First we only suppose the basic assumptions (A1)—(A3). In the second case we make some additional regularity assumptions with respect to the initial value b and the time dependence of the operator A.

Arbitrary dates

Theorem 4.1 [7]. *For an arbitrary parameter* $r \in \left]0, \frac{2m}{M^2}\right[$ *the iteration sequence* (v_i) *defined by*

$$rv_i' + L^*Lv_i = L^*Lv_{i-1} - rL^*ALv_{i-1}, \qquad v_i(0) = b, \qquad v_i \in Z,$$
$$i = 1, 2, \ldots, \qquad v_0 \in L^2(S; V) \quad \text{arbitrary,} \tag{4.1}$$

converges in Z *to* v^*.

Theorem 4.2 [9]. *For an arbitrary parameter* $r \in \left]0, \frac{2m^*}{M_*^2}\right[$ *the iteration sequence* (u_i) *defined by*

$$u_i' + rLL^*u_i = u_{i-1}' - rA^{-1}u_{i-1}', \qquad u(0) = a, \qquad u_i \in W, \tag{4.2}$$
$$i = 1, 2, \ldots, \qquad u_0' \in L^2(S; Y) \quad \text{arbitrary,}$$

converges in W *to* u_*.

Remark 4.1. If A (and therefore A^{-1}) is a potential operator (see e.g. [11]) then in Theorems 4.1 and 4.2 the intervals $\left]0, \frac{2m}{M^2}\right[$ and $\left]0, \frac{2m_*}{M_*^2}\right[$ may be replaced by $\left]0, \frac{2}{M}\right[$ and $]0, 2m[$, respectively. A similar statement holds for the Theorems 4.5, 4.6, 4.13 and 4.15 formulated below.

In order to formulate convergence results for projection methods we need some preparations.

Definition 4.1. A sequence (X_n) of finite dimensional subspaces of a Hilbert space X is said to be a *basis* in X if

$$X_n \subset X_{n+1},\ n = 0, 1, 2, \ldots, \bigcup_{n=0}^{\infty} X_n \text{ is dense in } X.$$

Definition 4.2. Let X be a Hilbert space with $X \subset X^*$ and let J_X be the duality mapping from X onto X^*. A basis (X_n) in X is said to be *special* if $J_X X_n = X_n$, $n = 0, 1, 2, \ldots$

The following Lemma gives a sufficient condition for the existence of special bases in V and U.

Lemma 4.1. *Let* V *and* Y *be separable. Moreover let the imbedding from* V *in* H *be completely continuous. Then there are special bases in* V *and* U.

Proof. It is well known (see e.g. [14]) that under our hypotheses there exists in V a complete system (g_n) of eigenfunctions of the duality mapping $J_V = L^*L \in (V \to V^*)$. Let (d_n) be an arbitrary increasing sequence of natural numbers with $\lim_{n\to\infty} d_n = \infty$. We set $V_n = \operatorname{span}(g_1, \ldots, g_{d_n})$, $n = 0, 1, \ldots$

Then, obviously, (V_n) is a special basis in V.

The space Y has the orthogonal decomposition

$$Y = Y_1 + Y_2 \qquad \text{with} \qquad Y_1 = \operatorname{im} L = \{y \in Y \mid \exists\, x \in V, y = Lx\},$$

$$Y_2 = \operatorname{cer} L^* = \{y \in Y \mid L^*y = 0\}.$$

Let (Y_{2n}) be an arbitrary basis in Y_2. Then, obviously, (U_n) with $U_n = L\{V_n\} \cup Y_{2n}$ is a basis in U. On the other hand we have

$$J_U\{L\{V_n\}\} = L\{V_n\} + LL^*\{L\{V_n\}\} = L\{V_n\} + L\{J_V\{V_n\}\}$$
$$= L\{V_n\} + L\{V_n\} = L\{V_n\},$$
$$J_U\{Y_{2n}\} = Y_{2n} + LL^*\{Y_{2n}\} = Y_{2n}$$

and therefore $J_U\{U_n\} = U_n$, i.e., (U_n) is a special basis in U. Lemma 4.1 is proved.

In the following Theorems 4.3—4.6 let (V_n) and (U_n) be special bases in V and U, respectively. Moreover let (b_n) and (a_n) be sequences such that

$$b_n \in V_n, \quad b_n \to b \text{ in } H, \quad a_n \in U_n, \quad a_n \to a \text{ in } G.$$

Theorem 4.3 [8]. *The Galerkin sequence (v_n) defined by*

$$(v_n' + L^*ALv_n, g)_S = 0 \quad \forall g \in L^2(S; V_n), \; v_n(0) = b_n, \; v_n \in H^1(S; V_n) \tag{4.3}$$

converges in Z to v_*.

Theorem 4.4 [10]. *The Galerkin sequence (u_n) defined by*

$$\big((A^{-1}u_n' + LL^*u_n, h)\big)_S = 0 \;\forall h \in L^2(S; U_n), \; u_n(0) = a_n, \; u_n \in H^1(S; U_n) \tag{4.4}$$

converges in W to u_*.

Theorem 4.5 [8]. *For an arbitrary parameter $r \in \left]0, \frac{2m}{M^2}\right[$ the projection-iteration sequence (z_n) defined by*

$$(rz_n' + L^*Lz_n, g)_S = (L^*Lz_{n-1} - rL^*ALz_{n-1}, g)_S \quad \forall g \in L^2(S; V_n)$$
$$z_n(0) = b_n, \quad z_n \in H^1(S; V_n), \quad z_0 \in L^2(S; V_0) \qquad \textit{arbitrary},$$

converges in Z to v_*.

Theorem 4.6 [10]. *For an arbitrary parameter $r \in \left]0, \frac{2m_*}{M_*^2}\right[$ the projection-iteration sequence (w_n) defined by*

$$\big((w_n' + rLL^*w_n, h)\big)_S = \big((w_{n-1}' - rA^{-1}w_{n-1}', h)\big)_S \quad \forall h \in L^2(S; U_n),$$
$$w_n(0) = a_n, \quad w_n \in H^1(S; U_n), \quad w_0 \in L^2(S; U_0) \qquad \textit{arbitrary},$$

converges in W to u_*.

Regular dates

Up to the end of this section we make in addition to (A1)—(A3) the following assumptions:

(A4) The operator A has the representation

$$(Au)(t) = A(t)\, u(t) \quad \text{where} \quad A(t) \in (Y \to Y), \quad \forall t \in S, \quad \forall u \in L^2(S; Y).$$

Moreover there exist a function $\varphi \in H^1(S, Y)$ and a constant ϱ such that

$$|||A(t)\, y - A(s)\, y||| \leqq \varrho\big(|||\varphi(t) - \varphi(s)||| + |t - s|\, (1 + |||y|||)\big)$$

$\forall t, \forall s \in S, \forall y \in Y$.

(A5) $b \in V, \quad L^*A(0)\, Lb \in H$.

In order to apply some regularity results from [9, 10] we need the following

Lemma 4.2. The operator A^{-1} has the representation

$$(A^{-1}u)(t) = A^{-1}(t)\, u(t) \qquad \forall t \in S, \quad \forall u \in L^2(S; Y)$$

and it holds $\forall t, \forall s \in S, \forall y \in Y$

$$|||A^{-1}(t)\, y - A^{-1}(s)\, y||| \leqq \varrho_*\big(|||\varphi(t) - \varphi(s)||| + |t - s|\, (1 + |||y|||)\big) \tag{4.7}$$

where $\varrho_* = \dfrac{\varrho}{m} \max\left(\dfrac{1}{m}, 1 + \dfrac{\varrho}{m}\left(2|||\varphi|||_C + T\left(1 + |||A^{-1}(0)\, 0|||\right)\right) + |||A^{-1}(0)\, 0|||\right)$.

Proof. Because of (A2) and (A4) we have (see e.g. [11], Kap. VI, Lemma 2.2 and Lemma 3.3) $\forall t \in S, \forall y_1, \forall y_2 \in Y$

$$\big((A(t)\, y_1 - A(t)\, y_2, y_1 - y_2)\big) \geqq m|||y_1 - y_2|||^2, \quad |||A(t)\, y_1 - A(t)\, y_2||| \leqq M|||y_1 - y_2|||.$$

From this the existence, strong monotonicity, Lipschitz continuity of $A^{-1}(t)$ and the given representation of $A^{-1} \in \big(L^2(S; Y) \to L^2(S; Y)\big)$ follow. To prove (4.7), we set $y_t = A^{-1}(t)\, y$ and $y_s = A^{-1}(s)\, y$. Then we obtain

$$|||y_t - y_s||| \leqq \frac{1}{m}\, |||A(t)\, y_t - A(t)\, y_s||| = \frac{1}{m}\, |||A(t)\, y_t - A(s)\, y_s + A(s)\, y_s - A(t)\, y_s|||$$

$$= \frac{1}{m}\, |||A(t)\, y_s - A(s)\, y_s||| \leqq \frac{\varrho}{m}\left(|||\varphi(t) - \varphi(s)||| + |t - s|\left(1 + |||A^{-1}(s)\, y|||\right)\right)$$

$$\leqq \frac{\varrho}{m}\left(|||\varphi(t) - \varphi(s)||| + |t - s|\left(1 + \frac{1}{m}\, |||y||| + |||A^{-1}(s)\, 0|||\right)\right).$$

Hence we find, setting $y = 0$,

$$|||A^{-1}(s)\,0||| \leqq |||A^{-1}(s)\,0 - A^{-1}(0)\,0||| + |||A^{-1}(0)\,0|||$$

$$\leqq \frac{\varrho}{m}\left(2\,|||\varphi|||_C + T\left(1 + |||A^{-1}(0)\,0|||\right)\right) + |||A^{-1}(0)\,0|||,$$

and the Lemma is proved.

Theorem 4.7 [7]. *The solution* v_* *of* (1.2) *has the regularity properties* $v_* \in C(S; V)$ *and* $v_*' \in Z$.

Theorem 4.8 [9]. *The solution* u_* *of* (2.2) *has the regularity properties* $u_* \in C(S; U)$ *and* $u_*' \in W$.

Theorem 4.9 [7]. *Suppose that the starting element* v_0 *of the iteration sequence* (v_i) *defined by* (4.1) *satisfies the conditions* $v_0(0) = b$, $v_0' \in L^2(S; V)$. *Then* $v_i \to v_*$ *in* Z *and* $v_i' \to v_*'$ *in* $L^2(S; H)$.

Theorem 4.10 [9]. *Suppose that the starting element* u_0 *of the iteration sequence* (u_i) *defined by* (4.2) *satisfies the conditions* $u_0'(0) = A(0)\,Lb$ *and* $u_0'' \in L^2(S; Y)$. *Then* $u_i \to u_*$ *in* W *and* $u_i' \to u_*'$ *in* $L^2(S; G)$.

In the following theorems let (V_n) and (U_n) be arbitrary bases in V and U, respectively. Further let (b_n) and (a_n) be again sequences with $b_n \in V_n$, $b_n \to b$ in H, $a_n \in U_n$, $a_n \to a$ in G.

Theorem 4.11 [8]. *Suppose that the sequence* (b_n) *satisfies the additional condition*

$$\sup_n |L^*A(0)\,La_n| < \infty .$$

Then for the Galerkin sequence (v_n) *defined by* (4.3) *it holds*

$$v_n \to v_* \text{ in } Z \quad \text{and} \quad v_n' \to v_*' \text{ in } L^2(S; H).$$

Theorem 4.12 [10]. *Suppose that the following additional conditions are satisfied:*

$$A(0)\,Lb \in U_0, \quad \left((LL^*a_n, h)\right) = \left((LL^*a, h)\right) \qquad \forall h \in U_n .$$

Then for the Galerkin sequence (u_n) *defined by* (4.4) *it holds*

$$u_n \to u_* \text{ in } C(S; G) \quad \text{and} \quad u_n' \to u_*' \text{ in } L^2(S; G).$$

Theorem 4.13 [8]. *Suppose that*

$$b_n = b \in V_0, \qquad n = 0, 1, \ldots, \qquad z_0(0) = b, \qquad z_0' \in L^2(S; V_0).$$

Then for the projection-iteration sequence (z_n) *defined by* (4.5) *it holds:*

$$z_n \to v_* \text{ in } Z \quad \text{and} \quad z_n' \to v_*' \text{ in } L^2(S; H).$$

Theorem 4.14 [10]. *Suppose (4.8) and suppose that the starting element w_0 of the projection-iteration sequence (w_n) defined by (4.6) satisfies the conditions $w_0'(0) = ALb$ and $w_0'' \in L^2(S; U_0)$. Then $w_n \to u_*$ in $C(S; G)$ and $w_n' \to u_*'$ in $L^2(S; G)$.*

Remark 4.2. In the papers [7—10] further convergence results on iteration, Galerkin and projection-iteration methods are proved.

5. Examples

Let $D \subset R^n$ be a bounded domain with sufficiently smooth boundary Γ.
First we consider the initial boundary value problem

$$\left.\begin{aligned} &\frac{\partial v}{\partial t} - \operatorname{div}\big(g(|\operatorname{grad} v|)\operatorname{grad} v\big) = 0 \text{ in } D, \\ &v(0, x) = b(x), \qquad x \in D, \qquad v|_\Gamma = 0. \end{aligned}\right\} \tag{5.1}$$

We suppose with regard to hypothesis (A2) that the function $\psi(\alpha) = g(\alpha) \cdot \alpha$ satisfies the conditions

$$\psi(\alpha) - \psi(\beta) \geqq m(\alpha - \beta), \quad |\psi(\alpha) - \psi(\beta)| \leqq M|\alpha - \beta| \quad \forall \alpha, \forall \beta \in R^1,$$
$$\alpha \geqq \beta \geqq 0.$$

We set:

$$V = H_0^1(D), \quad H = L^2(D), \qquad V^* = H^{-1}(D), \qquad Y = \big(L^2(D)\big)^n,$$
$$L: v \to \operatorname{grad} v, \quad A: y \to g(|y|)\, y, \quad |y|^2 = \sum_{i=1}^n y_i^2, \quad L^*: y \to -\operatorname{div} y$$

and accordingly

$$G = X_1 + X_2, \qquad U = X_1 + X_3$$

where:
$$X_1 = \{y \in \big(L^2(D)\big)^n \mid \operatorname{div} y = 0\},$$
$$X_2 = \{y \in \big(L^2(D)\big)^n \mid \exists w \in H_0^1(D) \cap H^2(D), \quad y = \operatorname{grad} w\},$$
$$X_3 = \{y \in \big(L^2(D)\big)^n \mid \exists w \in H_0^1(D),\ \Delta w \in H_0^1(D), \quad y = \operatorname{grad} w\}.$$

Then we may understand (1.2) as an abstract version of (5.1) and the conjugate problem (2.2) as an abstract version of the initial boundary value problem

$$f(|u'|)\, u' - \operatorname{grad} \operatorname{div} u = 0 \quad \text{in} \quad D,$$
$$u(0, x) = a(x), \qquad x \in D, \qquad \operatorname{div} u|_\Gamma = 0$$

where $\varphi(\alpha) = f(\alpha)\, \alpha$ is the inverse function of ψ defined as usual by $\varphi\big(\psi(\alpha)\big) = \alpha$, $\alpha \in R^1$, and the initial value a is chosen such that $\operatorname{div} a = b$. $\Big($For example $a = \Big(\int_0^1 b(t, x_2, \ldots, x_n) \times dt, 0, \ldots, 0\Big).\Big)$

Now we consider the following nonlinear diffusion problem

$$g(c)\frac{\partial c}{\partial t} - \operatorname{div}\big(f(c)\operatorname{grad} c\big) = 0 \quad \text{in} \quad D,$$

$$c(0, x) = c_0(x), \qquad x \in D, \qquad c|_\Gamma = 0.$$

We suppose that the functions f and g are locally integrable and satisfy the conditions

$$0 < f_1 \leqq f(\alpha) \leqq f_2 < \infty, \quad 0 < g_1 \leqq g(\alpha) \leqq g_2 < \infty, \qquad \forall \alpha \in R^1.$$

Setting

$$p(s) = \int_0^s f(\alpha)\, d\alpha, \qquad q(s) = \int_0^s g(\alpha)\, d\alpha,$$

we can rewrite (5.2) in the form

$$\frac{\partial}{\partial t} q(c) - \Delta p(c) = 0, \qquad c(0, x) = c_0(x), \qquad x \in D, \qquad c|_\Gamma = 0.$$

Substituting $v = q(c)$ and $b = q(c_0)$, we obtain

$$\frac{\partial v}{\partial t} - \Delta p\big(q^{-1}(v)\big) = 0 \text{ in } D, \qquad v(0, x) = b, \qquad x \in D, \qquad v|_\Gamma = 0. \tag{5.3}$$

We define

$$V = L^2(D), \qquad H = H^{-1}(D), \qquad V^* = \big(L^2(D)\big)^*, \qquad Y = L^2(D),$$

$L =$ canonical isomorphism between V and Y,

$$A : y \to p\big(q^{-1}(y)\big), \qquad L^* = -\Delta$$

and accordingly

$$G = H_0^1(D), \qquad U = H_0^1(D) \cap H^2(D).$$

Then we may understand (1.2) as an abstract version of (5.3) and the conjugate problem (2.2) as an abstract version of the initial boundary value problem

$$q\big(p^{-1}(u')\big) - \Delta u = 0 \text{ in } D, \qquad u(0, x) = a(x), \qquad x \in D, \qquad u|_\Gamma = 0$$

where $a = (-\Delta)^{-1} b$.

Remark 5.1. It is easy to see that in the considered examples the hypotheses (A1)—(A5) are fulfilled provided the initial values b and c_0 are sufficiently regular. We note also that in both examples the operators A are potential ones.

References

[1] Aubin, J. P., Burchard, H. G., Some aspects of the method of the hypercircle applied to elliptic variational problems. *In:* Numerical Solution of Partial Differential Equations II (SYNSPADE 1970), 1–67. Academic Press, New York–London 1971.

[2] Bittner, L., Abschätzungen bei Variationsmethoden mit Hilfe von Dualitätssätzen I. Numer. Math. **11**, 129–143 (1968).

[3] —, Abschätzungen bei Variationsmethoden mit Hilfe von Dualitätssätzen II. Numer. Math. **16**, 285–303 (1971).

[4] Diaz, J. B., Weinstein, A., Schwarz' inequality and the methods of Rayleigh-Ritz and Trefftz. J. Math. Phys. **26**, 133–136 (1947).

[5] Friedrichs, K. O., Ein Verfahren der Variationsrechnung, das Minimum eines Integrals als das Maximum eines anderen Ausdrucks darzustellen. Nachr. Akad. Wiss. Gött. Math. Phys. Kl. 1929, 13–20 (1929).

[6] Gajewski, H., Gröger, K., Konjugierte Operatoren und a posteriori Fehlerabschätzungen. Math. Nachr. **73**, 315–333 (1976).

[7] —, —, Zur Konvergenz eines Iterationsverfahrens für Evolutionsgleichungen. Math. Nachr. **68**, 331–343 (1975).

[8] —, —, Ein Projektions-Iterationsverfahren für Evolutionsgleichungen. Math. Nachr. **72**, 119–136 (1976).

[9] —, —, Zur Konvergenz eines Iterationsverfahrens für Gleichungen der Form $Au' + Lu = f$. Math. Nachr. **69**, 329–341 (1975).

[10] —, —, Ein Projektions-Iterationsverfahren für Gleichungen der Form $Au' + Lu = f$. Math. Nachr. **73**, 249–267 (1976).

[11] —, — und Zacharias, K., Nichtlineare Operatorgleichungen und Operatordifferentialgleichungen. Berlin 1974.

[12] Hlaváček, I., On a conjugate semi-variational method for parabolic equations. Aplikace Matematiky **18**, 434–444 (1973).

[13] Kato, T., On some approximate methods concerning the operators T^*T. Math. Ann. **126**, 253–262 (1953).

[14] Michlin, S. G., Variationsmethoden der mathematischen Physik. Berlin 1962.

[15] Payne, L. E., Weinberger, H. F., Bounds for solutions of second order elliptic equations in terms of arbitrary vector fields. Arch. Rat. Mech. Anal. **20**, 95–106 (1965).

[16] Prager, W., Synge, J. L., Approximations in elasticity based on the concept of function space. Quart. Appl. Math. **5**, 241–269 (1947).

[17] Trefftz, E., Ein Gegenstück zum Ritzschen Verfahren. *In:* Proc. 2nd Intern. Congress Appl. Mech. Zürich (1926).

Address of the author:

Akademie der Wissenschaften der DDR
Zentralinstitut für Mathematik und Mechanik
108 Berlin
Mohrenstr. 39
DDR

On a projection-iteration method for some evolution equations related to Navier-Stokes equations

Herbert Gajewski and Konrad Gröger, Berlin

Introduction

In this paper we shall prove that it is possible to solve initial value problems for two dimensional Navier-Stokes equations and related equations by a projection-iteration method which reduces the original problem to a sequence of initial value problems for finite systems of linear ordinary differential equations. We shall consider initial value problems which can be written in the form

$$u' + Au + B(u, u) = 0, \qquad u(0) = a, \tag{0}$$

where A and B are in general nonlinear operators satisfying certain monotonicity and continuity conditions. The formulation of these conditions given below (see (1)—(3)) seems to be useful for two reasons:

1. The conditions are sufficiently simple. It is easily proved that they are satisfied in the case of two dimensional Navier-Stokes equations (with various types of boundary conditions).

2. The conditions are of such a generality that the problem (0) includes many initial value problems describing processes of heat and mass transfer coupled with fluid motions.

The paper consists of four sections. In the first section we shall introduce some notations and formulate precisely the problem we are interested in. In the second section we shall prove by methods which are standard in the theory of Navier-Stokes equations an existence and uniqueness result and a theorem on the convergence of Galerkin's method. The third section contains the main result on the convergence of a certain projection-iteration method. In the last section we shall consider an example. We shall show that it is possible to apply the results of the previous sections to the so called Marangoni equations. These equations describe the heat transfer across a fluid interface in the case that the interface tension depends on the temperature.

We remark that most of the results formulated here were already proved by Gajewski [1] in the case of Boussinesq equations which can be considered as another example of the problems studied in this paper.

1. Preliminaries

Let V and H be real separable Hilbert spaces such that V is densely and continuously imbedded into H. Furthermore, let the imbedding operator from V into H be compact. Identifying H and its dual space we obtain $V \subset H \subset V^*$, where V^* denotes the dual space of V. The norms in V, H and V^* we shall denote by $\|.\|$, $|.|$ and $\|.\|_*$, respectively. The pairing between V^* and V as well as the inner product in H will be denoted by $(.,.)$.

Let $S = [0, T]$ be a finite interval of the real axis. If E is a Hilbert space we shall denote by $L^2(S; E)$ the Hilbert space of all quadratically integrable functions on S with values in E (provided with the usual inner product) and by $C(S; E)$ the space of all continuous mappings from S into E with the supremum norm. For the sake of brevity we introduce the following notations:

$$X = L^2(S; V), \qquad X^* = L^2(S; V^*);$$

$$\langle f, u\rangle = \int_S \big(f(t), u(t)\big)\, dt \qquad \forall f \in X^*, \qquad \forall u \in X;$$

$J =$ duality map from V onto V^* characterized by

$$(Jx, x) = \|x\|^2 = \|Jx\|_*^2 \qquad \forall x \in V;$$

$L =$ duality map from X onto X^* characterized by

$$\langle Lu, u\rangle = \|u\|_X^2 = \|Lu\|_{X^*}^2 \qquad \forall u \in X;$$

$$Y = X \cap C(S; H), \qquad \|u\|_Y = \|u\|_X + \|u\|_{C(S;H)} \qquad \forall u \in Y;$$

$u' =$ derivative of $u \in X$ in the sense of distributions on $]0, T[$ with values in V^*;

$$W = \{u \mid u \in X,\ u' \in X^*\}, \qquad \|u\|_W^2 = \|u\|_X^2 + \|u'\|_{X^*}^2 \qquad \forall u \in W.$$

To formulate the announced initial value problem we assume that we are given operators A and B with the following properties:

$$\left.\begin{array}{l} A \in (Y \to X^*), \quad (Au)(t) = A(t)\,u(t) \text{ for a.e. } t \in S, \qquad \forall u \in X, \\ A(t) \in (V \to V^*), \quad \big(A(t)\,x - A(t)\,y,\, x - y\big) \geqq m\|x - y\|^2 - \big(k(t)\,|x - y|\big)^2, \\ m > 0, \qquad k \in L^2(S), \\ \|A(t)\,x - A(t)\,y\|_* \leqq M\|x - y\| + k(t)\,|x - y| \quad \forall x, \forall y \in V, \quad \forall t \in S, \end{array}\right\} \tag{1}$$

$$B \in (Y \times Y \to X^*), \quad \langle B(u, v), w\rangle = \int_S b\big(u(t), v(t), w(t)\big)\, dt, \tag{2}$$

where

$$\left.\begin{array}{l} b \in (V \times V \times V \to R) \text{ 3-linear}, \quad b(x, y, y) = 0 \qquad \forall x, \forall y \in V, \\ |b(x, y, z)| \leqq M_1(|x|\,\|x\|\,|y|\,\|y\|)^{\frac{1}{2}}\,\|z\| \qquad \forall x, \forall y, \forall z \in V, \\ |b(x, y, z)| \leqq K(z)\,|x|\,|y| \quad \forall x, \forall y \in V, \quad \forall z \in D, \quad D \text{ dense in V}. \end{array}\right\} \tag{3}$$

Remark 1. In view of (1) we have

$$\|Au - Av\|_{X^*} \leqq \left(\int_S \left(M\,\|u(t) - v(t)\| + k(t)\,|u(t) - v(t)|\right)^2 dt\right)^{\frac{1}{2}}$$

$$\leqq \text{const}\,\|u - v\|_Y \qquad \forall u, \forall v \in Y. \tag{4}$$

Similarly, (2) and (3) give

$$\|B(u, v)\|_{X^*} \leqq M_1 \left(\int_S |u(t)|\,\|u(t)\|\,|v(t)|\,\|v(t)\|\,dt\right)^{\frac{1}{2}}$$

$$\leqq M_1\,\|u\|_Y\,\|v\|_Y \qquad \forall u, \forall v \in Y. \tag{5}$$

In the following sections we shall consider the initial value problem

$$u' + Au + B(u, u) = 0, \qquad u \in W, \qquad u(0) = a, \tag{6}$$

where a is a given element of H.

Remark 2. In the case of two dimensional Navier-Stokes equations $B(u, u)$ represents the inertial forces and Au the viscosity and body forces.

Remark 3. The space W is continuously imbedded into $C(S; H)$ (see e.g. [4], Satz 1.17, Kap. IV). Therefore, the initial condition $u(0) = a \in H$ in (6) makes sense.

2. Uniqueness, existence, Galerkin's method

Theorem 1. *Let* (1)—(3) *be satisfied. Then the initial value problem* (6) *has at most one solution.*

Proof. Let u and v be solutions of (6) and let $w = u - v$. In view of (3) we have

$$b(x, x, x - y) - b(y, y, x - y) = b(x - y, x, x - y) \qquad \forall x, \forall y \in V.$$

Consequently,

$$0 = \int_0^t \left(\left(u' + Au + B(u, u) - v' - Av - B(v, v)\right)(s), w(s)\right) ds$$

$$= \int_0^t \left(\left(w' + Au - Av + B(w, u)\right)(s), w(s)\right) ds$$

$$\geqq \frac{1}{2}\,|w(t)|^2 + \int_0^t \Big\{m\|w(s)\|^2 - |k(s)|^2\,|w(s)|^2$$

$$- M_1\left(|u(s)|\,\|u(s)\|\,|w(s)|\right)^{\frac{1}{2}}\,\|w(s)\|^{\frac{3}{2}}\Big\}\,ds$$

$$\geqq \frac{1}{2}\,|w(t)|^2 - c\int_0^t \left(|u(s)|^2\,\|u(s)\|^2 + |k(s)|^2\right)|w(s)|^2\,ds, \qquad c = \text{const}.$$

Using Gronwall's Lemma we obtain from this inequality the desired result $w = 0$.

Theorem 2. *Let* (1)—(3) *be satisfied. Then the initial value problem* (6) *has a solution for every* $a \in H$.

We shall prove Theorem 2 by Galerkin's method. Let (V_n), $n = 1, 2, \ldots$, be a sequence of finite dimensional subspaces of V such that

$$V_n \subset V_{n+1}, \quad n = 1, 2, \ldots, \quad \bigcup_n V_n \text{ is dense in } V,$$

and let (a_n) be a sequence of elements of H such that

$$a_n \in V_n, \quad a_n \to a \text{ in } H. \tag{8}$$

For the sake of brevity we shall denote the space $L^2(S; V_n)$ by X_n.

Lemma 1. *Let* (1)—(3) *be satisfied. Then for every n the problem*

$$\begin{aligned} &\langle u_n' + Au_n + B(u_n, u_n), v\rangle = 0 \\ &\forall v \in X_n, \quad u_n \in X_n, \quad u_n' \in X_n, \quad u_n(0) = a_n, \end{aligned} \tag{9}$$

has a unique solution u_n.

Proof. The uniqueness can be proved in the same way as Theorem 1. Because V_n is finite dimensional, (9) is equivalent to an initial value problem for a system of ordinary differential equations. The existence of a solution u_n follows from Carathéodory's Theorem and the a priori estimates we shall establish. Let $t \in S$ be fixed and

$$v(s) = \begin{cases} u_n(s) & \text{if } 0 \leqq s \leqq t, \\ 0 & \text{if } t < s \leqq T. \end{cases}$$

Using this function v in (9) we obtain

$$0 = \int_0^t \Big(\big(u_n' + Au_n + B(u_n, u_n)\big)(s), u_n(s)\Big)\, ds \geqq \frac{1}{2}|u_n(t)|^2 - \frac{1}{2}|a_n|^2$$

$$+ \int_0^t \Big\{m\, \|u_n(s)\|^2 - \big(k(s)\, |u_n(s)|\big)^2 - \|A(s)\, 0\|_*\, \|u_n(s)\|\Big\}\, ds.$$

Thus,

$$|u_n(t)|^2 + m \int_0^t \|u_n(s)\|^2\, ds \leqq c + c \int_0^t |k(s)|^2\, |u_n(s)|^2\, ds, \tag{10}$$

where $c = \text{const}$ is independent of n and t. It follows from (10) by Gronwall's Lemma that

$$\sup_n \|u_n\|_Y < \infty. \tag{11}$$

This completes the proof of Lemma 1.

Proof of Theorem 2. Let u_n be the solution of (9). From (1), (4), (5) and (11) follow

$$\sup_{v \in X_n} \langle u_n', v \rangle = \sup_{v \in X_n} \langle -Au_n - B(u_n, u_n), v \rangle$$

$$\leq \|Au_n + B(u_n, u_n)\|_{X^*} \|v\|_X \leq \text{const } \|v\|_X .$$

Using this result and (11) we obtain that the sequence (u_n) is compact in $L^2(S; H)$ (cf. Gröger [5]). Hence, there exists a subsequence (u_{n_j}) of (u_n) such that

$$u_{n_j} \rightharpoonup u \text{ in } X, \tag{12}$$

$$u_{n_j}(T) \rightharpoonup z \text{ in } H, \tag{13}$$

$$u_{n_j} \to u \text{ in } L^2(S; H), \tag{14}$$

$$u_{n_j}(t) \to u(t) \text{ in } H \text{ for a.e. } t \in S, \tag{15}$$

$$Au_{n_j} \rightharpoonup f \text{ in } X^*, \tag{16}$$

$$B(u_{n_j}, u_{n_j}) \rightharpoonup g \text{ in } X^*. \tag{17}$$

Since $u'_{n_j} \to u'$ in the sense of distributions, (9), (16) and (17) give

$$u' + f + g = 0.$$

Consequently, we have $u' \in X^*$ and $u \in W \subset Y$. It follows from this result by standard arguments that $u(0) = a$ and $u(T) = z$ (see e.g. [4], p. 207).

We shall now show that $g = B(u, u)$. Let $\varphi \in \mathscr{D}(]0, T[)$ and $x \in D$ (cf. (3)). Using (2), (3) and (14) we obtain

$$\langle B(u_{n_j}, u_{n_j}) - B(u, u), \varphi x \rangle = \langle B(u_{n_j}, u_{n_j} - u) + B(u_{n_j} - u, u), \varphi x \rangle$$

$$= \int_S \varphi(t) \left\{ b\big(u_{n_j}(t), u_{n_j}(t) - u(t), x\big) + b\big(u_{n_j}(t) - u(t), u(t), x\big) \right\} dt$$

$$\leq \sup_{t \in S} |\varphi(t)| \, K(x) \int_S \big(|u_{n_j}(t)| + |u(t)|\big) \, |u_{n_j}(t) - u(t)| \, dt \to 0 \quad \text{if } n \to \infty .$$

In view of (17) this yields

$$\langle g - B(u, u), \varphi x \rangle = 0 \qquad \forall \varphi \in \mathscr{D}(]0, T[), \quad \forall x \in D. \tag{18}$$

The set D is dense in V by assumption. Therefore, the set of all functions of the form $\sum_{i=1}^{m} \varphi_i x_i$, $\varphi_i \in \mathscr{D}(]0, T[)$, $x_i \in D$, is dense in X, and from (18) follows the desired result $g = B(u, u)$.

Now we shall prove $f = Au$. Let $v \in \bigcup_n X_n \cap W$. Then, using the previous results we find

$$\langle f - Av, u - v\rangle = \langle -u' - B(u, u) - Av, u - v\rangle$$

$$= \frac{1}{2}|a|^2 - \frac{1}{2}|z|^2 + \big(z, v(T)\big) - \big(a, v(0)\big) - \langle v', u\rangle + \langle B(u, u), v\rangle - \langle Av, u - v\rangle$$

$$\geqq \varliminf_{j\to\infty} \Big\{ \frac{1}{2}|a_{n_j}|^2 - \frac{1}{2}|u_{n_j}(T)|^2 + \big(u_{n_j}(T), v(T)\big) - \big(a_{n_j}, v(0)\big) - \langle v', u_{n_j}\rangle$$

$$+ \langle B(u_{n_j}, u_{n_j}), v\rangle - \langle Av, u_{n_j} - v\rangle\Big\}$$

$$= \varliminf_{j\to\infty} \langle -u'_{n_j} - B(u_{n_j}, u_{n_j}) - Av, u_{n_j} - v\rangle$$

$$= \varliminf_{j\to\infty} \langle Au_{n_j} - Av, u_{n_j} - v\rangle$$

$$\geqq \lim_{j\to\infty} \int_S \big\{-\big(k(s)\,|u_{n_j}(s) - v(s)|\big)^2\big\}\, ds$$

$$= -\int_S \big(k(s)\,|u(s) - v(s)|\big)^2\, ds,$$

where the last equality follows from Lebesgue's Theorem. Because $\bigcup_n X_n \cap W$ is dense in W (see e.g. [4], Lemma 1.5, Kap. VI), the inequality

$$\langle f - Av, u - v\rangle \geqq -\int_S \big(k(s)\,|u(s) - v(s)|\big)^2\, ds$$

holds for every $v \in W$. Choosing $v = u - hw$, $w \in W$ arbitrary, $h > 0$, we obtain

$$\langle f - A(u - hw), w\rangle \geqq -h \int_S \big(k(s)\,|w(s)|\big)^2\, ds.$$

Letting $h \to 0$ we find

$$\langle f - Au, w\rangle \geqq 0 \qquad \forall w \in W.$$

Hence, $f = Au$. This completes the proof of Theorem 2.

Theorem 3. *Let* (1)—(3) *be satisfied. Then the sequence* (u_n) *of Galerkin approximations of* (6), *i.e. the sequence of solutions of* (9) *for* $n = 1, 2, \ldots$ *converges in* Y *to the solution* u *of* (6).

Proof. Let (v_n) be a sequence such that

$$v_n \in C(S; V_n), \qquad v_n' \in C(S; V_n), \qquad v_n \to u \text{ in } W. \tag{19}$$

The existence of a sequence satisfying (19) is proved e.g. in [4] (Lemma 1.5, Kap. VI).

Using (1)—(5) and (19) and denoting $u_n - u$ by w_n we obtain

$$0 = \int_0^t \Big(\big(u_n' + Au_n + B(u_n, u_n) - u' - Au - B(u, u)\big)(s), u_n(s) - v_n(s)\Big)\, ds$$
$$= \int_0^t \Big(\big(w_n' + Au_n - Au + B(u_n, u_n) - B(u, u)\big)(s), u(s) - v_n(s)\Big)\, ds$$
$$+ \int_0^t \Big(\big(w_n' + Au_n - Au + B(u_n, u_n) - B(u, u)\big)(s), w_n(s)\Big)\, ds$$
$$\geqq -c_1 \|u - v_n\|_W + \frac{1}{2} |w_n(t)|^2 - \frac{1}{2} |a_n - a|^2$$
$$+ \int_0^t \Big\{m\|w_n(s)\|^2 - \big(k(s)\, |w_n(s)|\big)^2 - M_1\big(|u(s)|\, \|u(s)\|\, |w_n(s)|\big)^{\frac{1}{2}}$$
$$\times \|w_n(s)\|^{\frac{3}{2}}\Big\}\, ds.$$

Thus,

$$|w_n(t)|^2 + m \int_0^t \|w_n(s)\|^2\, ds \leqq \varepsilon_n + \int_0^t \big(|k(s)|^2 + c_2\, |u(s)|^2\, \|u(s)\|^2\big)\, |w_n(s)|^2\, ds,$$

where $\varepsilon_n = |a_n - a|^2 + 2c_1 \|u - v_n\|_W \to 0$ if $n \to \infty$. In view of Gronwall's Lemma this implies

$$|w_n(t)|^2 + m \int_0^t \|w_n(s)\|^2\, ds \leqq c\varepsilon_n,$$

where c is independent of n and t. This completes the proof of Theorem 3.

3. The projection-iteration method

In this section we shall prove that it is possible to approximate the solution of (6) by solving a sequence of linear initial value problems. We assume that we have sequences (V_n) and (H_n), $n = 1, 2, \ldots$, of subspaces of V and H, respectively, with the following properties:

$$V_n \subset V_{n+1}, \quad H_n \subset H_{n+1}, \quad n = 1, 2, \ldots;$$
$$\bigcup_n V_n \text{ dense in } V, \quad V_n \text{ dense in } H_n. \tag{20}$$

Furthermore, let (a_n) be a sequence of elements of H such that

$$a_n \in H_n, \qquad a_n \to a \text{ in } H. \tag{21}$$

Identifying the space H_n and its dual we find

$$V_n \subset H_n \subset V_n^*,$$

where V_n^* denotes the dual of V_n. In this section V_n is not necessarily finite dimensional. We shall use the following notations:

$$X_n = L^2(S; V_n), \qquad X_n^* = L^2(S; V_n^*), \qquad Y_n = X_n \cap C(S; H_n),$$

$$W_n = \{u \mid u \in X_n, u' \in X_n^*\} \subset Y_n,$$

I_n = imbedding operator from X_n into X,

I_n^* = adjoint operator of I_n (from X^* onto X_n^*),

L_n = duality map from X_n onto X_n^* characterized by

$$\langle L_n u, u\rangle = \|u\|_X^2 = \|L_n u\|_{X_n^*}^2 \qquad \forall u \in X_n.$$

Moreover, we define $A_n \in (Y_n \to X_n^*)$ and $B_n \in (Y_n \times Y_n \to X_n^*)$ by

$$A_n u = I_n^* A u, \qquad B_n(u, v) = I_n^* B(u, v) \qquad \forall u, \forall v \in Y_n. \tag{22}$$

We shall consider the following sequence of initial value problems:

$$\left.\begin{aligned} & r\big(u_n' + B_n(u_{n-1}, u_n)\big) + L_n u_n = L_n u_{n-1} - r A_n u_{n-1}, \\ & u_n \in W_n, \quad u_n(0) = a_n, \quad n = 1, 2, \ldots, \quad u_0 \in Y_1 \text{ arbitrary}, \end{aligned}\right\} \tag{23}$$

where r is a strictly positive real number.

Theorem 4. *Let* (1)—(3) *and* (20)—(22) *be satisfied. Then for fixed n the problem* (23) *has a unique solution u_n if $u_{n-1} \in Y_n$ is given. If $0 < r < \dfrac{2m}{M^2}$ it follows that $u_n \to u$ in Y, where u denotes the solution of* (6).

To prove this theorem we need the following lemma.

Lemma 2. *Let* (1) *be satisfied and $0 < r < \dfrac{2m}{M^2}$. Then*

$$\left.\begin{aligned} & \big\|\big(J - rA(t)\big)x - \big(J - rA(t)\big)y\big\|_*^2 \le p_r\|x - y\|^2 + q_r\big(k(t)\,|x - y|\big)^2, \\ & \forall x, \forall y \in V, \quad \forall t \in S, \quad 0 < p_r < 1, \end{aligned}\right\} \tag{24}$$

where p_r and q_r do not depend on x, y and t.

Proof. Using (1) and the properties of J we find

$$\begin{aligned} \big\|\big(J - rA(t)\big)x - \big(J - rA(t)\big)y\big\|_*^2 &= \|x - y\|^2 - 2r\big(A(t)x - A(t)y, x - y\big) \\ &\quad + r^2\|A(t)x - A(t)y\|_*^2 \\ &\le \|x - y\|^2(1 - 2rm + r^2M^2) + 2r\big(k(t)\,|x - y|\big)^2 \\ &\quad + 2r^2M|k(t)|\,\|x - y\|\,|x - y| + r^2\big(k(t)\,|x - y|\big)^2 \\ &\le \|x - y\|^2(1 - 2rm + r^2M^2 + \delta r^4M^2) + \left(2r + r^2 + \frac{1}{\delta}\right)\big(k(t)\,|x - y|\big)^2, \end{aligned}$$

where $\delta > 0$ is arbitrary. Because $0 < r < \dfrac{2m}{M^2}$ implies $1 - 2rm + r^2M^2 < 1$ this inequality proves the assertion of Lemma 2.

Proof of Theorem 4. The unique solvability of (23) for fixed n can be obtained easily from Theorems 1 and 2 (with 0 instead of B). We shall prove next that the sequence (u_n) defined by (23) is bounded in Y. From (23) follows

$$\int_0^t \big((ru_n' + rB(u_{n-1}, u_n) + L_n u_n)(s), u_n(s)\big)\, ds = \int_0^t \big((L_n u_{n-1} - rA_n u_{n-1})(s), u_n(s)\big)\, ds.$$

Hence,

$$\frac{r}{2}|u_n(t)|^2 - \frac{r}{2}|a_n|^2 + \int_0^t \|u_n(s)\|^2\, ds$$

$$= \int_0^t \big((Lu_{n-1} - rAu_{n-1} + rA0 - rA0)(s), u_n(s)\big)\, ds$$

$$\leq \frac{1}{2}\int_0^t \|(Lu_{n-1} - rAu_{n-1} + rA0)(s)\|_*^2\, ds + \frac{1}{2}\int_0^t \|u_n(s)\|^2\, ds - \int_0^t \big(rA(s)\,0, u_n(s)\big)\, ds.$$

Evidently, $(Lu_{n-1} - rAu_{n-1})(s) = Ju_{n-1}(s) - rA(s)\,u_{n-1}(s)$. Therefore, using (24) we obtain

$$r|u_n(t)|^2 + \int_0^t \|u_n(s)\|^2\, ds \leq c + \int_0^t \big\{p_r' \|u_{n-1}(s)\|^2 + q_r'\big(k(s)\,|u_{n-1}(s)|\big)^2\big\}\, ds, \tag{25}$$

where $p_r < p_r' < 1$ and c is a constant not depending on n. We put

$$g(t) = \frac{q_r'}{rp_r'}|k(t)|^2 \exp\left(-\frac{q_r'}{rp_r'}\int_0^t |k(s)|^2\, ds\right)$$

Multiplying (25) by $g(t)$ and integrating with respect to t from 0 to T we find

$$\int_0 g(t)\left\{r|u_n(t)|^2 + \int_0^t \|u_n(s)\|^2\, ds\right\} dt$$

$$\leq c_1 + \int_S g(t) \int_0^t \big\{p_r' \|u_{n-1}(s)\|^2 + q_r'\big(k(s)\,|u_{n-1}(s)|\big)^2\big\}\, ds\, dt$$

$$= c_1 + p_r' \int_S g(t) \int_0^t \|u_{n-1}(s)\|^2\, ds\, dt + q_r' \int_S \big(k(s)\,|u_{n-1}(s)|\big)^2 \int_s^T g(t)\, dt\, ds$$

$$\leq c_1 + p_r' \int_S g(t) \int_0^t \|u_{n-1}(s)\|^2\, ds\, dt$$

$$+ q_r' \int_S \big(k(s)\,|u_{n-1}(s)|\big)^2 \exp\left(-\frac{q_r'}{rp_r'}\int_0^s |k(\sigma)|^2\, d\sigma\right) ds$$

$$= c_1 + p_r' \int_S g(t)\left\{\int_0^t \|u_{n-1}(s)\|^2\, ds + r|u_{n-1}(t)|^2\right\} dt, \qquad c_1 = \text{const.}$$

Since $p_r' < 1$ from this inequality follows easily by induction

$$\int_S g(t) \left\{ r\,|u_n(t)|^2 + \int_0^t \|u_n(s)\|^2\,ds \right\} dt \leqq c_2,$$

where c_2 is independent of n. Using the definition of g we find

$$\sup_n \int_S \left(k(t)\,|u_n(t)|\right)^2 dt = c_3 < \infty. \tag{26}$$

Substituting this result into (25) (in the case $t = T$) we obtain

$$\int_S \|u_n(t)\|^2\,dt \leqq c + q_r' c_3 + p_r' \int_S \|u_{n-1}(t)\|^2\,dt.$$

Again by induction it follows that

$$\sup_n \|u_n\|_X < \infty. \tag{27}$$

The relations (25), (26) and (27) show that (u_n) is bounded also in $C(S; H)$. Thus, we have established that the sequence (u_n) defined by (23) is bounded in Y.

Now we are going to prove the convergence statement of Theorem 4. Let (v_n) be a sequence of functions such that

$$v_n \in C(S; V_n), \qquad v_n' \in C(S; V_n), \qquad v_n \to u \text{ in } W. \tag{28}$$

Let $w_n = u_n - u$. From (23), (6) and (24) follows

$$\begin{aligned}
0 &= \int_0^t \Big(\big(rw_n' + rB(u_{n-1}, u_n) - rB(u, u) + Lw_n - Lw_{n-1}\\
&\quad + rAu_{n-1} - rAu\big)(s), u_n(s) - v_n(s)\Big)\,ds\\
&= \int_0^t \Big(\big(rw_n' + rB(u_{n-1}, u_n) - rB(u, u) + Lw_n - Lw_{n-1}\\
&\quad + rAu_{n-1} - rAu\big)(s), u(s) - v_n(s)\Big)\,ds\\
&\quad + \int_0^t \Big(\big(rw_n' + rB(w_{n-1}, u) + Lw_n - Lw_{n-1} + rAu_{n-1} - rAu\big)(s), w_n(s)\Big)\,ds\\
&\geqq -c\,\|u - v_n\|_W + \frac{r}{2}\,|w_n(t)|^2 - \frac{r}{2}\,|a_n - a|^2\\
&\quad + \int_0^t \Big\{ -rM_1\big(|w_{n-1}(s)|\,\|w_{n-1}(s)\|\,|u(s)|\,\|u(s)\|\big)^{\frac{1}{2}}\,\|w_n(s)\| + \|w_n(s)\|^2\\
&\quad - \Big(p_r\,\|w_{n-1}(s)\|^2 + q_r\,\big(k(s)\,|w_{n-1}(s)|\big)^2\Big)^{\frac{1}{2}}\,\|w_n(s)\| \Big\}\,ds\\
&\geqq \frac{r}{2}\,|w_n(t)|^2 + \int_0^t \Big\{ \|w_n(s)\|^2\,(1 - \delta) - \|w_{n-1}(s)\|^2\,(p_r + \delta)\\
&\quad - c_1\big(|u(s)|^2\,\|u(s)\|^2 + |k(s)|^2\big)\,|w_{n-1}(s)|^2 \Big\}\,ds - \varepsilon_n,
\end{aligned} \tag{29}$$

where $\varepsilon_n = c\|u - v_n\|_W + \frac{r}{2}|a_n - a|^2 \to 0$ if $n \to \infty$ and $\delta > 0$ arbitrary. Let

$p = \frac{p_r + \delta}{1 - \delta} < 1$ and $q = 1 - \delta$. We put

$$k_1(s) = c_1\big(|u(s)|^2 \|u(s)\|^2 + |k(s)|^2\big)$$

and

$$g(t) = \frac{2}{rp} k_1(t) \exp\left(-\frac{2}{rp}\int_0^t k_1(s)\,ds\right).$$

Multiplying the last inequality by $g(t)$ and integrating with respect to t from 0 to T we obtain

$$\int_0 g(t)\left\{\frac{r}{2}|w_n(t)|^2 + q\int_0^t \|w_n(s)\|^2\,ds\right\}dt$$

$$\leqq \int_S g(t)\int_0^t \big(pq\,\|w_{n-1}(s)\|^2\,ds + k_1(s)\,|w_{n-1}(s)|^2\big)\,ds\,dt + c_2\varepsilon_n$$

$$= \int_S k_1(s)\,|w_{n-1}(s)|^2\int_s^T g(t)\,dt\,ds + pq\int_S g(t)\int_0^t \|w_{n-1}(s)\|^2\,ds\,dt + c_2\varepsilon_n$$

$$\leqq p\int_S g(t)\left\{\frac{r}{2}|w_{n-1}(s)|^2 + q\int_0^t \|w_{n-1}(s)\|^2\,ds\right\}dt + c_2\varepsilon_n.$$

From this inequality it follows easily by induction

$$\int_S g(t)\left\{\frac{r}{2}|w_n(t)|^2 + q\int_0^t \|w_n(s)\|^2\,ds\right\}dt \leqq c_3 p^n + c_2\sum_{i=1}^{n} \varepsilon_i p^{n-1}.$$

Since we have $p < 1$, the right hand side of this inequality tends to 0 if $n \to \infty$. Using this result and the definition of g we find

$$\lim_{n\to\infty} \int_S k_1(t)\,|w_n(t)|^2\,dt = 0. \tag{30}$$

Furthermore, we obtain (cf. (29))

$$q\int_S \|w_n(t)\|^2\,dt \leqq pq\int_S \|w_{n-1}(t)\|^2\,dt + \varepsilon_n + \int_S k_1(t)\,|w_{n-1}(t)|^2\,dt.$$

Again by induction it follows that

$$\lim_{n\to\infty} \|w_n\|_X = 0 \tag{31}$$

The relations (29), (30) and (31) imply

$$w_n \to 0 \quad \text{in} \quad C(S; H).$$

Thus, we have also

$$w_n \to 0 \qquad \text{in } Y.$$

This completes the proof of Theorem 4.

Remark 4. If we choose $V_n = V$, $H_n = H$ and $a_n = a$, then $A_n = A$, $B_n = B$, $L_n = L$ and (23) becomes a pure iteration method, which reduces the nonlinear problem (6) to a sequence of linear problems. If all the spaces V_n (and therefore also the spaces H_n) are finite dimensional, then (23) is a so called projection-iteration method. Each step of this method requires the solution of an initial value problem for a system of linear ordinary differential equations.

Remark 5. If $B = 0$, then Theorem 4 reduces to a slight generalization of known results on projection-iteration methods for evolution equations (see Gajewski-Gröger [2, 3]). In this case it is not necessary to assume the compactness of the imbedding operator from V into H.

Remark 6. If $A = L + C$, where

$$\left.\begin{aligned} &(Cu)(t) = C(t)\, u(t) \qquad \forall u \in Y, \forall t \in S, C(t) \in (V \to V^*), \\ &\|C(t)\, x - C(t)\, y\|_* \leqq \left(|k(t)|\, \|x - y\|\, |x - y|\right)^{\frac{1}{2}} \leqq \frac{\delta}{2}\, \|x - y\| + \frac{1}{2\delta}\, |k(t)|\, |x - y|, \end{aligned}\right\} \tag{32}$$

then the method (23) works for $r \in \,]0, 2[$. If $r = 1$, we can write (23) as follows:

$$\left.\begin{aligned} &u_n' + B_n(u_{n-1}, u_n) + L_n u_n + C_n u_{n-1} = 0, \qquad u_n \in W_n, \\ &u_n(0) = a_n, \qquad n = 1, 2, \ldots, \qquad u_0 \in Y_1 \text{ arbitrary.} \end{aligned}\right\} \tag{33}$$

Remark 7. Let $C \in (Y \to X^*)$ be an operator satisfying (32). Then Theorem 4 remains true, if we replace A by $A + C$ in (6) and $A_n u_{n-1}$ by $A_n u_{n-1} + C_n u_n$ in (23). This can be shown by some obvious modifications in the proof of this theorem. This is of interest especially if C is linear because in this case (23) (in its modified form) is again a sequence of linear problems.

Remark 8. In the case of m-dimensional Navier-Stokes equations, $m \geqq 3$, the operator B defined by

$$B(u, v) = u \operatorname{grad} v$$

does not satisfy (3). Nevertheless one can prove the convergence of a projection-iteration method analogous to (23) provided the existence of a solution

$$u \in L^p\left(S; \left(L^q(G)\right)^m\right), \quad \frac{2}{p} + \frac{m}{q} = 1, \qquad p \geqq 2,$$

is known. In this case the solution u is unique (cf. Serrin [7]).

4. An example

In this section we consider a simple two dimensional version of the so called Marangoni equations which describe the heat transfer across an interface between two fluids and the motion of these fluids in the case that the interface tension depends on the temperature.

Let d_1, d_2^-, d_2^+ be constants such that $d_1 > 0$, $d_2^- < 0 < d_2^+$ and let G be the domain in the x_1,x_2-plane defined by

$$G =]0, d_1[\times]d_2^-, d_2^+[.$$

Furthermore, let

$$G^+ = G \cap \{x_2 > 0\}, \qquad G^- = G \cap \{x_2 < 0\}$$

(domains occupied by the two fluids)

and

$$\Gamma = G \cap \{x_2 = 0\} \qquad \text{(interface).}$$

The Marangoni equations can be written formally as follows (see e.g. Wilke [8]):

$$\left.\begin{aligned} &\varrho \frac{\partial u_i}{\partial t} - \mu \Delta u_i + \varrho(u_1, u_2) \operatorname{grad} u_i = -\frac{\partial p}{\partial x_i}, \qquad i = 1, 2, \\ &\varrho c_p \frac{\partial u_3}{\partial t} - \lambda \Delta u_3 + \varrho c_p(u_1, u_2) \operatorname{grad} u_3 = -\frac{\beta}{\lambda} \varrho c_p u_2, \\ &\operatorname{div}(u_1, u_2) = 0 \qquad \text{in } G^+ \text{ and } G^-, \qquad 0 \leqq t \leqq T; \end{aligned}\right\} \tag{34}$$

$$\left.\begin{aligned} &u = (u_1, u_2, u_3) = 0 \quad \text{if} \quad x_2 = d_2^- \quad \text{or} \quad x_2 = d_2^+, \\ &u_1 = \frac{\partial u_2}{\partial x_1} = \frac{\partial u_3}{\partial x_1} = 0 \quad \text{if} \quad x_1 = 0 \quad \text{or} \quad x_1 = d_1, \\ &u_2 \Big|_{x_2=0} = [u_1] = [u_3] = \left[\lambda \frac{\partial u_3}{\partial x_2}\right] = 0, \quad \left[\mu \frac{\partial u_1}{\partial x_2}\right] - \xi \frac{\partial u_3}{\partial x_1}\Big|_{x_2=0}, \\ &\text{where} \quad [v] := \lim_{x_2 \downarrow 0} v(x_2) - \lim_{x_2 \uparrow 0} v(x_2); \end{aligned}\right\} \tag{35}$$

$$u \Big|_{t=0} = a. \tag{36}$$

The notations used in this formulation have the following meaning:

u_1, u_2 components of velocity,

u_3 modified temperature (the temperature is $\theta = u_3 + \alpha + \frac{\beta}{\lambda} \sigma_2$, where α, β are constants depending on the boundary values of θ),

p	pressure,
t	time,
a	initial value of $u = (u_1, u_2, u_3)$,
ζ	constant rate of interface tension with respect to temperature,
ϱ	density,
μ	viscosity,
λ	thermal conductivity,
c_p	specific heat.

We assume that

$$\varrho = \begin{cases} \varrho^+ & \text{in } G^+, \\ \varrho^- & \text{in } G^-, \end{cases} \qquad \varrho^+, \varrho^- \text{ constant,}$$

and that corresponding relations hold for μ, λ and c_p.

To formulate the problem (34)—(36) precisely we introduce the following spaces:

$$D = \{u \in (\mathcal{D}(\bar{G}))^3 \mid u_1(0, x_2) = u_1(d_1, x_2) = 0 \qquad \forall x_2 \in [d_2^-, d_2^+],$$

$$u(x_1, d_2^-) = u(x_1, d_2^+) = 0, \quad u_2(x_1, 0) = 0 \quad \forall x_1 \in [0, d_1],$$

$$\operatorname{div}(u_1, u_2) = 0\};$$

H = closure of D in $(L^2(G))^3$ with the inner product

$$(u, v) = \int_G \sum_{i=1}^{3} \varrho_i u_i v_i \, dx \qquad \forall u, \forall v \in H, \quad \text{where} \quad \varrho_1 = \varrho_2 = \varrho,$$

$$\varrho_3 = k^2 \varrho c_p \qquad (k \text{ a constant to be determined later});$$

V = closure of D in $(H^1(G))^3$ with the inner product

$$((u, v)) = \int_{\mathring{G}} \sum_{i=1}^{3} \mu_i \operatorname{grad} u_i \operatorname{grad} v_i \, dx \qquad \forall u, \forall v \in V,$$

where $\mu_1 = \mu_2 = \mu, \qquad \mu_3 = k^2 \lambda.$

The corresponding norms in H and V we shall denote as in the general case by $|.|$ and $\|.\|$, respectively. Evidently, V is densely imbedded into H and the imbedding operator is completely continuous by Rellich's Theorem. As in the previous sections let $S = [0, T]$ be a finite (time) interval and $X = L^2(S; V)$, $Y = X \cap C(S; H)$.

Let us assume for the time being that u is a sufficiently smooth solution of (34), (35).

Then for every $v \in X$ such that $v(t) \in D$ for a.e. $t \in S$ we obtain

$$0 = \int_S \int_G \left\{ \sum_{i=1}^{3} \left(\varrho_i u_i' - \mu_i \Delta u_i + \varrho_i (u_1, u_2) \operatorname{grad} u_i \right) v_i + \frac{\beta}{\lambda} \varrho_3 u_2 v_3 \right\} dx\, dt$$

$$= \int_S \left\{ (u', v) + ((u, v)) + \int_\Gamma \zeta \frac{\partial u_3}{\partial x_1} v_1 \, dx_1 \right.$$

$$\left. + \int_G \left(\sum_{i=1}^{3} \varrho_i \left((u_1, u_2) \operatorname{grad} u_i \right) v_i + \frac{\beta}{\lambda} \varrho_3 u_2 v_3 \right) dx \right\} dt \cdot$$

$$= \int_S \left\{ (u', v) + ((u, v)) + \int_{G^+} \zeta \left(\frac{\partial u_3}{\partial x_2} \frac{\partial v_1}{\partial x_1} - \frac{\partial u_3}{\partial x_1} \frac{\partial v_1}{\partial x_2} \right) dx \right.$$

$$\left. + \int_G \left(\sum_{i=1}^{3} \varrho_i \left((u_1, u_2) \operatorname{grad} u_i \right) v_i + k^2 \frac{\beta}{\lambda} \varrho c_p u_2 v_3 \right) dx \right\} dt. \tag{37}$$

This equation suggests to introduce an operator $A_0 \in (V \to V^*)$ and a 3-linear form $b \in (V \times V \times V \to R)$ by the following relations:

$$(A_0 u, v) = ((u, v)) + \int_{G^+} \zeta \left(\frac{\partial u_3}{\partial x_2} \frac{\partial v_1}{\partial x_1} - \frac{\partial u_3}{\partial x_1} \frac{\partial v_1}{\partial x_2} \right) dx$$

$$+ \int_G k^2 \frac{\beta}{\lambda} \varrho c_p u_2 v_3 \, dx \qquad \forall u, \forall v \in V, \tag{38}$$

$$b(u, v, w) = \int_G \sum_{i-1}^{3} \varrho_i \left((u_1, u_2) \operatorname{grad} v_i \right) w_i \, dx \qquad \forall u, \forall v, \forall w \in V. \tag{39}$$

The operator A_0 is evidently linear. From (38) follows

$$(A_0 u, u) \geqq \|u\|^2 - \frac{\zeta}{k \sqrt{\mu^+ \lambda^+}} \left(\int_{G^+} \mu_3 \operatorname{grad}^2 u_3 \, dx \right)^{\frac{1}{2}} \left(\int_{G^+} \mu_1 \operatorname{grad}^2 u_1 \, dx \right)^{\frac{1}{2}}$$

$$- kc \, |u|^2 \geqq \|u\|^2 \left(1 - \frac{\zeta}{k \sqrt{\mu^+ \lambda^+}} \right) - kc \, |u|^2 \qquad \forall u \in V, \qquad c = \text{const}, \tag{40}$$

and

$$(A_0 u, v) \leqq \|u\| \, \|v\| \left(1 + \frac{\zeta}{k \sqrt{\mu^+ \lambda^+}} \right) + kc \, |u| \, |v| \qquad \forall u, \forall v \in V. \tag{41}$$

In what follows we shall assume that

$$k > \frac{\zeta}{\sqrt{\mu^+ \lambda^+}}. \tag{42}$$

Using div $(u_1, u_2) = 0$ we obtain from (39) by partial integration

$$b(u, v, v) = 0 \qquad \forall u, \forall v \in V. \tag{43}$$

Furthermore, we find (also by partial integration)

$$|b(u, v, w)| \leqq c \, \|w\|_{(C^1(\bar{G}))^2} \, |u| \, |v| \qquad \forall w \in D, \quad \forall u, \forall v \in V, \quad c = \text{const.} \tag{44}$$

As a consequence of well known imbedding theorems (see e.g. Ladyshenskaja-Uralzewa [6]) we have

$$|b(u, v, w)| \leqq M_1 \left(|u| \, \|u\| \, |v| \, \|v\|\right)^{\frac{1}{2}} \|w\| \qquad \forall u, \forall v, \forall w \in V. \tag{45}$$

We define now $A \in (Y \to X^*)$ and $B \in (Y \times Y \to X^*)$ by

$$(Au)(t) = A_0 u(t) \qquad \forall t \in S, \quad \forall u \in Y, \tag{46}$$

$$\langle B(u, v), w \rangle = \int_S b\big(u(t), v(t), w(t)\big) \, dt \qquad \forall u, \forall v \in Y, \quad \forall w \in X. \tag{47}$$

It follows from (37) that a smooth solution of (34)—(36) is a solution of the problem

$$u' + Au + B(u, u) = 0, \qquad u \in W, \quad u(0) = a. \tag{48}$$

We shall consider (48) as the precise formulation of (34)—(36). It is easy to see (cf. (40)—(47)) that the operators A and B satisfy the conditions (1)—(3). Therefore, the problem (48) is uniquely solvable and the solution can be approximated as described in Theorem 4 (cf. also Remarks 6 and 7).

Remark 9. If the fluids in the preceding example are non-Newtonian, then we have to replace $-\mu \Delta u_i$ in (34) by

$$-\sum_{j=1}^{2} \frac{\partial}{\partial x_j} \left\{ \mu \left(\sum_{k,l} \left(\frac{\partial u_k}{\partial x_l} + \frac{\partial u_l}{\partial x_k} \right)^2 \right) \left(\frac{\partial u_i}{\partial x_j} + \frac{\partial u_j}{\partial x_i} \right) \right\},$$

where now μ is a function. Under reasonable assumptions on μ the corresponding (nonlinear) operator A again allows to apply the results of the previous sections.

Bibliography

[1] Gajewski, H., Zur iterativen Lösung der zweidimensionalen Boussinesq-Gleichungen. Z. angew. Math. Mech. (in print).

[2] Gajewski, H., Gröger, K., Ein Iterationsverfahren für Gleichungen mit einem maximal monotonen und einem stark monotonen Lipschitz-stetigen Operator. Math. Nachr. **69**, 307—317 (1975).

[3] —, Ein Projektions-Iterationsverfahren für Evolutionsgleichungen. Math. Nachr. **72**, 119 bis 136 (1976).

[4] Gajewski, H., Gröger, K., Zacharias, K., Nichtlineare Operatorgleichungen und Operatordifferentialgleichungen. Berlin 1974.

[5] Gröger, K., Ein Kompaktheitssatz für Folgen von Näherungslösungen von Evolutionsgleichungen. Math. Nachr. 72, 15—19 (1976).
[6] Ладыженская, О. А., Уральцева, Н. Н., Линейные и квазилинейные уравнения эллиптического типа. Москва 1973.
[7] Serrin, J., The initial value problem for the Navier-Stokes equations. *In:* Non linear Problems, ed. by R. E. Langer 1963, 69—98.
[8] Wilke, H., Ein numerisches Verfahren zur Behandlung grenzflächendynamischer Probleme beim gekoppelten Impuls- und Wärmetransport über fluide Phasengrenzen. Thesis, Berlin 1973.

Address of the authors:

Akademie der Wissenschaften der DDR
Zentralinstitut für Mathematik und Mechanik
108 Berlin
Mohrenstr. 39
DDR

Nonlinear elliptic problems in unbounded domains

Peter Hess, Zürich

1. About twelve years ago the question of solvability of variational boundary value problems for quasilinear elliptic equations in divergence form

$$\mathscr{A}u = \sum_{|\alpha| \leq m} (-1)^{|\alpha|} D^\alpha A_\alpha(x, u, \ldots, D^m u) = f, \tag{0}$$

considered in a bounded domain in $\boldsymbol{R}^N$, started to attract remarkable attention. For it had been realized that the theory of monotone mappings could be brought into play in a most powerful way. In connection with a different dependence of $\mathscr{A}$ on its top order part ($|\alpha| = m$) and its lower order terms, the concept of monotonicity was subsequently weakened, and various classes of nonlinear operators, which we summarize as "mappings of monotone type", were introduced. In that context the Sobolev embedding theorem, which guarantees certain compactness properties of the abstract mappings determined, was of utmost importance.

If the underlying domain is unbounded, these compactness assertions however fail to hold in general, and additional nontrivial difficulties arise. In some recent papers Berger & Schechter [1, 2], Edmunds & Evans [4] and Edmunds & Webb [7] overcame these obstacles by discovering a substitute for these embedding theorems. They proved that multiplication by certain functions defines a compact mapping of some (weighted) Sobolev space to an appropriate L^q-space. This result was used by them to establish an existence theory for the Dirichlet problem for coercive equations of the form (0) in unbounded domains.

The purpose of this paper is to propose a new approach to boundary value problems in unbounded domains which seems to be considerably simpler and does not require any special knowledge of Sobolev spaces over unbounded domains.

2. Let Ω denote a — possibly unbounded — domain in the Euclidean space $\boldsymbol{R}^N$ ($N \geqq 1$). We assume that Ω can be represented in the form $\Omega = \bigcup_k \omega_k$, where (ω_k) is an increasing sequence of *bounded* domains as follows:

(i) for each $k \in \boldsymbol{N}$ and $1 < q < \infty$, the embedding $W^{1,q}(\omega_k) \hookrightarrow L^q(\omega_k)$ is compact;

(ii) to each k there exists a function $\psi_k \in C_0^\infty(\boldsymbol{R}^N)$ with values in [0, 1], such that $\psi_k(x) = 1 \; \forall x \in \omega_k$ and supp $\psi_k \subset \omega_{k+1}$.

This can always be realized if the boundary of Ω is piecewise sufficiently smooth.

Let $\mathscr{A}$ be the quasilinear elliptic second order differential operator in divergence form:

$$(\mathscr{A}u)(x) = -\sum_{i=1}^{N} \frac{\partial}{\partial x_i} A_i(x, u(x), \nabla u(x)), \qquad \text{a.a. } x \in \Omega.$$

On the functions A_i $(i = 1, \ldots, N)$ the following standard conditions of Leray-Lions type are imposed.

(A1) Each $A_i: \Omega \times \mathbf{R} \times \mathbf{R}^N \to \mathbf{R}$ satisfies the Carathéodory conditions (i.e. $A_i(x, t, \xi)$ is measurable in $x \in \Omega$ for all fixed $(t, \xi) \in \mathbf{R} \times \mathbf{R}^N$ and continuous in (t, ξ) for a.a. fixed x). There exist constants $q: 1 < q < \infty$, $c_0 \geqq 0$ and a function $k_0 \in L^{q'}(\Omega)$ $(q' = q/q - 1)$ such that

$$|A_i(x, t, \xi)| \leqq k_0(x) + c_0(|t|^{q-1} + |\xi|^{q-1}), \qquad i = 1, \ldots, N,$$

for a.a. $x \in \Omega$, $\forall (t, \xi) \in \mathbf{R} \times \mathbf{R}^N$.

$$\text{(A 2)} \qquad \sum_{i=1}^{N} \big(A_i(x, t, \xi) - A_i(x, t, \xi^*)\big)(\xi_i - \xi_i^*) > 0$$

for a.a. $x \in \Omega$, $\forall t \in \mathbf{R}$, $\forall \xi, \xi^* \in \mathbf{R}^N$ with $\xi \neq \xi^*$.

$$\text{(A 3)} \qquad \sum_{i=1}^{N} A_i(x, t, \xi)\, \xi_i \geqq \gamma\, |\xi|^q \qquad (\gamma > 0),$$

for a.a. $x \in \Omega$, $\forall (t, \xi) \in \mathbf{R} \times \mathbf{R}^N$.

As a consequence of (A1) the semi-linear form a:

$$a(u, v) = \sum_{i=1}^{N} \int_\Omega A_i(., u, \nabla u) \frac{\partial v}{\partial x_i}\, dx$$

is defined on $W^{1,q}(\Omega) \times W^{1,q}(\Omega)$.

Let further the function p: $\Omega \times \mathbf{R} \times \mathbf{R}^N \to \mathbf{R}$ satisfy the Carathéodory conditions. For a function v on Ω we set

$$(Pv)(x) = p(x, v(x), \nabla v(x)), \qquad x \in \Omega.$$

Let V denote the closed subspace of $W^{1,q}(\Omega)$ containing the test functions, which determines the stable boundary conditions. For given $f \in V^*$ we ask for a solution of the boundary value problem

$$\text{(BVP)} \quad \begin{cases} \mathscr{A}u + Pu = f & \text{in } \Omega, \\ u \text{ satisfies the boundary conditions implied by } V. \end{cases}$$

Definition 1. A function u is *weak solution* of the (BVP) provided

$$u \in V,\ Pu \in L^{q'}(\Omega),$$

$$a(u, v) + \int_\Omega Puv\, dx = (f, v) \qquad \forall v \in V.$$

Theorem 1. *Suppose* $Pv \in L^{q'}(\Omega)$ $\forall v \in V$, *and that there exist a number* $\delta < \gamma$ *and a function* $r \in L^1(\Omega)$ *such that*

$$p(x, t, \xi)\, t \geqq -\delta\, |\xi|^q - r(x)$$

for a.a. $x \in \Omega$, $\forall (t, \xi) \in \boldsymbol{R} \times \boldsymbol{R}^N$. *Suppose further that*

$$\frac{1}{\|v\|_{1,q}} \left\{ a(v, v) + \int_\Omega Pvv\, dx \right\} \to +\infty \qquad (\|v\|_{1,q} \to \infty, v \in V). \tag{1}$$

Then the (BVP) *is weakly solvable for each* $f \in V^*$.

Theorem 1 generalizes results of [2, 4, 7], where either the domain Ω was assumed not to differ too much from a bounded domain, or else more restrictive growth conditions on the coefficient functions had to be imposed. We note that our result can easily be extended to problems of higher order (Hess [8]).

In case the coercivity condition (1) is not satisfied, the method of lower and upper solutions has shown to be most fruitful. We now extend that method, which we only know from the literature to hold for problems in *bounded* domains. Suppose in the following that V has the property

$$v \in V \Rightarrow v^+ \in V$$

(here $v^+ = \max\{v, 0\}$ denotes the positive part of the function v).

Definition 2. We say that a function Φ is *weak upper solution* of the (BVP) if

$$\Phi \in W^{1,q}(\Omega), \qquad P\Phi \in L^{q'}(\Omega),$$

$$(v - \Phi)^+ \in V \qquad \forall v \in V,$$

$$a(\Phi, v) + \int_\Omega P\Phi v\, dx \geqq (f, v) \qquad \forall v \in V \text{ with } v \geqq 0 \text{ in } \Omega.$$

This is a natural extension of the classical concept of upper solution. Similarly a weak lower solution φ is characterized by the reversed inequality sign and the hypothesis that $(v - \varphi)^- \in V$ $\forall v \in V$.

Theorem 2. *Suppose* φ, Φ *are weak lower and upper solution of the* (BVP), *respectively, with* $\varphi \leqq \Phi$ a.e. *in* Ω. *Suppose further that, with a constant* c_1 *and some function* $k_1 \in L^{q'}(\Omega)$,

$$|p(x, t, \xi)| \leqq k_1(x) + c_1 |\xi|^{q-1} \tag{2}$$

for a.a. $x \in \Omega$, $\forall \xi \in \boldsymbol{R}^N$, $\forall t\colon \varphi(x) \leqq t \leqq \Phi(x)$.

Then the (BVP) *admits a weak solution* u *with* $\varphi \leqq u \leqq \Phi$ a.e. *in* Ω.

Remarks. 1) Condition (2) guarantees that $Pv \in L^{q'}(\Omega)$ for all functions $v \in V$ with $\varphi \leqq v \leqq \Phi$ a.e. in Ω.

2) It is not known whether, in the general situation of Theorem 2, there exist a minimal and a maximal solution of the (BVP) between φ and Φ. If however the coefficient functions A_i are independent of t, if p does not depend on ξ, and if there exists

a number $\lambda > 0$ such that for a.a. $x \in \Omega$ the function $t \mapsto p(x, t) - \lambda t$ is nonincreasing, the classical monotone iteration scheme (e.g. [10]) can be adapted and yields minimal and maximal solutions.

3. We indicate now the simple proof of Theorem 1. Let χ_{ω_n} denote the characteristic function of ω_n. For each n we define

$$a_{\omega_n}(u, v) = \sum_{i=1}^{N} \int_\Omega A_i(\cdot, \chi_{\omega_n} u, \nabla u) \frac{\partial v}{\partial x_i} dx,$$

$$b_{\omega_n}(u, v) = \int_\Omega \chi_{\omega_n} Pu\, v\, dx,$$

$\forall u, v \in V$.

Lemma. *The mapping $T_n: V \to V^*$ determined by*

$$(T_n u, v) = a_{\omega_n}(u, v) + b_{\omega_n}(u, v) \qquad \forall u, v \in V,$$

is pseudo-monotone and continuous.

Its proof is identical with that for bounded domains; for the latter we refer to Lions [9, Chap. II.2]. Due to the modifications in the lower order terms, there may be some loss of coercivity. This will be compensated by the introduction of the semi-linear form c:

$$c(u, v) = \int_\Omega |u|^{q-2} uv\, dx \qquad (u, v \in V),$$

which defines a continuous monotone operator from V to V^*. It follows from a well-known surjectivity result for coercive, continuous pseudo-monotone mappings that (with a fixed positive λ) to each n there exists $u_n \in V$:

$$a_{\omega_n}(u_n, v) + b_{\omega_n}(u_n, v) + \lambda c(u_n, v) = (f, v) \qquad \forall v \in V. \tag{3}$$

Letting $n \to \infty$ we first note that $\|u_n\|_{1,q} \leqq \text{const}\ \forall n$. By the reflexivity of V we thus may pass to a subsequence (without changing the notation) such that

$$u_n \rightharpoonup u \text{ in } V \qquad (n \to \infty).$$

For a fixed $k \in \mathbf{N}$ set $v = \psi_k(u_n - u)$ in (3). Since the Sobolev embedding theorem is assumed to hold for the bounded domain ω_{k+1}, we may suppose (if necessary by passing to a further subsequence) that

$$u_n \to u \quad (n \to \infty) \quad \text{in } L^q(\omega_{k+1}) \text{ and a.e. in } \omega_{k+1}.$$

We then conclude from (3) that

$$\sum_{i=1}^{N} \int_\Omega \psi_k A_i(\cdot, \chi_{\omega_n} u_n, \nabla u_n) \frac{\partial}{\partial x_i}(u_n - u)\, dx \to 0 \qquad (n \to \infty).$$

Since also

$$\sum_{i=1}^{N} \int_\Omega \psi_k A_i(\cdot, \chi_{\omega_n} u_n, \nabla u) \frac{\partial}{\partial x_i}(u_n - u)\, dx \to 0 \qquad (n \to \infty),$$

we obtain

$$\sum_{i=1}^{N} \int_{\Omega} \psi_k [A_i(., \chi_{\omega_n} u_n, \nabla u_n) - A_i(., \chi_{\omega_n} u_n, \nabla u)] \frac{\partial}{\partial x_i} (u_n - u)\, dx \to 0 \qquad (n \to \infty),$$

and hence in particular

$$\sum_{i=1}^{N} \int_{\omega_k} [A_i(., u_n, \nabla u_n) - A_i(., u_n, \nabla u)] \frac{\partial}{\partial x_i} (u_n - u)\, dx \to 0 \qquad (n \to \infty). \tag{4}$$

From (4) we derive, employing an argument introduced by Leray-Lions (e.g. [9, Lemma 2.2, p. 184]), that

$$\nabla u_n \to \nabla u \qquad \text{a.e. in } \omega_k.$$

Further we already know that

$$u_n \to u \qquad \text{a.e. in } \omega_k.$$

We now set $k = 1, 2, \ldots$ and run through a diagonal process to obtain a sequence such that

$$\left. \begin{array}{l} u_n \to u \\ \nabla u_n \to \nabla u \end{array} \right\} \text{ a.e. in } \Omega.$$

Passage to the limit $n \to \infty$ in (3) is then simple and shows that

$$a(u, v) + \int_{\Omega} Puv\, dx + \lambda c(u, v) = (f, v) \qquad \forall v \in V.$$

(We hereby use the well-known fact that weak limit and pointwise limit a.e. coincide in an L^q-space.)

The limiting argument for $\lambda \to 0^+$ is now similar and completes the proof of Theorem 1.

Remarks. 1) The same method permits to attack "strongly non-linear" problems (cf. also Edmunds-Moscatelli-Webb [6], Webb [11]) as well as "semi-coercive" problems (cf. Edmunds & Moscatelli [5]) in unbounded domains.

2) It is open whether our new approach is also successful in the study of variational inequalities in unbounded domains.

The proof of Theorem 2 is identical with that given by Deuel & Hess [3] for the case of bounded domains. It consists in modifying the given problem outside the convex set of "interesting" functions $v: \varphi \leqq v \leqq \Phi$, in such a way to obtain a coercive problem. The existence of a solution of this altered problem then follows from a trivial variant of Theorem 1 (instead of the Nemytskii operator P one considers its composition with a truncation mapping). Finally one proves that any solution of the modified problem necessarily lies between φ and Φ.

References

[1] Berger, M. S., M. Schechter, L^p embedding and nonlinear eigenvalue problems for unbounded domains. Bull. Amer. Math. Soc. **76**, 1299–1302 (1970).
[2] —, Embedding theorems and quasi-linear elliptic boundary value problems for unbounded domains. Trans. Amer. Math. Soc. **172**, 261–278 (1973).
[3] Deuel, J., P. Hess, A criterion for the existence of solutions of nonlinear elliptic boundary value problems. Proc. Royal Soc. Edinburgh **74**, 49–54 (1975).
[4] Edmunds, D. E., W. D. Evans, Elliptic and degenerate-elliptic operators in unbounded domains. Ann. Scuola Norm. Sup. Pisa **27**, 591–640 (1973).
[5] Edmunds, D. E., V. B. Moscatelli, Semi-coercive nonlinear problems. Boll. Un. Mat. Ital. **11**, 144–153 (1975).
[6] Edmunds, D. E., V. B. Moscatelli, J. R. L. Webb, Strongly nonlinear elliptic operators in unbounded domains. Publ. Math. Bordeaux **4**, 6–32 (1974).
[7] Edmunds, D. E., J. R. L. Webb, Quasilinear elliptic problems in unbounded domains. Proc. Royal Soc. London A, **334**, 397–410 (1973).
[8] Hess, P., Problèmes aux limites non linéaires dans des domaines non bornés. C. R. Acad. Sci. Paris Sér. A **281**, 555–557 (1975).
[9] Lions, J. L., Quelques méthodes de résolution des problèmes aux limites non linéaires. Dunod, Gauthier-Villars, Paris 1969.
[10] Sattinger, D. H., Monotone methods in nonlinear elliptic and parabolic boundary value problems. Indiana Univ. Math. J. **21**, 979–1000 (1972).
[11] Webb, J. R. L., On the Dirichlet problem for strongly nonlinear elliptic operators in unbounded domains. J. London Math. Soc. **10**, 163–170 (1975).

Bibliographical remarks added in proof

(i) Similar problems in unbounded domains have been treated recently also by Simader („Über schwache Lösungen des Dirichletproblems für streng nichtlineare elliptische Differentialgleichungen“, Math. Z. **150**, 1–26 (1976)) by a truncation method; at the decisive steps his arguments differ essentially from ours.

(ii) The present method has been carried over by Mustonen („An elliptic boundary value problem for strongly nonlinear equations in unbounded domains“, to appear in Proc. Royal Soc. Edinburgh) to the setting of Orlicz-Sobolev spaces over unbounded domains, for wich no results on compactness of embeddings seem to be known.

(iii) By the same method Mahler (Thesis ETH Zürich, publications to appear) has handled variational inequalities and parabolic problems over unbounded domains.

Address of the author:

Mathematics Institute
University of Zürich
Freiestr. 36
8032 Zürich
Switzerland

Monotone Operatoren und irreguläre Variationsprobleme

Rolf Klötzler, Leipzig

Abstract. This traitment is an approach to problems with partially monotone or pseudomonotone operators and its application on differential equations and variational problems in Sobolev spaces. There are stated existence theorems and some examples: a nonlinear eigenvalue problem, surfaces with maximal area and a new conception of the plasticity theory.

1. Einleitung

Die Theorie der monotonen Operatoren und die direkten Methoden der Variationsrechnung sind wohlbekannte wesentliche Elemente zum Lösbarkeitsnachweis nichtlinearer Gleichungen oder Ungleichungen (vgl. z. B. [8]). Wir beschränken uns hier auf Gleichungen des Typs

$$Au = g, \tag{1}$$

wobei A ein beschränkter Operator sei, welcher den separablen reflexiven und reellen Banachraum X in seinen konjugierten Raum X^* abbilde; g sei ein vorgegebenes Element aus X^*. Falls A speziell ein Potentialoperator und damit Gradient eines reellen endlichen Funktionals $J(u)$ ist, wollen wir das formal zugeordnete Variationsproblem mit

$$J_0(w) = J(w) - \langle g, w \rangle \to \text{Min auf } X \tag{2}$$

bezeichnen. Aus der Existenz einer Minimalstelle u von (2) resultiert das Erfülltsein von (1) (vgl. etwa [3], S. 97).

Für die oben genannten Elemente der Existenztheorie zu (1) und (2) spielen Grundvoraussetzungen über die Monotonie oder Pseudomonotonie des Operators A eine sehr bedeutende Rolle bzw. Konvexitätseigenschaften des zugeordneten Funktionals $J(w)$ von (2).

Der Gegenstand nachfolgender Ausführungen ist darauf gerichtet, einen Zugang zu Lösbarkeitsfragen von (1) und (2) auch in solchen Fällen zu geben, in denen eben erwähnte Grundvoraussetzungen nicht generell erfüllt sind. Dazu führen wir einige neue Begriffe ein.

Definition. Ein Operator $A \in (X \to X^*)$ heißt *partiell monotoner* (oder *partiell pseudomonotoner*) *Operator* genau dann, wenn ein monotoner (bzw. pseudomonotoner)

Operator $\hat{A} \in (X \to X^*)$ existiert, für den auf einer nichtleeren Teilmenge $\hat{\mathfrak{D}} \subset X$ gilt $Aw = \hat{A}w \, \forall w \in \hat{\mathfrak{D}}$. $\hat{A}$ nennen wir einen *korrespondierenden* monotonen (bzw. pseudomonotonen) Operator zu A. Wir bezeichnen weiterhin A als *koerzitiv* und *hemistetig*, wenn $\hat{A}$ die entsprechenden Eigenschaften besitzt.

Mit diesen Begriffsbildungen folgern wir aus den Hauptsätzen der Theorie monotoner bzw. pseudomonotoner Operatoren (vgl. J. L. Lions [8], S. 170 und 180) unmittelbar die beiden nachstehenden Aussagen.

Hilfssatz 1. *Ist A ein koerzitiver, hemistetiger partiell monotoner* (oder *pseudomonotoner*) *Operator, dann existiert für alle $g \notin \hat{A}(X \setminus \hat{\mathfrak{D}})$ eine Lösung u von* (1).

Hilfssatz 2. *Erfüllt A die Voraussetzungen von Hilfssatz* 1 *und sind die Wertebereiche $A(X) = \hat{A}(X)$, dann existiert für alle $g \in X^*$ eine Lösung u von* (1).

Denn infolge unserer Voraussetzungen existiert zunächst für alle $g \in X^*$ ein $u \in X$ mit $\hat{A}u = g$. Ist $g \notin \hat{A}(X \setminus \hat{\mathfrak{D}})$, muß das Urbild u zu g in $\hat{\mathfrak{D}}$ liegen. Hierfür ist aber $Au = \hat{A}u = g$, also Hilfssatz 1, bestätigt. Hilfssatz 2 ist insofern eine triviale Folgerung, als $\hat{A}(X) = X^*$ ist und damit $A(X) = X^*$.

2. Eine Anwendung partiell pseudomonotoner Operatoren in Sobolewräumen

Die in den Hilfssätzen 1 und 2 verankerte elementare Idee wenden wir jetzt auf spezielle Operatoren in Sobolewräumen $W_0^{m,p}(\Omega)$ an. Es sei hier $X = W_0^{m,p}(\Omega)$ zu einem regulären beschränkten Gebiet $\Omega \subset R^{N_0}$. Als Element von X^* definieren wir den Operatorwert $A(u)$ mittels des in w linearen stetigen Funktionals

$$a(u, w) = \sum_{|\alpha| \leq m} \int_\Omega A_\alpha(x, \delta u, D^m u)\, D^\alpha w \, dx = \langle A(u), w \rangle \text{ auf } X. \tag{3}$$

Dabei ist unter Verwendung allgemein gebräuchlicher Bezeichnungen (vgl. [8], S. 182) und Multiindizes $\alpha = \alpha_1, \ldots, \alpha_k$ $D^\alpha u = \dfrac{\partial^k u}{\partial x_{\alpha_1} \cdots \partial x_{\alpha_k}}$ und $D^k u$ das System aller (verallgemeinerten) partiellen Ableitungen k-ter Ordnung von $u \in X$; d. h., $D^k u = \{D^\beta u \,|\, |\beta| = k\}$. Weiterhin ist $\delta u = \{u, Du, \ldots, D^{m-1}u\}$ und $u(x), w(x), A_\alpha(x, \eta, \xi) \in R^r \, \forall (x, \eta, \xi) \in \Omega \times R^{N_1} \times R^{N_2}$.

Als *Grundvoraussetzungen* stellen wir mit Lions die Forderungen

$$\left.\begin{array}{l}
\forall x \in \Omega \text{ ist die Abbildung } (\eta, \xi) \to A_\alpha(x, \eta, \xi) \text{ stetig;} \\
\forall (\eta, \xi) \in R^{N_1} \times R^{N_2} \text{ ist die Abbildung } x \to A_\alpha(x, \eta, \xi) \text{ meßbar;} \\
\exists k(x) \in L^{p'}(\Omega) \text{ und Konstante } c > 0 \text{ mit} \\
|A_\alpha(x, \eta, \xi)| \leq c[|\eta|^{p-1} + |\xi|^{p-1} + k(x)] \quad \text{für} \quad 1 < p < \infty; \\
\forall u, w \in X \text{ ist die Abbildung } x \to A_\alpha(x, \delta u, D^m w) \in L^{p'}(\Omega) \\
(1/p + 1/p' = 1).
\end{array}\right\} \tag{4}$$

Dann gilt zunächst nach H. Brezis [1] und Lions [8], S. 183, folgender Satz.

Satz 1. *A sei ein durch* (3) *definierter Operator, der die Grundvoraussetzungen* (4) *erfüllt.*

Außerdem sei

a) $a(w, w)/\|w\| \to \infty$ *für* $\|w\| \to \infty$;

b) $\sum_{|\alpha|=m} A_\alpha(x, \eta, \xi)\, \xi_\alpha/(|\xi| + |\xi|^{p-1}) \to \infty$ *für* $|\xi| \to \infty$, *beliebige* η *und fast alle* $x \in \Omega$;

c) $\sum_{|\alpha|=m} \big(A_\alpha(x, \eta, \xi) - A_\alpha(x, \eta, \xi^*)\big)\, (\xi_\alpha - \xi_\alpha^*) > 0$ *für* $\xi \neq \xi^*$, *beliebige* η *und fast alle* $x \in \Omega$.

Dann ist A ein koerzitiver pseudomonotoner Operator auf $X = W_0^{m,p}(\Omega)$, *so daß für alle* $g \in X^*$ *eine Lösung u von* (1) *existiert.*

Ein sorgfältiges Studium des Beweises von Satz 1 (vgl. [8], S. 183—187) läßt leicht erkennen, daß die gleichen Ergebnisse bereits unter etwas schwächeren Voraussetzungen erhalten werden können; nämlich schon dann, wenn Bedingung c) durch nachstehende Voraussetzung c*) ersetzt wird:

c*) $\sum_{|\alpha|=m} \big(A_\alpha(x, \eta, \xi) - A_\alpha(x, \eta, \xi^*)\big)\, (\xi_\alpha - \xi_\alpha^*) \geqq 0 \quad \forall \xi, \xi^*, \eta$ *und fast alle* $x \in \Omega$, *wobei das Gleichheitszeichen dann und nur dann gelte, falls entweder* $\xi = \xi^*$ *oder* $A_\alpha(x, \eta, \xi) = A_\alpha(x, \eta, \xi^*)$ *für alle* α *mit* $|\alpha| = m$ *ist.*

Wir benennen das erweiterte Theorem zu Satz 1, bei dem die Voraussetzung c) durch c*) ersetzt worden ist, als Satz 1*.

Erfüllt A jedoch nicht die Voraussetzungen von Satz 1*, können wir über die Hilfssätze 1 und 2 versuchen, eine Lösung von (1) zu gewinnen. Dazu benötigen wir die Eigenschaft von A, ein koerzitiver partiell pseudomonotoner Operator zu sein. Wir geben nachfolgend einige mit Satz 1* einfach zu überprüfende hinreichende Kriterien dafür an.

Hilfssatz 3. *Es sei* $N_0 = 1$, $r = 1$ *und A ein Operator, der die Grundvoraussetzungen* (4) *erfüllt, hingegen die Bedingungen* a) b) *und* c*) *von Satz* 1* *nur für* $|\xi|, |\xi^*| \geqq K$ *mit* $K = \text{const} > 0$. *Dann ist A ein koerzitiver partiell pseudomonotoner Operator mit dem korrespondierenden Operator* $\hat{A}$, *definiert über* (3) *durch*

$$\hat{A}_\alpha(x, \eta, \xi) = A_\alpha(x, \eta, \xi) \qquad \forall |\alpha| < m$$

$$\hat{A}_\alpha(x, \eta, \xi) = \sup_{\zeta \leqq \xi} A_\alpha(x, \eta, \zeta) \quad \textit{für} \quad |\alpha| = m.$$

Hilfssatz 4. *Es sei* $A = \operatorname{grad} J(u)$ *ein Potentialoperator, der die Grundvoraussetzungen* (4) *und Bedingungen* a) *und* b) *von Satz* 1* *erfüllt. Weiterhin sei* $J(u) = \int_\Omega f(x, \delta u, D^m u)\, dx$ *mit* $f \in C^1$; *die Abbildung* $\xi \to f(x, \eta, \xi)$ *sei für alle* ξ *mit* $|\xi| \geqq K$ *zu beliebigen festen* x, η *konvex. Dann ist A ein koerzitiver partiell pseudomonotoner Operator mit dem korrespondierenden Operator* $\hat{A} = \operatorname{grad} \hat{J}(u)$. *Dabei ist* $\hat{J}(u) = \int_\Omega \hat{f}(x, \delta u, D^m u)\, dx$ und $\hat{f}(x, \eta, \xi) = f^{**}(x, \eta, \xi)$ *die doppeltkonjugierte Funktion zu f bezüglich der Variablen* ξ (vgl. z. B. [3], S. 108).

Wir nehmen im weiteren vereinfachend an, daß $m = 1$, $p > N_0$ und $X = W_0^{1,p}(\Omega)$ ist und daß der Operator $A \in (X \to X^*)$ ein koerzitiver partiell pseudomonotoner Operator ist, dessen korrespondierender Operator $\hat{A}$ den Grundvoraussetzungen (4) und Bedingungen a), b), c*) von Satz 1* genügt. Hier ist A (und in entsprechender Weise $\hat{A}$) charakterisiert durch

$$\langle A(u), w\rangle = \int_\Omega [A_i(x, u, u_x)\, w^i + A_i{}^\alpha(x, u, u_x)\, w^i_{x_\alpha}]\, dx, \tag{5}$$

wobei über doppelt auftretende Indizes summiert wird ($i = 1, \ldots, r$; $\alpha = 1, \ldots, N_0$). Weiterhin spezialisieren wir uns auf lineare Funktionale $g(w)$ der Gestalt

$$g(w) = \int_\Omega g_i(x)\, u^i(x)\, dx, \qquad g_i(x) \in L^{p'}(\Omega). \tag{6}$$

Um mittels der Hilfssätze 1 und 2 die Lösbarkeit von (1) folgern zu können, bedarf es noch gewisser Kriterien, aus denen die Voraussetzungen $g \notin \hat{A}(X \setminus \mathfrak{D})$ bzw. $A(X) = \hat{A}(X)$ ersichtlich werden. Dazu definieren wir die mengenwertige Funktion

$$\mathfrak{M}(x, \eta) = \{\xi \in R^{N_2} \mid A_i{}^\alpha(x, \eta, \xi) = \hat{A}_i{}^\alpha(x, \eta, \xi)\ \forall \alpha, i\}, \tag{7}$$

mit deren Hilfe wir folgende zwei einschränkende Bedingungen an A stellen:

1. Die *Planar-Bedingung* (kurz P-Bedingung): (P).

Für jedes $\xi^0 \in R^{N_2} \setminus \mathfrak{M}(\mathring{x}, \mathring{\eta})$ und willkürlich fixierte $\mathring{x}, \mathring{\eta}$ existiere eine Umgebung $U_\varrho(\mathring{x}, \mathring{\eta}, \mathring{\xi}) \subset R^{N_0+N_1+N_2}$ mit dem Radius $\varrho = \varrho(\mathring{x}, \mathring{\eta}, \mathring{\xi})$, in der

$$\hat{A}_i{}^\alpha(x, \eta, \mathring{\xi}) = \hat{A}_i{}^\alpha(x, \eta, \xi) \quad \forall \alpha, i \qquad \text{und} \qquad (x, \eta, \xi) \in U_\varrho \tag{8}$$

ist. Für $\xi \in \mathfrak{M}(x, \eta)$ sei hingegen neben $A_i{}^\alpha(x, \eta, \xi) = \hat{A}_i{}^\alpha(x, \eta, \xi)$ auch $A_i(x, \eta, \xi) = \hat{A}_i(x, \eta, \xi)$.

Die geometrische Bedeutung von (P) ist besonders in dem durch Hilfssatz 4 umrissenen Spezialfall von Potentialoperatoren A leicht zu überschauen. Hier ist $\hat{A}_i{}^\alpha(x, \eta, \xi) = f^{**}_{\xi_\alpha{}^i}(x, \eta, \xi)$, $\hat{A}_i(x, \eta, \xi) = f^{**}_{\eta^i}(x, \eta, \xi)$, und ($P$) bedeutet, daß der Teilgraph von $\hat{f}(x, \eta, \xi)$ (zu festem x, η), welcher keine Punkte mit dem Graphen von $f(x, \eta, \xi)$ gemeinsam hat, aus Ebenenstücken besteht.

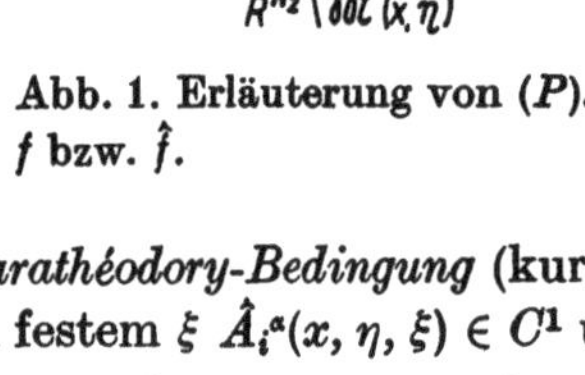

Abb. 1. Erläuterung von (P). Die ausgezogene (gestrichelte) Linie ist der Graph von f bzw. $\hat{f}$.

2. Die *Carathéodory-Bedingung* (kurz C-Bedingung): (C).[1]

Es sei zu festem ξ $\hat{A}_i{}^\alpha(x, \eta, \xi) \in C^1$ und die Abbildung

$$x \to \hat{A}^\alpha_{ix_\alpha}(x, w, w_x), \quad \hat{A}^\alpha_{i\eta^j}(x, w, w_x) \in L^{p'}(\Omega) \qquad \forall w \in X.$$

[1] Erstmalig von C. Carathéodory [2] beim Studium von Anfangswertaufgaben irregulärer Variationsprobleme für $m = 1$, $N_0 = 1$, $r = 1$ eingeführt.

Für alle η und fast alle $x \in \Omega$ gelte

$$\mathfrak{v}(x, \eta, \xi) = (v_1, \ldots, v_r) \neq 0 \quad \text{für} \quad \xi \in R^{N_1} \setminus \mathfrak{M}(x, \eta) \tag{9}$$

mit $v_i = (x, \eta, \xi) = \hat{A}^\alpha_{ix_\alpha} + \hat{A}^\alpha_{i\eta^j}\xi_\alpha{}^j - \hat{A}^i + g_i$.

Die formal analytische Bedeutung von $\mathfrak{v}(x, \eta, \xi)$ besteht darin, Lagrangescher Differentialausdruck zur Grundfunktion

$$\Phi(x, w, w_x) = \hat{A}_i(x, z, z_x)\, w^i + \hat{A}_i{}^\alpha(x, z, z_x)\, w^i_{x\alpha} - g_i w^i$$

in *solchen* Punkten $x \in \Omega$ zu sein, wo mit $z(x) \in C^1$, $\eta = z$, $\xi = z_x$ die Bedingung $z_x(x) \in R^{N_1} \setminus \mathfrak{M}(x, z(x))$ gilt. In der Tat ist

$$\frac{\partial \Phi w^i_{x\alpha}}{\partial x_\alpha} - \Phi_{w^i} = \frac{\partial \hat{A}_i{}^\alpha(x, z, z_x)}{\partial x_\alpha} - [\hat{A}_i(x, z, z_x) - g_i(x)]$$

und wegen (P) $= \hat{A}^\alpha_{ix_\alpha} + \hat{A}^\alpha_{i\eta^j} z^j_{x_\alpha} - \hat{A}_i + g_i = v_i(x, z, z_x)$.

Satz 2. *Der durch* (5) *definierte Operator* A *auf* $X = W_0{}^{1,p}(\Omega)$ $(p > N_0)$ *sei partiell pseudomonoton und sein korrespondierender Operator* $\hat{A}$ *erfülle die Voraussetzungen von Satz* 1*, *die P-Bedingung und nebst* g *von* (6) *die C-Bedingung. Dann existiert eine Lösung* $u \in X$ *von* $\hat{A}u = g$. *Diese erfüllt für* $N_0 = 1$ *auch die Gleichung* $Au = g$; *für* $N_0 > 1$ *gilt Entsprechendes, falls* $u(x)$ *für fast alle* x_0 *aus* $\mathfrak{N}(u) = \{x \in \Omega \mid u_x \notin \mathfrak{M}(x, u(x))\}$ *eine Bedingung der Gestalt*

$$\left|\hat{A}_i{}^\alpha(x, u(x), u_x(x)) - \hat{A}_i{}^\alpha(x, u,(x), u_x(x_0))\right| \leqq M(x_0)\, |x - x_0| \tag{10}$$

erfüllt.

Beweis. Daß unter den Voraussetzungen von Satz 2 eine Lösung u von $\hat{A}u = g$ existiert, ist eine unmittelbare Folge von Satz 1*. Falls mes $\mathfrak{N}(u) = 0$, ist fast überall in Ω $u_x(x) \in \mathfrak{M}(x, u(x))$ und somit unter Beachtung von (5), (8) und (P) zugleich $Au = \hat{A}u = g$, also u Lösung von (1). Wir werden nachfolgend zeigen, daß außerdem nicht mes $\mathfrak{N}(u) > 0$ sein kann. Damit ist dem Vorhergehenden zufolge die Lösbarkeit von (1) nachgewiesen.

Wir nehmen also an, es sei mes $\mathfrak{N}(u) > 0$. Wegen $\hat{A}u = g$ ist nach (5)

$$\int\limits_\Omega [\hat{A}_i(x, u, u_x)\, w^i + \hat{A}_i{}^\alpha(x, u, u_x)\, w^i_{x_\alpha} - g_i w^i]\, dx = 0 \quad \forall w \in X. \tag{11}$$

Wir geben uns zunächst irgendeinen später noch näher festzulegenden Punkt $x_0 \in \mathfrak{N}$ vor, in welchem $u_x(x_0)$ definiert ist (als eigentlicher endlicher Wert der summierbaren Funktion $u_x(x)$). Durch Ergänzung der Glieder $\hat{A}_i{}^\alpha(x, u(x), u_x(x_0))\, w^i_{x_\alpha}$ erhält (11) nach Anwendung des Gaußschen Integralsatzes (bzw. partieller Integration für $N_0 = 1$) die Gestalt

$$\int\limits_\Omega \Big\{ \Delta \hat{A}_i{}^\alpha(x, x_0)\, w^i_{x_\alpha} + \left[\hat{A}_i(x, u(x), u_x(x)) - g_i(x) - \frac{\partial \hat{A}_i{}^\alpha(x, u(x), u_x(x_0))}{\partial x_\alpha}\right] w_i \Big\}\, dx = 0 \tag{12}$$

$\forall w \in X$. Dabei ist $\Delta \hat{A}_i^\alpha(x, x_0) = \hat{A}_i^\alpha\big(x, u(x), u_x(x)\big) - \hat{A}_i^\alpha\big(x, u(x), u_x(x_0)\big)$. Der Ausdruck in der eckigen Klammer von (12) werde abkürzend mit $B_i(x, x_0)$ bezeichnet.

Fall 1: $N_0 = 1$.

Hier resultiert aus (12) nach dem Lemma von Du Bois-Reymond (vgl. [10], S. 28) die Relation

$$\Delta \hat{A}_i^1(x, x_0) = \int_{x_0}^{x} B_i(\bar{x}, x_0)\, d\bar{x} + c_i \quad \text{mit} \quad c_i = \text{const.} \tag{13}$$

Für fast alle $x \in \Omega$ existiert (unabhängig von der Wahl des x_0) nach Lebesgue die Ableitung der auf der rechten Seite von (12) stehenden Funktion von x in der Gestalt $B(x, x_0)$. Die Menge dieser x nennen wir Ω_0. Zugleich ist nach einer bekannten Folgerung zum maßtheoretischen Satz von Lusin (vgl. [11], S. 266) jede auf Ω definierte meßbare und fast überall endliche Funktion in fast allen Punkten x von Ω approximativ stetig, also hat auch $u' = u_x(x)$ diese Eigenschaft. Analytisch heißt das, es existiert zu fast allen $x^* \in \Omega$ eine Teilmenge $E(x^*) \subset \Omega$, auf welcher u' in x^* stetig ist, mit

$$\lim_{\varrho \to 0} \frac{\operatorname{mes}\big(E(x^*) \cap \mathfrak{K}(x^*, \varrho)\big)}{\operatorname{mes} \mathfrak{K}(x^*, \varrho)} = 1; \tag{14}$$

$\mathfrak{K}(x^*, \varrho)$ ist dabei ϱ-Umgebung von x^* im R^{N_0}. Aus diesem Grunde ist auch u' in fast allen Punkten x^* von $\mathfrak{N}(u)$ approximativ stetig. Wegen der Annahme $\operatorname{mes} \mathfrak{N} > 0$ gibt es somit einen Punkt $x_0 = x^* \in \Omega_0 \cap \mathfrak{N}$, in dem u' approximativ stetig ist. Infolge der P-Bedingung und der Stetigkeit von $u(x)$ (wegen $p > N_0$) existiert somit ein $\delta > 0$, so daß für alle $x \in \mathfrak{K}(x_0, \delta) \cap E(x_0)$ $\Delta \hat{A}_i^1(x, x_0) = 0$ ist. Wählen wir speziell aus $\mathfrak{K}(x_0, \delta) \cap E(x_0)$ eine Folge x_k mit $\lim_{k \to \infty} x_k = x_0$ aus, ergibt (13) $c_i = 0$ und nach Division durch $x_k - x_0$

$$0 = \frac{\Delta \hat{A}_i^1(x_k, x_0)}{x_k - x_0} = \frac{1}{x_k - x_0} \int_{x_0}^{x_k} B_i(\bar{x}, x_0)\, d\bar{x}.$$

Nach Ausführung des Grenzübergangs $k \to \infty$ erhalten wir hieraus $0 = B_i(x_0, x_0) = -v_i\big(x_0, u(x_0), u'(x_0)\big)$, $i = 1, \ldots, r$. Dieses Ergebnis steht im Widerspruch zur Bedingung (9). Also muß doch $\operatorname{mes} \mathfrak{N}(u) = 0$ sein.

Fall 2: $N_0 > 1$.

Hier wenden wir ähnliche Schlußweisen an. Aus (12) resultiert zunächst nach dem verallgemeinerten Haarschen Lemma (vgl. C. B. Morrey [9] und R. Klötzler [6]) an Stelle von (13)

$$\int_{\partial \mathfrak{K}(x_0, \varrho)} \Delta \hat{A}_i^\alpha(x, x_0) \cos(\mathfrak{n}, x_\alpha)\, do = \int_{\mathfrak{K}(x_0, \varrho)} B_i(\bar{x}, x_0)\, d\bar{x} \tag{15}$$

für fast alle $\varrho > 0$. Dabei ist $\mathfrak{n}$ der äußere Normaleneinheitsvektor des Flächenelements do. Für fast alle Punkte $x_0 \in \Omega$ existiert nach Lebesgue wiederum der Grenzwert

$$\lim_{\varrho \to 0} \left[\int_{\mathfrak{K}(x_0, \varrho)} B_i(x, x_0)\, dx / \operatorname{mes} \mathfrak{K}(x_0, \varrho) \right] = B_i(x_0, x_0). \tag{16}$$

Die Menge dieser x_0 nennen wir Ω_0.

Entsprechend obiger Betrachtungen ist auch $u_x(x)$ für fast alle Punkte x^* von Ω approximativ stetig, also auch für fast alle Punkte von $\mathfrak{N}(u)$. Unter der Annahme mes $\mathfrak{N} > 0$ gibt es somit einen Punkt $x_0 = x^* \in \Omega_0 \cap \mathfrak{N}$, in dem u_x approximativ stetig ist. Infolge der P-Bedingung und der Stetigkeit von $u(x)$ (wegen $p > N_0$) existiert ein $\delta > 0$, so daß

$$\Delta \hat{A}_i^\alpha(x, x_0) = 0 \quad \text{für alle} \quad x \in \mathfrak{K}(x_0, \delta) \cap E(x_0) \tag{17}$$

ist. Aus Bedingung (14) (bezüglich u_x statt u') entnimmt man leicht unter Anwendung des Satzes von Fubini die Existenz einer Nullfolge ϱ_k mit

$$\lim_{k\to\infty} \frac{\text{mes}^* \big(E(x_0) \cap \partial\mathfrak{K}(x_0, \varrho_k)\big)}{\text{mes}^* \, \partial\mathfrak{K}(x_0, \varrho_k)} = 1\,, \tag{18}$$

wobei mes* das $(N_0 - 1)$-dimensionale Maß auf $\partial\mathfrak{K}(x_0, \varrho_k)$ charakterisiert mit der Eigenschaft mes* $\partial\mathfrak{K}(x_0, \varrho_k) = \int\limits_{\partial\mathfrak{K}(x_0,\varrho_k)} do$.

Wegen (17) ist

$$\left| \frac{1}{\varrho_k} \int\limits_{\partial\mathfrak{K}(x_0,\varrho_k)} \Delta \hat{A}_i^\alpha(x, x_0) \cos(\mathfrak{n}, x_\alpha)\, do \right| = \left| \frac{1}{\varrho_k} \int\limits_{\partial\mathfrak{K}\setminus E(x_0)} \Delta \hat{A}_i^\alpha(x, x_0) \cos(\mathfrak{n}, x_\alpha)\, do \right|$$

und infolge der Zusatzbedingung von Satz 2 für $N_0 > 1$

$$\leqq M(x_0)\, N_0 \int\limits_{\partial\mathfrak{K}(x_0,\varrho_k)\setminus E(x_0)} do = M(x_0)\, N_0 \text{ mes}^* \big(\partial\mathfrak{K}(x_0, \varrho_k) \setminus E(x_0)\big).$$

Somit erhalten wir unter Beachtung der Relation

$$\text{mes } \mathfrak{K}(x_0, \varrho_k) = C_0 \varrho_k \text{ mes}^* \, \partial\mathfrak{K}(x_0, \varrho_k) \quad \text{mit} \quad C_0 = \text{const} > 0$$

und (18) schließlich mit (16) und (15)

$$\lim_{k\to\infty} \left(\int\limits_{\partial\mathfrak{K}(x_0,\varrho_k)} \Delta \hat{A}_i^\alpha(x, x_0) \cos(\mathfrak{n}, x_\alpha)\, do \,/\, \text{mes } \mathfrak{K}(x_0, \varrho_k) \right)$$
$$= 0 = B_i(x_0, x_0) = -v_i\big(x_0, u(x_0), u_x(x_0)\big), \qquad i = 1, \ldots, r.$$

Dieses Resultat steht aber im Widerspruch zur C-Bedingung. Also muß mes $\mathfrak{N}(u) = 0$ sein.

3. Erweiterungen und Ergänzungen zu Satz 2

a) Der durch Satz 2 gegebene Existenzsatz hat zweifellos für $N_0 > 1$ noch eine wesentliche Schwäche in der Bedingung (10), deren Erfülltsein Anforderungen an g stellt, die im allgemeinen bisher noch nicht geklärt werden konnten. Bei speziellen Problemklassen (vgl. unten Abschnitt b)!) ist es hingegen bereits gelungen, diese Einschränkung (10) zu überwinden. Aus dem Beweisablauf ersieht man übrigens, daß Satz 2 auch dann noch gültig bleibt, wenn man statt (10) die wesentlich schwächere Bedingung

$|\Delta A_i^\alpha(x, x_0)| \leqq M(x) \, |x - x_0|$ fordert, wobei $M(x)$ für alle ϱ eine auf $\partial\mathfrak{K}(x_0, \varrho)$ summierbare Funktion mit

$$\lim_{\varrho \to 0} \left[\int_{\partial\mathfrak{K}(x_0,\varrho)\setminus E(x_0)} M(x) \, do \,/\, \text{mes}^* \, \partial\mathfrak{K}(x_0, \varrho) \right] = 0$$

ist.

b) Für koerzitive partiell pseudomonotone Potentialoperatoren A im Sinne von Hilfssatz 4 sind durch Untersuchungen von R. Klötzler [4] im Falle $r = m = 1$ Existenzsätze ähnlicher Art wie Satz 2 gewonnen worden, in denen auf die Bedingung (10) verzichtet wird. In geringfügig erweiterter Form lauten diese Ergebnisse auf unsere hier gewählten Bezeichnungen übertragen:

Satz 3. *Es sei A ein koerzitiver partiell pseudomonotoner Potentialoperator auf $X = W_0^{1,p}(\Omega)$ im Sinne von Hilfssatz 4, dessen korrespondierender Operator $\hat{A}$ nebst g (bis auf Bedingung (10)) den Voraussetzungen von Satz 2 genüge. Es gebe weiterhin zwei Konstanten $C_1 > 0$ und C_2 mit*

$$f(x, \eta, \xi) \geqq C_1|\xi|^p + C_2 \qquad \forall (x, \eta, \xi) \in \Omega \times R^{N_1} \times R^{N_2}. \tag{19}$$

Dann existiert stets eine Lösung $u \in X$ des Variationsproblems $\hat{J}_0(w) = \hat{J}(w) - \langle g, w \rangle \to \text{Min}$ auf X und damit der Gleichung $\hat{A}u = g$. Für $N_0 = 1$ ist u zugleich Lösung des Variationsproblems (2) und der Gleichung (1). Für $N_0 > 1$ gilt die entsprechende Eigenschaft, falls u lokal lipschitzstetig ist.

Auch hier in Satz 3 wird für $N_0 > 1$ eine a-priori-Eigenschaft von u gefordert. Durch Resultate von Ladyshenskaja und Uralzewa ([7], S. 121 und 297) sind jedoch grundsätzlich berechenbare Bedingungen gegeben worden, aus denen die lokale Hölder-Stetigkeit von $u(x)$ und speziell dessen Lipschitzstetigkeit gefolgert werden kann.

Satz 3'. *Die Existenzaussage von Satz 3 gilt auch dann, wenn in der C-Bedingung die Forderung $\mathfrak{v} \not\equiv 0$ von (9) durch $\mathfrak{v} \equiv 0$ ersetzt wird.* (Vgl. dazu [4].)

c) Erweiterungen von Satz 3 und Satz 3' auf Variationsprobleme (2) in $X = W_0^{1,p}(\Omega)$ unter Nebenbedingungen des Typs $w_x(x) \in \mathfrak{B} = \overline{\mathfrak{B}} \subset R^N$ sind bei analogen Voraussetzungen in [5] dargelegt. Hier ist lediglich im Unterschied zu oben bei der Konstruktion von $J(w)$ und $A(u)$ gemäß Hilfssatz 4 $\hat{f}(x, \eta, \xi)$ als doppeltkonjugierte Funktion zu

$$f_0(x, \eta, \xi) = \begin{cases} f(x, \eta, \xi) & \forall \xi \in \mathfrak{B} \\ \infty & \forall \xi \notin \mathfrak{B} \end{cases} \tag{20}$$

zu beliebig fixierten x, η aufzufassen. Mit diesen Resultaten werden zugleich Existenzaussagen für Variationsungleichungen des Typs

$$\langle A(u), w - u \rangle \geqq \langle g, w - u \rangle \qquad \forall w \in V \tag{21}$$

geschaffen. Dabei ist $V = \{w \in X \mid w_x(x) \in \mathfrak{B} \text{ für fast alle } x \in \Omega\}$. Präzisiert lautet das einschlägige Theorem:

Satz 4. *Es sei A ein koerzitiver partiell pseudomonotoner Potentialoperator auf $X = W_0^{1,p}(\Omega)$ zu $r = 1$ im Sinne von Hilfssatz 4. Sein korrespondierender Operator werde mittels $\hat{J}(u) = \int_\Omega f_0^{**}(x, u, u_x) \, dx$ durch $\hat{A} = \text{grad} \, \hat{J}(u)$ erzeugt und genüge nebst g von (6)* *auf* $\hat{V} = \{w \in X \mid w_x(x) \in \operatorname{conv} \mathfrak{B}$ *für fast alle* $x \in \Omega\}$ *den Voraussetzungen von Satz* 2. *Außerdem gelte* (19). *Dann existiert stets eine Lösung u des Variationsproblems* $\hat{J}_0(w) = \hat{J}(w) - \langle g, w\rangle \to \mathrm{Min}$ *auf* $\hat{V}$ *und damit der Variationsungleichung* $\langle \hat{A}(u), w - u\rangle \geqq \langle g, w - u\rangle$ $\forall w \in \hat{V}$. *Für* $N_0 = 1$ *ist u zugleich Lösung des Variationsproblems* $J_0(w) \to \mathrm{Min}$ *auf* V *und der Variationsungleichung* (21). *Für* $N_0 > 1$ *gilt Entsprechendes, falls u lokal lipschitzstetig ist, insbesondere wenn* $\mathfrak{B}$ *beschränkt ist.*

Die gleiche Aussage gilt auch unter der in Satz 3' *formulierten Modifikation der C-Bedingung.*

4. Beispiele und Anwendungen

a) Nichtlineare Eigenwertprobleme

Fragestellungen des hier behandelten Typs treten häufig bei Eigenwertproblemen im $W_0^{1,p}(\Omega)$ der Gestalt

$$L_1[u] = \lambda L_2[u]$$

auf, wenn die L_i Differentialoperatoren gleicher Ordnung sind. Wir demonstrieren dies an dem einfachen Beispiel

$$(4u'^3)' = \lambda(2u')' \quad \text{in} \quad \Omega = [0, 1] \tag{22}$$

unter Randbedingungen $u(0) = u(1) = 0$ und $\lambda \in R$. Aufgabe (22) fassen wir auf als Gleichung des Typs $Au = 0$ im $W_0^{1,4}(0, 1)$ mit $A = \operatorname{grad} J(u)$ und $J(u) = \int\limits_0^1 [(u')^4 - \lambda(u')^2]\, dx$. Zu festem $\lambda \in R$ und $p = 4$ sind offenbar (4) und (19) erfüllt. Aus der

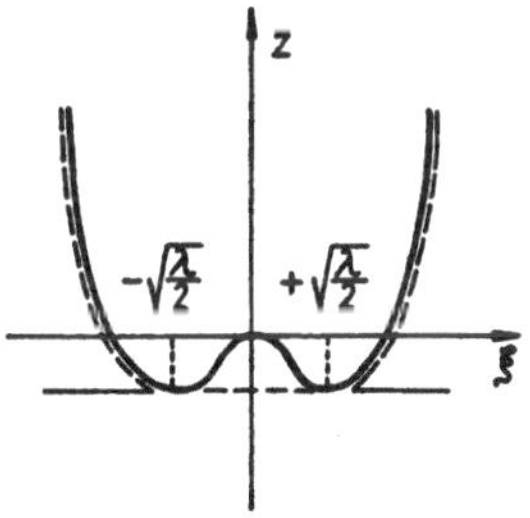

Abb. 2. Graph von f (ausgezogen) und von $\hat{f}$ (gestrichelt)

Gestalt des abgebildeten Graphen von $f(\xi) = \xi^4 - \lambda\xi^2$ lesen wir in Anlehnung an Hilfssatz 4 unmittelbar den Graphen von $\hat{f}(\xi)$ ab und das Erfülltsein der P-Bedingung. Die C -Bedingung ist hier in der modifizierten Form $\mathfrak{v} \equiv 0$ von Satz 3' erfüllt. Auch die Bestätigung seiner weiteren Voraussetzungen bereitet keine Mühe. Somit existiert eine Lösung u von (22). In der Tat ist jede Funktion $u \in W_0^{1,4}(0, 1)$ mit $u'(x) = \pm\sqrt{\frac{\lambda}{2}}$, deren Graph also eine Zickzack-Linie dargelegten Anstiegs beschreibt, eine Lösung von (22). Bei zusätzlicher Vorgabe einer Normierung $\int\limits_0^1 (u')^2\, dx = c > 0$ berechnen wir daraus durch Einsetzen von u' den zugeordneten Eigenwert $\lambda = 2c$. $\lambda_0 = 0$ ist folglich Verzweigungspunkt von (22).

b) Maximalflächen unter Anstiegsbeschränkungen

Wir studieren das Variationsproblem

$$J(w) = \int_\Omega \left(-\sqrt{1 + w_{x_1}^2 + w_{x_2}^2}\right) dx \to \text{Min} \quad \text{im } W_0^{1,p}(\Omega) \tag{23}$$

zu $\Omega \subset R^2$, $p > 2$ unter Nebenbedingungen $w_{x_1}^2 + w_{x_2}^2 \leqq 1$. Es beinhaltet anschaulich die Frage nach der von $\partial\Omega$ aufgespannten Maximalfläche unter vorgegebener Anstiegsbeschränkung. Wir wenden hierauf Satz 4 an mit $\mathfrak{B}$ als Einheitskreis. Hier ist $\hat{f}(\xi) \equiv -1 \ \forall \xi \in \mathfrak{B}$, also die P-Bedingung erfüllt und $\mathfrak{v} \equiv 0$. Nach Satz 4 existiert dann zu jedem regulären beschränkten Gebiet $\Omega \subset R^2$ eine Lösung des Variationsproblems (23).

Bemerkenswert ist, daß eine entsprechende Aussage nicht gilt, wenn die Nebenbedingung die abgeänderte Gestalt $(aw_{x_1})^2 + (bw_{x_2})^2 \leqq 1$ mit $a, b > 0$ und $a \neq b$ hat. Hier ist die P-Bedingung nicht erfüllt, wie man elementargeometrisch aus dem Graphen von f_0 aus (20) zu (23) unmittelbar erkennt. Wählen wir z. B. $\Omega = \{x \in R^2 \mid |x_i| \leqq 1, i = 1, 2\}$ und $a < b$, so ist für alle zulässigen w zu (23) $J(w) \geqq -\int_\Omega \sqrt{1 + a^{-2}}\, dx = -4\sqrt{1 + a^{-2}}$.

Jede Funktionenfolge w_n zulässiger Funktionen mit $w_n(x) = \frac{1}{a}\left|x_1 - \frac{k}{n}\right|$ für alle x_1 mit $\left|x_1 - \frac{k}{n}\right| \leqq \frac{1}{2n} (k = -n, \ldots, 0, \ldots, n)$ und $|x_2| \leqq 1 - \frac{b}{2na}$ (und geeignet fortgesetzt in den übrigen Punkten von Ω), deren Graph offenkundig eine „waschbrettartige“ Struktur besitzt, hat die Eigenschaft $\lim\limits_{n\to\infty} J(w_n) = -4\sqrt{1 + a^{-2}}$, also die Eigenschaft einer Minimalfolge. Dennoch gibt es keine zulässige Funktion u mit $J(u) = -4\sqrt{1 + a^{-2}}$. Für sie müßte fast überall $|u_{x_1}| = a^{-1}$, $u_{x_2} = 0$ gelten. Aus der letzten Bedingung und Randbedingung $u(x_1, \pm 1) = 0$ würde somit $u(x) \equiv 0$ folgen im Widerspruch zu $|u_{x_1}| = a^{-1}$.

c) Anwendungen zur Plastizitätstheorie

Die vorliegende Theorie dürfte eines ihrer Hauptanwendungsgebiete in der Plastizitätstheorie finden. Wir beschränken uns dabei auf solche elastisch-plastische Vorgänge, bei denen in Anlehung an K. Washizu [12] die Formänderungsenergie eines den Bereich $\Omega \subset R^3$ (vor der Formänderung) ausfüllenden festen Körpers durch ein Integral der Gestalt

$$\int_\Omega \varphi\big(e_{ij}(u_x)\big)\, dx$$

zum Deformationstensor $e_{ij} = \frac{1}{2}(u_{x_j}^i + u_{x_i}^j + u_{x_i}^k u_{x_j}^k)$ und gegebenem Verschiebungsvektor $u = (u^1, u^2, u^3)$ beschrieben wird. Axiomatisch deduktiv charakterisieren wir zu vorgegebenen äußeren Volumenkräften $P(x) = (P_1, P_2, P_3)$ und Randkräften $\hat{P} = (\hat{P}_1, \hat{P}_2, \hat{P}_3)$ einen stabilen Gleichgewichtszustand u des elastisch-plastischen Körpers auf der Grundlage des *Dirichletschen Prinzips* als Lösung des Variationsproblems

$$J_0(w) = \int_\Omega \varphi\big(e_{ij}(w_x)\big)\, dx - \int_\Omega P_i(x)\, w^i\, dx - \int_{\partial\Omega} \hat{P}_i w^i\, do \to \text{Min} \tag{24}$$

für alle zulässigen (kleinen) Verschiebungen $w(x)$. Wir erheben dieses Variationsprinzip auch in solchen Fällen zum Grundprinzip einer energetischen Charakterisierung von u, wenn lediglich (entsprechend Hilfssatz 4) für sämtliche $|w_x| \geqq K$ die positive Regularität des Variationsproblems (24) bzw. die Elliptizität des zugeordneten Potentialoperators $A(w)$ erfüllt ist. Wir gehen dabei von der dynamischen Vorstellung aus, daß bei langsamer Eintragung der Kraft P ein möglicher Fließvorgang im Körper nur ein latentes Übergangsstadium zwischen zwei stabilen Zuständen bildet, das im Endzustand $u(x)$ durch mes $\mathfrak{N}(u) = 0$ (vgl. Satz 2) überwunden ist.

Die Bestätigung der Richtigkeit einer solchen Vorgehensweise kann freilich nur durch zahlreiche Einzeluntersuchungen und Praxisvergleiche an Spezialfällen erreicht werden. Wir beschränken uns nachfolgend auf ein einfaches Beispiel dieser Art.

Longitudinale Belastung eines homogenen elastisch-plastischen Fadens

Im unbelasteten Zustand fülle der Faden das Intervall $\Omega = [0, 1]$ aus; er sei bei $x = 0$ starr befestigt, am anderen Ende bei $x = l$ wirke in Richtung der x-Achse eine Randkraft $\hat{P}$. Dann lautet hier das Dirichletsche Prinzip (24) mit $\varphi\left(\frac{1}{2}\,u'\right) = \bar{\varphi}(u')$

$$J_0(w) = \int_0^l \bar{\varphi}(w')\,dx - \hat{P}w(l) \to \text{Min}$$

unter der Randbedingung $w(0) = 0$ bzw.

$$J_0(w) = \int_0^l [\bar{\varphi}(w') - \hat{P}w']\,dx \to \text{Min im } W^{1,p}(0, 1) \tag{25}$$

mit $w(0) = 0$.[2]

Das bedeutet mit unseren früheren Bezeichnungen $g \equiv 0$, $f(\xi) = \bar{\varphi}(\xi) - \hat{P}\xi$, $A_1^1 = \bar{\varphi}'(\xi) - \hat{P}$, $A_1 \equiv 0$. Wegen $N_0 = 1$, $r = 1$ ist in diesem Beispiel die P-Bedingung trivialerweise erfüllt. Die C-Bedingung ist hier in der modifizierten Form von Satz 3' mit $\mathfrak{v} \equiv 0$ befriedigt. Die übrigen Voraussetzungen von Satz 3' nehmen wir in den Eigenschaften von $\bar{\varphi}$ als erfüllt an. Somit existiert eine Lösung u von (25), sie genügt notwendig der zugeordneten Euler-Lagrangeschen Differentialgleichung in integrierter Form

$$\bar{\varphi}'(u') - \hat{P} = \text{const} \qquad \forall x \in [0, l]$$

und dynamischen Randbedingung $\bar{\varphi}'(u') - \hat{P}\,|_{x=l} = 0$. Somit ist

$$\bar{\varphi}'\big(u'(x)\big) - \hat{P} = 0 \qquad \forall x \in [0, l]. \tag{26}$$

$\bar{\varphi}'\big(u'(x)\big)$ ist sachgemäß nach [12] als Produkt der wahren Spannung $\sigma\big(u'(x)\big)$ und des Fadenquerschnitts $q\big(u'(x)\big)$ nach der Deformation zu interpretieren bzw. als Produkt der scheinbaren Spannung $\sigma_0\big(u'(x)\big)$ (bezogen auf den Ausgangsquerschnitt $q_0 = q(0)$) und q_0. (26) beinhaltet damit die generelle Forderung an den Gleichgewichtszustand u:

[2] Damit ordnet sich eigentlich unser Beispiel nicht ganz in unsere vorangehende Theorie ein, die allseitige Randbedingungen auf $\partial\Omega$ fordert. Die dargelegte Theorie ist jedoch mühelos auch auf unsere Randbedingung von (25) übertragbar.

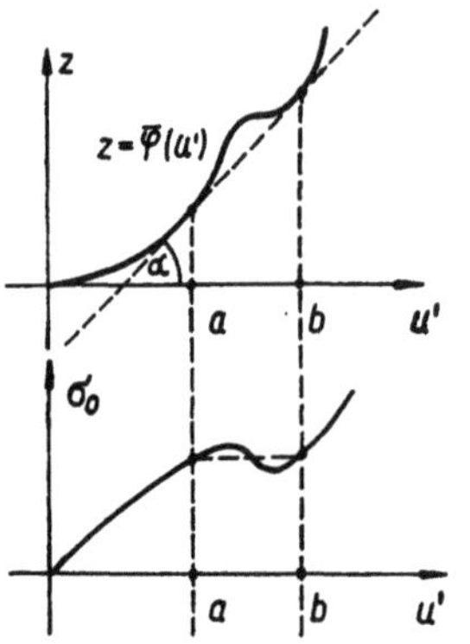

Abb. 3. Fadendeformation

$\hat{P} = \sigma(u'(x))\, q(u'(x)) = \sigma_0(u'(x))\, q_0$. Das heißt unter der vereinfachenden Annahme, daß hier $R^1 \setminus \mathfrak{M}(x, \eta)$ nur aus *einem* konstanten Intervall $[a, b]$ mit $0 < a$ besteht, ist für $\hat{P} < \tan\alpha = \bar{\varphi}'(a) = \bar{\varphi}'(b)$ die Deformation $u'(x) = \text{const} < a$ eindeutig bestimmt; das Gleiche gilt für $\hat{P} > \tan\alpha$ mit $u'(x) = \text{const} > b$. Lediglich im Falle $\hat{P} = \tan\alpha$ ist $u'(x) = a$ oder b. Das heißt, wir erhalten nur bei dieser speziellen Belastung im Gleichgewichtszustand zwei verschiedene Dehnungstypen und demzufolge zwei mögliche Querschnittsgrößen $q(b) < q(a)$, deren Lage freilich nicht eindeutig determiniert ist.

Uns erscheint an diesem Beispiel das Dirichletsche Prinzip (24) als geeignet zur Beschreibung des stabilen Gleichgewichtszustandes.

Literatur

[1] Brezis, H., Equations et inéquations non-linéaires dans les espaces vectoriels en dualité. Ann. Inst. Fourier 18, 115–176 (1968).

[2] Carathéodory, C., Über die diskontinuierlichen Lösungen in der Variationsrechnung. Gesammelte Mathematische Schriften I, 3–79, C. H. Beck, München 1954.

[3] Gajewski, H., Gröger, K., Zacharias, K., Nichtlineare Operatorgleichungen und Operatordifferentialgleichungen. Akademie-Verlag, Berlin 1974.

[4] Klötzler, R., Direkte Methoden bei irregulären Variationsproblemen. Beiträge zur Analysis 5, 1–17 (1973).

[5] Klötzler, R., On the existence of optimal processes. *In:* Banach Center Publications vol. 1, Mathematical Control Theory, 125–130. PWN, Warszawa 1976.

[6] Klötzler, R., Mehrdimensionale Variationsrechnung. VEB Deutscher Verlag der Wissenschaften, Berlin 1969.

[7] Ладыженская, О. А., Уральцева, Н. Н., Линейные и квазилинейные уравнения эллиптического типа. Узд. „Наука", Москва 1973.

[8] Lions, J. L., Quelques méthodes de résolution des problèmes aux limites non linéaires. Dunod, Gauthier-Villars, Paris 1969.

[9] Morrey, C. B., Multiple integrals in the calculus of variations and related topics. Univ. of California Press, 1, 1–130 (1943).

[10] Morrey, C. B., Multiple integrals in the calculus of variations. Springer-Verlag, Berlin—Heidelberg—New York 1966.
[11] Natanson, I. P., Theorie der Funktionen einer reellen Veränderlichen. Akademie-Verlag, Berlin 1954.
[12] Washizu, K., Variational methods in elasticity and plasticity. Pergamon Press, 1968.

Anschrift des Verfassers:

Karl-Marx-Universität Leipzig
Sektion Mathematik
701 Leipzig
Karl-Marx-Platz
DDR

Approximation methods for the determination of parameters in nonlinear problems

Reinhard Kluge, Berlin

In this paper we explain a general approach to a great variety of problems of parameter determination (inverse problems, deterministic identification, optimal control) in various nonlinear problems. It is based on the qualitative theory and numerical analysis for constrained minimum problems on sets with a convergence notion developed by the author in [12].

The exact statements concerning the abstract scheme (existence and numerical approximation of "optimal" parameters) are contained in [12]. For the determination of "optimal" parameters in minimum problems and variational inequalities, see Kluge [10]. Substantial complements to the theory of iteration methods in [12] are contained in the paper [16] by Kluge and Telschow (in these Proceedings) that deals with the iterative solution of variational inequalities in a Hilbert space with possibly multivalued mappings in the cases of unique as well as of non-unique solvability.

The present paper consists of four parts:

1. Parameter determination in nonlinear problems
2. Existence of "optimal" parameters
3. Approximation methods using sequences of minimum problems
4. Iteration methods.

1. Parameter determination in nonlinear problems

Let be Ω a bounded domain of the n-dimensional Euclidean space with boundary $d\Omega$, $s \in \Omega$, $x(s)$ a function defined on Ω ($x: \Omega \to R^1$), $\alpha = (\alpha_1, \ldots, \alpha_n)$ a multi-index, $|\alpha| = \alpha_1 + \cdots + \alpha_n$,

$$D^\alpha x(s) = \frac{\partial^{|\alpha|} x(s)}{\partial s_1^{\alpha_1} \cdots \partial s_n^{\alpha_n}}$$

the generalized derivative of order $|\alpha|$ of $x(s)$ in the sense of Sobolev, $D^r x = \{D^\alpha x: |\alpha| = r\}$, $\delta = \{\delta_\alpha: |\alpha| \leqq l\}$, $\delta(x) = (x, D^1 x, \ldots, D^l x)$, and let $A_\alpha: \Omega \times R^N \to R^1$ with $N = \dim \delta(x)$, $|\alpha| \leqq l$, denote functions that satisfy the Carathéodory condition and certain growth conditions such that the superposition operator generated by A_α transforms a certain Lebesgue space $L_{p_\alpha}(\Omega)$ into another Lebesgue space $L_{q_\alpha}(\Omega)$ that corresponds to the problem formulated.

Instead of Lebesgue spaces, also Orlicz spaces or the space L_∞ and the corresponding spaces on an unbounded domain may be considered. To simplify matters we restrict our investigations to the first case mentioned above.

We consider nonlinear differential operators in divergence form,

$$A_l(x) = \sum_{|\alpha| \leq l} (-1)^{|\alpha|} D^\alpha A_\alpha\big(s, \delta(x)\big).$$

Let be $\bar{g} : \Omega \to R^1$. We look for solutions of the equation

$$A_l(x)(s) = \bar{g}(s), \quad s \in \Omega, \tag{1}$$

that satisfy certain additional conditions for $x(s)$. We mention some possibilities:

(α) classical boundary value problems (linear equations for x on the boundary $d\Omega$, e.g., the Dirichlet problem);

(β) non-classical boundary value problems (e.g. unilateral boundary value problems or nonlinear equations for x on $d\Omega$) [3, 5, 19].

The problem (1), (α) can often be transformed, via the equation in variations for (1), into an equation

$$Ax = g \tag{1a}$$

with a nonlinear operator A mapping a subspace Y of the Sobolev space $W_p^l(\Omega)$ into its dual space Y^*. The problem (1), (β) may lead to a variational inequality

$$(g - Ax, x - y) \geqq 0, \qquad y \in X \subset Y \tag{1b}$$

with the constraint set X given by (β). We also arrive at (1b) when we start from (1a) and impose additional conditions (barriers in parts of Ω, elastic-plastic torsion, etc.). Here the solutions of (1a), (1b) are considered to be the solutions of the original problems (1), (α) and (1), (β), respectively.

Frequently the solution of these problems is not possible because of the difficulty that there is incomplete information, i.e., some of the parameters in (1), (α), (β) are unknown. These parameters may be numbers as well as functions. Then first of all these parameters have to be determined according to certain optimality criteria (inverse problems, deterministic identification). The field of parameter determination also encloses the problems of optimal control: here some parameters have to be determined (according to an optimality criterion) in order to achieve a certain goal.

The problems of determinations of parameters in the right-hand side of an equation (1) (sources, external forces, etc.) and of determination of boundary values are well known (cf. [18]). We mention some other interesting parameters that may be contained in the functions A_α. Let be

$$A_\alpha\big(s, \delta(x)\big) = \sum_{|\beta| \leq l} u_{\alpha\beta}\big(s, \delta(x)\big) D^\beta x, \qquad |\alpha| \leqq l.$$

We consider the following cases:

(i) $u_{\alpha\beta}\big(s, \delta(x)\big) = u_{\alpha\beta}(s), \qquad s \in \Omega,\ |\alpha|, |\beta| \leqq l.$

(ii) $u_{\alpha\beta}\big(s, \delta(x)\big) = u_{\alpha\beta}(s, x), \qquad s \in \Omega, \quad x \in R^1, \quad |\alpha|, |\beta| \leqq l.$

(iii) $u_{\alpha\beta}\big(s, \delta(x)\big) = u_{\alpha\beta}(s, D^l x), \qquad |\alpha|, |\beta| \leqq l.$

In (i) the parameters $u_{\alpha\beta}$ are defined on bounded domains. Problems of determination of diffusion coefficients are included here [4]. The parameters may also be scalars.

In (ii) the parameters are defined on unbounded domains. Problems of determination of diffusion coefficients that explicitly depend on the solution are included [4].

In (iii), again the parameters are defined on unbounded domains. This case includes, among others, the determination of material properties in problems of continuum mechanics.

Instead of (1 a), in the cases (i)—(iii) parametric equations of the form

$$A(x, u) = g$$

are given. The same holds for (1 b).

Another interesting situation is the following one: a system of equations or inequalities, for instance describing a multiphase problem, is given, and the coupling terms connecting the variables x^i are to be determined, e.g.,

$$A_i^i(x^i) = \bar{g}^i + u^i(x^1, \ldots, x^m), \qquad i = 1, \ldots, m.$$

Here the parameters u^i, $i = 1, \ldots, m$, also are defined on unbounded domains.

So we have to deal with very different types of parameters. The corresponding parameter spaces V will, as a rule, be provided with very different convergence notions since, for instance, different convergence definitions have to be used in spaces of functions that are defined on bounded domains and in spaces of functions that are defined on unbounded domains. Moreover, several types of convergence can be used in each parameter space (e.g. strong, weak and weak * convergence in a Banach space), and different types of parameters may appear at the same time.

So, for general investigations, it makes sense to assume the parameter space V to be a set with a (sequential) convergence notion τ, i.e., a set V where the convergence τ of countable sequences of elements is defined: $v_n \xrightarrow{\tau} v$, v_n, $v \in V$, $n = 1, 2, \ldots$ We suppose this convergence to satisfy the following condition: if $v_n \xrightarrow{\tau} v$ then $v_{n(k)} \xrightarrow{\tau} v$ for each subsequence $\{v_{n(k)}\}$ of $\{v_n\}$.

The same situation appears with respect to the sets Y in which we look for the solutions x of the original equations or inequalities. In dependence on the aims of investigation (existence of solutions or their numerical approximation) often different types of convergence are used (for instance, weak convergence for the existence proof, strong convergence for the investigation of convergence of approximation methods).

Moreover, the closedness with respect to a convergence notion in $Y \times V$ of a set of points $(x, u) \in Y \times V$ that satisfy a certain parametric relation is of greatest importance for the determination of parameters according to optimality criteria.

If Y is equipped with the convergence σ then we can introduce the "coordinatewise" defined convergence $\varrho = (\sigma, \tau)$ in $Y \times V = M$. In that way the variety of possible convergence notions is further enlarged. An additional motivation for our point of view arises from the inclusion of functionals into the problems, for also in dependence of the properties of these functionals different types of convergence of solutions of approximation problems towards solutions of the parameter determination problem (that still has to be defined more precisely) can be shown to hold.

Until now we have restricted our considerations to stationary problems (1). Of course analogous situations also arise in nonstationary problems. We will shortly mention this point. Let t be the time variable ($t \in [0, T]$, $T > 0$ fixed). Then, instead of (1), nonstationary equations of the form

$$\frac{\partial}{\partial t} A_l^1(x)(s, t) + A_l^2(x)(s, t) = \bar{g}(s, t) \tag{2}$$

("pseudo-parabolic" equations or equations of Sobolev-Galpern type of visco-elasticity and visco-plasticity theory) or of the form

$$\frac{\partial}{\partial t} x(s, t) + A_l(x)(s, t) = \bar{g}(s, t) \tag{3}$$

(equations of evolution type as, e.g., the equations of heat and mass transfer) can be given. Instead of the boundary value problems (α) and (β) let here the corresponding initial-boundary value problems (α) und (β), respectively, be posed.

Then we frequently obtain in case (2), (α) ordinary differential equations in infinite dimensional spaces, in case (2), (β) the corresponding ordinary differential inequalities (cf. [14]), in case (3), (α) evolution equations and in case (3), (β) evolution inequalities [3, 5, 19].

Possibly in these cases the functions A_α still depend on t. In general the parameter determination problems arising here are similar to those in the stationary case (1). But in addition, among others, also the initial values may act as parameters. So further spaces with different convergence notions appear.

Until now we have only dealt with partial differential equations and with systems of such equations (stationary and nonstationary equations may appear in one system). But a number of problems are formulated in the shape of variational problems.

Let be $J: \Omega \times R^N \to R^1$ and

$$j(x) = \int_\Omega J\big(s, \delta(x)(s)\big)\, ds .$$

We seek for points where j attains its minimum (possibly subject to additional constraints) in a corresponding function space. If the derivatives $\partial J/\partial \delta_\alpha$, $|\alpha| \leqq l$, exist then in the case $X = Y$ the corresponding Euler-Lagrange equation has the form (1) with $A_\alpha = \partial J/\partial \delta_\alpha$ and $\bar{g} = 0$. In the case of a convex set $X \neq Y$ as a necessary condition we obtain the corresponding partial differential inequality (in variations).

Of course the variational problems may also be handled in their original form. Here also, like in differential equations, parametric problems occur. These parametric problems may be transformed to the form

$$j(x, u) \leqq 0, \qquad (x, u) \in G_0 \subset Y \times V. \tag{4}$$

Let us return once more to the remaining problems. In the case of a convex and closed constraint set X in a reflexive Banach space Y the variational inequalities, ordinary differential inequalities and evolution inequalities can be transformed into operator equations, ordinary differential equations or evolution equations with multi-valued mappings. Taking into account that initial value problems for ordinary differential equations and evolution equations may be interpreted as operator equations, it is possible to write down all these problems ((1)—(3), (α), (β)) (and also parametric boundary value and initial-boundary value problems), under consideration of the "open" parameters, in the form of a parametric equation

$$A(x, u) \ni 0, \qquad (x, u) \in G_0 \subset Y \times V \tag{5}$$

with a multivalued mapping $A: Y \times V \to Z$ (frequently $Z = Y$ or $Z = Y^*$). We only mention that also integral equations and classical ordinary differential equations can be transformed to (5).

Because of their generality and practical importance we shall restrict our investigations here to the parametric relations (4) and (5).

To each $u \in V$ we relate the set Su of all solutions of $j(x, u) \leqq 0$, $(x, u) \in Y \times V$. Let $G(S)$ denote the graph of the solution operator S in $Y \times V$. Then we take

$$G = G(S) \cap G_0.$$

In the same manner we proceed in the case (5).

2. Existence of "optimal" parameters

In the above mentioned parameter determination problems (inverse problems, identification, optimal control) we assume the existence of optimality criteria given by certain functionals $F \in (V \to R^1)$ and $H \in (Y \to R^1)$ by means of which the desired (optimal) parameters have to be determined.

The *general parameter determination problem* is defined to be the following extremal problem:

$$H(x) + F(u) = \inf_{(x,u) \in G}. \tag{6}$$

Every pair $(x^*, u^*) \in G$ satisfying

$$H(x^*) + F(u^*) = \inf_{(x,u) \in G} [H(x) + F(u)]$$

is said to be a *solution* of (6). The element u^* is called an *"optimal" parameter*.

For H and F in an obvious manner the (sequential) *lower* σ- and τ-*semicontinuity* on Y and V, respectively, are defined (cf. [12]). If both properties hold then the functional $f(x, u) = H(x) + F(u)$ is *lower* ϱ-*semicontinuous* on $M = Y \times V$ with $\varrho = (\sigma, \tau)$.

The existence of solutions to (6) can be investigated making use of the Weierstraß Theorem on sets with a convergence notion. Here the ϱ-closedness and ϱ-compactness of the set G or the behaviour of the functional f outside some ϱ-compact set (and in connection with this possibly certain properties of the solution operator or of its inverse) are of particular importance (cf. the explanations in [12]).

We quote a general theorem.

Theorem 1. *Assume that f is lower ϱ-semicontinuous and, moreover, there exists a $b_0 \in G$ such that the set $\{b \in G: f(b) \leqq f(b_0)\}$ is ϱ-closed and ϱ-compact. Then there exists at least one solution of* (6), *and the set of all solutions of* (6) *is ϱ-closed and ϱ-compact.*

We shall give now examples for the ϱ-closedness of $G(S)$ in two important and general cases.

If j is lower ϱ-semicontinuous then $G(S)$ is ϱ-closed (cf. [12]).

Let Y be a reflexive Banach space, $X \subset Y$ convex and closed, $T: Y \to 2^{Y^*}$ maximally monotone, $P: X \to 2^{Y^*}$ pseudo-monotone, and let one of the conditions (i), (ii) be satisfied:

(i) $\operatorname{int} D(T) \cap X \neq \emptyset$;

(ii) $D(T) \cap \operatorname{int} X \neq \emptyset$.

Moreover, let be V another reflexive Banach space, $U \subset V$ convex and closed, and in case (ii) let be $\operatorname{int} U \neq \emptyset$ and the mapping $P^1: X \times U \to 2^{Y^* \times V^*}$ with range $R(P^1) \subset Y^*$ be pseudomonotone. Then the set of all points $(x, u) \in X \times U$ that satisfy the variational inequality

$$(Tx + Px + P^1(x, u), y - x) \geqq 0, \qquad y \in X \tag{7}$$

is weakly (i.e. (weakly, weakly)) closed in $Y \times V$. Because of space limitations we can not go into the details.

(7) includes stationary as well as nonstationary problems with single valued or multivalued mappings in the cases (α) and (β).

We still mention that in (6) also the cases $H = 0$ and $F = 0$ frequently occur and that the solution operator S may be multivalued and need not be Fréchet differentiable.

3. Approximation methods using sequences of minimum problems

In general (6) can not be solved directly. Therefore methods for the approximate determination of the solutions of (6) are of great importance. Often these approximation methods are combinations of approximation methods for the parametric relations with approximation methods for the (constrained) minimum problems. The convergence results for methods for the solution of (6) depend on convergence results for the individual "basic methods". Therefore a maximum of information concerning the solvability and approximate solvability of the relations (4) and (5) for fixed parameters and the dependence of solutions or methods on the parameters is of particular importance for the treatment of the parametric problems, even before the problems with the "desired" parameter values are to be solved.

We replace (6) by

$$H_n(x) + F_n(u) = \inf_{(x,u) \in G_n}, \tag{6n}$$

where $G_n \subset Y_n \times V_n = M_n$ is the set of all points $(x, u) \in G_0^n \subset M_n$ that satisfy the relation

$$A_n(x, u) \ni 0 \qquad (A_n: Y_n \times V_n \to 2^{Z_n}) \tag{5n}$$

or

$$j_n(x, u) \leqq 0 \qquad \big(j_n \in (Y_n \times V_n \to R^1)\big), \tag{4n}$$

respectively. The sets Y_n, V_n need not be subsets of Y, V.

We denote the solution operator by S_n. Consequently $G_n = G(S_n) \cap G_0^n$. Here S_n can already be single valued and even Fréchet differentiable. That is because the problems (4n), (5n) may be regularization or penalty methods for (4) or (5). Furthermore the spaces Y_n and V_n may be finite dimensional and the solution operators S_n may be connected with iteration methods for (4) or (5).

So the methods (4n), (5n) may include approximation methods (difference and projection methods, finite element methods, spline approximations), regularization methods, penalty methods and iteration methods as well as their various (possibly parameter dependent) combinations (cf., e.g., [6—9]). With respect to the functionals in (6n) also approximation methods, regularization methods and penalty methods as well as their combinations may occur, such that (6n), compared to (6), contains considerably simpler problems.

Let be $p_n^V \in (V_n \to V)$, $p_n^Y \in (Y_n \to Y)$ and $p_n = (p_n^V, p_n^Y.)$ For statements on the convergence of the solutions (x_n^*, u_n^*) of (6n) towards the solutions of (6) the condition

$$\varrho\text{-}\overline{\lim}\, p_n G_n \subset G \subset \varrho\text{-}\underline{\lim}\, p_n G_n \tag{8}$$

is of decisive importance (cf. [12]). Moreover, some approximation properties of the functionals H_n, F_n with respect to H, F have to be assumed. For details we refer to [12]. For optimal control with minimum problems and variational inequalities see, e.g., [10] (cf. also [15]).

In some cases there exist estimations between the elements (x_n^*, u_n^*) and (x^*, u^*) (cf. [13]) which can be interpreted as a quasi-optimality of the method (6n) like in the case considered by Aubin [1]. So these methods do not converge "more slowly" than the corresponding approximating spaces Y_n and V_n converge to Y and V. In this way results on the speed of convergence of (6n) can be obtained.

4. Iteration methods

In many cases the problems (6n), in spite of all simplifications, may still be too complicated because of the combination of x and u in (6n). Therefore we consider iteration methods which give the possibility to separate the variables and in addition to simplify the problems by means of a certain linearization.

The iteration methods can be used for the approximate solution of the approximation problems (6n). But by suitably connecting the number n of (6n) with the number of iteration also "diagonalized" methods for the direct approximate solution of (6) can be obtained which include combinations of the methods (6n) with iteration procedures and reduce, in every step, to the solution of some relatively easily solvable problems.

Let be F_n a real Hilbert space, I_n the identity mapping in F_n, $P_n(C_n)$ the projection operator from F_n onto a convex and strongly closed subset $C_n \subset F_n$, $n = 1, 2, \ldots$ Let be given single valued mappings $p_n \in (F_n \to F)$ and $r_n \in (F \to F_n)$, $n = 1, 2, \ldots$ In what follows both cases $F_n = Y_n \times V_n$ and $F_n = V_n$ are possible. On F_n let be given a strongly convex functional K_n $\big(K_n \in (F_n \to R^1)\big)$ with gradient K_n'. We consider iteration methods of the form

$$w_n = P_n(C_n)\,(I_n - t_n K_n')^{\alpha_n}\, r_n p_{n-1} w_{n-1}, \tag{9}$$

$n = 1, 2, \ldots,$ $w_0 \in F,$ $p_0 = I,$ with real numbers $t_n > 0$ and positive integers α_n, $n = 1, 2, \ldots$

The convergence of (9) has been investigated in [12] using results from [11]. Combining the results of [12] with those of the papers [11, 16] one obtains various statements on the convergence and speed of convergence of the method (9) (where the α_n can frequently be chosen to be equal to 1 or can be used for the "optimization" of the method with respect to the speed of convergence).

Moreover, similar results can be obtained based on [16] for the case of subgradients (multivalued mappings) for the K_n. Details will be given elsewhere. In the special case $F_n = F$, $n = 1, 2, \ldots$ ($F = Y \times V$ or $F = V$) these results are contained in [16]. Here use is made of the connection between constrained optimization problems and variational inequalities and of the fact that even in the case of essentially nonlinear solution operators the functions $f(x, u)$, considered as functions in u $\big(\hat{f}(u) = f(Su, u)\big)$, may be convex (cf., e.g., [17]).

Still some words to (9). In the case $F_n = V_n$ with the single valued and Fréchet differentiable solution operator S_n one calculates $K_n' = F_n'(.) + S_n'(.)^* H_n'\big(S_n(.)\big)$, $n = 1, 2, \ldots$ The question of numerical realization of (9) in this case has been discussed in [12].

Here we will still present an iteration method for (6) that is different from (9). Let be $A_0 \in (Y \to Y)$ a linear continuous and positive definite operator mapping the Hilbert space Y into itself, B_0 a linear continuous operator mapping the Hilbert space V into Y, $G_0 = Y \times V$, $H(x) = \frac{1}{2}\|x - y_0\|_Y^2$, $F(u) = \frac{1}{2}\|u - v_0\|_V^2$ and $A(x, u) = A_0x + B_0u = 0$. We consider (6), (5) and the following iteration method in $Y \times V$

$$\begin{pmatrix} z_{i+1} \\ w_{i+1} \end{pmatrix} = \begin{pmatrix} z_i \\ w_i \end{pmatrix} - t_0 \begin{pmatrix} \varepsilon_i z_i + A_0^*(y_i) \\ \varepsilon_i w_i + B_0^*(y_i) \end{pmatrix} \tag{10}$$

with $y_i = A_0 z_i + B_0 w_i$, $0 < t_0 < 2/(2 + \|C_0\|)$, $\varepsilon_i = \sqrt{1/i}$ and $C_0 = \begin{pmatrix} A_0^*A_0, & A_0^*B_0 \\ B_0^*A_0, & B_0^*B_0 \end{pmatrix}$, $i = 1, 2, \ldots$ Then

$$\|(x^*, u^*) - (z_i, w_i)\|_{V \times Y} = O\big(\sqrt{1/i}\big),$$

where $\|.\|_{Y\times V}^2 = \|.\|_Y^2 + \|.\|_V^2$ and (x^*, u^*) is the unique solution of (6), (5).

For the proof we remark that, according to [2] and [20], the penalty method

$$H(x) + F(u) + \frac{1}{2\varepsilon}\|A_0x + B_0u\|^2 \to \inf_{(x,u)\in Y\times V}$$

upon which the method (10) is based has a speed of convergence of order $O(\varepsilon)$. We apply Theorem 11 from [16] and obtain the assertion. In order to assure for (10) a linear speed of convergence $\left(O(k^n), k < 1\right)$ it is sufficient to take the operator

$$I - t_0 \left(\varepsilon_i I + \begin{pmatrix} A_0^*A_0, & A_0^*B_0 \\ B_0^*A_0, & B_0^*B_0 \end{pmatrix}\right)$$

on the right-hand side of (10) with power α_i, where

$$\alpha_i \geqq -\frac{\ln k}{\varepsilon_i}.$$

In method (10), compared to (9), parametric relations (for the calculation of S) no longer have to be solved. But the advontage of an easier numerical realization is paid for by a lower speed of convergence.

If one combines (10) with the Galerkin method for (5) then the corresponding iteration method

$$\begin{pmatrix} z_{i+1} \\ w_{i+1} \end{pmatrix} = \begin{pmatrix} z_i \\ w_i \end{pmatrix} - t_0 \begin{pmatrix} \varepsilon_i z_i + P_i^Y A_0^*(y_i) \\ \varepsilon_i w_i + P_i^V B_0^*(y_i) \end{pmatrix} \tag{11}$$

with $y_i = P_i^Y A_0 z_i + P_i^V B_0 w_i$, $i = 1, 2, \ldots$, still converges strongly in $Y \times V$ to (x^*, u^*).

References

[1] Aubin, J. P., Approximation of elliptic boundary-value problems. Wiley-Interscience, New York 1972.

[2] —, Estimate of the error in the approximation of optimization problems with constraints by problems without constraints. *In:* Control Theory and the Calculus of Variations. Ed. A. V. Balakrishnan, 153—173, Academic Press, New York 1969.

[3] Brezis, H., Problèmes unilateraux. J. Math. pures appl. 51, 1—168 (1972).

[4] Chavent, G., and P. Lemonnier, Identification de la nonlinéarité d'une équation parabolique quasilinéaire. IRIA Laboria. Rapport de Recherche N°. 45, 1973.

[5] Duvaut, C., and J. L. Lions, Les inéquations en méchanique et en physique. Dunod, Paris 1972.

[6] Kluge, R., Zur approximativen Lösung nichtlinearer Variationsungleichungen. Dissertation B. Berlin 1970.

[7] —, Näherungsverfahren zur approximativen Lösung nichtlinearer Variationsungleichungen. Math. Nachr. 51, 343—356 (1971).

[8] —, Näherungsverfahren für einige nichtlineare Probleme. *In:* Theory of Nonlinear Operators. Proc. Summer School Neuendorf/Hiddensee (GDR) (1972), 133—146. Akademie-Verlag, Berlin 1974.

[9] —, Approximation methods for nonlinear problems with constraints in form of variational inequalities. Banach Center Publications, vol. 1, 131—138 (1976).

[10] —, Optimal control with minimum problems and variational inequalities. *In:* Optimization Techniques. IFIP Technical Conference Novosibirsk 1974, Lecture Notes in Computer Science, vol. 27, 377—382. Springer-Verlag, Berlin 1975.

[11] —, On a class of iteration methods. Math. Nachr. 73, 7—18 (1976).

[12] —, Über eine Klasse von Minimumproblemen mit Nebenbedingungen. Math. Nachr. 78, 37—64 (1977).

[13] —, Über Näherungsverfahren zur Parameteroptimierung in Operatorgleichungen (Manuscript).
[14] Kluge, R., und G. Bruckner, Über einige Klassen nichtlinearer Differentialgleichungen und -ungleichungen im Hilbert-Raum. Math. Nachr. **64**, 5—32 (1974).
[15] Kluge, R., E. Krauß und R. Nürnberg, Zur Optimierung in Aufgaben mit Operatorgleichungen und Evolutionsgleichungen.
[16] Kluge, R., and G. Telschow, On the convergence and speed of convergence of some iteration methods for variational inequalities. I. These Proceedings, 135—163.
[17] Langenbach, A., Konvexe Steuerungsprobleme. These Proceedings, 171—182.
[18] Lions, J. L., Contrôle optimal de systèmes gouvernés par des équations aux dériveés partielles. Dunod, Gauthier-Villars, Paris 1968.
[19] —, Quelques méthodes de résolution des problemes aux limites non lineaires. Dunod, Gauthier-Villars, Paris 1969.
[20] Поляк, Б. Т., О скорости сходимости метода штрафных функций. Ж. выч. мат. и мат. физ. **11**, 3—11 (1971).

Address of the author:

Akademie der Wissenschaften der DDR
Zentralinstitut für Mathematik und Mechanik
108 Berlin
Mohrenstr. 39
DDR

On the convergence and speed of convergence of some iteration methods for variational inequalities. I

Reinhard Kluge and Gerhard Telschow, Berlin

We consider here iteration methods for the solution of variational inequalities with single-valued and multivalued monotone mappings in the cases of unique solvability as of non-unique solvability.

In special cases these methods turn out to be combinations of projection or regularization methods with iteration procedures.

In Chapter I, besides of convergence results, above all estimations for the speed of convergence for these iteration methods are given.

The proofs rest upon results on sequences of real numbers that are solutions of special sequences of scalar inequalities, which are derived in Chapter II.

For convergence results for some special methods see the papers [2, 3, 10, 14—25, 27, 33, 34].

I. Iteration methods

I.1. The problem

Let H denote a real Hilbert space with the norm $\|.\|$ and the scalar product $(.,.)$ and $P(K)$ the projector from H onto a closed and convex subset $K \subset H$.

Let $\{H_n\}$ be a sequence of subspaces of H. We denote $P(H_n)$ by P_n $(n = 1, 2, \ldots)$.

By A, A_n we denote multivalued mappings from H into H with domains of definition $D(A)$, $D(A_n)$ respectively. By Ax, for shortness, we denote either an arbitrary element of Ax or a fixed element of Ax. The same applies to A_nx. Let be $\{C_n\}$ a sequence of non-empty, convex and closed subsets C_n of H_n, $n = 1, 2, \ldots$ We state the following conditions:

(A_0) *For every n, A_n is strongly monotone, i.e. $(A_nx - A_ny, x - y) \geqq c_n\|x - y\|^2$ for all $x, y \in D(A_n)$ with $c_n = c(A_n) > 0$.*

(A_1) $C_n \subset D(A_n)$, $n = 1, 2, \ldots$

$y_n \in C_n$ is said to be a *solution* of the variational inequality

$$(A_ny, z - y) \geqq 0, \qquad z \in C_n, \tag{1}$$

if there exists an element $y_n^* \in A_ny_n$ such that

$$(y_n^*, z - y_n) \geqq 0 \qquad \text{for all } z \in C_n.$$

Sufficient conditions for the existence of a solution of variational inequalities with single-valued hemicontinuous monotone mappings are contained in [4, 6, 13, 17, 18, 26, 28, 33], and with possibly multivalued, maximal monotone mappings in [7, 30, 31].

(1) can be considered as an approximation method for the solution of variational inequalities of the form

$$(Ay, z - y) \geqq 0, \qquad z \in C, \tag{1a}$$

with a nonempty, convex and closed set $C \subset H$ and a mapping A that is either single valued, hemicontinuous and monotone or possibly multivalued and maximally monotone. It contains projection methods, the method of elliptic regularization, the penalty method and combinations of these methods. Conditions for the convergence of these methods are contained in [14—20, 28, 33] in the case of single-valued mappings. In the case of multivalued mappings one can obtain similar results from the papers [7, 8, 30—32].

We state the conditions

(A_2) *For every n there exists a solution y_n of* (1).

(A_3) (A_2) *is fulfilled. Furthermore, there exists an element $x^* \in H$ such that*

$$\lim_{n\to\infty} \|y_n - x^*\| = 0.$$

Let be $\{Q_n\}$ a sequence of single-valued mappings from H into itself with $D(Q_n) = H$ and $\{t_n\}$ a sequence of positive real numbers.

We state the following two iteration methods:

(A_4) *Let $\{x_n\}$ be a sequence with $x_n \in D(A_n)$ and*

$$x_{n+1} = Q_n(x_n - t_n A_n x_n), \qquad n = 1, 2, \ldots \tag{2}$$

(A_5) *Let $\{x_n\}$ be a sequence with $x_n \in D(A_n)$ and*

$$x_{n+1} = P(C_n)\, P_n \left(x_n - t_n \frac{A_n x_n}{\|A_n x_n\|}\right), \qquad n = 1, 2, \ldots \tag{3}$$

In the next sections we are going to investigate the convergence of the methods (2) and (3).

I.2. The convergence of iteration methods (part 1)

In this section we shall prove the convergence of (2) making use of the technique developed by Halpern [12] and Bruck [10].

Definition. The sequences $\{t_n\}$ and $\{e_n\}$ of non-negative numbers are said to be *admissibly paired* if the sequence $\{e_n\}$ is monotone nonincreasing and there is a strictly increasing sequence $\{n(k)\}$ of positive integers such that

$$\liminf e_{n(k)} \sum_{j=n(k)}^{n(k+1)-1} t_j > 0, \tag{4}$$

$$\lim \, (e_{n(k)} - e_{n(k+1)}) \sum_{j=n(k)}^{n(k+1)-1} t_j = 0, \tag{5}$$

$$\lim \sum_{j=n(k)}^{n(k+1)-1} t_j^2 = 0. \tag{6}$$

Bruck in [10] stated the following examples for admissibly paired sequences:

1. $t_n = \frac{1}{n}$, $\quad e_n = \frac{1}{\log \log n}$ and $n(k) = k^k$.

2. $t_n = \frac{1}{n^p}$, $\quad 0 < p < 1$, $\quad e_n = \frac{1}{n^q}$, $\quad q < \min (p, 1 - p)$,

$$0 < \left(1 - \frac{q}{1-p}\right)^{-1} < d < \begin{cases} \min \left(\frac{1-p}{1-2p}, 2\left(1 - \frac{q}{1-p}\right)^{-1}\right) & \text{if } p < \frac{1}{2} \\ 2 \cdot \left(1 - \frac{q}{1-q}\right)^{-1} & \text{if } p \geqq \frac{1}{2} \end{cases}$$

and $n(k) \sim k^{\frac{d}{1-p}}$.

Theorem 1. *Let the conditions (A_0), (A_1), (A_2), (A_3) and (A_4) be fulfilled. Furthermore assume:*

1. *The sequences $\{t_n\}$ and $\{e_n\}$ are admissibly paired.*
2. *$A_n = S + e_nT$, $n = 1, 2, \ldots$, where S is a monotone operator and T is a strongly monotone operator, i.e. $c_n = e_n c(T)$.*
3. *$Q_n = P(C_{n(k)})\, P_{n(k)}$ for $n(k) \leq n < n(k+1)$.*
4. *The sequences $\{x_n\}$, $\{Sx_n\}$ and $\{Tx_n\}$ are bounded.*

Then the iteration method (2) converges strongly to x^*.

Proof. Let be $n > n(1)$ arbitrary. There exists an integer k such that $n(k) < n \leqq n(k+1)$. Then because of $y_{n(k)} = Q_n y_{n(k)}$, the inequality

$$\begin{aligned} \|x_n - y_{n(k)}\|^2 &\leqq \|x_{n-1} - y_{n(k)} - t_{n-1}(Sx_{n-1} + e_{n-1}Tx_{n-1})\|^2 \\ &= \|x_{n-1} - y_{n(k)}\|^2 - 2t_{n-1}(Sx_{n-1} + e_{n(k)}Tx_{n-1}, x_{n-1} - y_{n(k)}) \\ &\quad + 2t_{n-1}(e_{n(k)} - e_{n-1})\,(Tx_{n-1}, x_{n-1} - y_{n(k)}) + t_{n-1}^2 \|Tx_{n-1} + e_{n-1}Sx_{n-1}\|^2 \end{aligned}$$

holds. Furthermore, we have

$$\begin{aligned} 0 &\leqq (Sx_{n-1} - Sy_{n(k)}, x_{n-1} - y_{n(k)}) = (Sx_{n-1} + e_{n(k)}Ty_{n(k)} - Sy_{n(k)} \\ &\quad - e_{n(k)}Ty_{n(k)}, x_{n-1} - y_{n(k)}) \leqq (Sx_{n-1} + e_{n(k)}Ty_{n(k)}, x_{n-1} - y_{n(k)}) \end{aligned}$$

and therefore

$$\begin{aligned} &(Sx_{n-1} + e_{n(k)}Tx_{n-1}, x_{n-1} - y_{n(k)}) = (Sx_{n-1} + e_{n(k)}Ty_{n(k)}, x_{n-1} - y_{n(k)}) \\ &+ e_{n(k)}(Tx_{n-1} - Ty_{n(k)}, x_{n-1} - y_{n(k)}) \geqq c_{n(k)}\|x_{n-1} - y_{n(k)}\|^2 \end{aligned}$$

and consequently

$$\|x_n - y_{n(k)}\|^2 \leqq (1 - 2t_{n-1}c_{n(k)}) \|x_{n-1} - y_{n(k)}\|^2$$

$$+ 2t_{n-1}(e_{n(k)} - e_{n-1}) (Tx_{n-1}, x_{n-1} - y_{n(k)}) + t_{n-1}^2 \|Sx_{n-1} + e_{n-1}Tx_{n-1}\|^2.$$

With

$$C = \sup \{2(Tx_{n-1}, x_{n-1} - y_{n(k)}), \|Sx_{n-1} + e_{n-1}Tx_{n-1}\|^2\} < +\infty$$

and making use of $1 - x \leqq e^{-x}$ for $x \geqq 0$ we conclude

$$\|x_n - y_{n(k)}\|^2 \leqq \exp(-2t_{n-1}c_{n(k)}) \|x_{n-1} - y_{n(k)}\|^2$$

$$+ Ct_{n-1}(e_{n(k)} - e_{n-1}) + Ct_{n-1}^2.$$

By induction we get

$$\|x_n - y_{n(k)}\|^2 \leqq \exp\left(-2c_{n(k)} \sum_{j=n(k)}^{n-1} t_j\right) \|x_{n(k)} - y_{n(k)}\|^2$$

$$+ C \sum_{j=n(k)}^{n-1} (e_{n(k)} - e_j)\, t_j + C \sum_{j=n(k)}^{n-1} t_j^2$$

and, because of the monotonicity of $\{e_n\}$,

$$\|x_n - y_{n(k)}\|^2 \leqq \exp\left(-2c_{n(k)} \sum_{j=n(k)}^{n-1} t_j\right) \|x_{n(k)} - y_{n(k)}\|^2$$

$$+ C(e_{n(k)} - e_n) \sum_{j=n(k)}^{n-1} t_j + C \sum_{j=n(k)}^{n-1} t_j^2. \tag{7}$$

In the particular case $n = n(k + 1)$ we obtain

$$\|x_{n(k+1)} - y_{n(k)}\|^2 \leqq \exp\left(-2c_{n(k)} \sum_{j=n(k)}^{n(k+1)-1} t_j\right) \|x_{n(k)} - y_{n(k)}\|^2$$

$$+ C(e_{n(k)} - e_{n(k+1)}) \sum_{j=n(k)}^{n(k+1)-1} t_j + C \sum_{j=n(k)}^{n(k+1)-1} t_j^2.$$

We conclude from (4), (5) and (6) that there is a number $\gamma \in (0, 1)$ such that

$$\limsup \|x_{n(k+1)} - y_{n(k)}\|^2 \leqq \gamma \limsup \|x_{n(k)} - y_{n(k)}\|^2. \tag{8}$$

On the other hand, $\lim y_{n(k)} = x^*$ yields

$$\limsup \|x_{n(k+1)} - y_{n(k)}\|^2 = \limsup \|x_{n(k+1)} - x^*\|^2 = \limsup \|x_{n(k)} - y_{n(k)}\|^2,$$

and making use of (8) we get $\lim \|x_{n(k)} - x^*\|^2 = 0$, i.e. $\lim x_{n(k)} = x^*$.

Let n be an arbitrary integer and k such that $n(k) < n \leqq n(k+1)$. Then from (7)

$$\|x_n - y_{n(k)}\|^2 \leqq \|x_{n(k)} - y_{n(k)}\|^2 + C(e_{n(k)} - e_n) \sum_{j=n(k)}^{n-1} t_j + C \sum_{j=n(k)}^{n-1} t_j^2$$

$$\leqq \|x_{n(k)} - y_{n(k)}\|^2 + C(e_{n(k)} - e_{n(k+1)}) \sum_{j=n(k)}^{n(k+1)-1} t_j + C \sum_{j=n(k)}^{n(k+1)-1} t_j^2 \tag{9}$$

follows, and consequently $\lim x_n = x^*$ holds.

Remark. Theorem 1 contains the main result of Bruck [10] in the special case that T is the (single-valued) identity in the space H and $C_n = H_n = H$ $(n = 1, 2, \ldots)$. Furthermore Theorem 1 contains iterated penalty methods for the variational inequality

$$(Tx, y - x) \geqq 0 \qquad \text{for all } y \in K,$$

where K is the set of all solutions of the variational inequality

$$(Sy, z - y) \geqq 0 \qquad \text{for all } z \in C.$$

Also the combination of these iteration methods with projection methods is contained.

I.3. The convergence of iteration methods (part 2)

In this and in the next section we consider methods of the form (2). We shall get statements about boundedness, convergence and speed of convergence of these methods under different assumptions. Some results of Chapter II concerning special sequences will be used here.

Theorem 2. *Let the assumptions* (A_0), (A_1), (A_2), (A_3) *and* (A_4) *be fulfilled with* $Q_n = P(C_n)\,P_n$. *Furthermore we suppose the following conditions:*

1. $C_{n+1} \supset C_n$ *for all* $n \geqq 1$.
2. *One of the conditions* (i) *or* (ii) *holds:*
   (i) $C_n \subset C_0$, $n = 1, 2, \ldots$; C_0 *is bounded; and the sequence* $\{A_n\}$ *is such that for every sequence* $\{u_n\}$ *with* $u_n \in C_n$, $n = 1, 2, \ldots$, *the sequence* $\{A_n u_n\}$ *is bounded.*
   (ii) $\|A_n x\| \leqq K\|x\| + L$ *for all* $x \in C_n$ *with some constants* $K, L \geqq 0$.
3. *It holds* $t_n, c_n > 0$, $\{t_n c_n\}$ *is bounded*, $\lim \dfrac{t_n}{c_n} = 0$, $\sum_{n=1}^{+\infty} t_n c_n = +\infty$ *and*
   $\lim \{\|y_n - x^*\|/c_n\} = 0$.

Then for the method (2) $\lim \|x_n - x^*\| = 0$ *holds.*

Proof. First we show that the sequences $\{x_n\}$ and $\{A_n x_n\}$ are bounded. This is evident if (i) holds. In the case (ii) we get:

$$\|A_n x_n\| \leqq K\|x_n\| + L \leqq K\|x_n - x^*\| + K\|x^*\| + L$$

and

$$\|A_n x_n\|^2 \leqq 3K^2\|x_n - x^*\|^2 + 3K^2\|x^*\|^2 + 3L^2.$$

Using the properties of the projector $P(C_n)$ we get

$$\begin{aligned}
&(x_{n+1} - x_n, x_{n+1} - y_n) = -(x_n - t_n A_n x_n - x_{n+1}, x_{n+1} - y_n) - t_n(A_n x_n, x_{n+1} - y_n)\\
&\leqq -t_n(A_n x_n, x_{n+1} - y_n) = -t_n(A_n x_n, x_{n+1} - x_n) - t_n(A_n x_n, x_n - y_n)\\
&\leqq +t_n\|A_n x_n\|\,\|x_{n+1} - x_n\| - t_n(A_n x_n - A_n y_n, x_n - y_n) - t_n(A_n y_n, x_n - y_n)\\
&\leqq t_n\|A_n x_n\|\,\|x_{n+1} - x_n\| - t_n c_n\|x_n - y_n\|^2\\
&= t_n\|A_n x_n\|\,\|P(C_n)(x_n - t_n A_n x_n) - P(C_n)\,x_n\| - t_n c_n\|x_n - y_n\|^2\\
&\leqq t_n^2\|A_n x_n\|^2 - t_n c_n\|x_n - y_n\|^2\\
&= t_n^2\|A_n x_n\|^2 - t_n c_n\|x_n - x^*\|^2 - 2t_n c_n(x_n - x^*, x^* - y_n) - t_n c_n\|y_n - x^*\|^2\\
&\leqq t_n^2\|A_n x_n\|^2 - t_n c_n\|x_n - x^*\|^2 + 2t_n c_n\|x_n - x^*\|\,\|y_n - x^*\|\\
&\leqq t_n^2(3K^2\|x_n - x^*\|^2 + 3K^2\|x^*\|^2 + 3L^2) - t_n c_n\|x_n - x^*\|^2\\
&\quad + 2t_n c_n\|x_n - x^*\|\,\|y_n - x^*\| = (3K^2 t_n^2 - t_n c_n)\,\|x_n - x^*\|^2\\
&\quad + 2t_n c_n\|x_n - x^*\|\,\|y_n - x^*\| + 3t_n^2(K^2\|x^*\|^2 + L^2)
\end{aligned}$$

and consequently

$$\begin{aligned}
&\|x_{n+1} - x^*\|^2 = \|x_{n+1} - x_n\|^2 + 2(x_{n+1} - x_n, x_n - x^*) + \|x_n - x^*\|^2\\
&\leqq \|x_n - x^*\|^2 + t_n^2\,\|A_n x_n\|^2 + 2(x_{n+1} - x_n, x_{n+1} - y_n) + 2(x_{n+1} - x_n, x_n - x_{n+1})\\
&\quad + 2(x_{n+1} - x_n, y_n - x^*)\\
&\leqq \|x_n - x^*\|^2 + t_n^2(3K^2\,\|x_n - x^*\|^2 + 3K^2\|x^*\|^2 + 3L^2) + (6K^2 t_n^2 - 2t_n c_n)\,\|x_n - x^*\|^2\\
&\quad + 4t_n c_n\|x_n - x^*\|\,\|y_n - x^*\| + 6t_n^2(K^2\,\|x^*\|^2 + L^2) + 2\,\|x_{n+1} - x_n\|\,\|y_n - x^*\|\\
&\leqq \|x_n - x^*\|^2 + (9K^2 t_n^2 - 2t_n c_n)\,\|x_n - x^*\|^2 + 4t_n c_n\|x_n - x^*\|\,\|y_n - x^*\|\\
&\quad + 9t_n^2(K^2\,\|x^*\|^2 + L^2) + 2t_n\|A_n x_n\|\,\|y_n - x^*\|\\
&\leqq \|x_n - x^*\|^2 - \left(1 - \frac{9K^2 t_n}{2c_n}\right)\cdot 2t_n c_n\|x_n - x^*\|^2 + 4t_n c_n\|x_n - x^*\|\,\|y_n - x^*\|\\
&\quad + 9t_n^2(K^2\|x^*\|^2 + L^2) + 2t_n(K\|x_n - x^*\| + K\|x^*\| + L)\,\|y_n - x^*\|\\
&= \|x_n - x^*\|^2 - \left(1 - \frac{9K^2 t_n}{2c_n}\right) 2t_n c_n\|x_n - x^*\|^2 + (4t_n c_n + 2t_n K)\,\|x_n - x^*\|\,\|y_n - x^*\|\\
&\quad + 9t_n^2(K^2\|x^*\|^2 + L^2) + t_n(2K\|x^*\| + 2L)\,\|y_n - x^*\|.
\end{aligned}$$

Because of $\lim \frac{t_n}{c_n} = 0$ there is an integer n_0 such that $1 - \frac{9Kt_n}{2c_n} \geqq \frac{1}{2}$ for all $n \geqq n_0$. We define $C = \max\{4,\ 2K,\ 9(K^2\|x^*\|^2 + L^2),\ 2(K\|x^*\| + L)\}$, $b_n = \|x_n - x^*\|^2$, $c = 1$, $k_n = 1 - t_nc_n$, $\delta_n = C(t_nc_n + t_n)\,\|y_n - x^*\|$ and $\gamma_n = C(t_n^2 + t_n\|y_n - x^*\|)$. With these notations we obtain $b_{n+1} \leqq k_nb_n + \delta_n \cdot \sqrt{b_n} + \gamma_n$, and consequently we get the boundedness of $\{b_n\}$ by Corollary 5 of Section II.1. From this the boundedness of $\{A_nx_n\}$ follows by (ii).

Now we can prove the convergence of (2). In both cases (i) and (ii) we get

$$\|x_{n+1} - x^*\|^2 = \|x_n - x^*\|^2 + \|x_{n+1} - x_n\|^2 + 2(x_{n+1} - x_n, x_n - x^*)$$
$$\leqq \|x_n - x^*\|^2 + t_n^2\,\|A_nx_n\|^2 + 2(x_{n+1} - x_n, x_{n+1} - y_n) + 2(x_{n+1} - x_n, x_n - x_{n+1})$$
$$+ 2(x_{n+1} - x_n, y_n - x^*) \leqq \|x_n - x^*\|^2 + t_n^2\|A_nx_n\|^2 - 2(x_n - t_nA_nx_n - x_{n+1},$$
$$x_{n+1} - y_n) - 2t_n(A_nx_n, x_{n+1} - y_n) + 2\,\|x_{n+1} - x_n\|\,\|y_n - x^*\|$$
$$\leqq \|x_n - x^*\|^2 + t_n^2\|A_nx_n\|^2 - 2t_n(A_nx_n, x_{n+1} - y_n) + 2t_n\|A_nx_n\|\,\|y_n - x^*\|$$
$$= \|x_n - x^*\|^2 + t_n^2\|A_nx_n\|^2 - 2t_n(A_nx_n, x_{n+1} - x_n) - 2t_n(A_nx_n, x_n - y_n)$$
$$+ 2t_n\|A_nx_n\|\,\|y_n - x^*\| \leqq \|x_n - x^*\|^2 + t_n^2\|A_nx_n\|^2 + 2t_n\|A_nx_n\|\,\|x_{n+1} - x_n\|$$
$$- 2t_n(A_nx_n - A_ny_n, x_n - y_n) - 2t_n(A_ny_n, x_n - y_n) + 2t_n\|A_nx_n\|\,\|y_n - x^*\|$$
$$\leqq \|x_n - x^*\|^2 + t_n^2\|A_nx_n\|^2 + 2t_n^2\|A_nx_n\|^2 - 2t_nc_n\|x_n - y_n\|^2 + 2t_n\|A_nx_n\|\,\|y_n - x^*\|$$
$$= \|x_n - x^*\|^2 + 3t_n^2\|A_nx_n\|^2 - 2t_nc_n\|x_n - x^*\|^2 - 4t_nc_n(x_n - x^*, x^* - y_n)$$
$$- 2t_nc_n\|y_n - x^*\|^2 + 2t_n\|A_nx_n\|\,\|y_n - x^*\|$$
$$\leqq \|x_n - x^*\|^2 - 2t_nc_n\|x_n - x^*\|^2 + (4t_nc_n\|x_n - x^*\| + 2t_n\|A_nx_n\|)\,\|y_n - x^*\|$$
$$+ 3t_n^2\|A_nx_n\|^2.$$

Because of the boundedness of $\{x_n\}$ and $\{A_nx_n\}$ we have

$$C = \sup_{n=1,2,\ldots} \{4\|x_n - x^*\|, 2\|A_nx_n\|, 3\|A_nx_n\|^2\} < +\infty.$$

We define $b_n = \|x_n - x^*\|^2$, $c = 2$, $\alpha_n = t_nc_n$, $\mu_n = 1$ and $a_n = Ct_nc_n\|y_n - x^*\| + Ct_n\|y_n - x^*\| + Ct_n^2$. We get

$$b_{n+1} \leqq b_n - \frac{c\alpha_n}{\mu_n}\, b_n + a_n, \tag{10}$$

and by Theorem 1 of Chapter II we obtain the convergence of $\{b_n\}$ to 0 and consequently $\lim \|x_n - x^*\| = 0$. This proves the theorem.

Theorem 3. *Let the conditions (A_0), (A_1), (A_2), (A_3) and (A_4) be fulfilled. We suppose furthermore that the assumptions 1 and 2 from Theorem 2 hold. Additionally we assume: $Q_n = P(C_n)\,P_n$, $c_n \geqq c > 0$, c fixed, $\{c_n\}$ is bounded and*

$$\text{(iii)} \quad \sum_{n=1}^{+\infty} t_n = +\infty, \qquad \lim t_n = 0.$$

Then the iteration method (2) converges to x^*.

If, moreover, $t_n = n^{-p}$, $0 < p \leqq 1$, *and* $\limsup \|y_n - x^*\| n^r < +\infty$ *hold then* $\limsup \|x_n - x^*\|^2 n^q < +\infty$ *takes place with* $q \leqq \min(r, p)$ *and in addition* $q < 2c$ *if* $p = 1$.

Proof. The convergence follows directly from Theorem 2. With the notations from Theorem 2 we obtain $a_n \leqq K_1 t_n \|y_n - x^*\| + C t_n^2$ with $K_1 = C\left(1 + \sup\limits_{n=1,2,\dots} c_n\right)$ and consequently $\limsup a_n n^\alpha < +\infty$ with $\alpha = p + \min(p, r)$. We introduce the notations $b_n = \|x_n - x^*\|^2$, $k_n = 1 - 2ct_n$ and $K_2 = \sup\limits_{n=1,2,\dots} a_n n^\alpha < +\infty$ and get from (10)

$$b_{n+1} \leqq k_n b_n + c_n n^{-\alpha}.$$

From Example 3 of Chapter II (II.2) we conclude the assertion.

Theorem 4. *Let the conditions* (A_0), (A_1), (A_2), (A_3) *and* (A_4) *be fulfilled with* $Q_n = P(C_n) P_n$. *We assume conditions* 1 *and* 2 *from Theorem* 2. *Furthermore we assume*

(iv) $t_n, c_n > 0$, $\limsup t_n c_n < +\infty$, $\sum\limits_{n=1}^{+\infty} t_n c_n = +\infty$,

$\lim (t_n/c_n) = 0$ *and* $\lim (\|y_{n-1} - y_n\|/t_n c_n) = 0$.

Then the iteration method (2) *converges to* x^*.

Proof. First we prove the boundedness of $\{x_n\}$ and $\{A_n x_n\}$. This is evident if (i) holds. In case (ii), because of $y_n = Q_n y_n$,

$$\begin{aligned}
\|x_{n+1} - y_n\|^2 &\leqq \|x_n - y_n - t_n A_n x_n\|^2 \\
&= \|x_n - y_n\|^2 - 2t_n(A_n x_n, x_n - y_n) + t_n^2 \|A_n x_n\|^2 \\
&= \|x_n - y_n\|^2 + t_n^2\|A_n x_n\|^2 - 2t_n(A_n x_n - A_n y_n, x_n - y_n) - 2t_n(A_n y_n, x_n - y_n) \\
&\leqq \|x_n - y_n\|^2 - 2t_n c_n \|x_n - y_n\|^2 + t_n^2 \|A_n x_n\|^2 \\
&\leqq \|x_n - y_n\|^2 - 2t_n c_n \|x_n - y_n\|^2 + t_n^2 (K\|x_n - y_n\| + K\|y_n\| + L)^2 \\
&\leqq \|x_n - y_n\|^2 - 2t_n c_n \|x_n - y_n\|^2 + 3t_n^2 K^2 \|x_n - y_n\|^2 + 3t_n^2 K^2 \|y_n\|^2 + 3t_n^2 L^2 \\
&= \|x_n - y_n\|^2 - 2t_n c_n \left(1 - \frac{3t_n K^2}{2c_n}\right) \|x_n - y_n\|^2 + 3t_n^2 (K^2 \|y_n\|^2 + L^2).
\end{aligned}$$

There is an integer n_0 with $1 - \dfrac{3t_n K^2}{2c_n} \geqq \dfrac{1}{2}$ for all $n \geqq n_0$ because of $\lim \dfrac{t_n}{c_n} = 0$. With $C = 3K^2 \sup\limits_{n=1,2,\dots} \|y_n\|^2 + 3L^2$ we get

$$\begin{aligned}
&\|x_{n+1} - y_n\|^2 \leqq \|x_n - y_n\|^2 - t_n c_n \|x_n - y_n\|^2 + C t_n^2 = \|x_n - y_{n-1}\|^2 \\
&+ \|y_{n-1} - y_n\|^2 + 2(x_n - y_{n-1}, y_{n-1} - y_n) - t_n c_n \|x_n - y_{n-1}\|^2 - t_n c_n \|y_n - y_{n-1}\|^2 \\
&- 2t_n c_n (x_n - y_{n-1}, y_{n-1} - y_n) + C t_n^2 \leqq \|x_n - y_{n-1}\|^2 - t_n c_n \|x_n - y_{n-1}\|^2 \\
&+ 2\|y_{n-1} - y_n\| \, \|x_n - y_{n-1}\| + 2t_n c_n \|y_{n-1} - y_n\| \, \|x_n - y_{n-1}\| + \|y_{n-1} - y_n\|^2 + C t_n^2.
\end{aligned}$$

We denote $k_n = 1 - t_n c_n$, $b_n = \|x_n - y_{n-1}\|^2$, $\delta_n = 2(1 + t_n c_n) \|y_{n-1} - y_n\|$ and $\gamma_n = \|y_{n-1} - y_n\|^2 + Ct_n^2$ and get

$$b_{n+1} \leqq k_n b_n + \delta_n \sqrt{b_n} + \gamma_n.$$

The boundedness of $\{x_n\}$ and $\{A_n x_n\}$ follows from Corollary 5 of Chapter II and from (ii), respectively.

Now we prove the convergence.

$$\|x_{n+1} - y_n\|^2 \leqq \|x_n - y_n - t_n A_n x_n\|^2 = \|x_n - y_n\|^2 - 2t_n(A_n x_n, x_n - y_n) + t_n^2\|A_n x_n\|^2$$
$$= \|x_n - y_n\|^2 + t_n^2\|A_n x_n\|^2 - 2t_n(A_n x_n - A_n y_n, x_n - y_n) - 2t_n(A_n y_n, x_n - y_n)$$
$$\leqq \|x_n - y_n\|^2 + t_n^2\|A_n x_n\|^2 - 2t_n c_n\|x_n - y_n\|^2 \leqq \|x_n - y_{n-1}\|^2 + \|y_{n-1} - y_n\|^2$$
$$- 2(x_n - y_{n-1}, y_n - y_{n-1}) - 2t_n c_n\|x_n - y_{n-1}\|^2 - 2t_n c_n\|y_{n-1} - y_n\|^2$$
$$+ 4t_n c_n(x_n - y_{n-1}, y_n - y_{n-1}) + t_n^2\|A_n x_n\|^2 \leqq \|x_n - y_{n-1}\|^2$$
$$- 2t_n c_n\|x_n - y_{n-1}\|^2 + \big(\|y_{n-1} - y_n\| + 2(1 + 2t_n c_n) \|x_n - y_{n-1}\|\big) \|y_n - y_{n-1}\|$$
$$+ t_n^2\|A_n x_n\|^2.$$

We denote $c = 2$, $\alpha_n = t_n c_n$, $\mu_n = 1$, $b_n = \|x_n - y_{n-1}\|^2$, $C = \sup\limits_{n=1,2,\dots} [\max \{2(1 + 2t_n c_n) \times \|x_n - y_{n-1}\| + \|y_{n-1} - y_n\|, \|A_n x_n\|^2\}] < +\infty$ and $a_n = C\|y_{n-1} - y_n\| + Ct_n^2$ and get

$$b_{n+1} \leqq b_n - \frac{c\alpha_n}{\mu_n} b_n + a_n.$$

The assertion follows from Theorem 1 of Chapter II.

Lemma 5. *Let be $A_n = S + e_n T$ with S monotone and T strongly monotone, $\{e_n\}$ monotone nonincreasing, $C_n = C$ and $\{Ty_n\}$ bounded. Then under the assumptions (A_0). (A_1), (A_2)*

$$\limsup \|y_n - y_{n-1}\| \frac{e_n}{e_{n-1} - e_n} < +\infty.$$

Proof. Since y_n is a solution of (1) and $C_n = C$, we have

$$e_n c(T)\|y_n - y_{n-1}\|^2 \leqq e_n(Ty_n - Ty_{n-1}, y_n - y_{n-1})$$
$$= (Sy_n + e_n Ty_n, y_n - y_{n-1}) + (Sy_{n-1} + e_{n-1}Ty_{n-1}, y_{n-1} - y_n)$$
$$+ (Sy_{n-1} - Sy_n, y_n - y_{n-1}) + (e_{n-1} - e_n) (Ty_{n-1}, y_n - y_{n-1})$$
$$\leqq (e_{n-1} - e_n) \|Ty_{n-1}\| \|y_n - y_{n-1}\|.$$

Therefore

$$\|y_n - y_{n-1}\| \left(\frac{e_n}{e_{n-1} - e_n}\right) \leqq \frac{\|Ty_{n-1}\|}{c(T)}.$$

Theorem 6. *Let the assumptions* (A_0), (A_1), (A_2), (A_3), (A_4) *be fulfilled with* $A_n = S + e_nT$, $D(A_n) = D(S) \cap D(T)$, $C_n = C$ *and* $Q_n = P(C)$. *We suppose that* S *is monotone and* T *is strongly monotone. Furthermore we suppose either* (i)$_1$ *or* (ii)$_2$:

(i)$_1$ C *is bounded, and* S *and* T *are bounded on* C.

(ii)$_2$ *There are non-negative constants* K *and* L *with* $\max (\|Sx\|, \|Tx\|) \leqq K \|x\| + L$ *for all* $x \in C$.

and (v):

(v) $t_n = n^{-s}, \quad e_n = n^{-r}, \quad 0 < r < s, \quad r + s < 1.$

Then (2) *converges to* x^*.

If, in addition, $\lim \|y_n - x^*\| n^t < +\infty$, $t > 0$, *holds then we have* $\limsup \|x_n - x^*\|^2 \times n^q < +\infty$ *with* $q \leqq \min \big(1 - (r + s), s - r, t\big)$.

Proof. The convergence was proved in [3]. From Lemma 5 we obtain $\lim \|y_n - y_{n-1}\| \times n^{r+s} = 0$ (cf. Example 4 of Chapter II). Therefore the convergence follows from Theorem 4, too.

We introduce the notations $b_n = \|x_n - y_{n-1}\|^2$, $k_n = 1 - 2t_ne_nc(T)$, $a_n^1 = \|y_{n-1} - y_n\|$, $a_n^2 = t_n^2$ and $a_n = C(a_n^1 + a_n^2)$ and get (cf. the proof of Theorem 4) $b_{n+1} \leqq k_nb_n + a_n$. We take $d_n = n^{-q}$. Because of Lemma 5 we get

$$\limsup \frac{a_n^1}{d_{n+1} - k_nd_n} < +\infty$$

if $q \leqq 1 - (r + s)$ (cf. Example 4 of Chapter II) and

$$\limsup \frac{a_n^2}{d_{n+1} - k_nd_n} < +\infty$$

if $q \leqq 2s - (r + s) = s - r$ (cf. Example 3 of Chapter II — II.2). Thus we have

$$\limsup \frac{a_n}{d_{n+1} - k_nd_n} < +\infty$$

if $q \leqq \min \big(1 - (r + s), r - s\big)$. From Theorem 7 of Chapter II (Section II.1) we obtain $\limsup b_nn^q < +\infty$. Consequently we get the assertion.

I.4. The convergence of iteration methods (part 3)

Let A be a single-valued mapping from H into H with $D(A) = H$. A is said to be *s-Lipschitz continuous in a point* $x \in H$ if there exists a monotone non-decreasing function $L(A, x, .)$: $[0, s] \to [0, +\infty)$ such that $\|Ax - Ay\| \leqq L(A, x, s) \|x - y\|$ for all $y \in H$ with $\|x - y\| \leqq s$. If $L(A, x, s) < 1$, A is said to be an *s-contraction.*

Lemma 7. *If* A *is strongly monotone and s-Lipschitz continuous in* $x \in H$ *then* $A_t = I - tA$ *is an s-contraction if* $0 < t < \dfrac{2c(A)}{L(A, x, s)^2}$ *with*

$$L(A_t, x, s) = \sqrt{1 - 2tc(A) + t^2L(A, x, s)^2}.$$

Proof. cf. [24]. If A is lipschitzian, cf. [5, 9].

Remark. $L(A_t, x, s)$ is minimal if $t = t_0 = \dfrac{c(A)}{L(A, x, s)^2}$ with

$$L(A_{t_0}, x, s) = \sqrt{1 - \frac{c(A)^2}{L(A, x, s)^2}}.$$

Lemma 8. *If A is the gradient of a functional and is strongly monotone and lipschitzian, then A_t is a contraction if $0 < t \leqq \dfrac{2}{L(A) + c(A)}$ with $L(A_t) = 1 - tc(A)$. $L(A_t)$ is minimal if $t = t_0 = \dfrac{2}{L(A) + c(A)}$ with $L(A_{t_0}) = \dfrac{L(A) - c(A)}{L(A) + c(A)}$.*

Proof. cf. [11] (p. 165—166).

Theorem 9. *Let (A_0), (A_1), (A_2), (A_3) and (A_4) be fulfilled with $Q_n = P(C_n) P_n$. Let A_n be single-valued and s_n-Lipschitz continuous in y_n with $s_n \geqq \|y_n - x_n\|$. Furthermore we suppose*

(vi) $\sum\limits_{n=1}^{\infty} (1 - k_n) = +\infty, \qquad \lim \dfrac{k_n \|y_n - y_{n-1}\|}{1 - k_n} = 0$ *with*

$k_n = L\big((A_n)_{t_n}, y_n, s_n\big)$.

Then the iteration method (2) *converges to x^*.*

Proof. Because of (1) $y_n = P(C_n) P_n(y_n - t_n A_n y_n)$ takes place. Consequently

$$\begin{aligned}
\|x_{n+1} - y_n\| &= \|P(C_n) P_n(x_n - t_n A_n x_n) - P(C_n) P_n(y_n - t_n A_n y_n)\| \\
&\leqq \|x_n - y_n - t_n(A_n x_n - A_n y_n)\| \\
&\leqq k_n \|x_n - y_n\| \\
&\leqq k_n \|x_n - y_{n-1}\| + k_n \|y_{n-1} - y_n\|.
\end{aligned}$$

With $b_n = \|x_n - y_{n-1}\|$ and $a_n = k_n \|y_{n-1} - y_n\|$ the assertion follows from Theorem 1 of Chapter II (Section II.1).

Corollary 10. *Let the suppositions of Theorem 9 hold. Instead of* (vi) *assume* (vii):

(vii) $k_n \leqq k < 1, n = 1, 2, \ldots, \lim \|y_n - y_{n-1}\| = 0$ *with*

$k_n = L\big((A_n)_{t_n}, y_n, s_n\big)$.

Then the iteration method (2) *converges to x^*.*

If in addition $\limsup \|y_n - x^*\| p^{-n} < +\infty$ *holds with $p < 1$, then* $\limsup \|x_n - x^*\| p^{-n} < +\infty$ *if $p > k$ and* $\limsup \|x_n - x^*\| \dfrac{p^{-n}}{n} < +\infty$ *if $p = k$ takes place.*

If $\limsup \|y_n - x^*\| n^q < +\infty$ *with $q > 0$ holds, then* $\limsup \|x_n - x^*\| n^q < +\infty$ *holds.*

Proof. The convergence follows from Theorem 9. The first assertion about the speed of convergence follows from Examples 1 and 2 of Chapter II (Section II.2).

To prove the last assertion we take $d_n = n^{-q}$. We get $k_n \frac{d_n}{d_{n+1}} \leqq k\left(\frac{n+1}{n}\right)^q < 1$ if n is sufficiently large,

$$\sum_{n=1}^{+\infty}\left(1 - k_n \frac{d_n}{d_{n+1}}\right) \geqq \sum_{n=1}^{+\infty}\left(1 - k\left(\frac{n+1}{n}\right)^q\right) = +\infty.$$

Furthermore (cf. the proof of Theorem 9)

$$\begin{aligned}
\|x_{n+1} - x^*\| &\leqq \|x_{n+1} - y_n\| + \|y_n - x^*\| \\
&\leqq k_n\|x_n - y_{n-1}\| + k_n\|y_{n-1} - y_n\| + \|y_n - x^*\| \\
&\leqq k_n\|x_n - x^*\| + k_n\|y_{n-1} - x^*\| + k_n\|y_{n-1} - x^*\| \\
&\quad + k_n\|y_n - x^*\| + \|y_n - x^*\| \\
&\leqq k_n\|x_n - x^*\| + 2\|y_{n-1} - x^*\| + 2\|y_n - x^*\|.
\end{aligned}$$

With $b_n = \|x_n - x^*\|$ and $a_n = 2\|y_{n-1} - x^*\| + 2\|y_n - x^*\|$ the assertion follows from Theorem 7 of Chapter II (Section II.1).

Remarks. 1. Let A_n be lipschitzian in y_n with $L(A_n) = L$ and strongly monotone with $c(A_n) = c$. Then $k_n = \sqrt{1 - 2t_nc + t_n^2L^2} \leqq k < 1$ if $0 < \varepsilon_1 \leqq t_n \leqq \varepsilon_2 < \frac{2c}{L^2}$, with $k = \max\left\{\sqrt{1 - 2\varepsilon_1c + \varepsilon_1^2L^2}, \sqrt{1 - 2\varepsilon_2c + \varepsilon_2^2L^2}\right\}$.

Proof. $f(t) = \sqrt{1 - 2tc + t^2L^2}$ is convex in the interval $\left(0, \frac{2c}{L^2}\right)$ because of $f''(t) = \frac{L^2 - c^2}{(1 - 2tc + t^2L^2)^{3/2}} \geqq 0$. Consequently we get $k_n = f(t_n) \leqq k$ if $t_n \in [\varepsilon_1, \varepsilon_2]$.

2. Let, in addition to the assumptions of remark 1, A_n be the gradient of some functional. Then $k_n \leqq k < 1$ if $0 < \varepsilon \leqq t_n \leqq \frac{2}{L + c}$ with $k_n = 1 - t_nc$ and $k = 1 - \varepsilon c$.

Proof. $k_n = 1 - t_nc \leqq 1 - \varepsilon c = k < 1$ if $t_n \geqq \varepsilon$.

3. For the case of locally Lipschitz continuous mappings A_n Theorem 9 contains the results of [25].

Theorem 11. *Let the conditions* (A_0), (A_1), (A_2), (A_3) *and* (A_4) *hold with* $A_n = S + e_nT$, $C_n = C$ *and* $Q_n = P(C)$, $n = 1, 2, \ldots$ *We suppose that* $D(S) = D(T) = H$, *S and T are lipschitzian, S monotone and T strongly monotone. In addition let* (viii) *be fulfilled:*

(viii) *T and S are gradients of functionales,* $t_n = t_0n^{-s}$, $e_n = n^{-r}$, $r > 0$, $r + s < 1$ *and*

$$0 < t_0 < \frac{2}{L(S) + L(T) + c(T)}.$$

Then the iteration method (2) *converges to* x^*. *If in addition* $\limsup \|y_n - x^*\|\, n^t < +\infty$ *holds with* $t > 0$ *then we get* $\limsup \|x_n - x^*\|\, n^q < +\infty$ *with* $q = \min\left(1 - (r + s), t\right)$.

Proof. The convergence follows because of $k_n = 1 - t_ne_nc(T)$, (viii), Lemma 5 and Example 4 of Chapter II (Section II.2) from Theorem 9 (cf. also [3]). The assertions about the speed of convergence follow from Example 4 of Chapter II (Section II.2) using the inequality from the proof of Corollary 10.

Theorem 12. *Let the suppositions of Theorem* 11 *be fulfilled. Instead of* (viii) *assume* (ix):

$$\text{(ix)}\ e_n = n^{-r},\ 0 < r < \frac{1}{2},\ t_n = t_0 e_n,\ 0 < t_0 < \frac{2c(T)}{L^2},\ L = L(S) + L(T).$$

Then the iteration method (2) *converges to* x^*. *If in addition* $\limsup \|y_n - x^*\|\, n^t < +\infty$ *holds with* $t > 0$ *we get* $\limsup \|x_n - x^*\|\, n^q < +\infty$ *with* $q = \min(1 - 2r, t)$.

Proof. The convergence follows because of (ix), Lemma 5 and Example 4 of Chapter II (Section II.2) from Theorem 9 (cf. also [3]). Furthermore we have

$$k_n = \sqrt{1 - 2t_n c_n + t_n^2 L(A_n)^2} \leqq \sqrt{1 - t_0 e_n^2 \left(2c(T) - t_0 L^2\right)}$$
$$\leqq 1 - t_0 e_n^2 \left(c(T) - \frac{t_0}{2} L^2\right).$$

From Example 4 of Chapter II (Section II.2) $\limsup \|x_n - y_{n-1}\|\, n^q < +\infty$ with $q = 1 - 2r$ follows using the inequalities from the proof of Theorem 9. From this we get the assertion.

I.5. The convergence of the iteration methods (part 4)

In this section we consider the iteration method (3).

Theorem 13. *Let the assumptions* (A_0), (A_1), (A_2), (A_3) *and* (A_5) *be fulfilled. Furthermore we suppose:*

1. $C_{n+1} \supset C_n, \quad n = 1, 2, \ldots$
2. *For all bounded sequences* $\{z_n\}$ *with* $z_n \in C_n$ *let* $\{A_n z_n\}$ *be bounded, and let either* (x) *or* (xi) *be satisfied:*
   (x) $C_n \subset C_0, \quad n = 1, 2, \ldots, \quad C_0$ *bounded.*
   (xi) $\sum_{n=1}^{\infty} t_n^2 < +\infty$ *and* $\sum_{n=1}^{\infty} t_n \|y_n - x^*\| < +\infty$.
3. $c_n \leq \text{const} < +\infty, \quad \sum_{n=1}^{\infty} t_n c_n = +\infty, \quad \lim \frac{t_n}{c_n} = 0$ *and*
   $\lim \frac{\|y_n - x^*\|}{c_n} = 0.$

Then the iteration method (3) *converges to* x^*.

Proof. First we prove the boundedness of $\{x_n\}$ and $\{A_n x_n\}$. If (x) holds this assertion is evident.

Assume (xi). We get $\left(x_n - t_n \frac{A_n x_n}{\|A_n x_n\|} - x_{n+1}, x_{n+1} - y_n\right) \geq 0$ because of (3) and therefore $(x_{n+1} - x_n, x_{n+1} - y_n) \leq \frac{-t_n}{\|A_n x_n\|} (A_n x_n, x_{n+1} - y_n)$. We conclude

$$\|x_{n+1} - x^*\|^2 = \|x_{n+1} - x_n\|^2 + \|x_n - x^*\|^2 + 2(x_{n+1} - x_n, x_n - x^*)$$

$$\leqq t_n^2 + \|x_n - x^*\|^2 + 2(x_{n+1} - x_n, x_{n+1} - y_n) + 2(x_{n+1} - x_n, x_n - x_{n+1})$$
$$+ 2(x_{n+1} - x_n, y_n - x^*)$$
$$\leqq t_n^2 + \|x_n - x^*\|^2 - \frac{2t_n}{\|A_nx_n\|}(A_nx_n, x_{n+1} - y_n) + 2t_n\|y_n - x^*\|$$
$$\leqq \|x_n - x^*\|^2 - \frac{2t_n}{\|A_nx_n\|}(A_nx_n - A_ny_n, x_n - y_n) - \frac{2t_n}{\|A_nx_n\|}(A_ny_n, x_n - y_n)$$
$$- \frac{2t_n}{\|A_nx_n\|}(A_nx_n, x_{n+1} - y_n) + 2t_n\|y_n - x^*\| + t_n^2$$
$$\leqq \|x_n - x^*\|^2 - \frac{2c_nt_n}{\|A_nx_n\|}\|x_n - y_n\|^2 + 2t_n\|x_{n+1} - x_n\| + 2t_n\|y_n - x^*\| + t_n^2$$
$$\leqq \|x_n - x^*\|^2 + 3t_n^2 + 2t_n\|y_n - x^*\| - \frac{2t_nc_n}{\|A_nx_n\|}\|x_n - y_n\|^2.$$

So we get

$$\|x_{n+1} - x^*\|^2 \leqq \|x_n - x^*\|^2 + 3t_n^2 + 2t_n\|y_n - x^*\|,$$

and the boundedness of $\{x_n\}$ and $\{A_nx_n\}$ follows from Lemma 2 of Chapter II (Section II.1) and assumption 2.

We define $K = \sup_{n=1,2,\dots} [\max(\|x_n - x^*\|, \|A_nx_n\|)] < +\infty$. From the inequality above we conclude

$$\|x_{n+1} - x^*\|^2 \leqq \|x_n - x^*\|^2 - \frac{2t_nc_n}{K}\|x_n - y_n\|^2 + 3t_n^2 + 2t_n\|y_n - x^*\|$$
$$= \|x_n - x^*\|^2 - \frac{2t_nc_n}{K}\|x_n - x^*\|^2 - \frac{2t_nc_n}{K}\|y_n - x^*\|^2 - \frac{4t_nc_n}{K}(x_n - x^*, x^* - y_n)$$
$$+ 3t_n^2 + 2t_n\|y_n - x^*\| \leqq \|x_n - x^*\|^2 - \frac{2t_nc_n}{K}\|x_n - x^*\|^2 + 4t_nc_n\|y_n - x^*\|$$
$$+ 3t_n^2 + 2t_n\|y_n - x^*\| = \|x_n - x^*\|^2 - \frac{2t_nc_n}{K}\|x_n - x^*\|^2$$
$$+ (4c_n + 2)\, t_n\|y_n - x^*\| + 3t_n^2.$$

The convergence follows from Theorem 1 of Chapter II (Section II.1).

Corollary 14. *Let the suppositions of Theorem 13 hold. Instead of assumption 3 assume*

$$\text{(xii)} \quad c_n = c > 0, \qquad \lim t_n = 0 \quad \text{and} \quad \sum_{n=1}^{\infty} t_n = +\infty.$$

Then the iteration method (3) converges to x^*.

If, in addition, $t_n = n^{-p}$, $0 < p < 1$ and $\limsup \|y_n - x^*\|\, n^r < +\infty$ with $r > 1 - p$ hold, we get $\limsup \|x_n - x^*\|^2 n^q < +\infty$ with $q = \min(p, r)$.

Proof. The convergence follows from Theorem 13 directly. Using the last inequality from the proof of Theorem 13 we get the assertion about the speed of convergence by Example 3 of Chapter II (Section II.2).

Theorem 15. *Let the conditions* (A_0), (A_1), (A_2), (A_3) *and* (A_5) *hold. We suppose that* $C_0 \supset C_{n+1} \supset C_n$, $n = 1, 2, \ldots$, *and for every sequence* $\{x_n\}$ *with* $x_n \in C_n$, $\{x_n\}$ *and* $\{A_n x_n\}$ *are bounded,* $t_n c_n \leqq \text{const} < +\infty$, $\sum\limits_{n=1}^{\infty} t_n c_n = +\infty$, $\lim \frac{t_n}{c_n} = 0$ *and* $\lim \frac{\|y_{n-1} - y_n\|}{t_n c_n} = 0$.

Then the iteration method (3) *converges to* x^*.

Proof. We define $C = \sup\limits_{n=1,2,\ldots} \|A_n x_n\| < +\infty$. We get

$$\|x_{n+1} - y_n\|^2 \leqq \|x_n - y_n\|^2 - \frac{2t_i}{\|A_n x_n\|} (A_n x_n, x_n - y_n) + t_n^2$$

$$\leqq \|x_n - y_n\|^2 + t_n^2 - \frac{2t_n}{\|A_n x_n\|} (A_n x_n - A_n y_n, x_n - y_n) - \frac{2t_n}{\|A_n x_n\|} (A_n y_n, x_n - y_n)$$

$$\leqq \|x_n - y_n\|^2 + t_n^2 - \frac{2t_n c_n}{\|A_n x_n\|} \|x_n - y_n\|^2$$

$$\leqq \left(1 - \frac{2t_n c_n}{C}\right) \|x_n - y_n\|^2 + t_n^2$$

$$= \left(1 - \frac{2t_n c_n}{C}\right) \|x_n - y_{n-1}\|^2 + \left(1 - \frac{2t_n c_n}{C}\right) \|y_{n-1} - y_n\|^2$$

$$+ 2\left(1 - \frac{2t_n c_n}{C}\right) \cdot (x_n - y_{n-1}, y_{n-1} - y_n) + t_n^2$$

$$\leqq \left(1 - \frac{2t_n c_n}{C}\right) \|x_n - y_{n-1}\|^2 + \|y_{n-1} - y_n\|^2 + 2\left(1 + \frac{2t_n c_n}{C}\right) \|x_n - y_{n-1}\| \, \|y_{n-1} - y_n\|$$

$$+ t_n^2 \leqq \left(1 - \frac{2t_n c_n}{C}\right) \|x_n - y_{n-1}\|^2 + K\|y_{n-1} - y_n\| + t_n^2$$

with

$$K = \sup_{n=1,2,\ldots} \left(\|y_{n-1} - y_n\| + 2\left(1 + \frac{2t_n c_n}{C}\right) \|x_n - y_{n-1}\|\right) < +\infty.$$

The assertion follows from Theorem 1 of Chapter II (Section II.1).

II. The convergence of special sequences of real numbers

In this chapter we consider sequences $\{b_n\}$ that satisfy a set of inequalities

$$b_{n+1} \leqq b_n - \frac{c\alpha_n}{\mu_n} b_n + a_n, \qquad n = 1, 2, \ldots, \tag{1}$$

where c is a given real number and $\{\alpha_n\}$, $\{\mu_n\}$ and $\{a_n\}$ are given sequences.

We shall give conditions for the convergence of $\{b_n\}$ to 0 and assertions on the speed of convergence. In II.2 we apply the results to particular cases of (1).

II.1. General assertions

We state the conditions

(V_0) *Let* (1) *be fulfilled with* $a_n, b_n, \alpha_n \geqq 0$, $\mu_n > 0$, $\quad n = 1, 2, \ldots$

and

(V_1) *Let* (V_0) *be fulfilled,* $\{\mu_n\}$ *be bounded and* $\sum_{n=1}^{\infty} \alpha_n = +\infty$. (2)

Theorem 1. *Let* (V_1) *and one of the conditions* (i), (ii) *be satisfied,*

(i) $\sum_{n=1}^{\infty} a_n < +\infty$

(ii) $\alpha_n > 0$, $\lim \frac{a_n}{\alpha_n} = 0$ *and, if* $\{\alpha_n\}$ *is unbounded then* $\lim a_n = 0$.

Then the sequence $\{b_n\}$ *converges to* 0.

Proof. The case (i) was proved in [2]. For a special case, cf. [34].

Let (ii) be fulfilled. We define $\delta_n = \frac{a_n \mu_n}{c \alpha_n}$, $n = 1, 2, \ldots$ We have $\lim \delta_n = 0$ because of (V_1) and (ii). We conclude from (1)

$$b_{n+1} \leqq b_1 - c \sum_{i=1}^{n} \frac{\alpha_i}{\mu_i} (b_i - \delta_i).$$

First we want to prove $\liminf b_n = 0$. For a proof from the contrary we suppose $\liminf b_n > 0$. Then we can find an integer $N > 0$ with $b_n > \delta_n$ for all $n \geqq N$ because of $\lim \delta_n = 0$. Consequently we get

$$\lim_{n \to \infty} \sum_{i=N}^{n} \frac{\alpha_i}{\mu_i} (b_i - \delta_i) \leqq \frac{b_1}{c} - \sum_{i=1}^{N-1} \frac{\alpha_i}{\mu_i} (b_i - \delta_i) < +\infty.$$

Because of (2) we conclude $\liminf \frac{b_i - \delta_i}{\mu_i} = 0$. Thus we have $\liminf b_i = 0$ because $\{\mu_n\}$ is bounded and $\lim \delta_i = 0$. This is a contradiction to the assumption $\liminf b_n > 0$. Consequently we can find a subsequence $\{b_{n_k}\}$ with $\lim b_{n_k} = 0$.

Let $\varepsilon > 0$ be arbitrary. Then there is an integer k_0 with

$$b_{n(k_0)} \leqq \varepsilon \text{ and } \frac{a_n \mu_n}{c \alpha_n} \leqq \frac{\varepsilon}{2}, \; a_n \leqq \frac{\varepsilon}{2} \text{ for all } n \geqq n(k_0). \tag{3}$$

Let $n \geqq n(k_0)$ be arbitrary.

If we suppose $b_n \leqq \varepsilon$ we get:

$$\text{if } 0 \leqq b_n \leqq \frac{\varepsilon}{2} \text{ then } b_{n+1} \leqq b_n - \frac{c \alpha_n}{\mu_n} b_n + a_n \leqq b_n + a_n \leqq \varepsilon,$$

and

$$\text{if } \frac{\varepsilon}{2} < b_n \leqq \varepsilon \text{ then } b_{n+1} \leqq b_n - \frac{c\alpha_n}{\mu_n} \frac{a_n \mu_n}{c\alpha_n} + a_n = b_n \leqq \varepsilon .$$

Thus we have proved the assertion.

Remark. Between conditions (i) and (ii) there is the following connection: Let $\{a_n\}$ be decreasing. Then we can conclude from (i) $\lim a_n \cdot n = 0$. Consequently (ii) holds with $\delta_n = a_n \cdot n$ and $\alpha_n = \frac{1}{n}$.

In the sequel we are interested in conditions for the boundedness of the sequence $\{b_n\}$ from which we can derive estimates for the speed of convergence of $\{b_n\}$ to zero.

Lemma 2. *If* (V_0) *and* $\sum_{n=1}^{\infty} a_n < +\infty$ *hold the sequence* $\{b_n\}$ *is bounded and the inequalities*

$$b_n \leqq b_1 + \sum_{i=1}^{\infty} a_i \quad \text{and} \quad \sum_{i=1}^{\infty} \frac{\alpha_i}{\mu_i} b_i < +\infty$$

are fulfilled.

Proof. From (1) we conclude $b_{n+1} \leqq b_1 - \sum_{i=1}^{n} c \frac{\alpha_i}{\mu_i} b_i + \sum_{i=1}^{n} a_i$, and from this we get the assertions.

Corollary 3. *Let* (V_0) *be fulfilled. If there is a sequence* $\{d_n\}$ *with* $d_n > 0$, $\sum_{i=1}^{\infty} \frac{a_i}{d_{i+1}} < +\infty$ *and* $\mu_i(d_i - d_{i+1}) \leqq c d_i \alpha_i$ *then we get:* $\left\{\frac{b_n}{d_n}\right\}$ *is bounded,* $b_n \leqq \left(\frac{b_1}{d_1} + \sum_{i=1}^{\infty} \frac{a_i}{d_{i+1}}\right) d_n$ *and* $\sum_{i=1}^{\infty} \frac{\mu_i(d_{i+1} - d_i) + c d_i \alpha_i}{d_{i+1}\mu_i} \frac{b_i}{d_i} < +\infty$.

Proof. From (1) we get

$$\frac{b_{n+1}}{d_{n+1}} \leqq \frac{b_n}{d_{n+1}} - \frac{c\alpha_n}{\mu_n d_{n+1}} b_n + \frac{a_n}{d_{n+1}} - \frac{d_n}{d_{n+1}} \frac{b_n}{d_n} - \frac{c\alpha_n d_n}{\mu_n d_{n+1}} \frac{b_n}{d_n} + \frac{a_n}{d_{n+1}}$$

$$= \frac{b_n}{d_n} - \left[1 - \left(1 - \frac{c\alpha_n}{\mu_n}\right) \frac{d_n}{d_{n+1}}\right] \frac{b_n}{d_n} + \frac{a_n}{d_{n+1}} . \tag{4}$$

From Lemma 2 we conclude the assertions.

Lemma 4. *Let* $\{a_n\}$, $\{b_n\}$, $\{c_n\}$, $\{k_n\}$ *and* $\{l_n\}$ *be sequences of non-negative real numbers with* $k_n, l_n < 1$, *and*

$$l_n b_n + c_n \leqq b_{n+1} \leqq k_n b_n + a_n , \qquad n = 1, 2, \ldots \tag{5}$$

Then

$$\min\left\{b_1, \min_{i \leq n} \frac{c_i}{1 - l_i}\right\} \leqq b_{n+1} \leqq \max\left\{b_1, \max_{i \leq n} \frac{a_i}{1 - k_i}\right\}, \qquad n = 1, 2, \ldots$$

Proof. We define $\delta_n = \dfrac{c_n}{1 - l_n}$ and $\gamma_n = \dfrac{a_n}{1 - k_n}$. Then follows from (5)

$$l_n b_n + (1 - l_n)\,\delta_n \leqq b_{n+1} \leqq k_n b_n + (1 - k_n)\,\gamma_n$$

and consequently

$$\min\{b_n, \delta_n\} \leqq b_{n+1} \leqq \max\{b_n, \gamma_n\}.$$

The assertion follows.

Remark. If the assumptions of Lemma 4 and $\sup\limits_{i=1,2,\dots} \left\{\dfrac{a_i}{1 - k_i}\right\} < +\infty$ hold then $\{b_n\}$ s bounded.

Corollary 5. *Let $\{b_n\}$, $\{k_n\}$, $\{\delta_n\}$ and $\{\gamma_n\}$ be sequences of non-negative real numbers with $k_n < 1$ such that there is a number $\delta_0 > 0$ with*

a) $\dfrac{\delta_n}{1 - k_n} \leqq 1 - \delta_0$ *for sufficiently large integers n,*

b) $\sup \dfrac{\gamma_n}{1 - k_n} < +\infty$ *and*

c) $b_{n+1} \leqq k_n b_n + \delta_n \sqrt{b_n} + \gamma_n, \quad n = 1, 2, \dots$

Then $\{b_n\}$ is bounded.

Proof. With

$$k_n' = \begin{cases} k_n & \text{if } b_n \leqq 1 \\ k_n + \delta_n & \text{if } b_n > 1 \end{cases}$$

and

$$a_n' = \begin{cases} \delta_n + \gamma_n & \text{if } b_n \leqq 1 \\ \gamma_n & \text{if } b_n > 1 \end{cases}$$

we obtain $b_{n+1} \leqq k_n' b_n + a_n'$. In the case $b_n \leqq 1$ we get

$$k_n' = k_n < 1 \quad \text{and} \quad \frac{a_n'}{1 - k_n'} = \frac{\delta_n + \gamma_n}{1 - k_n} \leqq 1 - \delta_0 + \frac{\gamma_n}{1 - k_n}.$$

In the case $b_n > 1$ we get $k_n' = k_n + \delta_n \leqq k_n + (1 - k_n)(1 - \delta_0) < 1$ and $\dfrac{a_n'}{1 - k_n'}$ $= \dfrac{\gamma_n}{1 - k_n - \delta_n} = \dfrac{\gamma_n}{1 - k_n} \dfrac{1 - k_n}{1 - k_n - \delta_n} \leqq \dfrac{\gamma_n}{1 - k_n} \cdot \dfrac{1}{\delta_0}$. Consequently $\sup \dfrac{a_n'}{1 - k_n'} < +\infty$.
Using Lemma 4 we get the assertions.

Remarks. 1. Let the assumption c) of Corollary 5 and the assumptions $k_n = k < 1$, $n = 1, 2, \dots$, $\lim \delta_n = 0$ and $\sup\limits_{n=1,2,\dots} \gamma_n < +\infty$ be fulfilled. Then the sequence $\{b_n\}$ is bounded.

2. Let the assumption c) of Corollary 5 and the assumptions $k_n \leqq 1 + \varepsilon_n$, $\sum_{n=1}^{\infty} \varepsilon_n < +\infty$, $\sum_{n=1}^{\infty} \delta_n < +\infty$ and $\sum_{n=1}^{\infty} \gamma_n < +\infty$ hold. Then the sequence $\{b_n\}$ is bounded (cf. [2]).

Corollary 6. *Let* $\{a_n\}$, $\{b_n\}$, $\{c_n\}$, $\{d_n\}$, $\{k_n\}$ *and* $\{l_n\}$ *be sequences of non-negative real numbers with* (5), $d_n > 0$ *and*

$$k_n, l_n < \frac{d_{n+1}}{d_n}, \qquad n = 1, 2, \ldots \tag{6}$$

Then

$$\min\left\{\frac{b_1}{d_1}, \min_{i \leqq n} \frac{c_i}{d_{i+1} - l_i d_i}\right\} \leqq \frac{b_{n+1}}{d_{n+1}} \leqq \max\left\{\frac{b_1}{d_1}, \max_{i \leqq n} \frac{a_i}{d_{i+1} - k_i d_i}\right\}.$$

Proof. From (5) we get

$$\frac{l_n d_n}{d_{n+1}} \cdot \frac{b_n}{d_n} + \left(1 - \frac{l_n d_n}{d_{n+1}}\right) \frac{c_n}{d_{n+1} - l_n d_n} \leqq \frac{b_{n+1}}{d_{n+1}} \leqq \frac{k_n d_n}{d_{n+1}} \cdot \frac{b_n}{d_n} + \left(1 - \frac{k_n d_n}{d_{n+1}}\right) \frac{a_n}{d_{n+1} - k_n d_n}.$$

So we can apply Lemma 4 to the sequence $\left\{\frac{b_n}{d_n}\right\}$.

Remarks. 1. Assume in addition to the assumptions of Corollary 6

$$a = \limsup \frac{a_n}{d_{n+1} - k_n d_n} < +\infty.$$

Then

$$\limsup \frac{b_n}{d_n} \leqq \max\left\{\frac{b_1}{d_1}, a\right\}.$$

Consequently, $\{b_n\}$ has at least the same speed of convergence as $\{d_n\}$.

If, otherwise, $b = \liminf \frac{c_n}{d_{n+1} - l_n d_n} > 0$ *holds*, we get $\liminf \frac{b_n}{d_n} \geqq \min\left\{\frac{b_1}{d_1}, b\right\} > 0$, and $\{b_n\}$ converges not faster than $\{d_n\}$.

2. From (6) follows $d_{n+1} \geqq d_1 \prod_{j=1}^{n} k_j$ and

$$d_{n+1} \geqq d_1 \prod_{j=1}^{n} l_j, \qquad n = 1, 2, \ldots$$

3. In addition to the assumptions of Corollary 6 assume $k_n = k < 1$, $d_n \leqq d_0 d_{n+1}$, $n = 1, 2, \ldots$, $k \cdot d_0 < 1$ and $\sup \frac{a_n}{d_{n+1}} = a_0 < +\infty$. Then

$$\frac{b_{n+1}}{d_{n+1}} \leqq \max\left\{\frac{b_1}{d_1}, \frac{a_0}{1 - k d_0}\right\}.$$

Indeed, it holds

$$\frac{a_n}{d_{n+1} - k_n d_n} = \frac{\dfrac{a_n}{d_{n+1}}}{1 - k_n \dfrac{d_n}{d_{n+1}}} \leqq \frac{a_0}{1 - kd_0}.$$

It is easy to show that the inequalities

$$c_n + b_1 \prod_{j=1}^{n} l_j + \sum_{i=1}^{n-1} c_i \prod_{j=i+1}^{n} l_j \leqq b_{n+1} \leqq a_n + b_1 \prod_{j=1}^{n} k_j + \sum_{i=1}^{n-1} a_i \prod_{j=i+1}^{n} k_j \tag{7}$$

follow from (5).

With the aid of (7) we get better results than in Corollary 6:

Theorem 7. *Let the assumptions of Corollary* 6 *hold.* (6) *has to hold only for sufficiently arge n. In addition assume*

$$\sum_{n=1}^{\infty} \left(1 - l_n \frac{d_n}{d_{n+1}}\right) = +\infty \quad \text{and} \quad \sum_{n=1}^{\infty} \left(1 - k_n \frac{d_n}{d_{n+1}}\right) = +\infty.$$

Then we get

$$\begin{aligned} &\liminf \frac{c_n}{d_{n+1}} + \left(\liminf \frac{c_n}{d_{n+1} - l_n d_n}\right) \left(\liminf \frac{l_n d_n}{d_{n+1}}\right) \\ &\leqq \liminf \frac{b_n}{d_n} \leqq \limsup \frac{b_n}{d_n} \\ &\leqq \limsup \frac{a_n}{d_{n+1}} + \left(\limsup \frac{a_n}{d_{n+1} - k_n d_n}\right) \left(\limsup \frac{k_n d_n}{d_{n+1}}\right). \end{aligned} \tag{8}$$

Proof. We will consider only the right-hand side of the inequalities. The other side can be handled in the same way. From (7) we get

$$\frac{b_{n+1}}{d_{n+1}} \leqq \frac{a_n}{d_{n+1}} + \frac{b_1}{d_1} \prod_{j=1}^{n} k_j \frac{d_j}{d_{j+1}} + \sum_{i=1}^{n-1} \frac{a_i}{d_{i+1}} \prod_{j=i+1}^{n} k_j \frac{d_j}{d_{j+1}}.$$

We can choose the integer N so large that (6) holds for $n > N$. We get

$$\begin{aligned} \sum_{i=1}^{n-1} \frac{a_i}{d_{i+1}} \prod_{j=i+1}^{n} k_j \frac{d_j}{d_{j+1}} &= \sum_{i=1}^{n} \frac{a_i}{d_{i+1}} \prod_{j=i+1}^{n} k_j \frac{d_j}{d_{j+1}} \\ &+ \sum_{i=N+1}^{n-1} \frac{a_i}{d_{i+1}} \prod_{j=i+1}^{n} k_j \frac{d_j}{d_{j+1}} = \prod_{j=N+1}^{n} k_j \frac{d_j}{d_{j+1}} \sum_{i=1}^{N} \frac{a_i}{d_{i+1}} \prod_{j=i+1}^{N} k_j \frac{d_j}{d_{j+1}} \\ &+ \sum_{j=N+1}^{n-1} \frac{a_i}{d_{i+1} - k_i d_i} \left(1 - k_i \frac{d_i}{d_{i+1}}\right) \prod_{j=i+1}^{n} k_j \frac{d_j}{d_{j+1}} \end{aligned}$$

$$\leq \prod_{j=N+1}^{n} k_j \frac{d_j}{d_{j+1}} \sum_{i=1}^{N} \frac{a_i}{d_{i+1}} \prod_{j=k+1}^{N} k_j \frac{d_j}{d_{j+1}}$$

$$+ \sup_{i>N} \frac{a_i}{d_{i+1} - k_i d_i} \sum_{i=N+1}^{n-1} \left(\prod_{j=i+1}^{n} k_j \frac{d_j}{d_{j+1}} - \prod_{j=i}^{n} k_j \frac{d_j}{d_{j+1}} \right)$$

$$= \prod_{j=N+1}^{N} k_j \frac{d_j}{d_{j+1}} \sum_{i=1}^{N} \frac{a_i}{d_{j+1}} \prod_{j=i+1}^{N} k_j \frac{d_j}{d_{j+1}}$$

$$+ \sup_{i \geq N} \frac{a_i}{d_{i+1} - k_i d_i} \left(k_n \frac{d_n}{d_{n+1}} - \prod_{j=N+1}^{n} k_j \frac{d_j}{d_{j+1}} \right).$$

Consequently

$$\limsup_{n\to\infty} \frac{b_n}{d_n} \leq \limsup_{N\to\infty} \left[\limsup_{n\to\infty} \frac{a_n}{d_{n+1}} + \frac{b_1}{d_1} \limsup_{n\to\infty} \prod_{j=1}^{n} k_j \frac{d_j}{d_{j+1}} \right.$$

$$+ \sum_{i=1}^{N} \frac{a_i}{d_{i+1}} \prod_{j=i+1}^{N} k_j \frac{d_j}{d_{j+1}} \limsup_{n\to\infty} \prod_{i=N+1}^{n} k_j \frac{d_j}{d_{j+1}}$$

$$\left. + \sup_{i>N} \frac{a_i}{d_{i+1} - k_i d_i} \limsup_{n\to\infty} k_n \frac{d_n}{d_{n+1}} \left(1 - \lim_{n\to\infty} \prod_{j=N+1}^{n-1} k_j \frac{d_j}{d_{i+1}} \right) \right]$$

$$= \limsup_{n\to\infty} \frac{a_n}{d_{n+1}} + \left(\limsup_{n\to\infty} \frac{a_n}{d_{n+1} - k_n d_n} \right) \left(\limsup_{n\to\infty} k_n \frac{d_n}{d_{n+1}} \right).$$

This is the assertion.

Remark. Let the assumptions of Theorem 7, $k_n \leq k < 1, n = 1, 2, \ldots, \lim \frac{d_n}{d_{n+1}} = d_0$, $k \cdot d_0 < 1$ and $\lim \frac{a_n}{d_{n+1}} = a_0$ hold.

Then we get $\lim \frac{b_n}{d_n} \leq \frac{a_0}{1 - kd_0}$.

Indeed: we have

$$\limsup \frac{a_n}{d_{n+1} - k_n d_n} = \limsup \frac{\frac{a_n}{d_{n+1}}}{1 - k_n \frac{d_n}{d_{n+1}}} \leq \frac{a_0}{1 - kd_0}$$

and consequently

$$\limsup \frac{b_n}{d_n} \leq a_0 + \frac{a_0}{1 - kd_0} kd_0 = \frac{a_0}{1 - kd_0}.$$

We use now Lemma 4, Corollary 6 and Theorem 7 to get results for (1).

Corollary 8. *Let* (V_0), $\mu_n \leq k$ *and* $\alpha_n \leq \frac{k}{c}$ *hold. Then we get the assertions:*

1. *If* $\alpha_n > 0$ *and* $\sup \frac{a_n}{\alpha_n} < +\infty$ *hold then* $\{b_n\}$ *is bounded.*

2. *Let* $\{d_n\}$ *be a sequence of positive real numbers with* $\left(1 - \frac{c\alpha_n}{k}\right) \frac{d_n}{d_{n+1}} < 1$ *and*

$\sup \frac{a_n}{d_{n+1} - \left(1 - \frac{c\alpha_n}{k}\right) d_n} < +\infty$. *Then the sequence* $\frac{b_n}{d_n}$ *is bounded.*

3. *Let* $\{d_n\}$ *be a sequence of positive real numbers with* $\left(1 - \frac{c\alpha_n}{k}\right) \frac{d_n}{d_{n+1}} < 1$ *if* n *is sufficiently large and* $\sum_{n=1}^{\infty} \left(1 - \left(1 - \frac{c\alpha_n}{k}\right) \frac{d_n}{d_{n+1}}\right) = +\infty$. *Then we get*

$$\limsup_{n\to\infty} \frac{b_n}{d_n} \leqq \limsup \frac{a_n}{d_{n+1}} + \limsup \frac{a_n}{d_{n+1} - \left(1 - \frac{c\alpha_n}{k}\right) d_n} \limsup \left(1 - \frac{c\alpha_n}{k}\right) \frac{d_n}{d_{n+1}}.$$

Proof. From (1) follows

$$b_{n+1} \leqq \left(1 - \frac{c\alpha_n}{k}\right) b_n + a_n. \tag{9}$$

Assertion 1 follows from Lemma 4, assertion 2 from Corollary 6 and assertion 3 from Theorem 7.

Lemma 9. *Let be* $\{q_n\}$ *a sequence of positive numbers converging to zero,* $\{a_n\}$, $\{b_n\}$ *and* $\{L_n\}$ *sequences of non-negative numbers and* $\{p_n\}$ *a sequence of integers with* $a_n + b_n > 0$, $0 < L_n < 1$,

$$p_n \geqq \frac{\ln\left(\frac{q_n}{a_n + b_n}\right)}{\ln(L_n)}, \tag{10}$$

$k_n = L_n^{p_n}$ *and*

$$b_{n+1} \leqq k_n b_n + k_n a_n. \tag{11}$$

Then $b_{n+1} \leqq q_n$ *and* $\lim b_n = 0$.

Proof. Because of (10) and (11) we get

$$b_{n+1} \leqq k_n(b_n + a_n) = L_n^{p_n}(b_n + a_n) \leqq q_n.$$

Lemma 10. *Let* $\{a_n\}$, $\{b_n\}$, $\{k_n\}$, $\{l_n\}$, $\{o_n\}$ *and* $\{L_n\}$ *be sequences of non-negative reals and* $\{p_n\}$ *be a sequence of integers with*

$$0 < L_n < 1, a_n, l_n, o_n > 0, \quad p_n \geqq \max \left\{\frac{\ln\left(\frac{l_n o_n}{a_n}\right)}{\ln L_n}, \frac{\ln l_n}{\ln L_n}\right\},$$

$k_n = L_n^{p_n}$ *and* $b_{n+1} \leqq k_n l_n + k_n a_n$.

Then $b_{n+1} \leqq l_n b_n + l_n o_n$.

II.2. Examples

In this section we give some applications of the results of Section II.1.

Example 1. *Let be* $0 < p, q < 1$, $0 < k, l < r < 1$, $c_n = c_0 q^n$, $a_n = a_0 p^n$, $d_n = d_0 r^n$, $c_0, a_0, d_0 > 0$ *and*

$$lb_n + c_n \leqq b_{n+1} \leqq kb_n + a_n, \qquad n = 1, 2, \ldots$$

Then

$$\min\left\{\frac{b_1}{d_1}, \frac{c_0}{d_0(r-l)}\right\} \leqq \frac{b_{n+1}}{d_{n+1}} \leqq \max\left\{\frac{b_1}{d_1}, \frac{a_0}{d_0(r-k)}\right\} \text{ if } \quad p = q = r$$

and

$$\lim \frac{b_n}{d_n} = 0 \quad \text{if} \quad p < r.$$

Proof. We define $k_i = k$, $l_i = l$, $i = 1, 2, \ldots$ We get

$$\frac{a_n}{d_{n+1} - k_n d_n} = \frac{a_0 p^n}{d_0 r^{n+1} - k d_0 r^n} = \frac{a_0}{d_0(r-k)} \left(\frac{p}{r}\right)^n,$$

$$\frac{c_n}{d_{n+1} - l_n d_n} = \frac{c_0 q^n}{d_0 r^{n+1} - l d_0 r^n} = \frac{a_0}{d_0(r-1)} \left(\frac{q}{r}\right)^n,$$

$$k_n \frac{d_n}{d_{n+1}} = \frac{k}{r} < 1 \quad \text{and} \quad l_n \frac{d_n}{d_{n+1}} = \frac{l}{r} < 1.$$

Using Corollary 6 we get

$$\min\left\{\frac{b_1}{d_1}, \frac{c_0}{d_0(r-l)}\right\} \leqq \frac{b_{n+1}}{d_{n+1}} \leqq \max\left\{\frac{b_1}{d_1}, \frac{a_0}{d_0(r-k)}\right\}$$

if $p = q = r$, and using Theorem 7 we get $\lim \frac{b_n}{d_n} = 0$ if $p < r$.

Example 2. *Let be* $0 < k < 1$, $a_n = a_0 k^n$ *and*

$$b_{n+1} \leqq kb_n + a_n, \qquad n = 1, 2, \ldots$$

Then $b_{n+1} \leqq (b_1 + a_0 n)\, k^n$ *and* $\limsup \dfrac{b^n}{k^n(n-1)} \leqq \dfrac{a_0}{k}$.

Proof. First we define $d_n = k^{n-1}\big(b_1 + a_0(n-1)\big)$, $k_n = k$, $n = 1, 2, \ldots$ Then we have $k_n d_n = k^n\big(b_1 + a_0(n-1)\big) < d_{n+1}$,

$$\frac{a_n}{d_{n+1} - k_n d_n} = \frac{a_0 k^n}{k^n(b_1 + a_0 n) - k^n\big(b_1 + a_0(n-1)\big)} = 1 \quad \text{and} \quad \frac{b_1}{d_1} = 1.$$

Consequently, from Corollary 6 we get $b_{n+1} \leqq d_{n+1}$.

Now we define $d_n = k^n(n-1)$. Then we have

$$k_n \frac{d_n}{d_{n+1}} = \frac{kk^n(n-1)}{k^{n+1}n} = \frac{n-1}{n} < 1, \quad \lim k_n \frac{d_n}{d_{n+1}} = 1,$$

$$\sum_{n=1}^{\infty} \left(1 - k_n \frac{d_n}{d_{n+1}}\right) = \sum_{n=1}^{\infty} \left(1 - \frac{n-1}{n}\right) = \sum_{n=1}^{\infty} \frac{1}{n} = +\infty,$$

$$\lim \frac{a_n}{d_{n+1}} = \lim \frac{a_0 k^n}{k^{n+1}n} = 0$$

and

$$\lim \frac{a_n}{d_{n+1} - k_n d_n} = \lim \frac{a_0 k^n}{k^{n+1}n - kk^n(n-1)} = \frac{a_0}{k},$$

and the last assertion follows from Theorem 7.

Example 3. *Let be* $a_n = \frac{c_2}{n^\alpha}$, $k_n = 1 - \frac{2c}{n^p}$, $0 < p \leqq 1, \alpha > p, b_n \geqq 0$ *and* $b_{n+1} \leqq k_n b_n + a_n$. *Then*

$$\limsup b_n n^q \leqq \begin{cases} 0 & \text{if } q < \alpha - p \text{ and } 0 < p < 1 \\ & \text{or } q < \alpha - 1, \ p = 1 \text{ and } q < 2c \\ \frac{c_2}{2c} & \text{if } q = \alpha - p \text{ and } 0 < p < 1 \\ \frac{c_2}{2c - q} & \text{if } q = \alpha - 1, \ p = 1 \text{ and } \alpha - 1 < 2c. \end{cases}$$

Proof. We have

$$\lim \frac{a_n}{(n+1)^{-q}} = c_2 \lim \frac{(n+1)^q}{n^\alpha} = 0$$

because of $q \leqq \alpha - p < \alpha$. First we consider the cases $0 < p < 1$, $q = \alpha - p$ and $p = 1$, $q = \alpha - 1$, $\alpha - 1 < 2c$. Defining $d_n = n^{-q}$ we have

$$\lim \frac{a_n}{d_{n+1} - k_n d_n} = \lim \frac{c_2 n^{-\alpha}}{(n+1)^{-q} - (1 - 2cn^{-p})\, n^{-q}}$$

$$= c_2 \lim \frac{n^{q-\alpha}}{\left(\frac{n}{n+1}\right)^q - 1 + 2cn^{-p}} = c_2(q-\alpha) \lim \frac{n^{q-\alpha-1}}{qn^{q-1}(n+1)^{-q-1} - 2cpn^{-p-1}}$$

$$= c_2(q-\alpha) \lim \frac{n^{p+q-\alpha}}{qn^{p+q}(n+1)^{-p-1} - 2cp}$$

$$= c_2(q-\alpha) \lim \frac{n^{p+q-\alpha}}{qn^{p+q}(n+1)^{-p-1} - 2cp} = \begin{cases} \frac{c_2}{2c} & \text{if } p < 1 \\ \frac{c_2}{2c - q} & \text{if } p = 1. \end{cases}$$

Consequently we get $k_n \frac{d_n}{d_{n+1}} < 1$ if n is sufficiently large,

$$\sum_{n=1}^{\infty} \left(1 - k_n \frac{d_n}{d_{n+1}}\right) \geqq \text{const} \sum_{n=1}^{\infty} \frac{a_n}{d_{n+1}} = \text{const} \sum_{n=1}^{\infty} \frac{(n+1)^q}{n^\alpha} = +\infty,$$

and

$$\lim k_n \frac{d_n}{d_{n+1}} = \lim \left(1 - \frac{2c}{n^p}\right)\left(\frac{n+1}{n}\right)^q = 1.$$

Using Theorem 7 we get the assertions in the cases considered.

If in these cases $q < \alpha - p$ holds then we get

$$\limsup b_n n^q = \lim n^{q+p-\alpha} \limsup b_n n^{\alpha-p} = 0.$$

Now we consider the case $p = 1$, $q < 2c \leqq \alpha - 1$. We prove the inequality

$$0 < \lim \left(1 - k_n \frac{d_n}{d_{n+1}}\right) n < +\infty.$$

$$\lim \left(1 - k_n \frac{d_n}{d_{n+1}}\right) n = \lim \left(1 - \left(1 - \frac{2c}{n}\right)\left(\frac{n+1}{n}\right)^q\right) n$$

$$= \lim \frac{1 - \left(\frac{n+1}{n}\right)^q + 2c(n+1)^q\, n^{-q-1}}{n^{-1}}$$

$$= \lim \frac{q(n+1)^{q-1}\, n^{-q-1} + 2cq(n+1)^{q-1}\, n^{-q-1} - 2c(q+1)\,(n+1)^q\, n^{-q-2}}{-n^{-2}}$$

$$= \lim \left(-(2c+1)\, q(n+1)^{q-1} n^{-q+1} + 2c(q+1)\,(n+1)^q\, n^{-q}\right)$$

$$= 2c(q+1) - (2c+1)\, q = 2c - q.$$

Consequently we have

$$\lim \frac{a_n}{d_{n+1} - k_n d_n} = \frac{1}{2c - q} \lim \frac{a_n n}{d_{n+1}} = \frac{c_2}{2c - q} \lim \frac{(n+1)^q}{n^{\alpha-1}} = 0$$

because of $q < 2c \leqq \alpha - 1$, $k_n \frac{d_n}{d_{n+1}} < 1$ if n is sufficiently large, $\sum_{n=1}^{\infty} \left(1 - k_n \frac{d_n}{d_{n+1}}\right) = +\infty$ and $\lim k_n \frac{d_n}{d_{n+1}} = 1$. So we get the last assertion of Theorem 7.

Example 4. *Let be* $e_n = n^{-r}$, $t_n = n^{-s}$; $k_n = 1 - t_n e_n c_0$, $d_n = n^{-q}$, $a_n = c \frac{e_{n-1} - e_n}{e_n}$, $r, s \geqq 0$, $0 < r + s < 1$, $q \leqq 1 - (r + s)$ *and* $b_{n+1} \leqq k_n b_n + a_n$. *Then we have*

$$\limsup \frac{b_n}{d_n} \leqq \begin{cases} 0 & \text{if } q < 1 - (r+s) \\ \dfrac{cr}{c_0} & \text{if } q = 1 - (r+s). \end{cases}$$

Proof. We use the fact that $(1 + x)^r \leqq 1 + rx$ if $x \geqq 0$ and $r \leqq 1$. We have

$$a_n = c\,\frac{(n-1)^{-r} - n^{-r}}{n^{-r}} = c\left(\left(1 + \frac{1}{n-1}\right)^r - 1\right) \leqq c \cdot r\,\frac{1}{n-1},$$

Consequently $b_{n+1} \leqq k_n b_n + \dfrac{cr}{n-1}$. Similar calculations as in Example 3 give the assertion.

Example 5. *Let be* $a_n = \dfrac{c_2}{n^{2c+1}}$, $k_n = 1 - \dfrac{2c}{n}$, $b_n \geqq 0$ *and* $b_{n+1} \leqq k_n b_n + a_n$.

Then $\limsup b_n \dfrac{n^{2c}}{\ln n} \leqq c_2$.

Proof. We compute the following limits:

a) $$\lim n \ln\left(\frac{n+1}{n}\right) = \lim \frac{\ln(n+1) - \ln n}{n^{-1}} = \lim \frac{\frac{1}{n+1} - \frac{1}{n}}{-n^{-2}}$$

$$= \lim\left(n - \frac{n^2}{n+1}\right) = 1,$$

b) $$\lim \ln\left(\frac{n+1}{n}\right) = 0,$$

c) $$\lim n^2\left[\left(1 - \frac{1}{n+1}\right)^{2c} - \left(1 - \frac{2c}{n}\right)\right] = \lim \frac{\left(\frac{n}{n+1}\right)^{2c} - 1 + \frac{2c}{n}}{n^{-2}}$$

$$= c \lim \frac{1 - \left(\frac{n}{n+1}\right)^{2c+1}}{n^{-1}} = c \lim \frac{-(2c+1)\,\frac{n^{2c}}{(n+1)^{2c+2}}}{-n^{-2}} = c(2c+1),$$

d) $$\lim\left[n\left(1 - \frac{1}{n+1}\right)^{2c} - (n - 2c)\right] \ln(n+1)$$

$$= \lim n^2\left[\left(1 - \frac{1}{n+1}\right)^{2c} - \left(1 - \frac{2c}{n}\right)\right] \frac{\ln(n+1)}{n} = 0.$$

e) We use the notations $d_n = \dfrac{\ln n}{n^{2c}}$. Then we get

$$\lim \frac{a_n}{d_{n+1} - k_n d_n} = \lim \frac{\frac{c_2}{n^{2c+1}}}{\frac{\ln(n+1)}{(n+1)^{2c}} - \left(1 - \frac{2c}{n}\right)\frac{\ln n}{n^{2c}}}$$

$$= \lim \frac{c_2}{n\left(\frac{n}{n+1}\right)^{2c} \ln(n+1) - (n - 2c)\ln n}$$

$$= c_2 \frac{1}{\lim \left\{(n-2c)\ln(n+1) + \left[n\left(\frac{n}{n+1}\right)^{2c} - (n-2c)\right]\ln(n+1) - (n-2c)\ln n\right\}}$$

$$= c_2 \frac{1}{\lim (n-2c)\ln\left(\frac{n+1}{n}\right) + \lim\left[n\left(1-\frac{1}{n+1}\right)^{2c} - (n-2c)\right]\ln(n+1)} = c_2 .$$

f) $\sum_{n=1}^{\infty}\left(1 - k_n\frac{d_n}{d_{n+1}}\right) \geqq \text{const}\sum_{n=1}^{\infty}\frac{a_n}{d_{n+1}} = \text{const}\sum_{n=1}^{\infty}\frac{c_2(n+1)^{2c}}{n^{2c+1}\ln(n+1)}$

$$= \text{const}\sum_{n=1}^{\infty}\left(\frac{n+1}{n}\right)^{2c+1}\frac{1}{(n+1)\ln(n+1)} \geqq \text{const}\sum_{n=2}^{\infty}\frac{1}{n\ln n} = +\infty$$

g) $\lim k_n\frac{d_n}{d_{n+1}} = \lim\left(1-\frac{2c}{n}\right)\left(\frac{n+1}{n}\right)^{2c}\frac{\ln n}{\ln(n+1)} = 1 .$

From Corollary 6 we get the assertion.

Example 6. *Let be* $a_n = \frac{c_2}{n^\alpha}$, $k_n = 1 - \frac{2c}{n}$, $b_n \geqq 0$, $\alpha > 2c+1$ *and* $b_{n+1} \leqq k_n b_n + a_n$. *Then we have* $\limsup b_n n^{2c} < +\infty$.

Proof. We define $d_n = \frac{1}{n^{2c}}$, $\mu_n = \frac{1}{2}$, $\alpha_n = \frac{1}{n}$ and get

$$b_{n+1} \leqq b - \frac{c\alpha_n}{\mu_n} b_n + a_n .$$

We compute

$$\lim n^{2+2c}[\mu_n(d_n - d_{n+1}) - cd_n\alpha_n] = \lim n^{2+2c}\left[\frac{1}{2}\left(\frac{1}{n^{2c}} - \frac{1}{(n+1)^{2c}}\right) - c\frac{1}{n^{2c+1}}\right]$$

$$= \lim n^2\left[\frac{1}{2}\left(1 - \left(\frac{n}{n+1}\right)^{2c}\right) - c\frac{1}{n}\right] = \frac{1}{2}\lim\frac{1 - \left(\frac{n}{n+1}\right)^{2c} - \frac{2c}{n}}{n^{-2}}$$

$$= \frac{1}{2}\lim\frac{-2c\frac{n^{2c-1}}{(n+1)^{2c+1}} + \frac{2c}{n^2}}{-2n^{-3}} = \frac{c}{2}\lim\frac{\left(\frac{n}{n+1}\right)^{2c+1} - 1}{n^{-1}}$$

$$= \frac{c}{2}\lim\frac{(2c+1)\frac{n^{2c}}{(n+1)^{2c+2}}}{-n^{-2}} = -\frac{c(2c+1)}{2} < 0 .$$

Consequently we have $\mu_n(d_n - d_{n+1}) \leqq cd_n\alpha_n$ if n is sufficiently large. Otherwise we have

$$\sum_{n=1}^{\infty} \frac{a_n}{d_{n+1}} = \sum_{n=1}^{\infty} \frac{c_2(n+1)^{2c}}{n^{\alpha}} < +\infty$$

because of $\alpha - 2c > 1$. Using Corollary 3 we get the assertion.

Remark. The results of Examples 5 and 6 may also be applied to Chapter I without additional difficulties. So some of the results given there still can be complemented.

References

[1] Aubin, J. P., Estimate of the error in the approximation of optimization problems with constraints by problems without constraints. *In:* Control Theory and the Calculus of Variations. Ed. A. V. Balakrishnan, 153–173. Academic Press, N. Y. 1969.

[2] Auslender, A., M. Gourgand, A. Goillet, Résolution numérique d'inégalités variationelles. *In:* Analyse convexe et ses applications. Ed. J. P. Aubin. Lectures Notes in Economics and Mathematical Systems vol. 102, 1–40. Springer-Verlag, Berlin 1974.

[3] Бакушинский, А. Б., Поляк, Б. Т., О решении вариационных неравенств. Докл. АН СССР **219**, 1038–1041 (1974).

[4] Brezis, H., Équations et inéquations non linéaires dans les espaces vectoriels en dualité. Ann. Inst. Fourier (Grenoble) **18**, 115–175 (1968).

[5] Browder, F. E., Problèmes non-linéaires. Les Presses de l'Université de Montreal 1966.

[6] —, On the unification of the calculus of variations and the theory of monotone nonlinear operators in Banach spaces. Proc. Nat. Acad. Sci. U.S.A. **56**, 419–425 (1966).

[7] —, Nonlinear maximal monotone operators in Banach space. Math. Ann. **175**, 89–113 (1968).

[8] Browder, F. E., and P. Hess, Nonlinear mappings of monotone type in Banach spaces. J. Funct. Anal. **11**, 251–294 (1972).

[9] Browder, F. E., and W. V. Petryshyn, Construction of fixed points of nonlinear mappings in Hilbert space. J. Math. Anal. Appl. **20**, 197–228 (1967).

[10] Bruck, R. E., A strongly convergent iterative solution of $o \in U(x)$ for a maximal monotone operator U in Hilbert space. J. Math. Anal. Appl. **48**, 114–126 (1974).

[11] Gajewski, H., K. Gröger und K. Zacharias, Nichtlineare Operatorgleichungen und Operatordifferentialgleichungen. Akademie-Verlag, Berlin 1974.

[12] Halpern, B., Fixed points of nonexpanding maps. Bull. Amer. Math. Soc. **73**, 957–961 (1967).

[13] Hartmann, P. H., and Stampacchia, G., On some nonlinear elliptic differential funktional equations. Acta Math. **115**, 271–310 (1966).

[14] Kluge, R., Zur approximativen Lösung nichtlinearer Variationsungleichungen. Mber. Dt. Akad. Wiss. **12**, 120–134 (1970).

[15] —, Ein Projektions-Iterationsverfahren zur approximativen Lösung von Minimumproblemen mit Nebenbedingungen und Variationsungleichungen mit nichtlinearen Potentialoperatoren. Mber. Dt. Akad. Wiss., **12**, 721–734 (1970).

[16] —, Zur approximativen Lösung konvexer Minimumprobleme mit Nebenbedingungen und nichtlinearer Variationsungleichungen über Folgen linearer algebraischer Gleichungssysteme. Math. Nachr. **48**, 341–352 (1971).

[17] —, Dissertation B, Berlin 1970.

[18] —, Näherungsverfahren zur approximativen Lösung nichtlinearer Variationsungleichungen. Math. Nachr. **51**, 343–356 (1971).

[19] Kluge, R., Näherungsverfahren für einige nichtlineare Probleme. *In:* Theory of Nonlinear Operators, 133–146, Akademie-Verlag, Berlin 1974.
[20] —, Approximation methods for nonlinear problems with constraints in form of variational inequalities. Proc. of the Conference on "Mathematical Questions of Optimal Control" held at Zakopane, Poland, 7.–13. 1. 1974. Banach Center Publications, vol. 1, 131–138 (1976).
[21] —, Optimal control with minimum problems and variational inequalities. *In:* "Optimization Techniques. IFIP Technical Conference. Novosibirsk, July 1–7, 1974". Lecture Notes in Computer Sciences, vol. 27, 377–382. Springer-Verlag, Berlin 1975.
[22] —, On a class of iteration methods. Math. Nachr. **73**, 7–18 (1976).
[23] —, Über eine Klasse von Minimumproblemen mit Nebenbedingungen. Math. Nachr. **78**, 37–64 (1977).
[24] Kluge, R., und Bruckner, G., Iterationsverfahren bei Folgen von Abbildungen und nichtlineare Probleme mit monotonen Operatoren. Math. Nachr. **54**, 335–353 (1972).
[25] —, Iterationsverfahren für einige nichtlineare Probleme mit Nebenbedingungen. Math. Nachr. **56**, 345–369 (1973).
[26] Lions, J. L., Quelques méthodes de résolution des problèmes aux limites non linéaires. Dunod-Gauthier-Villars, Paris 1969.
[27] Langmach, H., and G. Telschow, The penalty function method and its iteration (to be published).
[28] Mosco, U., Convergence of convex sets and solutions of variational inequalities. Advances in Math. **3**, 510–585 (1969).
[29] Поляк, Б. Т., О скорости ходимости метода штрафных функций. Ж. вычисл. мат. и мат. физ. **11**, 3–11 (1971).
[30] Rockafellar, R. T., Convex functions, monotone operators and variational inequalities. *In:* Theory and Application of Monotone Operator, 35–65. "Oderisi", Gubbio 1969.
[31] —, On the maximality of sums of nonlinear monotone operators. Trans. Amer. Math. Soc. **149**, 75–93 (1970).
[32] —, On the maximal monotonicity of subdifferential mappings. Pacific J. Math. **33**, 209–216 (1970).
[33] Sibony, M., Sur l'approximation d'équations et inéquations aux derivées partielles non linéaires de type monotone. J. Math. Anal. Appl. **34**, 502–564 (1971).
[34] Telschow, G., Weak and strong convergence of approximation-iteration penalty methods (to be published).

Address of the authors:

Akademie der Wissenschaften der DDR
Zentralinstitut für Mathematik und Mechanik
108 Berlin
Mohrenstr. 39
DDR

Some properties of nonlinear operators. II

Josef Kolomý, Praha

This paper is a continuation of [13]. In the first part of this communication we shall give an extension of some results of [14, 15] concerning the invariance of domain theorems, while the second section deals with the differentiability and continuity properties of some classes of nonlinear operators. The results of this section are due partly to my student Ho Duc Viet and the author.

Let X, Y be normed linear spaces, X^*, Y^* their dual spaces, $\langle , \rangle$ a pairing between X^*, X. We use the symbols "$\rightarrow$", "$\xrightarrow{w}$" to denote strong and weak convergence. For a set $V \subset X$ ∂V denotes the boundary of V in X, $S_1 = \{u \in X: \|u\| = 1\}$, $S_1^* = \{u^* \in X^*: \|u^*\| = 1\}$. We recall some known notions, compare Cudia [6] and Golomb and Tapia [10]. We shall say that:

(i) X is *weakly uniformly convex* (WUC) if for each $\varepsilon > 0$ and $u^* \in X^*$, $\|u^*\| = 1$ there is $\delta(\varepsilon, u^*) > 0$ so that if $u, v \in X$, $u, v \in S_1$ and $1 - \left|\left\langle \frac{u+v}{2}, u^* \right\rangle\right| < \delta$, then $\|u - v\| < \varepsilon$;

(ii) X is *weakly uniformly convex in each direction* (WUCD) if given $\varepsilon > 0$ and $u^*, v^* \in S_1^*$ there exists $\delta > 0$ such that

$$\left|\left\langle u^*, \frac{u+v}{2} \right\rangle\right| > 1 - \delta(u, v \in S_1) \Rightarrow |\langle v^*, u - v\rangle| < \varepsilon;$$

(iii) *smooth* if the norm $\|.\|$ of X is Gâteaux differentiable at every point of X away the origin. Using [16, § 26] and [21, Thm. 2] we obtain: *If X is reflexive, then X is strictly convex $\Leftrightarrow$ X^* is smooth $\Leftrightarrow$ $X^{**} = X$ is* WUCD.

Example [7]. Let X_n $(n - 1, 2, \ldots)$ be Banach spaces, $1 < p < \infty$. By $l_p(X_n)$ we mean the space of all sequences $x = (x_n)$, where $x_n \in X_n$ such that $\sum_{n=1}^{\infty} \|x_n\|_{X_n}^p < +\infty$, and $l_p(X_n)$ is provided by the norm $\|x\| = \left(\sum_{n=1}^{\infty} \|x_n\|^p\right)^{1/p}$.

Let X_n $(n = 1, 2, \ldots)$ be strictly convex and reflexive. Then $l_p(X_n)$ is reflexive and strictly convex and therefore WUCD.

By a *gauge function* $\mu: E_1^+ \rightarrow E_1^+$ we mean a real-valued strictly increasing continuous function such that $\mu(0) = 0$ and $\lim_{t \to +\infty} \mu(t) = +\infty$. A function $J: X \rightarrow 2^{X^*}$ is a *duality mapping* of X into X^* with the gauge function μ if $J(0) = 0$ and for each $x \neq 0$, $x \in X$, $J(x) = \{x^* \in X^*: \langle x^*, x\rangle = \|x^*\| \|x\|, \|x^*\| = \mu(\|x\|)\}$. This concept has been introduced independently by Beurling-Livingstone [2] and Cudia [6] and intensively studied by many authors in connection with the theory of monotone operators and the structure of certain classes of Banach spaces. Let us recall that (i) for each $x \in X$, $J(x)$ is a non-void convex bounded closed subset of X; (ii) X is a smooth Banach space $\Leftrightarrow$ J is single valued $\Leftrightarrow$ X^* is WUCD (Tapia [21]); (iii) if X is reflexive and smooth, then J is demicontinuous and onto (Browder [4]). (iv) If X^* is a uniformly convex Banach space, then J is uniformly continuous on each bounded subset of X (Kato [12]); (v) J is a duality mapping if and only if $J(x)$ is a subgradient at x ($x \in X$) of the convex function $\Phi(\|x\|) = \int_0^{\|x\|} \mu(t)\, dt$, where μ is the gauge function corresponding to J (Asplund [1]).

Let J^* denote the duality map between X^*, X^{**}, τ a canonical (natural) mapping between X, X^{**}.

Definition (Golomb, Tapia [10]). Let X be a normed linear space, $f: X \to E_1$ a Fréchet-differentiable functional on X. A mapping $\nabla f: X \to 2^X$ defined by

$$x \to \nabla f(x) = \tau^{-1} J^* f'(x), \qquad x \in X$$

is called a *metric gradient* of f.

For each $x \in X$ (fixed) $\nabla f(x)$ a is closed convex bounded set in X. If X is reflexive, then $\nabla f(x) \neq \emptyset$, and if X is strictly convex then $\nabla f(x)$ is at most one point.

A mapping $F: X \to Y$ is said to be:

(a) *demicontinuous at* $u_0 \in X$ if for each $u_n \to u_0$, $F(u_n) \xrightarrow{w} F(u_0)$;

(b) *weakly continuous* if $u_n \xrightarrow{w} u_0 \Rightarrow F(u_n) \xrightarrow{w} F(u_0)$;

(c) *strongly continuous* if $u_n \xrightarrow{w} u_0 \Rightarrow F(u_n) \to F(u_0)$.

1. Invariance of domain theorems

Theorem 1. *Let X, Y be Banach spaces, X reflexive, $\emptyset \neq U \subset X$ an open subset, $F: U \to Y$ a weakly continuous mapping having the Gâteaux derivative on U. Assume that for each $u_0 \in U$ there exists a bounded convex open neighborhood $V(u_0)$ of u_0 such that $\overline{V(u_0)} \subset U$ and that*

(*) $a_{u_0} = \inf\{\|F(u) - F(u_0)\| : u \in \partial V(u_0)\} > 0$.

Suppose that one of the two following equivalent hypotheses is satisfied:

(a) *the nullspace of $[F'(u)]^*: Y^* \to X^*$ is trivial for each $u \in U$,*

(b) *$\overline{F'(u)X} = Y$ for each $u \in U$.*

Then $F(U)$ is open in Y.

Remark. Let us point out that the conditions on X and F can be weakened. The condition (*) one can replace by the following one:

For each $u_0 \in U$ there exist $r_{u_0} > 0$ and a function $\psi_{u_0}: E_1^+ \to E_1^+$ having the property that if $r > 0$ then $\psi_{u_0}(r) > 0$ and such that $Kr_{u_0}(u_0) = \{u \in X: \|u - u_0\| \leqq r_{u_0}\} \subset U$ and $u \in \partial Kr_{u_0}(u_0) \Rightarrow \|F(u) - F(u_0)\| \geqq \psi_{u_0}(\|u - u_0\|)$.

Corollary 1. *Let X, Y be Banach spaces, X reflexive, $\emptyset \neq U \subset X$ an open set, $F: U \to Y$ a weakly continuous mapping having the Gâteaux derivative on U. Assume that $\|F(u) - F(v)\| \geqq \psi(\|u - v\|)$ for each $u, v \in U$ and some function $\psi: E_1^+ \to E_1^+$ which is such that $r > 0 \Rightarrow \psi(r) > 0$. If* Ker $[F'(u)]^* = (0)$ *for each $u \in U$, then $F(U)$ is open.*

Remark. Under the assumptions of Corollary 1 assume that $U = X$ and that either (i) $\psi(r) \to 0_+ \Rightarrow r \to 0_+$, Y is complete, F is continuous, or (ii) $\psi(r) \to +\infty$ as $r \to +\infty$. Then $F(X) = Y$.

The following two results are also consequences of Theorem 1.

Theorem 2. *Let X, Y be Banach spaces, X reflexive, $\emptyset \neq U \subset X$ an open subset of X, $F: U \to Y$ a weakly continuous mapping having the Fréchet derivative $F'(u)$ on U. If $F'(u) \in$* Isom (X, Y) *for each $u \in U$, then $F(U)$ is open.*

Theorem 3. *Let X be a reflexive Banach space, $\emptyset \neq U \subset X$ an open set, $\Phi: X \to X$ a strongly continuous mapping having the Fréchet derivative $\Phi'(u)$ on X. If $F'(u) = \mathrm{id} - \Phi'(u)$ is $1 - 1$ for each $u \in U$, then $F(U)$ is open.*

Let us remark that the classical inverse theorems heavily depend on the continuity the of Gâteaux (or Fréchet) derivative. A number of general invariance of domain theorems have been obtained by W. V. Petryshyn [18], we refer also the reader to this paper for an extensive literature in these topics.

The proof of the above and other theorems will be given in a common paper by J. Daneš and the author under the title: Fixed point, surjectivity and invariance of domain theorems.

2. Gradients and subgradients of some functions

We start with the following extension of [24].

Let $B_\alpha(M)$ denote an α-neighborhood of M ($\alpha > 0$).

Theorem 4. *Let X, Y be real Banach spaces, X separable and reflexive, $M \subset X$ a bounded closed convex set, $F: B_\alpha(M) \to Y$ a weakly continuous mapping which is Fréchet-differentiable on M.*

Then the set $Z(M)$ of all points at which a mapping $u \to F'(u)$ is strongly continuous is weakly dense in M and of G_δ-type in the relative weak topology of M.

The following result extends Lahiri's result [17] concerning real functions in E_1: Let X, Y be real Banach spaces, M an open subset of X, $N \subset M$ a closed subset of M, $F: M \to Y$ a Fréchet differentiable mapping. Assume that the mapping $u \to F'(u)$ is locally bounded on M. Then the set of all points from N at which $u \to F'(u)$ is continuous is a G_δ-set dense in N.

Theorem 5. *Let X be a real strictly convex Banach space. Then X is reflexive $\Leftrightarrow J^*$ is single-valued and demicontinuous.*

Let us recall that the first part of the theorem follows at once from Browder's result [4], while the second one depends on the James characterization of reflexivity [11], Bishops-Phelps theorem [3] and the results of de Prima and Petryshyn [20].

Using the properties of duality mapping Petryshyn [18] has characterized strictly convex Banach spaces (see also [20] and Torrance [22]) and H-spaces, while uniformly convex B-spaces have been characterized by Bynum [5]. J. R. Giles [9] has proved that a Banach space X is reflexive if it has a support mapping which is a homeomorphism.

Using the geometry of Banach spaces [6], the properties of duality maps and the result of Trojanski [23], we have the following extension of the results of Giles [8, 9].

Theorem 6. *Let X be a WUC [or reflexive (H)] Banach space, $f: X \to E_1$ a convex Gâteaux differentiable weakly continuous functional on X.*

Then the metric gradient ∇f is strongly continuous on X.

Let us remark that demicontinuity (continuity) of ∇f in WUCD (WUC) spaces have been investigated in [10].

References

[1] Asplund, E., Positivity of duality mappings. Bull. Amer. Math. Soc. 73, 200—203 (1967).

[2] Beurling, A., A. E. Livingstone, A theorem on duality mappings in Banach spaces. Ark. Mat. 4, 405—411 (1962).

[3] Bishop, E., R. R. Phelps, A proof that every Banach space is subreflexive. Bull. Amer. Math. Soc. 67, 97—98 (1961).

[4] Browder, F. E., Multivalued monotone nonlinear mappings and duality mappings in Banach spaces. Trans. Amer. Math. Soc. 118, 338—351 (1965).

[5] Bynum, W. L., Characterization of uniform convexity. Pacific J. Math. 38, 577—581 (1971).

[6] Cudia, D. F., The geometry of Banach spaces. Smoothness. Trans. Amer. Math. Soc. 110, 284—314 (1964).

[7] Day, M. M., Reflexive Banach spaces not isomorphic to uniformly convex spaces. Bull. Amer. Math. Soc. 47, 313—317 (1941).

[8] Giles, J. R., On a characterization of differentiability of the norm of a normed linear space. J. Austral. Math. Soc. 12, 106—114 (1971).

[9] Giles, J. R., A Banach space with support homeomorphism is reflexive. Bull. Austral. Math. Soc. 5, 187—189 (1971).

[10] Golomb, M., R. A. Tapia, The metric gradient in normed linear spaces. Numer. Math. 20 115—124 (1972).

[11] James, R. C., Reflexivity and the supremum of linear functionals. Ann. of Math. 66, 159 to 169 (1957).

[12] Kato, T., Nonlinear semigroups and evaluation equation. J. Math. Soc. Japan 19, 508—519 (1967).

[13] Kolomý, J., Some properties of monotone (potential) operators. Beiträge zur Analysis 8, 77—84 (1976).

[14] Kolomý, J., Open mappings and solvability of nonlinear equations. Communicated on the summer-school on "Nonlinear functional analysis and mechanics", St. Lesná, 23—27 September, 1974, Czechoslovakia.

[15] Kolomý, J., Invariance of domain theorem for weakly continuous mapping. Math. Nachr. 58, 137—144 (1973).

[16] Köthe, G., Topologische lineare Räume I. Springer-Verlag, Berlin 1960.

[17] Lahiri, B. K., On nonuniform differentiability. Amer. Math. Monthly 67, 649—652 (1960).

[18] Petryshyn, W. V., Invariance of domain theorem for locally A-proper mappings and its application. J. Funct. Anal. 5, 137—159 (1970).

[19] Petryshyn, W. V., A characterization of strict convexity of Banach spaces and other uses of duality mappings. J. Funct. Anal. 6, 282—291 (1970).

[20] Prima, C. R. de, W. V. Petryshyn, Remarks on strict monotonicity and surjectivity properties of duality mappings defined on real normed linear spaces. Math. Z. **123**, 49–55 (1971).
[21] Tapia, R. A., A characterization of inner product spaces. Proc. Amer. Math. Soc. **41**, 569 to 574 (1973).
[22] Torrance, E., Strictly convex spaces via semi-inner-product orthogonality. Proc. Amer. Math. Soc. **26**, 108–110 (1970).
[23] Trojanski, S., On locally uniformly convex and differentiable norms in certain nonseparable Banach spaces. Studia Math. **37**, 173–180 (1971).
[24] Ho Duc Viet, Strong continuity of gradient mappings (to appear).

Address of the author:

Mathematical Institute of Charles University
Sokolovská 83
18600 Praha — Karlin
Czechoslovakia

Konvexe Steuerungsprobleme

Arno Langenbach, Berlin

Abstract. An optimal control problem in the shape of the restricted minimum problem (1.1), (1.2) is considered. Under suitable conditions it may be replaced by an equivalent convex or strongly convex minimum problem.

I. In dieser Arbeit betrachten wir das Minimum-Problem

$$f(\hat{x}, \hat{g}, \hat{u}) = \inf_{(x,g,u) \in \mathfrak{X} \times \mathfrak{G} \times \mathfrak{U}} f(x, g, u), \tag{1.1}$$

in welchem $\mathfrak{X}$, $\mathfrak{G} \subseteqq \mathfrak{B}$, $\mathfrak{U} \subseteqq \mathfrak{B}'$ und $\mathfrak{B}$, $\mathfrak{B}'$ Banachräume sind. Über die Elemente des „Steuerbereiches" $\mathfrak{G} \times \mathfrak{U}$ steuern wir die Elemente des „Phasenbereiches" $\mathfrak{X}$ mit Hilfe der Steuergleichung

$$x = g + K(x, u), \qquad K \in (\mathfrak{X} \times \mathfrak{U} \to \mathfrak{B}). \tag{1.2}$$

Das Zielfunktional $f \in (\mathfrak{B} \times \mathfrak{B} \times \mathfrak{B}' \to \mathfrak{R})$ in (1.1) sei konvex. Durch die Nebenbedingung (1.2) wird diese Konvexität jedoch gestört. Wir stellen uns die Aufgabe, ein zu (1.1), (1.2) äquivalentes konvexes Minimum-Problem anzugeben.

Speziell untersuchen wir die Steuerung durch Urysohn-Operatoren.

Für eine Menge $\Omega \subseteqq \mathfrak{R}^m$ endlichen Maßes wählen wir $\mathfrak{B} = L_p(\Omega)$, $\mathfrak{B}' = L_{p'}(\Omega)$, die Räume der zur Potenz $p > 1$ bzw. $p' > 1$ Lebesgue-integrierbaren Funktionen. Die Funktion $\varphi \in (\Omega \times \Omega \times \mathfrak{R} \times \mathfrak{R} \to \mathfrak{R})$ heißt *Carathéodory-Funktion,* wenn $\varphi(t, s, \bar{x}, \bar{u})$ für fixierte $\bar{x}, \bar{u}$ meßbar in $\Omega \times \Omega$ und $\varphi(\bar{t}, \bar{s}, x, u)$ für fast alle $(\bar{t}, \bar{s}) \in \Omega \times \Omega$ in $\mathfrak{R} \times \mathfrak{R}$ stetig ist. Mit einer geeigneten Carathéodory-Funktion als Kern erklären wir den *Urysohn-Operator*

$$K_0(x, u)\,(t) = \int_\Omega \varphi\big(t, s, x(s), u(s)\big)\, ds, \qquad t \in \Omega, \tag{1.3}$$

auf $L_p(\Omega) \times L_{p'}(\Omega)$, wobei der Kern φ so beschaffen sein soll, daß $K_0 \in \big(L_p(\Omega) \times L_{p'}(\Omega) \to L_p(\Omega)\big)$.

Ist der Operator $K \in (\mathfrak{X} \times \mathfrak{U} \to \mathfrak{B})$ bezüglich x, u linear, so erhalten wir aus der Gleichung (1.2) eine von der Steuerung (g, u) linear abhängige Phase $x[g, u]$ und ein konvexes Zielfunktional

$$F(g, u) = f(x[g, u], g, u). \tag{1.4}$$

Denn es gilt

$$\begin{aligned}
&F(\alpha g + (1-\alpha) h, \alpha u + (1-\alpha) v) \\
&= f(x[\alpha g + (1-\alpha) h, \alpha u + (1-\alpha) v], \alpha g + (1-\alpha) h, \alpha u + (1-\alpha) v) \\
&= f(\alpha x[g, u] + (1-\alpha) x[h, v], \alpha g + (1-\alpha) h, \alpha u + (1-\alpha) v) \\
&\leqq \alpha F(g, u) + (1-\alpha) F(h, v), \qquad \alpha \in [0, 1],
\end{aligned}$$

auf Grund der Konvexität von f.

Ist die Steuergleichung (1.2) nichtlinear, so ist das Zielfunktional (1.4) im allgemeinen nicht konvex, wodurch sich die konstruktive Lösung des Minimum-Problems (1.1), (1.2) schwierig gestalten kann. Unter Verwendung der in der konvexen Optimierung üblichen Halbordnungsstrukturen kann in einigen Fällen ein äquivalentes konvexes Minimum-Problem angegeben werden. Am Beispiel der Steuerung mit Volterra-Operatoren wurde dies in [1] gezeigt. Die dort entwickelte Methodik erfordert die eindeutige Auflösbarkeit der Steuergleichung für alle zulässigen Steuerungen aus einem recht allgemein gewählten Steuerungsbereich. Durch die Einschränkung des Minimum-Problems (1.1) auf eine Teilmenge des Graphen der Steuergleichung (1.2) können Schwierigkeiten mit der Auflösbarkeit der Steuergleichung teilweise umgangen werden.

Wir wollen in dieser Arbeit untersuchen, wie Minimum-Probleme auf Graphen zu konvexen Minimum-Problemen erweitert werden können.

Im folgenden sei der Banachraum $\mathfrak{B}$ halbgeordnet. Speziell den Raum $L_p(\Omega)$ versehen wir mit der Halbordnungsstruktur

$$u \prec v \Leftrightarrow u(x) \leqq v(x) \text{ für fast alle } x \in \Omega. \tag{1.5}$$

Als *Graphen* der Steuergleichung (1.2) bezeichnen wir die Menge

$$G = \{(x, g, u) \in \mathfrak{X} \times \mathfrak{B} \times \mathfrak{U};\ x = g + K(x, u)\}. \tag{1.6}$$

Definition 1.1. Ein Operator $K \in (\mathfrak{X} \times \mathfrak{U} \to \mathfrak{B})$ heiße *konvex* (*konkav*), wenn $\mathfrak{X} \times \mathfrak{U} \subseteqq \mathfrak{B} \times \mathfrak{B}'$ konvex und

$$K(\alpha x + (1-\alpha) y, \alpha u + (1-\alpha) v) \prec (\succ)\ \alpha K(x, u) + (1-\alpha) K(y, v). \tag{1.7}$$

Wir betrachten dann die Obermenge

$$\hat{G} = \{(x, g, u) \in \mathfrak{X} \times \mathfrak{B} \times \mathfrak{U};\quad \exists\, g' \in \mathfrak{B},\ g' \prec g,\ (x, g', u) \in G\}.$$

Lemma 1.1. *Die Mengen $\mathfrak{X}$, $\mathfrak{U}$ seien konvex; überdies sei der Operator $-K$ in (1.2) konvex. Dann ist die Menge $\hat{G}$ konvex.*

Beweis. Für (x, g, u), $(y, h, v) \in \hat{G}$ und $\alpha \in [0, 1]$ gilt

$$\begin{aligned}
&\alpha(x, g, u) + (1-\alpha)(y, h, v) \\
&= (\alpha x + (1-\alpha) y, \alpha g + (1-\alpha) h, \alpha u + (1-\alpha) v) \in \mathfrak{X} \times \mathfrak{B} \times \mathfrak{U}.
\end{aligned}$$

Überdies finden wir $g', h' \in \mathfrak{B}$, $g' < g$, $h' < h$ derart, daß

$$\begin{aligned} \alpha g' + (1-\alpha) h' &= \alpha x + (1-\alpha) y - \alpha K(x, u) - (1-\alpha) K(y, v) \\ &> \alpha x + (1-\alpha) y - K(\alpha x + (1-\alpha) y, \alpha u + (1-\alpha) v). \end{aligned} \tag{1.8}$$

Bezeichnen wir daher die rechte Seite in (1.8) mit k_α, so finden wir $k_\alpha < \alpha g + (1-\alpha) h$ und $(\alpha x + (1-\alpha) y, k_\alpha, \alpha u + (1-\alpha) v) \in G$, also $(\alpha x + (1-\alpha) y, \alpha g + (1-\alpha) h, \alpha u + (1-\alpha) v) \in \hat{G}$, q.e.d.

Satz 1.1. *Auf der abgeschlossenen konvexen Menge $\mathfrak{X} \times \mathfrak{B} \times \mathfrak{U}$ sei das Zielfunktional f stark konvex:*

$$f\left(\frac{x+y}{2}, \frac{g+h}{2}, \frac{u+v}{2}\right) + \gamma(\|x-y\|^2 + \|g-h\|^2 + \|u-v\|^2)$$
$$\leqq \frac{1}{2} f(x, g, u) + \frac{1}{2} f(y, h, v) \quad \text{für ein} \quad \gamma > 0,$$

unterhalbbeschränkt und unterhalbstetig. Überdies sei f monoton wachsend in $g \in \mathfrak{B}$:

$$f(x, g, u) \geqq f(x, g', u), \quad \text{falls} \quad x \in \mathfrak{X}, g, g' \in \mathfrak{B}, \; u \in \mathfrak{U}, \; g > g'.$$

Der Operator $K \in (\mathfrak{X} \times \mathfrak{U} \to \mathfrak{B})$ sei abgeschlossen und konkav. Dann besitzt das Minimum-Problem (1.1), (1.2) *für $\mathfrak{G} = \mathfrak{B}$ eine eindeutig bestimmte Lösung $(\hat{x}, \hat{g}, \hat{u})$; jede Minimalfolge $\{(x_n, g_n, u_n)\}$ von f auf $\hat{G}$ konvergiert stark gegen diese Lösung.*

Beweis. Betrachten wir eine beliebige Minimalfolge $\{(x_n, g_n, u_n)\}$ von f auf G. Da f monoton in $g \in \mathfrak{B}$ ist, gibt es eine Minimalfolge $\{(x_n, g_n', u_n)\} \subseteqq G \subseteqq \hat{G}$. Diese konvergiert wegen der starken Konvexität von f gegen ein Element $(\hat{x}, \hat{g}, \hat{u}) \in G$, da G überdies abgeschlossen ist, weil der Operator K abgeschlossen ist. Da jede Mischfolge der Minimalfolgen $\{(x_n, g_n, u_n)\}$ und $\{(x_n, g_n', u_n)\}$ wieder Minimalfolge in $\hat{G}$ ist und gleichzeitig Fundamentalfolge sein muß, konvergiert auch die Folge $\{(x_n, g_n, u_n)\} \subseteqq \hat{G}$ gegen $(\hat{x}, \hat{g}, \hat{u})$. Wegen der Unterhalbstetigkeit von f ist $(\hat{x}, \hat{g}, \hat{u})$ Lösung des Minimum-Problems für f auf G und auf $\hat{G}$. Ist $(\tilde{x}, \tilde{g}, \tilde{u})$ irgendeine andere Lösung des Minimum-Problems (1.1), (1.2), so bilden wir eine Mischfolge aus der Minimalfolge $\{(x_n, g_n, u_n)\}$ und der stationären Folge $\{(\tilde{x}, \tilde{g}, \tilde{u})\}$. Diese ist wiederum Minimalfolge für f. Es gilt daher $\tilde{x} = \hat{x}$, $\tilde{g} = \hat{g}$, $\tilde{u} = \hat{u}$, q.e.d.

Eine abweichende Methode kann angewandt werden, wenn der Operator K stetig ist. Wir können aus der Gleichung (1.2) nämlich $g = g[x, u]$ berechnen und über $f(x, g[x, u], u) = \Phi(x, u)$ auf $\mathfrak{X} \times \mathfrak{U}$ minimisieren.

Lemma 1.2 *$f \in (\mathfrak{X} \times \mathfrak{B} \times \mathfrak{U} \to \mathfrak{R})$ sei konvex, $f(x, g, u)$ monoton wachsend in g auf $\mathfrak{B}$. Ist der Operator $g[x, u]$, $(x, u) \in \mathfrak{X} \times \mathfrak{U}$, konvex, so ist auch $\Phi(x, u)$ konvex für $(x, u) \in \mathfrak{X} \times \mathfrak{U}$. Ist f überdies stark konvex in x und u:*

$$f\left(\frac{x+y}{2}, \frac{g+h}{2}, \frac{u+v}{2}\right) + \gamma(\|x-y\|^2 + \|u-v\|^2)$$
$$\leqq \frac{1}{2} f(x, g, u) + \frac{1}{2} f(y, h, v) \quad \text{für ein} \quad \gamma > 0,$$

so ist auch Φ stark konvex.

Beweis. Für $(x, u), (y, v) \in \mathfrak{X} \times \mathfrak{U}$ und $\alpha \in [0, 1]$ gilt

$$\begin{aligned}
&\Phi\big(\alpha x + (1-\alpha) y, \alpha u + (1-\alpha) v\big) \\
&= f\big(\alpha x + (1-\alpha) y, g[\alpha x + (1-\alpha) y, \alpha u + (1-\alpha) v], \alpha u + (1-\alpha) v\big) \\
&\leqq f\big(\alpha x + (1-\alpha) y, \alpha g[x, u] + (1-\alpha) g[y, v], \alpha u + (1-\alpha) v\big) \\
&\leqq \alpha f(x, g[x, u], u) + (1-\alpha) f\big(y, g[y, v], v\big) \\
&= \alpha \Phi(x, u) + (1-\alpha) \Phi(y, v).
\end{aligned}$$

Im Falle, daß f in x und u stark konvex ist, erhalten wir in der gleichen Weise

$$\begin{aligned}
&\Phi\left(\frac{x+y}{2}, \frac{u+v}{2}\right) \leqq f\left(\frac{x+y}{2}, \frac{g[x, u] + g[y, v]}{2}, \frac{u+v}{2}\right) \\
&\leqq \frac{1}{2} f(x, g[x, u], u) + \frac{1}{2} f(y, g[y, v], v) - \gamma(\|x-y\|^2 + \|u-v\|^2) \\
&= \frac{1}{2} \Phi(x, u) + \frac{1}{2} \Phi(y, v) - \gamma(\|x-y\|^2 + \|u-v\|^2), \text{ q.e.d}
\end{aligned}$$

Satz 1.2. *Für konvexe und abgeschlossene Bereiche* $\mathfrak{X} \subseteqq \mathfrak{B}$ *und* $\mathfrak{U} \subseteqq \mathfrak{B}'$ *genüge das Zielfunktional* $f \in (\mathfrak{X} \times \mathfrak{B} \times \mathfrak{U} \to \mathfrak{R})$ *allen Bedingungen von Lemma* 1.2, *sei überdies unterhalbbeschränkt und unterhalbstetig. Der Operator* $-K \in (\mathfrak{X} \times \mathfrak{U} \to \mathfrak{B})$ *in* (1.2) *sei stetig und konvex. Dann besitzt das Minimum-Problem* (1.1), (1.2) *für* $\mathfrak{G} = \mathfrak{B}$ *genau eine Lösung* $(\hat{x}, \hat{g}, \hat{u})$, *gegen die jede Minimalfolge des Funktionals* f *konvergiert.*

Beweis. Für $(x, u), (y, v) \in \mathfrak{X} \times \mathfrak{U}$ erhalten wir

$$\begin{aligned}
&g(\alpha x + (1-\alpha) y, \alpha u + (1-\alpha) v] = \alpha x + (1-\alpha) y \\
&- K\big(\alpha x + (1-\alpha) y, \alpha u + (1-\alpha) v\big) < \alpha\big(x - K(x, u)\big) \\
&+ (1-\alpha)\big(y - K(y, v)\big) = \alpha g[x, u] + (1-\alpha) g[y, v].
\end{aligned}$$

Der Operator $g[x, u]$, $(x, u) \in \mathfrak{X} \times \mathfrak{U}$ ist damit konvex. Lemma 1.2 gemäß ist das Funktional $\Phi(x, u)$, $(x, u) \in \mathfrak{X} \times \mathfrak{U}$ stark konvex. Ist nun $\{(x_n, g_n, u_n)\} \subseteqq \mathfrak{X} \times \mathfrak{B} \times \mathfrak{U}$ Minimalfolge von f mit $x_n = g_n + K(x_n, u_n)$, so ist $\{(x_n, u_n)\}$ eine Minimalfolge des Funktionals $\Phi(x, u)$. Es gibt daher ein Element $(\hat{x}, \hat{u}) \in \mathfrak{X} \times \mathfrak{U}$ derart, daß $(x_n, u_n) \to (\hat{x}, \hat{u})$. Der Operator K ist stetig; mit $\hat{g} = \hat{x} - K(\hat{x}, \hat{u})$ gilt dann $g_n \to \hat{g}$. Schließlich ist f unterhalbstetig und folglich $f(\hat{x}, \hat{g}, \hat{u}) = \inf\limits_{(x,g,u) \in G} f(x, g, u)$. Ist $(\tilde{x}, \tilde{g}, \tilde{u})$ irgendeine andere Lösung des Minimum-Problems (1.1), (1.2), so ist $\tilde{g} = \tilde{x} - K(\tilde{x}, \tilde{u})$. Bilden wir eine Mischfolge aus der stationären Folge $\{(\tilde{x}, \tilde{u})\}$ und der Folge $\{[x_n, u_n)\}$, so ist diese wiederum Minimalfolge für Φ. Es gilt dann $\tilde{x} = \hat{x}$ und $\tilde{u} = \hat{u}$, folglich auch $\tilde{g} = \hat{g}$, q.e.d.

Das letztere Verfahren zur Überführung in ein konvexes Minimum-Problem ist auch dann anwendbar, wenn das Zielfunktional (1.1) lediglich konvex ist. Wir erhalten nach Lemma 1.2 ein konvexes Minimum-Problem für $\Phi(x, u)$ auf $\mathfrak{X} \times \mathfrak{U}$. Lösungskriterien für dieses Minimum-Problem findet man beispielsweise in [2, 3].

Ergebnisse über stetige Urysohn-Operatoren $K_0 \in \big(L_p(\Omega) \times L_{p'}(\Omega) \to L_p(\Omega)\big)$ findet man bei M. A. Krasnoselski [4] und M. M. Wajnberg [5].

Satz 1.3. *Der Kern* $\varphi \in (\Omega \times \Omega \times \mathfrak{R} \times \mathfrak{R} \to \mathfrak{R})$ *in* (1.3) *sei eine Carathéodory-Funktion; überdies gelte*

$$|\varphi(t, s, x, u)| \leqq \alpha(t, s) + \beta\,|x| + \gamma\,|u|^{p'/p} \tag{1.9}$$

mit einer Funktion $\alpha \in L_p(\Omega \times \Omega)$ *und geeigneten Konstanten* $\beta, \gamma > 0$. *Dann ist der Operator* $K_0 \in \big(L_p(\Omega) \times L_{p'}(\Omega) \to L_p(\Omega)\big)$ *in* (1.3) *stetig.*

Beweis. Betrachten wir zunächst den Superpositionsoperator

$$\varphi \in \big(L_p(\Omega \times \Omega) \times L_{p'}(\Omega \times \Omega) \to L_p(\Omega \times \Omega)\big),$$

$\varphi(x, u)\,(t, s) = \varphi\big(t, s, x(t, s), u(t, s)\big)$ fast überall in $\Omega \times \Omega$. Für fixierte $x \in L_p(\Omega \times \Omega)$, $u \in L_{p'}/(\Omega \times \Omega)$ ist die Funktion $\varphi(x, u)\,(t, s)$ meßbar in $\Omega \times \Omega$. Ferner gilt auf Grund von (1.9) $|\varphi(x, u)\,(t, s)|^p \leqq |\alpha(t, s) + \beta|x(t, s)| + \gamma|u(t, s)|^{p'/p}|^p \leqq 3^p\{|\alpha(t, s)|^p + \beta^p \times |x(t, s)|^p + \gamma^p|u(t, s)|^{p'}\}$[1], wobei die rechte Seite in dieser Abschätzung integrierbar ist. φ bildet tatsächlich $L_p(\Omega \times \Omega) \times L_{p'}(\Omega \times \Omega)$ in $L_p(\Omega \times \Omega)$ ab und ist nach einem allgemeinen Resultat von M. M. Wajnberg [5] stetig. Betrachten wir nun Funktionen $x \in L_p(\Omega)$ als Elemente von $L_p(\Omega \times \Omega)$, so ist

$$\|x\|^p_{L_p(\Omega \times \Omega)} = \int_\Omega \int_\Omega |x(t)|^p\,dt\,ds = \operatorname{mes}(\Omega)\,\|x\|^p_{L_p(\Omega)}.$$

Der Superpositionsoperator $\tilde{\varphi} \in \big(L_p(\Omega) \times L_{p'}(\Omega) \to L_p(\Omega \times \Omega)\big)$, $\tilde{\varphi}(x, u)\,(t, s) = \varphi(t, s, x(s), u(s))$ f. ü. in $\Omega \times \Omega$ ist daher ebenfalls stetig.

Wir führen nun den Integraloperator $J \in \big(L_p(\Omega \times \Omega) \to L_p(\Omega)\big)$, $Jv(t) = \int_\Omega v(t, s)\,ds$ ein. Nach dem Satz von Fubini existiert dieses Integral für fast alle $t \in \Omega$, wenn $v \in L_p(\Omega \times \Omega) \subseteqq L_1(\Omega \times \Omega)$. Nach der Hölderungleichung gilt

$$\left|\int_\Omega v(t, s)\,ds\right|^p \leqq \int_\Omega |v(t, s)|^p\,ds\,[\operatorname{mes}(\Omega)]^{p/q}, \qquad \frac{1}{p} + \frac{1}{q} = 1.$$

Wiederum nach dem Satz von Fubini ist die rechte Seite dieser Abschätzung integrierbar und daher

$$\|Jv\|_{L_p(\Omega)} \leqq \|v\|_{L_p(\Omega \times \Omega)} \operatorname{mes}(\Omega)^{1/q}.$$

Der Operator J ist linear und folglich stetig. Als Komposition des stetigen Superpositionsoperators $\tilde{\varphi} \in \big(L_p(\Omega) \times L_{p'}(\Omega) \to L_p(\Omega \times \Omega)\big)$ aus (1.10) und des stetigen Integraloperators $J \in \big(L_p(\Omega \times \Omega) \to L_p(\Omega)\big)$ ist $K_0 = J\tilde{\varphi} \in \big(L_p(\Omega) \times L_{p'}(\Omega) \to L_p(\Omega)\big)$ stetig, q.e.d.

Für Urysohn-Operatoren mit stetigem Kern φ kann auf Wachstumsbedingungen der Form (1.9) verzichtet werden, wenn der Phasenbereich $\mathfrak{X}$ und die Komponente $\mathfrak{U}$ des Steuerbereichs aus beschränkten Funktionen bestehen. Sei beispielsweise $\varphi \in (\bar{\Omega} \times \bar{\Omega}$

---

[1] Aus der Ungleichung $\sum_{i=1}^{n} x_i y_i \leqq \left(\sum_{i=1}^{n} x_i^p\right)^{1/p} \left(\sum_{i=1}^{n} y_i^q\right)^{1/q}$, $\frac{1}{p} + \frac{1}{q} = 1$, $x_i, y_i \geqq 0$, $i = 1, 2, \ldots, n$, erhält man mit $y_i = 1$, $i = 1, 2, \ldots, n$, die elementare Ungleichung $\left(\sum_{i=1}^{n} x_i\right)^p \leqq \sum_{i=1}^{n} x_i^p n^{p/q}$.

$\times \mathfrak{R} \times \mathfrak{R} \to \mathfrak{R})$ stetig und $x \in \mathfrak{X} \Rightarrow |x(t)| \leqq a$, $u \in \mathfrak{U} \Rightarrow |u(t)| \leqq b$ für geeignete Konstanten $a, b > 0$ und fast alle $t \in \Omega$. Konvergieren die Folgen $\{x_n\} \subseteqq L_p(\Omega)$, $\{u_n\} \subseteqq L_{p'}(\Omega)$ gegen Elemente x bzw. u in ihren Räumen, so wählen wir Teilfolgen $\{x_n'\} \subseteqq \{x_n\}$, $\{u_n'\} \subseteqq \{u_n\}$, die gegen ihre Grenzwerte fast überall in Ω konvergieren. Wir finden dann für alle $t \in \Omega$ und fast alle $s \in \Omega$

$$\left|\varphi\big(t, s, x_n'(s), u_n'(s)\big)\right| \leqq \max_{(t,s,x,u) \in \bar{\Omega} \times \bar{\Omega} \times [-a,a] \times [-b,b]} |\varphi(t, s, x, u)| = M .$$

Nach dem Satz von Lebesgue können wir unter dem Integraloperator zum Grenzwert übergehen und erhalten

$$\lim_{n \to \infty} K_0(x_n', u_n')\,(t) = K_0(x, u)\,(t) \qquad \forall t \in \Omega .$$

Da überdies $|K_0(x_n', u_n')\,(t)| \leqq M \operatorname{mes}(\Omega)$, so konvergiert die Folge $\{K_0(x_n', u_n')\}$ auch in $L_p(\Omega)$ gegen $K_0(x, u)$. Aus diesen Überlegungen folgt in üblicher Weise, daß $K_0 \in (\mathfrak{X} \times \mathfrak{U} \to L_p)$ stetig ist.

Schließlich bemerken wir noch, daß der Urysohn-Operator K_0 konvex (konkav) ist, wenn sein Kern in entsprechender Weise konvex (konkav) ist. Gilt nämlich

$$\varphi\big(t, s, \alpha x + (1 - \alpha)\, y, \alpha u + (1 - \alpha)\, v\big) \leqq (\geqq)\ \alpha\varphi(t, s, x, u)$$
$$+ (1 - \alpha)\, \varphi(t, s, y, v) \ \text{ für fast alle } (t, s) \in \Omega \times \Omega ,$$

so finden wir für fast alle $t \in \Omega$

$$\begin{aligned} & K_0\big(\alpha x + (1 - \alpha)\, y, \alpha u + (1 - \alpha)\, v\big)\,(t) \\ & = \int_\Omega \varphi\big(t, s, \alpha x(s) + (1 - \alpha)\, y(s), \alpha u(s) + (1 - \alpha)\, v(s)\big)\, ds \\ & \leqq (\geqq) \int_\Omega \{\alpha\varphi\big(t, s, x(s), u(s)\big) + (1 - \alpha)\, \varphi\big(t, s, y(s), v(s)\big)\}\, ds \\ & = \alpha K_0(x, u)\,(t) + (1 - \alpha)\, K_0(y, v)\,(t) \end{aligned}$$

oder

$$K_0\big(\alpha x + (1 - \alpha)\, y, \alpha u + (1 - \alpha)\, v\big) < (>)\ \alpha K_0(x, u) + (1 - \alpha)\, K_0(y, v) .$$

II. Steuerung durch Volterra-Operatoren

In diesem Fall ist $\Omega = [0, T] \subseteqq \mathfrak{R}$, und die Steuergleichung lautet

$$x(t) = g(t) + \int_0^t \varphi\big(t, s, x(s), u(s)\big)\, ds \qquad t \in [0, T] \tag{2.1}$$

mit dem Kern $\varphi \in (\Delta \times \mathfrak{R} \times \mathfrak{R} \to \mathfrak{R})$, wobei $\Delta = \{(t, s) \in \mathfrak{R}^2;\ 0 \leqq t \leqq T,\ 0 \leqq s \leqq t\}$ ist.

Mit dem erweiterten Kern

$$\bar{\varphi}(t, s, x, u) = \begin{cases} \varphi(t, s, x, u) & \text{falls } (t, s, x, u) \in \Delta \times \mathfrak{R} \times \mathfrak{R} \\ 0 & \text{für } (t, s, x, u) \in \Delta' \times \mathfrak{R} \times \mathfrak{R} \end{cases}$$

mit $\Delta' = \{(t, s) \in \mathfrak{R}^2;\ 0 \leqq t \leqq T,\ t < s \leqq T\}$ erhält der Integraloperator in der Gleichung (2.1) die Form eines Urysohn-Operators. Offensichtlich ist $\bar{\varphi}$ bezüglich $[0, T] \times [0, T] \times \mathfrak{R} \times \mathfrak{R}$ eine Carathéodory-Funktion, wenn φ bezüglich $\Delta \times \mathfrak{R} \times \mathfrak{R}$ eine Carathéodory-Funktion ist. Überdies ist $\bar{\varphi}(t, s, x, u)$ nur für solche $t, s \in [0, T]$ in (x, u) nicht konvex, für die $\varphi(t, s, x, u)$ in (x, u) nicht konvex ist. In Anwendung von Satz 1.1 oder 1.2 kommen wir damit zu folgendem Ergebnis:

Satz 2.1. *Das Zielfunktional* $f \in \big(L_p(0, T) \times L_p(0, T) \times L_{p'}(0, T) \to \mathfrak{R}\big)$ *sei stark konvex, unterhalbstetig und unterhalbbeschränkt auf der konvexen, abgeschlossenen Teilmenge* $\mathfrak{X} \times L_p(0, T) \times \mathfrak{U}$. *Überdies sei* $f(x, g, u)$ *monoton wachsend in* $g \in L_p(0, T)$:

$$f(x, g, u) \geqq f(x, g', u), \quad \textit{falls} \quad x \in \mathfrak{X},\ g, g' \in L_p(0, T), \quad u \in \mathfrak{U}, \quad g > g'.$$

Der Kern $\varphi \in (\Delta \times \mathfrak{R} \times \mathfrak{R} \to \mathfrak{R})$ *der Gleichung* (2.1) *sei eine Carathéodory-Funktion, genüge der Abschätzung*

$$|\varphi(t, s, x, u)| \leqq \alpha(t, s) + \beta|x| + \gamma|u|^{p'/p}$$

mit einer Funktion $\alpha \in L_p(\Omega \times \Omega)$ *und geeigneten Konstanten* $\alpha, \beta > 0$, φ *sei überdies konkav in* (x, u): $\varphi\big(t, s, \alpha x + (1 - \alpha)\, y, \alpha u + (1 - \alpha)\, v\big) \geqq \alpha\varphi(t, s, x, u) + (1 - \alpha)\, \varphi(t, s, y, v)$ *für fast alle* $(t, s) \in \Delta$ *und alle* $\alpha \in [0, 1]$, $x, y, u, v \in \mathfrak{R}$. *Dann besitzt das Minimum-Problem* (1.1) *mit der Steuergleichung* (2.1) *genau eine Lösung, gegen die jede Minimalfolge stark konvergiert.*

Bei Steuergleichungen mit Volterra-Operatoren gibt es jedoch noch eine weitere Lösungsmöglichkeit, wenn $\mathfrak{X} = L_p(0, T)$ gesetzt werden kann. Wir lösen die Gleichung (2.1) bezüglich $x = x[g, u]$ auf und setzen in das Zielfunktional $f(x, g, u)$ ein. Dann erhalten wir das Minimum-Problem

$$F(\hat{g}, \hat{u}) = f(x[\hat{g}, \hat{u}], \hat{g}, \hat{u}) = \inf_{(g,u) \in \mathfrak{G} \times \mathfrak{U}} F(g, u). \tag{2.2}$$

Der Einfachheit halber setzen wir $p = p' = 2$, $\varphi \in (\Delta \times \mathfrak{R} \times \mathfrak{R} \to \mathfrak{R})$ sei stetig und

$$\begin{aligned} &|\varphi(t, s, x, u) - \varphi(t, s, y, v)| \leqq L(|x - y| + |u - v|) \quad \text{für} \quad (t, s) \in \Delta, \\ &x, y, u, v \in \mathfrak{R}. \end{aligned} \tag{2.3}$$

Auf $L_2(0, T)$ führen wir noch die Normen

$$\|x\|_{2k}^2 = \int_0^T |x(t)|^2\, e^{-2kt}\, dt, \qquad k = 1, 2, \ldots, \tag{2.4}$$

ein. Da

$$\int_0^T |x(t)|^2\, dt \geqq \int_0^T |x(t)|^2\, e^{-2kt}\, dt \geqq \int_0^T |x(t)|^2\, dt\, e^{-2kT},$$

sind die Normen (2.4) mit der Norm $\|x\|_{L_2(0,T)} = \|x\|_{20}$ topologisch äquivalent.

Den Raum $C([0, T])$ der in $[0, T]$ stetigen Funktionen betrachten wir als Teilmenge von $L_2(0, T)$. In $C[(0, T])$ führen wir ebenfalls eine Folge von gewichteten Normen

$$\|x\|_{Ck} = \max_{t\in[0,T]} \{e^{-kt} |x(t)|\}, \qquad k = 1, 2, \ldots, \tag{2.5}$$

ein. Diese Normen sind für alle k der üblichen Tschebyschew-Norm ($k = 0$) topologisch äquivalent.

Satz 2.2. *Der stetige Kern φ der Steuergleichung* (2.1) *genüge der Lipschitz-Bedingung* (2.3). *Dann existiert zu beliebigem $g, u \in L_2(0, T)$ eine eindeutig bestimmte Lösung $x[g, u] \in L_2(0, T)$ der Steuergleichung. Der Lösungsoperator $x \in \big(L_2(0, T) \times L_2(0, T) \to L_2(0, T)\big)$ ist Lipschitz-stetig. Für $(g, u) \in C([0, T]) \times C([0, T])$ gilt $x[g, u] \in C([0, T])$. Der Lösungsoperator $x \in \big(C([0, T]) \times C([0, T]) \to C([0, T])\big)$ ist ebenfalls Lipschitz-stetig.*

Beweis. Sei

$$K_0(x, u)(t) = \int_0^t \varphi\big(t, s, x(s), u(s)\big)\, ds. \tag{2.6}$$

Zerlegen wir den Kern in der Form

$$\varphi(t, s, x, u) = \varphi^*(t, s, x, u) + \varphi(t, s, 0, 0),$$
$$\varphi^*(t, s, x, u) = \varphi(t, s, x, u) - \varphi(t, s, 0, 0),$$

so ist mit der stetigen Funktion $\alpha(t, s) = \varphi(t, s, 0, 0)$ die Bedingung (1.9) mit $\alpha = \beta = L, p = p' = 2$ erfüllt. Die Beziehung (2.6) definiert nach Satz 1.3 einen stetigen Operator $K_0 \in \big(L_2(0, T) \times L_2(0, T) \to L_2(0, T)\big)$. Wir finden

$$\begin{aligned}
&\|K_0(x, u) - K_0(y, v)\|_{2k}^2 \\
&= \int_0^T e^{-2kt} \left(\int_0^t \left[\varphi\big(t, s, x(s), u(s)\big) - \varphi\big(t, s, y(s), v(s)\big)\right] ds\right)^2 dt \\
&\leqq L^2 \int_0^T e^{-2kt} \left(\int_0^t e^{ks} e^{-ks}[|x(s) - y(s)| + |u(s) - v(s)|]\, ds\right)^2 dt \\
&\leqq 2L^2[\|x - y\|_{2k}^2 + \|u - v\|_{2k}^2] \int_0^T e^{-2kt} \left(\int_0^t e^{2ks}\, ds\right) dt \\
&\leqq \frac{TL^2}{k} (\|x - y\|_{2k}^2 + \|u - v\|_{2k}^2).
\end{aligned} \tag{2.7}$$

Aus der Abschätzung (2.7) ergibt sich zunächst für fixiertes $u \in L_2(0, T)$, daß der Operator $K_0(\cdot, u) \in \big(L_2(0, T) \to L_2(0, T)\big)$ in einer Norm $\|\cdot\|_{2k}$ strikt kontraktiv ist. Die Gleichung (2.1) besitzt daher zu jedem Paar $(g, u) \in L_2(0, T) \times L_2(0, T)$ genau eine Lösung $x[g, u] \in L_2(0, T)$. Zeigen wir, daß diese Lösung von beiden Argumenten stetig abhängt. Bei fixiertem u ist $x[g, u] = g + K_0(x[g, u], u)$, $x[h, u] = h + K_0(x[h, u], u)$ und folglich

$$\|x[g, u] - x[h, u]\|_{2k} \leqq \|g - h\|_{2k} + L \sqrt{\frac{T}{k}}\, \|x[g, u] - x[h, u]\|_{2k},$$

also

$$\|x[g, u] - x[h, u]\|_{2k} \leqq \frac{1}{1 - L\sqrt{\frac{T}{k}}} \|g - h\|_{2k}.$$

Bei fixiertem g finden wir

$$\|x[g, u] - x[g, v]\|_{2k}^2 = \|K_0(x[g, u], u) - K_0(x[g, v], v)\|_{2k}^2$$
$$\leqq \frac{TL^2}{k} (\|x[g, u] - x[g, v]\|_{2k}^2 + \|u - v\|_{2k}^2)$$

und daher

$$\|x[g, u] - x[g, v]\|_{2k} \leqq \frac{L\sqrt{\frac{T}{k}}}{\sqrt{1 - L^2\frac{T}{k}}} \|u - v\|_{2k}.$$

Insgesamt erhalten wir daraus

$$\|x[g, u] - x[h, v]\|_{2k} \leqq \|x[g, u] - x[g, v]\|_{2k} + \|x[g, v] - x[h, v]\|_{2k}$$
$$\leqq \varkappa[\|g - h\|_{2k} + \|u - v\|_{2k}] \tag{2.8}$$

für eine geeignete Konstante $\varkappa > 0$, falls k hinreichend groß gewählt wird. Aus den Beziehungen der topologischen Äquivalenz für die Normen $\|\cdot\|_{2k}$ folgt dann die Lipschitz-Stetigkeit auch von $x \in \big(L_2(0, T) \times L_2(0, T) \to L_2(0, T)\big)$.

Betrachten wir die Gleichung (2.1) nun im Raum $C([0, T])$. Da der Kern $\varphi \in (\Delta \times \mathfrak{R} \times \mathfrak{R} \to \mathfrak{R})$ stetig vorausgesetzt ist, gilt auch $K_0 \in \big(C([0, T]) \times C([0, T]) \to C([0, T])\big)$. Ferner erhalten wir die Abschätzung

$$e^{-kt} |K_0(x, u)(t) - K_0(y, v)(t)| \leqq e^{-kt} \int_0^t \big|\varphi\big(t, s, x(s), u(s)\big)$$
$$- \varphi\big(t, s, y(s), v(s)\big)\big| \, ds \leqq L\, e^{-kt} \int_0^t [|x(s) - y(s)| + |u(s) - v(s)|] \, ds$$
$$\leqq L\, e^{-kt} \int_0^t e^{ks} \, ds \, [\|x - y\|_{Ck} + \|u - v\|_{Ck}] \leqq \frac{L}{k} [\|x - y\|_{Ck} + \|u - v\|_{Ck}]$$

und daher

$$\|K_0(x, u) - K_0(y, v)\|_{Ck} \leqq \frac{L}{k} [\|x - y\|_{Ck} + \|u - v\|_{Ck}]. \tag{2.9}$$

Aus dieser Abschätzung folgern wir wie aus (2.7) die strikte Konvexität von K_0 bei fixiertem $u \in C([0, T])$ und hinreichend großem k, damit auch die eindeutige Lösbarkeit der Gleichung (2.1) im Raum $C([0, T])$. Außerdem erhalten wir eine Abschätzung der Form (2.8), diesmal im Raum $C([0, T])$,

$$\|x[g, u] - x[h, v]\|_{C^k} \leqq \varkappa \|g - h\|_{C^k} + \|u - v\|_{C^k}, \tag{2.10}$$

und folgern schließlich aus der topologischen Äquivalenz der Normen $\|\cdot\|_{C^k}$, $k = 0, 1, \ldots$, die Lipschitz-Stetigkeit auch von $x \in \big(C([0, T]) \times C([0, T]) \to C([0, T])\big)$, q. e. d.

Für $(g, u) \in C([0, T]) \times C([0, T])$ lassen sich nun wie in [1] Bedingungen angeben, unter denen der Operator $x[g, u]$ konvex (konkav) ist. Da der Raum $C([0, T])$ in $L_2(0, T)$ dicht ist, erhalten wir aus der Ungleichung

$$x[\alpha g + (1 - \alpha) h, \alpha u + (1 - \alpha) v] (t) \leqq (\geqq) \alpha x[g, u] (t) + (1 - \alpha) x[h, v] (t) \tag{2.11}$$

für $t \in [0, T]$ und $\alpha \in [0, 1]$, $g, h, u, v \in C([0, T])$, mit dem Ergebnis von Satz 2.2, indem wir die entsprechenden Elemente $g, h, u, v \in L_2(0, T)$ durch fast überall in $[0, T]$ konvergente Folgen aus $C([0, T])$ approximieren, die entsprechenden Ungleichungen

$$\begin{aligned} &x[\alpha g + (1 - \alpha) h, \alpha u + (1 - \alpha) v] < (>) \alpha x[g, u] + (1 - \alpha) x[h, v], \\ &g, h, u, v \in L_2(0, T), \qquad \alpha \in [0, 1]. \end{aligned} \tag{2.12}$$

Widmen wir uns daher dem Nachweis der Ungleichungen (2.11).

Lemma 2.1. *Die Funktionen $x_\alpha, y, z \in C([0, T])$ mögen die Volterragleichungen*

$$x_\alpha(t) = \xi_\alpha(t) + \int_0^t \varphi_\alpha\big(t, s, x_\alpha(s)\big) \, ds, \tag{2.13}$$

$$\left.\begin{aligned} y(t) &= \eta(t) + \int_0^t \psi\big(t, s, y(s)\big) \, ds, \\ z(t) &= \zeta(t) + \int_0^t \chi\big(t, s, z(s)\big) \, ds \end{aligned}\right\} \tag{2.14}$$

mit stetigen Kernen $\varphi_\alpha, \psi, \chi \in (\Delta \times \mathfrak{R} \to \mathfrak{R})$ und $\xi_\alpha, \eta, \zeta \in C([0, T])$ erfüllen.

i) *Es sei $\xi_\alpha(t) < \alpha\eta(t) + (1 - \alpha) \zeta(t)$, $\alpha \in [0, 1]$ und $t \in [0, T]$. Aus $x_\alpha \leqq \alpha y + (1 - \alpha) z$, $\alpha \in [0, 1]$, möge $\varphi_\alpha(t, s, x_\alpha) \leqq \alpha\psi(t, s, y) + (1 - \alpha) \chi(t, s, z)$ für $(t, s) \in \Delta$ folgen. Dann gilt*

$$x_\alpha(t) < \alpha y(t) + (1 - \alpha) z(t) \quad \text{für} \quad \alpha \in [0, 1] \quad \text{und} \quad t \in [0, T]. \tag{2.15}$$

ii) *Ist die Gleichung (2.13) für alle $\xi_n(t) = \xi_\alpha(t) - \frac{1}{n}$, $t \in [0, T]$, $n = 1, 2, \ldots$, lösbar und konvergieren die entsprechenden Lösungen $x_{\alpha n} \in C([0, T])$ gegen $x_\alpha \in C([0, T])$, so folgt aus der schwächeren Voraussetzung $\xi_\alpha(t) \leqq \alpha\eta(t) + (1 - \alpha) \zeta(t)$ in* i)

$$x_\alpha(t) \leqq \alpha y(t) + (1 - \alpha) z(t) \quad \text{für} \quad \alpha \in [0, 1] \quad \text{und} \quad t \in [0, T]. \tag{2.16}$$

Beweis. i) Nehmen wir an, daß die Ungleichung (2.15) für ein $t_0 \in (0, T]$ verletzt ist (für $t = 0$ ist sie nach Voraussetzung richtig), so können wir auf Grund der Stetigkeit von x_α, y, z annehmen, daß t_0 die erste Stelle in $(0, T]$ ist. Entgegen dieser Annahme ist aber

$$\alpha y(t_0) + (1-\alpha) z(t_0) - x_\alpha(t_0) = \alpha\eta(t_0) + (1-\alpha)\zeta(t_0) - \xi_\alpha(t_0)$$
$$+ \int_0^{t_0} \{\alpha\psi(t_0, s, y(s)) + (1-\alpha)\chi(t_0, s, z(s)) - \varphi_\alpha(t_0, s, x_\alpha(s))\}\, ds > 0.$$

ii) Da für $x_{\alpha n} \in C([0, T])$ nach i)

$$x_{\alpha n}(t) < \alpha y(t) + (1-\alpha) z(t), \qquad \alpha \in [0, 1], \qquad t \in [0, T]$$

gilt, erhalten wir daraus die Ungleichung (2.16) durch Grenzübergang für $n \to \infty$.

Satz 2.3. *Der stetige Kern* $\varphi \in (\Delta \times \Re \times \Re \to \Re)$ *in der Gleichung* (2.1) *möge die Lipschitzbedingung* (2.3) *erfüllen, sei in* $x \in \Re$ *monoton wachsend:*

$$\varphi(t, s, x, u) \leqq \varphi(t, s, y, u) \quad \text{für} \quad (t, s) \in \Delta \quad \text{und} \quad x, y, u \in \Re,\ x \leqq y, \tag{2.17}$$

sowie in $(x, u) \in \Re \times \Re$ *konvex:*

$$\varphi(t, s, \alpha x + (1-\alpha) y, \alpha u + (1-\alpha) v) \leqq \alpha\varphi(t, s, x, u)$$
$$+ (1-\alpha)\varphi(t, s, y, v) \quad \text{für} \quad (t, s) \in \Delta, \quad x, y, u, v \in \Re, \quad \alpha \in [0, 1]. \tag{2.18}$$

Dann genügt der Lösungsoperator $x[g, u]$ *der Gleichung* (2.1) *für* $(g, u) \in C([0, T]) \times C([0, T])$ *der Konvexitätsbedingung* (2.11).

Beweis. Wir setzen in Lemma 2.1

$$x_\alpha(t) = x[\alpha g + (1-\alpha) h, \alpha u + (1-\alpha) v]\,(t),$$
$$y(t) = x[g, u]\,(t),$$
$$z(t) = x[h, v]\,(t), \qquad t \in [0, T], \qquad \alpha \in [0, 1].$$

Dann gelten die Beziehungen (2.13), (2.14) mit

$$\varphi_\alpha(t, s, x) = \varphi(t, s, x, \alpha u(t) + (1-\alpha) v(t)),$$
$$\psi(t, s, y) = \varphi(t, s, y, u(t)),$$
$$\chi(t, s, z) = \varphi(t, s, z, v(t)), \qquad \alpha \in [0, 1], \qquad (t, s) \in \Delta, \qquad x, y, z \in \Re,$$
$$\xi_\alpha(t) = \alpha g(t) + (1-\alpha) h(t), \qquad \eta(t) = g(t), \qquad \zeta(t) = h(t),$$

so daß die Aussage ii) in Betracht kommt, da der Lösungsoperator der ersten Gleichung (2.13) dann nach Satz 2.2 stetig vom Außenglied $\xi_\alpha \in C([0, T])$ abhängt.

Auf Grund der Bedingungen (2.17), (2.18) erhalten wir für $x_\alpha \leqq \alpha y + (1-\alpha) z$, $\alpha \in [0, 1]$

$$\varphi_\alpha(t, s, x) = \varphi(t, s, x, \alpha u(t) + (1-\alpha) v(t))$$
$$\leqq \varphi(t, s, \alpha y + (1-\alpha) z, \alpha u(t) + (1-\alpha) v(t))$$
$$\leqq \alpha\varphi(t, s, u(t)) + (1-\alpha)\varphi(t, s, v(t)) = \alpha\psi(t, s, y)$$
$$+ (1-\alpha)\chi(t, s, z), \qquad \alpha \in [0, 1], \qquad (t, s) \in \Delta, \qquad x, y, z \in \Re.$$

Die Aussage (2.16) von Lemma 2.1 geht nun in die Ungleichung (2.11) über, q. e. d.
Wir können nun Satz 1.2 sinngemäß auf das Minimum-Problem (2.2) übertragen.

Satz 2.4. *Für konvexe und abgeschlossene Bereiche* $\mathfrak{G}, \mathfrak{U} \subseteqq L_2(0, T)$ *sei das Zielfunktional* $f \in \left(L_2(0, T) \times \mathfrak{G} \times \mathfrak{U} \to \mathfrak{R}\right)$ *monoton wachsend in g:*

$$f(x, g, u) \leqq f(x, h, u) \quad \text{für} \quad x \in L_2(0, T), \quad g, h \in \mathfrak{G}, \quad u \in \mathfrak{U}, g < h,$$

und stark konvex bezüglich $(g, u) \in \mathfrak{G} \times \mathfrak{U}$:

$$f\left(\frac{x+y}{2}, \frac{g+h}{2}, \frac{u+v}{2}\right) + \gamma(\|g-h\|_{L_2}^2 + \|u-v\|_{L_2}^2)$$

$$\leqq \frac{1}{2} f(x, g, u) + \frac{1}{2} f(y, h, v) \quad \text{für} \quad x, y \in L_2(0, T), \quad g, h \in \mathfrak{G}, \quad u, v \in \mathfrak{U}$$

$$\text{und ein} \quad \gamma > 0.$$

Überdies sei f unterhalbstetig und unterhalbbeschränkt. Der Kern φ *des Integraloperators der Gleichung* (2.1) *genüge den Bedingungen in Satz* 2.3. *Dann besitzt das Minimum-Problem* (2.2) *genau eine Lösung, gegen die jede Minimalfolge stark konvergiert.*

Literatur

[1] Langenbach, A., Stark konvexe Steuerungsprobleme. Math. Nachr. (im Druck).

[2] Поляк, В. Т., Теоремы существования и сходимость минимизирующих последовательностей для задач на экстремум при наличии ограничений. ДАН СССР **166**, 287–290 (1966).

[3] Вайнберг, М. М., Вариационный метод и метод монотонных операторов. Москва 1972.

[4] Красносельский, М. А., Топологические методы в теории нелинейных интегральных уравнений. Москва 1956.

[5] Вайнберг, М. М., Вариационные методы исследования нелинейных операторов. Москва 1956.

Anschrift des Verfassers:

Humboldt-Universität Berlin
Sektion Mathematik
108 Berlin
Unter den Linden 6
DDR

Some quasi-variational inequalities arising in stochastic impulse control theory

Umberto Mosco, Roma

1. So called *quasi-variational inequalities* (QVI) have been recently introduced by A. Bensoussan and J. L. Lions in a dynamic programming approach to some impulse control problems. We refer to the series of recent notes of these authors in C. R. Acad. Sci. Paris or, for instance, to J. L. Lions [21, 22] or A. Bensoussan and J. L. Lions [4], where an extensive bibliography on the subject can be found.

In our lecture we will try to illustrate some special features of QVI's, by focusing our attention on a stationary example that seems to play an important role in the applications.

The QVI we are interested in can be stated as a system of partial differential inequalities in an open region $\mathcal{O}$ of the n-dimensional euclidean space R^n. We assume that $\mathcal{O}$ is bounded and has a smooth boundary Γ.

These inequalities involve a second order partial differential operator in divergence form

$$Lu = \sum_{i,j=1}^{n} \left(\frac{\partial}{\partial x_j} a_{ij} \frac{\partial}{\partial x_i} + b_j \frac{\partial}{\partial x_j} + c \right) u,$$

an operator M of the form

$$Mu = 1 + \inf_{\substack{\xi \geq 0 \\ x+\xi \in \mathcal{O}}} u(x + \xi), \qquad x \in \mathcal{O},$$

a given function f on $\mathcal{O}$, and can be formally written as

$$\left.\begin{aligned} &u \leq Mu \\ &Lu - f \leq 0 \\ &(u - Mu)(Lu - f) = 0 \end{aligned}\qquad \text{in } \mathcal{O}, \right\} \tag{1}$$

with, in addition, a

boundary condition on Γ, to be specified. (2)

The condition on Γ may be of Dirichlet or Neumann type, or a Signorini condition like

$$u \leq Mu, \quad \frac{\partial u}{\partial \nu_L} \leq 0, \quad (u - Mu) \frac{\partial u}{\partial \nu_L} = 0 \quad \text{on } \Gamma, \tag{2'}$$

where

$$\frac{\partial u}{\partial \nu_L} = \sum_{i,j=1}^{n} a_{ij} \frac{\partial u}{\partial x_i} \cos(\vec{\nu}, x_j)$$

is the conormal derivative of u on Γ associated with the operator L.

2. Before stating our problem more precisely, let us point out two important features of the system of inequalities (1) (2).

The first remark is that the constraint appearing in (1), that is

$$u \leqq Mu \qquad \text{in } \mathcal{O},$$

is an "implicit" constraint, in the following sense: It is the usual unilateral constraint of the theory of variational inequalities, namely

$$u \leqq \psi \qquad \text{in } \mathcal{O},$$

with, however, the basic difference now that the "obstacle" ψ is not given *ex ante* among the data of the problem, but it is only known *ex post*, via the map M, once the solution u itself has been found out. Let us also remark, incidentally, that such a kind of "circularity" between solution and constraints has been known since long time, for example, in many classical problems of the economic theory.

A second feature of the problem above, due to the specific form of the mapping M involved, is its *global* nature. This aspect of the problem, as we shall see, drastically affects the study of the regularity of the solution and makes the usual regularity techniques for variational inequalities inadequate in the case at hand.

On the other hand, as naturally suggested by the "circularity" mentioned above, we can actually deal with problem (1) (2) in two steps, the first of which relies on the basic existence results for variational inequalities, while the second one consists in setting up suitable fixed-point arguments.

In fact, we may first consider the "obstacle" Mu as "frozen", that is, as given by some *fixed* function u via the map M, and solve the inequalities which the system (1) (2) consequently reduce to, namely

$$\left.\begin{array}{l} v \leqq Mu \\ Lv - f \leqq 0 \qquad\qquad \text{in } \mathcal{O}, \\ (v - Mu)(Lv - f) = 0 \end{array}\right\} \tag{3}$$

with a

boundary condition on Γ corresponding to (2). (4)

Problem (3) (4) is indeed a *variational inequality* (VI) in the sense of Lions-Stampacchia [23] and can be dealt with by techniques by now well known, see also [20].

Once the VI (3) (4) has been solved for any given u, assuming the solution v exists and is unique, we end up with a mapping S, by setting

$$v = Su. \tag{5}$$

This map is called the *variational selection* of problem (1) (2).

The second, and last, step for solving problem (1) (2) consists then in finding a *fixed-point* of the selection S. Indeed, it follows by the very definition (5) of S that u is a solution of (1) (2) if and only if u verifies

$$u = Su. \tag{6}$$

Let us remark, finally, that despite our assumption that the VI (3) (4) has a unique solution v for each fixed u, the fixed-point (6) of S, hence the solution of QVI (1) (2), may well not be unique. We may say, in general, that further restrictions must be put on the solution of a QVI such as (1) (2) above, in order the solution be determined uniquely. Moreover, the proof of the uniqueness may be not so easy to achieve, as, instead, we know to be the case for a large class of variational inequalities.

3. We suppose that the coefficients of the operator L are bounded measurable real-valued functions defined in $\mathcal{O}$. This allows us to consider L as a continuous linear operator from the Sobolev space $H^1(\mathcal{O})$ to the dual V' of any given closed subspace V of $H^1(\mathcal{O})$, by means of the identity

$$(Lu, w) = a(u, w), \qquad u \in H^1(\mathcal{O}), \quad w \in V,$$

where

$$a(u, w) = \sum_{i,j=1}^{n} \int_{\mathcal{O}} (a_{ij} u_{x_i} w_{x_j} + b_j u_{x_j} w + cw)\, dx. \tag{7}$$

In the example discussed in section 4 below, we have

$$V = H^1(\mathcal{O}),$$

while in that of the following section 5 we have

$$V = H_0^1(\mathcal{O}),$$

where $H_0^1(\mathcal{O})$, as usual, denotes the subspace of all functions of $H^1(\mathcal{O})$ that vanish on the boundary Γ of $\mathcal{O}$.

In both cases, we assume that $c(x) \geqq 0$ a.e. in $\mathcal{O}$ and that the form (7) is *coercive* on V, that is, that there exists a constant $\gamma > 0$, such that

$$a(v, v) \geqq \gamma \|v\|^2 \qquad \text{for all } v \in V,$$

where $\|\cdot\|$ is the usual norm of the space $H^1(\mathcal{O})$.

The operator M is defined for any $u \in L^\infty(\mathcal{O})$ by

$$(Mu)(x) = 1 + \operatorname*{essinf}_{\substack{\xi \geqq 0 \\ x+\xi \in \mathcal{O}}} u(x + \xi), \qquad x \text{ a.e. in } \mathcal{O}. \tag{8}$$

Since the space $L^\infty(\mathcal{O})$ is a complete lattice under the a.e. ordering, see e.g. [25], $Mu \in L^\infty(\mathcal{O})$, therefore

$$M\colon L^\infty(\mathcal{O}) \mapsto L^\infty(\mathcal{O}). \tag{9}$$

Moreover, M is *increasing*, that is

$$u_1 \leqq u_2 \text{ a.e. in } \mathcal{O} \text{ implies } Mu_1 \leqq Mu_2 \text{ a.e. in } \mathcal{O}, \tag{10}$$

a property that will play a crucial role in what follows.

Let us notice that, as soon as $n > 1$, the operator M does not carry the space $H^1(\mathcal{O})$ into itself, as it has been shown by Temam with simple two-dimensional examples.

4. We now specify the boundary condition (2) to be the Signorini condition (2'). Then, as it is well known from the theory of variational inequalities, the system of inequalities (1) (2') admits the following weak formulation

$$\left.\begin{aligned} &u \in H^1(\mathcal{O}) \cap L^\infty(\mathcal{O}), \qquad u \leqq Mu \quad \text{a.e. in } \mathcal{O}, \\ &a(u, u - w) \leqq (f, u - w) \quad \forall w \in H^1(\mathcal{O}), \quad w \leqq Mu \quad \text{a.e. in } \mathcal{O}, \end{aligned}\right\} \tag{11}$$

where

$$(f, z) = \int_{\mathcal{O}} fz\,dx \qquad \text{for any } z \in L^2(\mathcal{O}),$$

see e.g. ref. [20].

We suppose in this section that

$$f \in L^\infty(\mathcal{O}) \qquad \text{and} \qquad f \geqq 0 \qquad \text{a.e. in } \mathcal{O}, \tag{12}$$

and we denote by $\bar{u}$ the solution of the following Neumann problem

$$\left.\begin{aligned} &\bar{u} \in H^1(\mathcal{O}) \\ &a(\bar{u}, w) = (f, w) \qquad \forall w \in H^1(\mathcal{O}). \end{aligned}\right\} \tag{13}$$

Note that, by (12), we have

$$\bar{u} \in H^1(\mathcal{O}) \cap L^\infty(\mathcal{O}) \qquad \text{and} \qquad \bar{u} \geqq 0 \qquad \text{a.e. in } \mathcal{O}. \tag{14}$$

The existence of a solution u of problem (11) has been proved by Bensoussan-Goursat-Lions [3] and, in a more abstract setting, by Tartar [26] who also gave conditions for the uniqueness of the solution. A direct proof of this uniqueness has been given by Laetsch [18].

Theorem 1. *Under the assumption* (12), *there exists a unique solution $u \geqq 0$ a.e. in $\mathcal{O}$ of problem* (11). *Moreover, $u \leqq \bar{u}$ a.e. in $\mathcal{O}$.*

The proof can be given along the lines of section 3. By using known results for variational inequalities, see the references quoted above, we prove first that for any u of the interval

$$0 \leqq u \leqq \bar{u} \qquad \text{a.e. in } \mathcal{O}, \tag{15}$$

$u \in L^\infty(\mathcal{O})$, there exists a unique solution

$$v = Su$$

of the problem

$$\left.\begin{aligned} &v \in H^1(\mathcal{O}), \qquad v \leqq Mu \qquad \text{a.e. in } \mathcal{O}, \\ &a(v, v - w) \leqq (f, v - w) \qquad \forall w \in H^1(\mathcal{O}), \quad w \leqq Mu \quad \text{a.e. in } \mathcal{O}. \end{aligned}\right\} \tag{16}$$

Note that, by (14), the interval (15) is not empty, the map M is, by (9), well defined on it and, moreover, the vector $w = 0 \in H^1(\mathcal{O})$ is always admissible in (16) since

$$Mu \geqq 0 \quad \text{a.e. in } \mathcal{O} \text{ for every } \quad u \geqq 0 \quad \text{a.e. in } \mathcal{O}. \tag{17}$$

We now consider the interval (15) as an interval in the space $L^2(\mathcal{O})$, which is also a complete lattice under the a.e. ordering, see ref. [25]. Note that, by (14), the functions u of $L^2(\mathcal{O})$ belonging to that interval are the same as those of the space $L^\infty(\mathcal{O})$.

We can also identify the space $H^1(\mathcal{O})$ with a subspace of $L^2(\mathcal{O})$, in the distribution sense. This makes $H^1(\mathcal{O})$ a sub-lattice of $L^2(\mathcal{O})$, see e.g. [19], however, not a complete one.

We are now in a position to consider the map S above as a map of the interval (15) of the space $L^2(\mathcal{O})$, a complete lattice, into the space $L^2(\mathcal{O})$ itself.

In consequence of property (10) of M and of the comparison results for variational inequalities, see e.g. ref. [20], the map S is *increasing* for the a.e. ordering,

$$u_1 \leqq u_2 \qquad \text{a.e. in } \mathcal{O} \text{ implies} \qquad Su_1 \leqq Su_2 \qquad \text{a.e. in } \mathcal{O}. \tag{18}$$

At this point we can appeal to classical fixed-point theorems, see [6], asserting that an increasing mapping of an interval $z_1 \leqq z_2$ of a complete lattice admits a maximum and a minimum fixed-point in $z_1 \leqq z_2$, provided z_1 is a *sub-solution* of S, i.e.,

$$z_1 \leqq Sz_1, \tag{19}$$

and z_2 is a *super-solution of* S, i.e.,

$$Sz_2 \leqq z_2. \tag{20}$$

Clearly, for an increasing S condition (19) and (20) together are the same thing as the property of the interval $z_1 \leqq z_2$ of being *stable* under S.

Now, by the comparison theorems, any solution v of VI (16) satisfies

$$v \leqq \bar{u} \qquad \text{a.e. in } \mathcal{O},$$

hence, in particular

$$Su \leqq \bar{u} \qquad \text{a.e. in } \mathcal{O}. \tag{21}$$

Similarly, since $f \geqq 0$ and $M0 \geqq 0$, we have

$$0 \leqq S0 \qquad \text{a.e. in } \mathcal{O}. \tag{22}$$

By remarking that the fixed-point u of S, which has been found in the space $L^2(\mathcal{O})$, is indeed an element of $H^1(\mathcal{O})$ for it can be identified as a distribution with Su (and $Su \in H^1(\mathcal{O})$), we conclude the proof of the existence of a solution of (11).

The uniqueness of the solution $u \geqq 0$ of problem (11), under the assumption (12), has been proved, as we said, by Laetsch [18], by relying only on properties (9), (10), (17) of the mapping M and, furthermore, on the following property

$$\text{For every } u \geqq 0 \text{ and every } \alpha \in [0, 1[, \text{ there exists some } \beta \in]\alpha, 1[\text{ such that } \beta Mu \leqq M(\alpha u), \tag{23}$$

which is not difficult to check on our mapping (8).

The existence proof sketched above is an adaptation of Tartar's proof (based on Zorn's lemma) of an abstract fixed-point theorem for increasing mappings defined in inductive ordered spaces which are not necessarily complete lattices, see [26].

The proof given by Bensoussan-Goursat-Lions is of a more constructive type and is based on the following iterative procedure

$$u_0 = \bar{u}, \qquad u_k = Su_{k-1}, \qquad k = 1, 2, \ldots \tag{24}$$

where $\bar{u}$ is, as above, the solution of (13). By using, in addition to (10) and (17), a mild continuity property of M, these authors show that the sequence u_k is non-increasing, bounded from below by 0 and weakly convergent to the solution $u \geqq 0$ of (11). It is not known at the present whether the non-decreasing sequence defined iteratively by

$$u_0' = 0, \qquad u_k' = Su'_{k-1}, \qquad k = 1, 2, \ldots \tag{25}$$

does converge (from below) to the solution of (11).

5. Let us now specify the boundary condition (2) to be the Dirichlet condition

$$u = 0 \qquad \text{a.e. on } \Gamma. \tag{2''}$$

The weak formulation of problem (1) (2″) is then the following

$$\left.\begin{aligned} &u \in H_0^1(\mathcal{O}) \cap L^\infty(\mathcal{O}), \quad u \leqq Mu \quad \text{a.e. in } \mathcal{O}, \\ &a(u, u - w) \leqq (f, u - w) \quad \forall w \in H_0^1(\mathcal{O}), \quad w \leqq Mu \quad \text{a.e. in } \mathcal{O}. \end{aligned}\right\} \tag{26}$$

If we assume, as in (12), that $f \in L^\infty(\mathcal{O})$ and $f \geqq 0$ a.e. in $\mathcal{O}$, it is easy to check that there exists a unique solution $u \geqq 0$ of (26), given by the VI

$$\left.\begin{aligned} &u \in H_0^1(\mathcal{O}), \quad u \leqq 1 \quad \text{a.e. in } \mathcal{O}, \\ &a(u, u - w) \leqq (f, u - w) \quad \forall w \in H_0^1(\mathcal{O}), \quad w \leqq 1 \quad \text{a.e. in } \mathcal{O}. \end{aligned}\right\} \tag{27}$$

In fact, if $u \geqq 0$ and (2″) holds, then $Mu \equiv 1$ in $\mathcal{O}$; on the other hand, by (12), the solution u of (27) belongs to $H_0^1(\mathcal{O}) \cap L^\infty(\mathcal{O})$ and is $u \geqq 0$; therefore, u is a solution of (26). Vice versa, if $u \geqq 0$ is a solution of (26), then $Mu = 1$ as remarked above and (26) reduces to (27).

If the assumption (12) is dropped, problem (26) is a "true" QVI. However, we cannot repeat the solution pattern followed in section 4. What goes wrong now is that (22) is not in general satisfied, that is, 0 no more is a sub-solution.

Under the rather drastic assumption that the operator L has *constant coefficients*, the proof of the existence of a *regular* solution of (26) has been given in [16, 17], under suitable hypotheses on the function f replacing (12).

We suppose that

$$f \in L^p(\mathcal{O}) \tag{28}$$

for some

$$p > \max\left\{1, \frac{n}{2}\right\}; \tag{29}$$

while we do not assume now that $f \geqq 0$.

For any such p, the Sobolev space $H^{2,p}(\mathcal{O})$ is continuously imbedded in the space of continuous functions on $\overline{\mathcal{O}}$, $C(\overline{\mathcal{O}})$. For any $u \in C_0(\overline{\mathcal{O}})$, the function Mu can be defined pointwise by

$$(Mu)(x) = 1 + \inf_{\substack{\xi \geqq 0 \\ x+\xi \in \overline{\mathcal{O}}}} u(x + \xi), \qquad x \in \overline{\mathcal{O}}, \tag{30}$$

and it belongs to $C(\overline{\mathcal{O}})$. Moreover, M is continuous on sequences of $C_0(\overline{\mathcal{O}})$.

As we know from the theory of variational inequalities, the natural regularity we may expect for the solution of problem (26), in consequence of the hypothesis (28), is that u belong to $H^{2,p}(\mathcal{O})$. We thus rewrite our problem as

$$\left.\begin{array}{l} u \in H_0^1(\mathcal{O}) \cap H^{2,p}(\mathcal{O}), \qquad u \leqq Mu \qquad \text{a.e. in } \mathcal{O}, \\ a(u, u - w) \leqq (f, u - w) \quad \forall w \in H_0^1(\mathcal{O}), \quad w \leqq Mu \quad \text{a.e. in } \mathcal{O}. \end{array}\right\} \tag{31}$$

A *necessary* condition for the existence of a solution u can be expressed in terms of the solution $\overline{u}$ of the Dirichlet problem

$$\left.\begin{array}{l} \overline{u} \in H_0^1(\mathcal{O}), \\ a(\overline{u}, w) = (f, w) \qquad \forall w \in H_0^1(\mathcal{O}), \end{array}\right\} \tag{32}$$

namely,

$$\overline{u} \geqq -1 \qquad \text{in } \mathcal{O}. \tag{33}$$

In fact, by the comparison theorems, $\overline{u} \geqq u$ if u is a solution of (31). On the other hand, we have $u \geqq 1$, for if $u(x_0)$ at some point $x_0 \in \overline{\mathcal{O}}$ (recall that u is continuous on $\overline{\mathcal{O}}$), then we could find some $x \in \Gamma$, such that $x_0 = x + \xi$ with $\xi \geqq 0$, hence we would have

$$u(x) \leqq (Mu)(x) = 1 + \inf_{\substack{\xi \geqq 0 \\ x+\xi \in \overline{\mathcal{O}}}} u(x + \xi) \leqq 1 + u(x_0) < 0,$$

whereas $u(x) = 0$ for all $x \in \Gamma$.

A *sufficient* condition can be given instead in terms of the solution $\underline{u}$ of the Dirichlet problem

$$\left.\begin{array}{l} \underline{u} \in H_0^1(\mathcal{O}), \\ a(\underline{u}, w) = (g, w) \qquad \forall w \in H_0^1(\mathcal{O}), \end{array}\right\} \tag{34}$$

involving the function g obtained from the given f by

$$g(x) = \inf\left\{0, \operatorname*{essinf}_{\substack{\xi \geqq 0 \\ x+\xi \in \mathcal{O}}} f(x + \xi)\right\}, \qquad x \text{ a.e. in } \mathcal{O}, \tag{35}$$

and that is supposed to satisfy the condition

$$g \in L^p(\mathcal{O}), \tag{36}$$

with the same p as in (28). Note that, if $f \in L^\infty(\mathcal{O})$, then both (28) and (35) hold for any p. The sufficient condition we are talking about, as Theorem 2 below shows, consists in assuming that

$$\underline{u} \geqq -1 \qquad \text{in } \mathcal{O}. \tag{37}$$

Note that this condition is actually stronger than the necessary condition (33), for we have

$$\overline{u} \geqq \underline{u} \qquad \text{in } \mathcal{O}, \tag{38}$$

always in consequence of the comparison results ($g \leqq f$ a.e. in $\mathcal{O}$).

Theorem 2. *Under the assumptions* (28), (29), (36) *and* (37), *there exists a solution* u *of problem* (31), *satisfying*

$$g \leqq Lu \leqq f \qquad \text{a.e. in } \mathcal{O}. \tag{39}$$

In particular, $\underline{u} \leqq u \leqq \overline{u}$, *where* $\overline{u}$ *and* $\underline{u}$ *are the solutions of the Dirichlet problems* (32) *and* (34), *respectively.*

6. The proof of Theorem 2, as that of Theorem 1, can be carried out in two steps. First we solve the VI

$$\left.\begin{array}{l} v \in H_0^1(\mathcal{O}), \qquad v \leqq Mu \qquad \text{a.e. in } \mathcal{O}, \\ a(v, v - w) \leqq (f, v - w) \quad \forall w \in H_0^1(\mathcal{O}), \quad w \leqq Mu \quad \text{a.e. in } \mathcal{O}, \end{array}\right\} \tag{40}$$

or any fixed u satisfying

$$u \in H_0^1(\mathcal{O}) \cap H^{2,p}(\mathcal{O}), \qquad u \geqq -1. \tag{41}$$

Since the last condition implies $Mu \geqq 0$, the set of the admissible w's in (40) is never empty.

The second step, indeed the difficult one in the present circumstances, is to show that the map

$$v = Su, \tag{42}$$

that associates the (unique) solution v of (40) to each u of (41), has a fixed-point.

Let us notice that the existence of such a fixed-point of S requires, "a fortiori", that, (40) admit a solution $v \in H^{2,p}(\mathcal{O})$, in correspondence with some u satisfying (41); a fact, this, that could not be said by a direct inspection of (40). In fact, we can only say that Mu is a continuous function, however not that it belongs to $H^{2,p}(\mathcal{O})$, as required by the regularity results for VI's such as (40) in order to conclude that the solution v itself is an element of $H^{2,p}(\mathcal{O})$.

The argument used to prove the existence of a fixed-point of S is, in the present case, a *topological* one, namely Schauder's fixed-point theorem. In order to apply this theorem, we have to find a closed convex subset of the space $H_0^1(\mathcal{O}) \cap H^{2,p}(\mathcal{O})$, which is bounded in the $H^{2,p}(\mathcal{O})$ norm and is carried into itself by the map S, continuously for the weak topology of $H^{2,p}(\mathcal{O})$.

It turns out that the set C_0 of all functions $u \in H_0^1(\mathcal{O}) \cap H^{2,p}(\mathcal{O})$ satisfying the inequality

$$f \geqq Lu \geqq g \qquad \text{a.e. in } \mathcal{O}, \tag{43}$$

meets all our requirements.

The boundedness of C_0 in the $H^{2,p}(\mathcal{O})$ norm follows from classical estimates for the Dirichlet problem, in consequence of our assumptions (28) and (36). Since C_0 is closed and convex, it is weakly compact in $H^{2,p}(\mathcal{O})$.

It is also not difficult to prove, even if it is not immediate, that S is continuous on sequences of C_0 in the weak topology of $H^{2,p}(\mathcal{O})$ (M behaves nicely on continuous functions), provided we know that S carries the set C_0 into itself. This is indeed the most difficult point of the proof. Let us state it as a formal proposition.

Proposition. *Under the assumptions of Theorem* 2, *if* $u \in H_0^1(\mathcal{O}) \cap H^{2,p}(\mathcal{O})$ *satisfies condition* (43), *then the solution* $v = Su$ *of problem* (40) *also verifies*

$$f \geqq Lv \geqq g \qquad \text{a.e. in } \mathcal{O}. \tag{44}$$

The proof is too technical to be given here in detail. However, it is perhaps worthwile to give even only a rough sketch of the proof, just to point out which are the main tools used and where the hypothesis that L has constant coefficients comes in.

The scheme of the proof is the following. The operator M, given by (30), is approximated by a directed family of mappings $\{M_F\}$, F being an arbitrary *finite* set of vectors $\xi \geqq 0$ of R^n, containing $0 \in R^n$, the family of all such F's being partially ordered by increasing inclusion. Accordingly, the selection S is approximated by the directed family of mappings S_F, each defined for any fixed F, by associating the solution v_F of the following VI

$$\left.\begin{aligned} &v_F \in H_0^1(\mathcal{O}), \qquad v_F \leqq M_F u \quad \text{a.e. in } \mathcal{O}, \\ &a(v_F, v_F - w) \leqq (f, v_F - w) \quad \forall w \in H_0^1(\mathcal{O}), \quad w \leqq M_F u \quad \text{a.e. in } \mathcal{O} \end{aligned}\right\} \tag{45}$$

with every $u \in C_0$,

$$v_F = S_F u, \qquad u \in C_0. \tag{46}$$

The operator M_F is defined as

$$M_F u = 1 + \inf\left\{0, \inf_{\xi \in F} (\tilde{u})_\xi\right\}, \qquad \text{in } \overline{\mathcal{O}}, \tag{47}$$

where we have used the notation

$$\tilde{u}(x) = u(x) \quad \text{if} \quad x \in \overline{\mathcal{O}}, \quad = 0 \quad \text{if} \quad x \notin \overline{\mathcal{O}}, \tag{48}$$

and

$$(\tilde{u})_\xi(x) = \tilde{u}(x + \xi), \qquad x \in R^n, \qquad \xi \in R^n. \tag{49}$$

Let us remark that a straightforward definition of M_F as

$$(M_F u)(x) = 1 + \inf_{\substack{\xi \in F \\ x+\xi \in \mathcal{O}}} u(x + \xi)$$

would bring very bad discontinuities in $M_F u$, even if u is smooth, as it can be easily illustrated by simple one-dimensional pictures.

Next step is to prove that C_0 is stable under each one of the mappings S_F. Proposition above will then follow by a suitable limit argument on the family $\{S_F\}$. We must then prove that the solution (46) verifies the inequality

$$f \geqq L v_F \geqq g, \tag{50}$$

as a *measure* of the dual $H^{-1}(\mathcal{O})$ of $H_0^1(\mathcal{O})$. Once (50) has been established, the additional property $v_F \in H^{2,p}(\mathcal{O})$ comes from the estimates mentioned above, in consequence of assumptions (28) and (36).

The main tool for the proof of (50) is the following estimate

$$f \geqq Lu \geqq \inf \{f, L\psi\} \tag{51}$$

of the solution u of the VI

$$\left.\begin{aligned} &u \in H_0^1(\mathcal{O}), \qquad u \leqq \psi \qquad \text{a.e. in } \mathcal{O}, \\ &a(u, u - w) \leqq (f, u - w) \qquad \forall w \in H_0^1(\mathcal{O}), \quad w \leqq \psi \ \text{ a.e. in } \mathcal{O}. \end{aligned}\right\} \tag{52}$$

Inequalities of this type were first considered by H. Lewy and G. Stampacchia [19] for *regular* classical solutions. The particular estimate needed in the case at hand, where u is a *weak* solution, can be stated as follows, see [24].

Lemma. *If $\psi \in H^1(\mathcal{O}) \cap C(\overline{\mathcal{O}})$, $\psi \geqq 0$ on Γ and $L\psi$ is a measure in $H^{-1}(\mathcal{O})$, then the solution u of the* VI (52) *satisfies the inequality* (51) *in the sense of measures.*

We are assuming that f is given as in (28), however, f could be any (signed) measure in $H^{-1}(\mathcal{O})$. In any case, the last term in (51) must be given the meaning $f + (Lu - f)^-$, where $(Lu - f)^-$ denotes the negative part of the (signed) measure $Lu - f$.

In order to apply the Lemma to problem (45), we only need to check, in addition to what we already know, that $L(M_F u)$ is a *measure* in $H^{-1}(\mathcal{O})$, for any $u \in C_0$. If so, we find that the solution v_F satisfies the inequality

$$f \geqq L v_F \geqq \inf \{f, L(M_F u)\} \tag{53}$$

and this will bring us to the inequality (50), provided we are able to *estimate* $L(M_F u)$ *from below* by

$$L(M_F u) \geqq g. \tag{54}$$

The proof of these two points is too technical to be reproduced here. We only remark that the following well known *lattice property* of the operator L comes in

$$L(\inf z_1, z_2) \geqq \inf \{L z_1, L z_2\},$$

together with the fact that L *commutes with the translations* occuring in the expression (47) of $M_F u$, since the operator L has constant coefficients.

7. Let us conclude with some additional remarks and some comments on open problems and extensions of the theory.

a) Laetsch's uniqueness result does not seem to have a direct application to the problem of section 5. However, the fact that the solution u of problem (31) is continuous makes it possible to interpret it, following Bensoussan-Lions, as the solution of a stochastic impulse control problem, for which the uniqueness of the solution is known. It would be nice to have a direct proof of the uniqueness of the solution of (31) and we might also ask whether the solution of (26) is unique.

b) The solution of problem (31) can be obtained, as that of problem (11) of section 4, by an iterative process such as (24). In fact, it is also proved in [17], under the same assumptions of Theorem 2, that if $\bar{u}$ and $\underline{u}$ are the solutions of the Dirichlet problems (32) and (34), then the sequences $\{u_k\}$, $\{u_k'\}$, defined iteratively by

$$u_0 = \bar{u}, \qquad u_k = S u_{k-1}, \qquad k = 1, 2, \ldots,$$

$$u_0' = \underline{u}, \qquad u_k' = S u_{k-1}', \qquad k = 1, 2, \ldots,$$

converge to the (unique) solution u of (31), weakly in $H^{2,p}(\mathcal{O})$. Moreover, $u_k' \leqq u \leqq u_k$ for every k; the sequence $\{u_k\}$ is pointwise non-increasing; the sequence $\{u_k'\}$ is pointwise non-decreasing, and both sequences converge uniformly to u on $\bar{\mathcal{O}}$. These results are more complete than those available for problem (11), since in that case, as already remarked in section 4, the convergence from below of an iterative process starting with a subsolution ($u = 0$ in that case, see (25)) has not yet been proved (and it may be indeed false, unless, perhaps, a proper initial subsolution is chosen).

c) The regularity result of Theorem 2 has been adapted to problem (11) by Hanouzet-Joly [12], by taking L of the form

$$L = -\Delta + b_1 \frac{\partial}{\partial x_1} + b_2 \frac{\partial}{\partial x_2} + c, \qquad b_1, b_2, c \text{ constant},$$

and the region $\mathcal{O}$ of rectangular type. The main tool is again an inequality of Lewy-Stampacchia type, see Hanouzet-Joly [12].

d) Estimates such as (51) have proven to be also useful in establishing the existence of *regular* strong solutions of *parabolic VI's*, as shown by Charrier-Troianiello [7], as well as in generalizing the regularity result described above to the corresponding *evolutionary QVI*, as done in a forthcoming paper by Charrier and M. A. Vivaldi [8].

e) Many QVI's have been considered up to now, involving different types of operators M and in connection with a variety of *free boundary problems*, we refer again to A. Bensoussan and J. L. Lions, loc. cit., and to A. Friedman, R. Jensen [9, 10], C. Baiocchi [1]. For *nonlinear* QVI's see also A. Bensoussan and J. L. Lions [3]. QVI's have been also studied in a general abstract framework, by relying on the relevant order and topological features shared by a wide class of "implicit" variational problems presenting the "circularity" mentioned in section 2. See [13, 14, 15], where QVI's involving *monotone operators* are also considered.

f) Inequalities of type (51) also occur in axiomatic potential theory and can be obtained by relying only on the lattice structure of the space $H^1(\mathcal{O})$ and the lattice properties of the operator L, see ref. [11] for further comments and references. We state here without proof an "abstract" estimate of this type for variational inequalities such as

$$\left.\begin{aligned} &u \in V, \quad u \leqq \psi, \\ &(Au, u - w) \leqq (f, u - w) \quad \forall w \in V, \quad w \leqq \psi, \end{aligned}\right\}$$

where A is a strictly T-monotone operator from a Banach lattice X to the dual V' of a subspace V of X, V being also a sublattice of X, A coercive and hemicontinuous on V, f is an element of the order dual V^* of V (see ref. [25]), and $\psi \in X$ is such that $\inf\{\psi, v\} \in V$ for every $v \in V$, and $A\psi$ belongs to V^*. We recall that $A: X \mapsto V'$ is said to be strictly T-monotone if $\left(Au - Av, (u-v)^+\right) \geqq 0$ for all $u, v \in X$ with $(u - v)^+ \in V$ and where $=$ holds iff $(u - v)^+ = 0$. Then, the solution u of the VI above satisfies the inequality

$$f \geqq Au \geqq \inf\{f, A\psi\}$$

in the sense of the dual order of V^*, the inf being also in the sense of the dual lattice structure of V^*.

References

[1] Baiocchi, C., C. R. Acad. Sci. Paris, série A **278**, (1974).
[2] Bensoussan, A., J. L. Lions, C. R. Acad. Sci. Paris, série A **276**, 1411—1415, 1189—1192, 1333—1338 (1973) and *ibidem*, **278**, 675—679, 747—751 (1974).
[3] Bensoussan, A., J. L. Lions, Trudy Mat. Inst. Steklov. **134**, 5—22 (1975).
[4] Bensoussan, A., J. L. Lions, Contrôle impulsionnel et temps d'arrêt: Inéquations variationnelles et quasi-variationnelles d'évolution. Lectures at the "Scuola di Matematica", Erice, March 1975.
[5] Bensoussan, A., M. Goursat, J. L. Lions, C. R. Acad. Sci. Paris, série A **276**, 1279—1284 (1973).
[6] Birkhoff, G., Lattice theory, Amer. Math. Soc. Coll. Publ. XXV, New York 1948.
[7] Charrier, P., G. Troianiello, C. R. Acad. Sci. Paris, série A **281**, 621—623 (1975).
[8] Charrier, P., M. A. Vivaldi, C. R. Acad. Sci. Paris, série A **283**, 465—467 (1976).
[9] Friedman, A., R. Jensen, Ann. Scuola Norm. Sup. Pisa (4) **2**, 421—468 (1975).
[10] Friedman, A., R. Jensen, Elliptic quasi-variational inequalities and application to a nonstationary problem in hydraulics (to appear).
[11] Hanouzet, B., J. L. Joly, C. R. Acad. Sci. Paris, série A **281**, 373—376 (1975).
[12] Hanouzet, B., J. L. Joly, C. R. Acad. Sci. Paris, série A **281**, 799—800 (1975).
[13] Joly, J. L., U. Mosco, C. R. Acad. Sci. Paris, série A **279**, 499—502 (1974).
[14] Joly, J. L., U. Mosco, *In:* Proc. Coll. IRIA LABORIA on Control Theory, June 1974, Lecture Notes in Economics and Mathematical Systems, Vol. 107, 625—642. Springer-Verlag Berlin 1975.
[15] Joly, J. L., U. Mosco, A própos de l'existence et de la régularité des solutions de certaines inéquations quasi variationnelles (to appear).
[16] Joly, J. L., U. Mosco, G. Troianiello, C. R. Acad. Sci. Paris, série A **279**, 937—940 (1974).

[17] Joly, J. L., U. Mosco, G. Troianiello, On the regular solution of a quasi variational inequality connected to a problem of stochastic impulse control (to appear).
[18] Laetsch, T., J. Funct. Anal. **18**, 286–287 (1975).
[19] Lewy, H., G. Stampacchia, J. Analyse Math. **23**, 227–236 (1970).
[20] Lions, J. L., Quelques méthodes de résolution des problèmes aux limites non linéaires. Dunod and Gauthiers-Villars ed., Paris 1969.
[21] Lions, J. L., Sur la théorie du contrôle. Int. Math. Congress, Vancouver 1974.
[22] Lions, J. L., On the numerical approximation of problems of impulse control. I. F. I. P. Meeting, Novosibirsk, July 1974.
[23] Lions, J. L., G. Stampacchia, Comm. Pure Appl. Math. **20**, 493–519 (1967).
[24] Mosco, U., G. Troianiello, Boll. Un. Mat. Ital. **8**, 57–67 (1973).
[25] Schaeffer, H., Topological vector spaces, Mac-Millan ed., New York 1966.
[26] Tartar, L., C. R. Acad. Sci. Paris, série A **278**, 1193–1196 (1974).

Address of the author:

Università di Roma
Istituto Matematico
00185 Roma
Italia

and

Centre de Recherche de Mathématiques de la Décision
Université Paris IX Dauphine
Place du Maréchal De Lattre de Tassigny
75775 Paris Cedex 16
France

Example of an irregular solution to a nonlinear elliptic system with analytic coefficients and conditions for regularity

Jindrich Nečas, Praha

§ 1. Introduction

The problem on the regularity of a weak solution to a nonlinear elliptic system of the second-order with analytic coefficients was negatively solved by the example of E. Giusti, M. Miranda [1] (see also other examples of the same type in V. G. Mazja [2], S. A. Arakčejev [3]) who have given for the dimension $n \geqq 3$, $0 \in \Omega \subset R_n$ and for the vector function $u = (u_1, \ldots, u_n)$ a functional

$$\Phi(u) = \int_{\Omega} F(u, \partial u)\, dx \tag{1.1}$$

with analytic function $F(u, \partial u)$, whose critical point is a discontinuous function with components $\frac{x_i}{r}$. The function $F(u, \partial u)$ depends on u in a substantial way and the functional (1.1) has not the second differential.

In this paper, in § 2, we give a functional for $n \geqq 3$, $0 \in \Omega \subset R_n$,

$$\Phi(u) = \int_{\Omega} F(\partial u)\, dx, \tag{1.2}$$

with an analytic function F, satisfying the growth conditions:

$$|D^l F| \leqq c_l \frac{|\partial u|^2}{(1 + |\partial u|)^l}, \quad l = 0, 1, \ldots, \tag{1.3}$$

where u is an n^2 dimensional vector function. It is $\frac{\partial F}{\partial \xi_k{}^{ij}}(0) = 0$ and for n enough large

$$\frac{\partial^2 F}{\partial \xi_k{}^{ij} \partial \xi_\gamma{}^{\alpha\beta}}(\xi)\, \eta_k{}^{ij} \eta_\gamma{}^{\alpha\beta} \geqq c|\eta|^2, \quad c > 0. \tag{1.4}$$

The functional (1.2) is on the Sobolev space $[W^{1,2}]^{n^2}$ twice continuously differentiable and for n enough large the second differential is positive definite. The critical point of (1.2) is the function u^0 with components $u^0_{ij} = \frac{x_i x_j}{r}$. We give also an example of a second-order elliptic system

$$\int_{\Omega} a_k{}^{ij}(\partial u) \frac{\partial \varphi_{ij}}{\partial x_k}\, dx = 0, \tag{1.5}$$

with analytic $a_k^{ij}(\partial u)$, $a_k^{ij}(0) = 0$, satisfying the ellipticity condition for $n \geqq 3$:

$$a_k^{ij}(\xi)\, \xi_k^{ij} \geqq c\, |\xi|^2, \qquad c > 0 \tag{1.6}$$

and the strong monotony condition for $n \geqq 5$:

$$\frac{\partial a_k^{ij}(\xi)}{\partial \xi_\gamma^{\alpha\beta}}\, \eta_k^{ij} \eta_\gamma^{\alpha\beta} \geqq c|\eta|^2, \qquad c > 0. \tag{1.7}$$

A weak solution of (1.5) is also the above defined function u^0.

In § 3, we give an "a priori" estimate for the regularity of a critical point to the functional

$$\Phi(u) = \int_\Omega [F(\partial u) - (f, \partial u)]\, dx, \tag{1.8}$$

where $\Omega \subset R_3$, $\partial\Omega$ is enough smooth, $u_1 = (u, u_2, \ldots, u_m)$ and ∂u is the jet of the first order derivatives. The function $F(\xi)$ is defined in a bounded ball K_a, $F \in C^{(4)}(\bar{K}_a)$. Let $f \in [W^{2,2}]^{3m}$, $\bar{u} \in [W^{3,2}]^m$ and let $u \in [W^{3,2}]^m$ be a critical point of (1.8), such that $\forall x \in \bar{\Omega}: |\partial u(x)| < r \leqq a$ and let $u = \bar{u}$ on $\partial\Omega$. Let us suppose for $|\xi| < r$:

$$\frac{\partial^2 F}{\partial \xi_i^r\, \partial \xi_j^s}(\xi)\, \eta_i^r \eta_j^s \geqq c(r)\, |\eta|^2, \qquad c(r) > 0, \tag{1.9}$$

$$\frac{\partial^4 F}{\partial \xi_i^r\, \partial \xi_j^s\, \partial \xi_k^t\, \partial \xi_l^v}\, \eta_i^r \eta_j^s \eta_k^t \eta_l^v \leqq T(r) \sum_{r,i} (\eta_i^r)^4, \tag{1.10}$$

$$c(r) - 3r^2 T(r) > 0 \quad \text{(the regularity condition)}, \tag{1.11}$$

then

$$\|u\|_{3,2} \leqq A(r)\left[1 + \|\bar{u}\|_{3,2}^2 + \|f\|_{2,2}^2\right].$$

This estimate enables us to find classical solutions of problems with weak solutions, or of problems where the existence of weak solutions is not proved, in the form of solution's path when the data's paths $f(t)$, $\bar{u}(t)$ are prescribed. For details see the paper by J. Nečas, J. Stará, R. Švarc [4].

In § 4, constructive aspects of the method are briefly discussed and the second problem to the finite elasticity is defined as an example, where the theory of monotone operators fail.

§ 2. Examples for irregular solutions

Let $0 \in \Omega \subset R_n$, where Ω is a bounded domain. Put $m = n^2$ and let us look for $u \in [W^{1,2}]^m$, such that

$$u - u^0 \in [W_0^{1,2}]^m \quad \text{(functions with zero traces on } \partial\Omega\text{)}, \tag{2.1}$$

where $u_{ij}^0 = \dfrac{x_i x_j}{r}$, $i, j = 1, 2, \ldots, n$, and where u is the weak solution to the elliptic system: $\forall \varphi \in [\mathscr{D}(\Omega)]^m$

$$\int_\Omega a_k^{ij}(\partial u)\, \frac{\partial \varphi_{ij}}{\partial x_k} = 0. \tag{2.2}$$

Let

$$\Phi(u) = \int_\Omega \left[\frac{1}{2}\frac{\partial u_{ij}}{\partial x_k}\frac{\partial u_{ij}}{\partial x_k} + \frac{\mu}{2}\frac{\partial u_{ij}}{\partial x_i}\frac{\partial u_{kk}}{\partial x_j} + \lambda \frac{\partial u_{ij}}{\partial x_i}\frac{\partial u_{k\alpha}}{\partial x_k}\frac{\partial u_{l\beta}}{\partial x_l}\frac{\partial u_{j\alpha}}{\partial x_\beta}\left(1 + \frac{\partial u_{ab}}{\partial x_c}\frac{\partial u_{ab}}{\partial x_c}\right)^{-1}\right] dx \equiv \int_\Omega F(\partial u)\, dx, \quad (2.3)$$

where $\lambda = 2\dfrac{n^3 - 1}{n(n-1)(n^3 - n + 1)}$, $\mu = -\dfrac{4 + \lambda n}{n^2 - n + 1}$.

We have (1.7), u^0 is a critical point of (2.3), $\dfrac{\partial F}{\partial \xi_k^{ij}}(0) = 0$, and for n enough large (1.4) holds. In order to see the last statement, put $\nu^i = \xi_j^{ji}$, $\nu_\eta^{\;i} = \eta_j^{ji}$, $M^i = \xi_i^{jj}$, $N_i = \eta_i^{jj}$ and note $V = \dfrac{\partial u_{ij}}{\partial x_i}\dfrac{\partial u_{k\alpha}}{\partial x_k}\dfrac{\partial u_{l\beta}}{\partial x_l}\dfrac{\partial u_{j\alpha}}{\partial x_\beta}$, $\dfrac{\partial u_{ab}}{\partial x_l}\dfrac{\partial u_{ab}}{\partial x_l} = U$, $\xi_k^{ij}\eta_k^{ij} = [\xi, \eta]$, $\nu^i\nu_\eta^{\;j} = (\nu, \nu_\eta)$. We get

$$\begin{aligned} D^2F(\xi, \eta, \eta) = {} & |\eta|^2 + \mu(\nu_\eta, M_\eta) \\ & + \lambda D^2V(\xi, \eta, \eta)(1 + U)^{-1} - 4\lambda DV(\xi, \eta)(1 + U)^{-2}[\xi, \eta] \\ & + 8\lambda V(1 + U)^{-3}[\xi, \eta]^2 - 2\lambda V(1 + U)^{-2}|\eta|^2. \end{aligned} \quad (2.4)$$

Because $|\nu| \leqq \sqrt{n}\,|\xi|$, $|\nu_\eta| \leqq \sqrt{n}\,|\eta|, \ldots$, we get from (2.4)

$$D^2F(\xi, \eta, \eta) \geqq \left(1 - \frac{c}{\sqrt{n}}\right)|\eta|^2. \quad (2.5)$$

The construction of the function F and the estimate (2.5) are crude. In order to get better estimations for the dimension where "the theorem on the regularity" is not valid, let us consider the elliptic system

$$\int_\Omega \left[\frac{\partial u_{ij}}{\partial x_k}\frac{\partial \varphi_{ij}}{\partial x_k} + a_1\frac{\partial u_{ik}}{\partial x_i}\frac{\partial \varphi_{jk}}{\partial x_j} + a_2\frac{\partial u_{ii}}{\partial x_k}\frac{\partial \varphi_{jj}}{\partial x_k} + \mu\frac{\partial u_{ij}}{\partial x_i}\frac{\partial \varphi_{kk}}{\partial x_j} + \lambda\frac{\partial u_{ij}}{\partial x_i}\frac{\partial u_{\alpha\beta}}{\partial x_\alpha}\frac{\partial u_{k\gamma}}{\partial x_k}\frac{\partial \varphi_{j\beta}}{\partial x_j}\left(1 + \frac{\partial u_{ab}}{\partial x_u}\frac{\partial u_{cb}}{\partial x_c}\right)^{-1}\right] dx \equiv \int_\Omega f(\partial u, \partial\varphi)\, dx. \quad (2.6)$$

Put $a_1 = \dfrac{n+1}{n}$, $a_2 = \dfrac{n+3}{n-1}$, $\lambda = 2\dfrac{(n+1)(n^2+1)}{n^3(n-1)}$, $\mu = -2\dfrac{n+3}{n(n-1)}$, $u_{ij}^0 = \dfrac{x_i x_j}{r}$. We get (2.6) for $u = u^0$. For $n \geqq 3$, we get

$$\begin{aligned} f(\xi, \xi) & \geqq |\xi|^2 + \frac{n+1}{n}|\nu|^2 + \frac{n+3}{n-1}|M|^2 - |\mu|\,|\nu|\,|M| - \lambda\frac{|\nu|^3\,|\xi|}{1 + |\nu|^2} \\ & \geqq |\xi|^2 + \frac{n^3 - 2n - 3}{n^2(n-1)}|\nu|^2 - \frac{2(n+1)(n^2+1)}{n^3(n-1)}|\nu|\,|\xi|. \end{aligned} \quad (2.7)$$

It follows that for $n \geqq 3$ the last form is positive definite.

To obtain (1.7), we can see that the integral $I = \int_\Omega \frac{\partial u_{ij}}{\partial x_i} \frac{\partial u_{\alpha\beta}}{\partial x_\alpha} \frac{\partial u_{k\gamma}}{\partial x_k} \frac{\partial \varphi_{j\beta}}{\partial x_\gamma}\, dx$ is independent with respect to a permutation of the indices j, β, γ. Put $\eta^{i,j,k} = \frac{1}{3}\left(\eta_k{}^{ji} + \eta_k{}^{ji} + \eta_j{}^{ik}\right)$, $V(\xi, \eta) = \xi_i{}^{ij}\xi_a{}^{a\beta}\xi_k{}^{k\gamma}\eta_\gamma{}^{j\beta}\left(1 + \xi_a{}^{ab}\xi_c{}^{ab}\right)^{-1}$. We have

$$DV(\xi, \eta, \eta) = 3\nu_\eta{}^j\nu^\beta\nu^\gamma\eta^{j\beta\gamma}(1 + |\nu|^2)^{-1} - 2\nu^j\nu^\beta\nu^\gamma\eta^{j\beta\gamma}(1 + |\nu|^2)^{-2}\,(\nu, \nu_\eta). \tag{2.8}$$

Put $\nu' = \frac{\nu}{(1 + |\nu|^2)^{\frac{1}{2}}}$, $\mu^j = \nu'^\beta\nu'^\gamma\eta^{j\beta\gamma}$.

We have to estimate the term

$$W \equiv |3(\nu_\eta, \eta) - 2(\nu', \mu)\,(\nu', \nu_\eta)|.$$

By the projection of the vector ν' on the plane generated by ν_η, μ (we let the details to the reader), we get finally

$$W \leqq 3\,|\eta|\,|\nu_\eta|. \tag{2.9}$$

Hence

$$Df(\xi, \eta, \eta) \geqq |\eta|^2 + \frac{n+1}{n}|\nu_\eta|^2 + \frac{n+3}{n-1}|M_\eta|^2 - |\mu|\,|\nu_\eta|\,|M_\eta| - 3\lambda\,|\nu_\eta|\,|\eta|$$

$$\geqq |\eta|^2 + \frac{n^3 - 2n - 3}{n^2(n-1)}|\nu_\eta|^2 - \frac{6(n+1)(n^2+1)}{n^3(n-1)}|\nu_\eta|\,|\eta|, \tag{2.10}$$

which implies (1.7) for $n \geqq 5$.

We have also

$$|f(\xi, \eta)| \leqq c|\xi|\,|\eta|, \tag{2.11}$$

for

$$k \geqq 1: |D_\xi{}^k f(\xi, \eta)| \leqq c_k|\eta|\,(1 + |\nu|)^{1-k}. \tag{2.12}$$

§ 3. Conditions for regularity

Let us consider $\Omega \subset R_3$ with $\partial\Omega$ enough smooth and let $F(\xi)$, where $\xi = \{\xi_i{}^r, r = 1, 2, \ldots, m, i = 1, 2, 3\}$, be a real function defined on the ball $K_a = \{\xi, |\xi| < a\}$. Let $F \in C^{(4)}(\overline{K}_a)$, $\frac{\partial F}{\partial \xi_i{}^r}(0) = 0$. Let $f \in [W^{2,2}]^{3m}$, $\tilde{u} \in [W^{3,2}]^m$. Let us have $u \in [W^{3,2}]^m$ such that $u = \tilde{u}$ on $\partial\Omega$, $\|u\|_{1,\infty} \equiv \max_{x\in\overline{\Omega}}\left(\frac{\partial u_r}{\partial x_i}(x), \frac{\partial u_r}{\partial x_i}(x)\right)^{\frac{1}{2}} < a$, and such that it is a solution to the elliptic system:

$$\forall \varphi \in [W_0{}^{1,2}]^m$$

$$\int_\Omega \frac{\partial F}{\partial \xi_i{}^r}(\partial u)\frac{\partial \varphi_r}{\partial x_i}\, dx = \int_\Omega f_i{}^r \frac{\partial \varphi_r}{\partial x_i}\, dx. \tag{3.1}$$

Theorem 3.1. *Let u be a solution of the system* (3.1), *satisfying the boundary condition $\tilde{u}$, let $\|u\|_{1,\infty} < r < \varrho \leqq a$ and let for $|\xi| < r$ the conditions* (1.9), (1.10), (1.11) *be valid. Then there exists a continuous function $d(r)$ for $0 \leqq r < \varrho$ such that*

$$\|u\|_{3,2} \leqq d(r) \left(1 + \|f\|_{2,2}^2 + \|\tilde{u}\|_{3,2}^2\right). \tag{3.2}$$

If $\lim_{r\to\varrho} c(r) = \alpha > 0$ and $\lim_{r\to\varrho} c(r) - 3T(r)\, r^2 = \beta > 0$, then $d(r)$ is bounded on the interval $[0, \varrho[$.

We refer the reader for details to the paper [4]. We sketch the proof in this paper for $\tilde{u} = 0$ and $\Omega = R_3^+$, the halfspace.

Theorem 3.2. *Let $\Omega = R_3^+$, let u be a solution of* (3.1) *with zero boundary data and let the conditions of Theorem* 3.1 *be satisfied. Then there exists a continuous function $c_1(r)$ in the interval $[0, \varrho]$, such that*

$$\|u\|_{3,2} \leqq c_1(r) \left(c(r)\right)^{-4} \left[c(r) - 3T(r)\, r^2\right]^{-1} \left(\|f\|_{2,2} + \|f\|_{2,2}^2\right). \tag{3.3}$$

Proof. For the halfspace R_3^+ we use the following norms for $W^{k,2} \cap W_0^{1,2}$ spaces $(k \geqq 1)$:

$$\|u\|_{k,2} = \left(\int_{R_3^+} \sum_{1\leqq|\alpha|\leqq k} (D^\alpha u)^2\, dx\right)^{\frac{1}{2}}.$$

Let $\varphi \in [\mathscr{D}]^m$ and let φ' be the derivative $\frac{\partial\varphi}{\partial x_1}$ or $\frac{\partial\varphi}{\partial x_2}$. We have

$$\int_{R_3^+} \frac{\partial F}{\partial \xi_i^r} \frac{\partial \varphi_r^{(4)}}{\partial x_i}\, dx = \int_{R_3^+} (f_i^r)'' \frac{\partial \varphi_r''}{\partial x_i}\, dx = \int_{R_3^+} \frac{\partial^2 F}{\partial \xi_i^r\, \partial \xi_j^s} \frac{\partial u_s''}{\partial x_j} \frac{\partial \varphi_r''}{\partial x_i}\, dx$$
$$+ \int_{R_3^+} \frac{\partial^3 F}{\partial \xi_i^r\, \partial \xi_j^s\, \partial \xi_k^t} \frac{\partial u_s'}{\partial x_j} \frac{\partial u_t'}{\partial x_k} \frac{\partial \varphi_r'}{\partial x_i}\, dx. \tag{3.4}$$

Passing to the corresponding limit, we can put $\varphi = u$. We first get

$$\int_{R_3^+} \frac{\partial^3 F}{\partial \xi_i^r\, \partial \xi_j^s\, \partial \xi_k^t} \frac{\partial u_r''}{\partial x_i} \frac{\partial u_s'}{\partial x_j} \frac{\partial u_t'}{\partial x_k}\, dx$$
$$= -\frac{1}{3} \int_{R_3^+} \frac{\partial^4 F}{\partial \xi_i^r\, \partial \xi_j^s\, \partial \xi_k^t\, \partial \xi_l^v} \frac{\partial u_r'}{\partial x_i} \frac{\partial u_s'}{\partial x_j} \frac{\partial u_t'}{\partial x_k} \frac{\partial u_v'}{\partial x_l}\, dx. \tag{3.5}$$

In the following we use a special case of the L. Nirenberg's interpolation inequality, see L. Nirenberg [5], which we can get by a simple integration by parts:

$$\text{for } v \in \mathscr{D}(-\infty, \infty), \quad \int_{-\infty}^{\infty} (v')^4\, dx \leqq 9 \max_{x\in R_1} |v(x)|^2 \int_{-\infty}^{\infty} (v'')^2\, dx. \tag{3.6}$$

From (3.6) and (1.10) we obtain

$$-\frac{1}{3}\int\limits_{R_3^+}\frac{\partial^4 F}{\partial\xi_i^r\,\partial\xi_j^s\,\partial\xi_k^t\,\partial\xi_l^v}\,\frac{\partial u_r'}{\partial x_i}\,\frac{\partial u_s'}{\partial x_j}\,\frac{\partial u_t'}{\partial x_k}\,\frac{\partial u_v'}{\partial x_l}\,dx$$

$$\geqq -3T(r)\,r^2\int\limits_{R_3^+}\frac{\partial u_r''}{\partial x_i}\,\frac{\partial u_r''}{\partial x_i}\,dx. \tag{3.7}$$

It follows from (3.4), (3.5), (3.7) and (1.11) that

$$\left(c(r) - 3T(r)\,r^2\right)\|u''\|_{1,2} \leqq \|f\|_{2,2}. \tag{3.8}$$

In virtue of (3.6), we get

$$\int\limits_{R_3^+}\sum_{r,i}\left(\frac{\partial u_r'}{\partial x_i}\right)^4 dx \leqq \frac{9r^2}{\left(c(r) - 3T(r)\,r^2\right)^2}\,\|f\|_{2,2}^2, \tag{3.9}$$

hence

$$\left\|\left(\frac{\partial F}{\partial\xi_i^r}\right)'\right\|_{0,4} \leqq \frac{M(r)}{\left(c(r) - 3T(r)\,r^2\right)^{\frac{1}{2}}}\,\|f\|_{2,2}^{\frac{1}{2}}. \tag{3.10}$$

But

$$\sup_{\varphi\in[\mathscr{D}]^s,\|\varphi\|_{0,\frac{4}{3}}\leq 1}\left|\int\limits_{R_3^+}\frac{\partial}{\partial x_3}\left(\frac{\partial F}{\partial\xi_3^r}\right)\varphi_r\,dx\right|$$

$$\leqq \sup\left[\left|\int\limits_{R_3^+}\frac{\partial}{\partial x_1}\left(\frac{\partial F}{\partial\xi_l^r}\right)\varphi_r\,dx\right| + \left|\int\limits_{R_3^+}\frac{\partial}{\partial x_2}\left(\frac{\partial F}{\partial\xi_2^r}\right)\varphi_r\,dx\right| + \left|\int\limits_{\Omega} f_i^r\,\frac{\partial\varphi_r}{\partial x_i}\,dx\right|\right]$$

$$\leqq \frac{N(r)}{\left(c(r) - 3T(r)\,r^2\right)^{\frac{1}{2}}}\,\|f\|_{2,2}^{\frac{1}{2}} + c\,\|f\|_{2,2}. \tag{3.11}$$

We can calculate the derivatives $\frac{\partial^2 u_s}{\partial x_3^2}$, $s = 1, 2, \ldots, m$, from the elliptic system

$$-\frac{\partial^2 F}{\partial\xi_i^r\,\partial\xi_j^s}\,\frac{\partial^2 u_s}{\partial x_i\,\partial x_j} = -\frac{\partial f_i^r}{\partial x_i}, \qquad r = 1, 2, \ldots, m \tag{3.12}$$

and we get

$$\left\|\frac{\partial^2 u_s}{\partial x_3^2}\right\|_{0,4} \leqq \frac{O(r)}{c(r)\,[c(r) - 3T(r)\,r^2]^{\frac{1}{2}}}\left[\|f\|_{2,2}^{\frac{1}{2}} + \|f\|_{2,2}\right]. \tag{3.13}$$

If we differentiate (3.12) with respect to x_1 or x_2, we obtain

$$-\frac{\partial^2 F}{\partial\xi_i^r\,\partial\xi_j^s}\,\frac{\partial^2 u_s'}{\partial x_i\,\partial x_j} = -\frac{\partial f_i^{r\prime}}{\partial x_i} + \frac{\partial^3 F}{\partial\xi_i^r\,\partial\xi_j^s\,\partial\xi_k^t}\,\frac{\partial u_t'}{\partial x_k}\,\frac{\partial^2 u_s}{\partial x_i\,\partial x_j}, \tag{3.14}$$

hence

$$\|u'\|_{2,2} \leqq \frac{Q(r)}{c(r)^3\,[c(r) - 3T(r)\,r^2]}\left(\|f\|_{2,2} + \|f\|_{2,2}^2\right). \tag{3.15}$$

Once more differentiating (3.12) with respect to x_3, we get the inequality (3.2), q.e.d.

Let data paths be given: $\tilde{u} \in C([0, T], [W^{3,2}]^m)$, $f \in C([0, T], [W^{2,2}]^{3m})$, $\tilde{u}(0) = 0$, $f(0) = 0$, and let us look for $u(t) \in [W^{3,2}]^m$ such that, $\forall t$ from the below precised interval, $u(t)$ solves the equation (3.1) with the boundary condition $\tilde{u}(t) = u(t)$ and such that $\|u(t)\|_{1,\infty} < \varrho \leqq a$. We have

Theorem 3.3. *Let the conditions* (1.9) *and* (1.11) *be satisfied for all* $r < \varrho$. *Then there exists a unique solution path* $u(t)$ *and we have:*

$$u \in C([0, T], [W^{3,2}]^m) \tag{3.16}$$

or

$$u \in C([0, t_{cr}[, [W^{3,2}]^m), \tag{3.17}$$

where $t_{cr} \leqq T$ *and, in the case* (3.17), *we have*

$$\lim_{t \to t_{cr}} \min_{x \in \bar{\Omega}} |\partial u(x, t) - \partial K_\varrho| = 0. \tag{3.18}$$

If $\lim_{r \to \varrho} c(r) = \alpha > 0$ *and* $\lim_{r \to \varrho} \left(c(r) - 3r^2 T(r)\right) = \beta > 0$, *then*

$$\sup_{t \in [0, t_{cr}[} \|u(t)\|_{3,2} < \infty. \tag{3.19}$$

The reader can easily see that Theorem 3.3 follows readily from Theorem 3.1 and from the implicit function theorem applied to our problem. We sketch the situation for $\Omega = R_3^+$:

Theorem 3.4. *Let* $u_0(t) \in [W^{3,2}(R_3^+)]^m \cap [W_0^{1,2}(R_3^+)]^m$ *be a solution of the system* (3.1) *for some* t_0 *and let* $U(u(t_0))$ *be some small neighbourhood of* $u(t_0)$. *On* $U(u(t_0)) \times (t_0 - \varepsilon, t_0 + \varepsilon)$, *let us define the operator* A *to* $[W^{2,2}(R_3^+)]^{3m}/S$

$$\int_{R_3^+} \frac{\partial F}{\partial \xi_i^r}\left(\partial u(t)\right) \frac{\partial \varphi_r}{\partial x_i}\, dx = \int_{R_3^+} g_i^r \frac{\partial \varphi_r}{\partial x_i}; \tag{3.20}$$

here

$$S = \left\{ g \,\middle|\, \int_{R_3^+} g_i^r \frac{\partial \varphi_r}{\partial x_i}\, dx = 0 \right\}.$$

Then the operator $A(u, t)$ *is continuous, has a continuous derivative* $A_u'(u, t)$, *invertible at the point* $[u(t_0), t_0]$.

Proof. We shall sketch the proof of the invertibility of A_u'. But this is equivalent to the proof that the system

$$\int_{R_3^+} \frac{\partial^2 F}{\partial \xi_i^r\, \partial \xi_j^s}\left(\partial u(t_0)\right) \frac{\partial v_s}{\partial x_j} \frac{\partial \varphi_r}{\partial x_i}\, dx = \int_{R_3^+} h_i^r \frac{\partial \varphi_r}{\partial x_i}\, dx \tag{3.21}$$

has a unique solution in $[W^{3,2}]^m \cap [W_0^{1,2}]^m$. But the condition (1.9) implies the unique solvability of (3.21) in $[W_0^{1,2}]^m$. We have to prove that such a solution $v \in [W^{3,2}]^m$. This we can do by investigating the integral

$$\int_{R_3^+} \frac{\partial^2 F}{\partial \xi_i^r \, \partial \xi_j^s} \left(\partial u(t_0)\right) \frac{\partial v_s}{\partial x_j} \frac{\partial \varphi_r^{(4)}}{\partial x_i} \, dx \tag{3.22}$$

and by following the analogous steps as in the proof of Theorem 3.2. For the usual $W^{l,2}$ norms $\|f\|_{l,2} = \left(\int_{R_3^+} \sum_{|\alpha| \le l} (D^\alpha f)^2 \, dx \right)^{\frac{1}{2}}$ we use the inequalities

$$\|f\|_{k,2} \le \varepsilon \|f\|_{l,2} + c_\varepsilon \|f\|_{0,2} \tag{3.23}$$

for $k < l$, which we get easily in R_3 by the Fourier transform and in R_3^+ by the extension of f from R_3^+ to R_3. For details see, for example, J. Nečas [6]. Using the inequality (3.6), we get first

$$\|u\|_{2,4} \le c \, \|u\|_{3,2} \tag{3.24}$$

and then

$$\|v''\|_{1,2} \le \varepsilon \, \|v\|_{3,2} + c_\varepsilon \|h\|_{2,2}. \tag{3.25}$$

The next step is the inequality

$$\|v\|_{2,4} \le \varepsilon \|v\|_{3,2} + d_\varepsilon \|h\|_{2,2} \tag{3.26}$$

and finally, we obtain the inequality

$$\|v\|_{3,2} \le c \|h\|_{2,2}. \tag{3.27}$$

(3.27) is an "a priori" inequality, whose validity for the solution v can be obtained by the standard homotopy argument.

§ 4. Constructive aspects and finite elasticity

Let t_0 be an attainable parameter from Theorem 3.2. The path $u(t)$ can be understood as an embedding process, connecting $t = 0$, where $u(0) = 0$, with $u(t_0)$. For details see H. Schwetlick [7]. Instead of looking for the whole path, we can calculate $u(0)$, $u(t_1)$, $u(t_2), \ldots, u(t_0)$ with $t_i - t_{i-1}$ enough small, using Newton's method. The conditions for the applicability of this method are, roughly speaking, the same as for the implicit function theorem, see Theorem 3.4, provided that the paths $\tilde{u}(t)$ and $f(t)$ are in $C^{(1)}$. Let us underline the fact that the conditions for the applicability of Newton's method are not satisfied for the weak solutions of nonlinear elliptic systems of the type (3.1).

Let us define the place boundary value problem for the finite elasticity. For details see K. Washizu [8].

Let $\Omega \subset R_3$ be a bounded domain with $\partial\Omega$ enough smooth, the considered body before the deformation Ω is supposed to play the role of a reference domain. Let u be the displacement vector and

$$e_{ij} = \frac{1}{2}\left(\frac{\partial u_i}{\partial x_j} + \frac{\partial u_j}{\partial x_i} + \frac{\partial u_k}{\partial x_i}\frac{\partial u_k}{\partial x_j}\right) \tag{4.1}$$

the finite strain-tensor. Let σ^i be the stress vectors acting on the surface, which before the deformation had the normal parallel to the x_i axis, and let σ^i be defined with respect to the unit area of the surface before the deformation. Let P be the vector of body forces with respect to the unit volume before the deformation. In every point of the body, we have the equations of equilibrium

$$\frac{\partial \sigma^i}{\partial x_i} + P = 0. \tag{4.2}$$

Let $y = x + u(x)$, and let us define a tensor $\sigma^{\lambda\mu}$ by

$$\sigma^\lambda = \sigma^{\lambda\mu}\frac{\partial y}{\partial x_\mu}. \tag{4.3}$$

The moment equation of equilibrium gives $\sigma^{\mu\lambda} = \sigma^{\lambda\mu}$.

In some domain of the strain tensor, let the Hooke law be defined by

$$\sigma^{\mu\lambda} = \frac{\partial A}{\partial e_{\mu\lambda}}. \tag{4.4}$$

On $\partial\Omega$ let the displacement $\tilde{u}$ be described. Let us suppose that the body forces depend only on x (and not on $u(x)$). From (4.2), (4.3) and (4.4) we then get the system of finite elasticity:

$$\frac{\partial}{\partial x_i}\left[\frac{\partial A}{\partial e_{ij}}\left(\delta_{\alpha j} + \frac{\partial u_\alpha}{\partial x_j}\right)\right] + P_\alpha = 0. \tag{4.5}$$

(4.5) are the Euler equations for the functional of the total potential energy

$$\Phi(u) = \int_\Omega A(e_{ij}(u))\,dx - \int_\Omega u_i P_i\,dx. \tag{4.6}$$

If we put $F(\partial u) = A(e_{ij}(u))$, we have an example of the system (3.1). The functional (4.6) is not positive definite and not lower weakly semicontinuous in the energy space (which is, for a linear Hooke law, $[W^{1,4}]^3$!), see the examples from the paper by J. Nečas [9]. The conditions (1.9) can be defined only locally in the space $\|u\|_{1,\infty}$ because $c(r) > 0$ gives the condition of the local stability of the system (and the instability can occur). The paths $\tilde{u}(t)$, $f(t)$ express the loading paths.

If the conditions (1.9), (1.11) are satisfied in the domain of definition for the function $F(\partial u)$, the solution path $u(t)$ exists up to a point where it loses its physical meaning.

Bibliography

[1] Giusti, E., M. Miranda, Un esempio di soluzioni discontinue per un problema di minimo relativo ad un integrale regolare del calcolo delle variazioni. Boll. Un. Mat. Ital. (4) 2, 1—8 (1969).

[2] Мазья, В. Г., Примеры нерегулярных решений квазилинейных эллиптических уравнений с аналитическими коэффициентами. Функц. Анализ **2**, 3, 53—57 (1968).
[3] Аракчеев, С. А., О гладкости обобщенных решений некоторого класса квазилинейных эллиптических уравнений. Вестник Московск. Унив. **30**, 1, 49—56 (1975).
[4] Nečas, J., J. Stará, R. Švarc, Classical solution to second-order nonlinear elliptic system in R_3 (to appear).
[5] Nirenberg, L., An extended interpolation inequality. Ann. Scuola Norm. Sup. Pisa **20**, 733 to 737 (1966).
[6] Nečas, J., Les méthodes directes en théorie des équations elliptiques. Academia, Prague 1967.
[7] Schwetlick, H., Continuation methods for solving nonlinear equations (to appear in the volume: Mathematical Models and Numerical Methods Semester, edited by S. Banach international mathematical center).
[8] Washizu, K., Variational methods in elasticity and plasticity. Pergamon Press, 1968.
[9] Nečas, J., Theory of locally monotone operators modeled on the finite displacement theory for hyperelasticity (to appear).

Address of the author:

Mathematical Institute of the Czechoslovak Academy of Sciences
Žitná 25
11567 Praha 1
Czechoslovakia

Some results on infinite dimensional extremal problems

Andrzej W. Olbrot, Warszawa

1. Introduction

The lecture is devoted to optimal control problems in Banach and Hilbert spaces and introduces some recent results obtained at Warsaw Technical University, Institute of Automatics. The results of section 2 are based on a paper of S. Kurcyusz [6] (to be published). In section 2 the fundamental question of nontriviality and normality of generalized Lagrange-Euler equation is examined and appropriate conditions (necessary and sufficient) are described. Some of them indicate how to choose the right topology in the space of controls and the constraint space.

These results show that if the reachable subspace of the linearized system is closed then the normal form of necessary optimality conditions can be established. The explicit algebraic criteria for the closedness of the reachable subspace for differential-difference systems are briefly summarized in section 3 (the full version will appear in J. Diff. Eqs.).

In section 4 necessary conditions for optimality of time-lag systems with a ball in function space as a target set are stated. Although such type of constraints results, as a rule, in the maximum principle of integral form — we are able to show the validity of results rather similar to classical ones.

2. On Lagrange multipliers for infinite dimensional optimization problems

Consider the following minimization problem.

Problem 2.1.

$$\begin{cases} \text{minimize } J(u) \\ \text{subject to } u \in M, \quad S(u) \in K. \end{cases}$$

It is assumed that the following holds.

Assumption 2.1. U, L are Banach spaces, $\mathcal{O} \subset U$ is an open set, $K \subset L$ is a closed convex cone with vertex at zero, $S: \mathcal{O} \mapsto L$ is a continuously Fréchet differentiable operator, $\hat{u}$ is a given point in $\mathcal{O}$ satisfying $S(\hat{u}) \in K$.

Assumption 2.2. $J: \mathcal{O} \to R$ is a Fréchet differentiable functional.

Assumption 2.3. M is a subset of U with $\hat{u} \in M$ possessing "a good conical approximation" at $\hat{u}$. More precisely, the open cone

$$M_{\hat{u}} = \{u \in U: \exists\, \varepsilon_0, r > 0, \forall\, 0 < \varepsilon \leqq \varepsilon_0, \forall\, \|v\| \leqq r \Rightarrow \hat{u} + \varepsilon(u + v) \in M\}^1$$

is convex.

The goal of this section is to investigate the conditions under which nontrivial Lagrange multipliers for 2.1 at $\hat{u}$ exist, provided $\hat{u}$ is a local solution to 2.1. Recall the following

Definition. By *nontrivial Lagrange multipliers for Problem* 2.1 *at* $\hat{u}$ we mean a pair $(\lambda_0, l^*) \in R \times L^*$ satisfying

$$(\lambda_0, l^*) \neq (0, 0),$$

$$\lambda_0 \geqq 0, \qquad l^* \in K^*, \qquad \langle l^*, S(\hat{u})\rangle = 0,$$

$$\lambda_0 J_{\hat{u}} - S_{\hat{u}}{}^* l^* \in W_{\hat{u}}{}^*.$$

A *normal Lagrange* multiplier is a vector $l^* \in L$ such that the above conditions hold with $\lambda_0 = 1$.

Remark 2.1. The assumptions imposed on S, J and M are rather strong. Notice, however, that we do not assume anything about K except that it is convex and closed. In particular cases (int $K \neq \emptyset$) one can easily relax the assumption on S [1]; it is also possible to discuss problems 2.1 for which Assumption 2.3 is not satisfied [2, 3]. Nevertheless we accept here assumptions 2.1—2.3 since they allow a rather full discussion of the problem.

Since the necessary optimality conditions for a nonlinear problem 2.1 are expressed in terms of linearizations $J_{\hat{u}}$, $S_{\hat{u}}$, $M_{\hat{u}}$ ($J_{\hat{u}}$, $S_{\hat{u}}$ being Fréchet derivatives) it is clear that we shall need 2.1 to be "well linearized". This is ensured by the so called Kuhn-Tucker regularity condition [2, 3]. Define first conical approximations of the sets $S^{-1}(K)$ at $\hat{u}$ and K at $S(\hat{u})$.

Definition. Denote

$$\big(S^{-1}(K)\big)_{\hat{u}} = \left\{\begin{matrix} u \in U: \exists\, \{\varepsilon_n\} \subset R^+,\ \varepsilon_n \to 0, \\ \exists\, \{u_n\} \subset S^{-1}(K), \end{matrix} \qquad u = \lim \frac{u_n - \hat{u}}{\varepsilon_n}\right\}^2$$

$$K_{S(\hat{u})} = K + \operatorname{Lin} S(\hat{u})^3.$$

Definition. We shall say that the *Kuhn-Tucker condition* (KT) holds if

$$S^{-1}\big((K)\big)_{\hat{u}} \supset S_{\hat{u}}^{-1} K_{S(\hat{u})}.$$

Condition (KT) is related to the nonlinearity of 2.1 regardless the dimensionality of U and L. If L is infinite dimensional, new conditions must be added which are automatically satisfied if dim $L < +\infty$.

[1] This is the cone of admissible directions in the terminology of [4].

[2] $(S^{-1}(K))_{\hat{u}}$ is called a sequential *tangent cone* to $S^{-1}(K)$ at $\hat{u}$ [11], [5].

[3] This is a *radial cone* to K at $S(\hat{u})$, [5].

Definition. We shall say that

(i) the *Farkas condition* (F) holds, if

$$(S_{\hat{u}}^{-1}K_{S(\hat{u})})^* = S_{\hat{u}}^*K_{S(\hat{u})}^*.$$

(ii) *condition* (NS 1) holds, if the set

$$\text{im}\, S_{\hat{u}} + K_{S(\hat{u})} = \text{im}\, S_{\hat{u}} + \text{Lin}\, S(\hat{u}) + K$$

is not dense in L.

(iii) the *nonsingularity condition* (NS) holds, if either (F) or (NS 1) is satisfied.

The following two theorems show the close relationship between the nonsingularity condition (NS) and the existence of Lagrange multipliers.

Theorem 2.1. *Suppose that Assumptions* 2.1, 2.2, 2.3 *are satisfied and $\hat{u}$ is a local solution to Problem* 2.1.

(i) *If either* (NS 1) *or both* (F) *and* (KT) *hold then there exist nontrivial Lagrange multipliers for* 2.1 *at $\hat{u}$.*

(ii) *If* (F) *and* (KT) *are satisfied and*

$$M_{\hat{u}} \cap S_{\hat{u}}^{-1}K_{S(\hat{u})} \neq \emptyset$$

then there exists a normal Lagrange multiplier for Problem 2.1 *at $\hat{u}$.*

Theorem 2.2. *Suppose that Assumption* 2.1 *is satisfied with $\mathcal{O} = U$, S affine and $M = U$. Consider the following statements.*

(i) *nontrivial Lagrange multipliers exist for Problem* 2.1 *at $\hat{u}$ for every J satisfying Assumption* 2.2 *such that $\hat{u}$ is a local solution to Problem* 2.1;

(ii) *a normal Lagrange multiplier exists for Problem* 2.1 *at $\hat{u}$ for every J as above;*

(iii) (NS) *holds;*

(iv) (F) *holds.*

Then (i) $\Leftrightarrow$ (iii) *and* (ii) $\Leftrightarrow$ (iv).

Now we specialize the preceeding theorems to the important case of $K = \{0\}$ (equality constraints) and then show how various known theorems guaranteeing the existence of Lagrange multipliers can be derived from Theorem 2.1.

Theorem 2.3. *Suppose Assumption* 2.1 *is satisfied with $K = \{0\}$ and also Assumptions* 2.2, 2.3 *hold. Let $\hat{u}$ be a local solution to Problem* 2.1.

(i) *If* $\text{im}\, S_{\hat{u}}$ *is not a proper subspace dense in L, then there exist nontrivial Lagrange multipliers for Problem* 2.1 *at $\hat{u}$.*

(ii) *If* $\text{im}\, S_{\hat{u}}$ *is a closed subspace of L,* (KT) *holds* (i.e. $(S^{-1}(0))_{\hat{u}} \supset \ker S_{\hat{u}}$) *and*

$$M_{\hat{u}} \cap \ker S_{\hat{u}} \neq \emptyset$$

then there exists a normal Lagrange multiplier for Problem 2.1 *at $\hat{u}$.*

Theorem 2.4. *Suppose Assumption 2.1 is satisfied with $K = \{0\}$, $\mathcal{O} = U$, S affine and $M = U$. Consider the following statements.*

(i) *nontrivial Lagrange multipliers exist for the problem*

$$\textbf{Problem 2.2} \quad \begin{cases} \textit{minimize } J(u) \\ \textit{subject to } S(u) = 0 \end{cases}$$

at fixed $\hat{u} \in S^{-1}(0)$ and for every Frechet differentiable J such that $\hat{u}$ is a local solution to Problem 2.2;

(ii) *a normal Lagrange multiplier exists for Problem 2.2 at $\hat{u}$ for every J as above;*

(iii) im $S_{\hat{u}}$ *is not a proper subspace dense in L;*

(iv) im $S_{\hat{u}}$ *is a closed subspace of L.*

Then (i) $\Leftrightarrow$ (iii) *and* (ii) $\Leftrightarrow$ (iv).

Remark 2.2. Theorem 2.2 *does not* say that the nonsingularity condition is necessary for the existence of nontrivial multipliers for any single problem. In fact, a normal Lagrange multiplier exists iff

$$J_{\hat{u}} \in S_{\hat{u}}{}^* K^*_{S(\hat{u})} + W_{\hat{u}}{}^*$$

regardless any regularity conditions.

Nevertheless, Theorem 2.2 asserts that the nonsingularity conditions — explicit or implicit — must be present in any statement dealing with the existence of multipliers for a larger class of problems 2.1 with various functionals J. In this sense the nonsingularity conditions can be viewed as the "weakest" regularity conditions for infinite dimensional problems with affine constraints.

Remark 2.3. If the nonsingularity condition NS is not satisfied then (as the examples show [6]) one may suspect that either the target space L or the constraint cone K are not appropriate for the problem. First of all, it seems that the space L is not well chosen if im $S_{\hat{u}}$ is not closed in the topology of L. This is the basic requirement for the problems with equality constraints (see Theorem 2.4) recognized by many authors [1, 7, 8, 9]. We shall attempt to give an informal justification of this — apparently — purely mathematical condition.

For simplicity, consider the problem

$$\begin{cases} \text{minimize } \|u\|^2 \\ \text{subject to } Su = l, \end{cases}$$

where $S \in \mathscr{L}(U, L)$ and U and L are assumed to be Hilbert spaces, so that their norms may be viewed as energies.

Suppose that im S is a *proper dense* subspace of L, so that the system described by the triple U, L, S is approximately and not exactly controllable. At first glance the situation is very advantageous, since — practically — any target $l \in L$ can be attained. Notice, however, that if the target is not exactly attainable, $l \notin$ im S, and the sequence $\{u_n\}$ is such that $l = \lim Su_n$, then — by the Eberlein-Smulian theorem — we must have

$$\|u_n\| \to \infty \quad \text{for} \quad n \to \infty.$$

Therefore the targets only approximately attainable can be "reached" at a price of high energy controls; moreover if we want to improve *slightly* a given approximation of l we must use controls which are *far* from the preceeding ones.

The conclusion which can be drawn from these considerations is that our notion of what are close targets is not compatible with the structure of the system. To choose the target space L for an optimization problem is essentially to choose its topology, i.e., to decide what will be meant by the words "close targets".

Remark 2.4. One can always define a norm topology on im S, turning this into a Banach space:

$$|||l||| = ||l||_L + \inf_{u \in S^{-1}(l)} ||u||_U, \qquad l \in \operatorname{im} S.$$

Usually, however, it is possible to find among commonly used function spaces a suitable one: one should find a space L with a sufficiently strong topology. Observe, however, that the stronger is the topology in L, the smaller is usually L as a set of functions and the larger is L^*. As a result, Lagrange multipliers are more and more irregular (see also [10]). It is therefore reasonable to avoid too strong a topology in L, if possible.

Remark 2.5. In connection with these problems, a new question arises in the controllability theory of various infinite dimensional linear systems U, L, S: to find out whether (when), im S is closed in L. It occurs that the investigation of the closedness property of im S may provide an interesting insight into the structure of the system (see [11]).

Remark 2.6. This problem seems to be also of strong practical interest. Experiments have been made with the application of penalty functional techniques for solving certain problems of type as in Remark 2.2 [12]. The penalty term is of the form $||Su - l||_L^2$ and therefore depends on the choice of the target space L. It occured that the use of an appropriate space L with im S closed in L resulted in a considerable reduction of computational effort.

3. The closure conditions for the attainable subspace of a time-lag system

Consider a difference-differential system

$$\left.\begin{array}{ll} \dot{x}(t) = A(t)\,x(t) + B(t)\,x(t-1) + C(t)\,u(t) & \text{a.e. in } [0, T], \\ x(t) = 0 & \forall t \in [-1, 0], \end{array}\right\} \tag{3.1}$$

where $T > 1$, $x(t) \in R^n$, $u(t) \in R^r$ and $A(\cdot)$, $B(\cdot)$, $C(\cdot)$ are matrix functions of appropriate dimensions.

A complete final state for such a system is a terminal piece of trajectory $x(\cdot)|_{[T-1,T]}$, often abbreviated by x_T.

The set of all complete final states which can be attained with controls $u(\cdot) \in L^p[0, T; R^r]$ clearly forms a vector space and will therefore be called the *attainable subspace* of the system (3.1). Any complete final state is absolutely continous; one may therefore consider the attainable subspace as a subset of different Banach spaces, for example $L^q[T-1; T; R^n]$, $1 \leqq q \leqq \infty$ or $C[T-1, T; R^n]$. The attainable subspace is, however, never closed in any of the above spaces if one excludes the trivial case $C(t) = 0$. What is more, if the system (3.1) is completely controllable, then its attainable subspace is a proper dense subset of L^q and C (approximate controllability). This situation is pathological and leads to difficulties in establishing the necessary optimality conditions in the optimal control problem for the system (3.1) with complete final state constraints. It has been shown in [6] that these difficulties are not related to the particular features of the system (3.1), but are a general rule for extremal problems with infinite dimensional operator constraints.

Before stating the main result concerning closedness of the attainable subspace in the topology of $W_1^q[T-1, T; R^n]$ — the Sobolev space of n-vector valued functions with derivative in $L^q[T-1, T; R^n]$ and the norm

$$\|x_T(\cdot)\| = \left(\|x(T)\|^2_{R^n} + \|\dot{x}_T(\cdot)\|^2_{L^q}\right)^{1/2} -$$

let us introduce the following notations:

$$\mathscr{A}_p = \{x_T(\cdot) : x(\cdot) \text{ satisfies (3.1) with } u \in L^p\},$$

C^+ is the generalized inverse of C,

$$L(t) = \frac{d}{dt} - A(t).$$

Theorem 3.1. *Suppose that in* (3.1) $A(\cdot)$, $B(\cdot)$, $C(\cdot)$ *are analytic in* $t \in [0, T]$, $C(t) \neq 0$. *Then the attainable subspace* $\mathscr{A}_p$ *is closed in* W_1^q *iff the following conditions are satisfied*

(i) $p = q$;

(ii) *the map* $t \to C^+(t)$ *is bounded on* $[T-1, T]$ *or, equivalently, the rank of* $C(t)$ *is constant in* $[T-1, T]$;

(iii) $\operatorname{im} B(t)\, L^i(t)\, C(t) \subset \operatorname{im} C(t)$, $i = 0, 1, \ldots, n-1$, *for all but isolated points in* $[0, T]$.

If the matrices A, B, C are constant, the above conditions take a much simpler form.

Corollary 3.1. *Suppose the matrices* A, B, C *are constant and* $C \neq 0$. *Then*

$$\mathscr{A}_p \text{ is closed in } W_1^q \text{ if } p = q \text{ and } \operatorname{im} BA^iC \subset \operatorname{im} C, \quad i = 0, 1, \ldots, n-1.$$

Using Kalman's decomposition theorem [13] one obtains another characterization of the closedness property for time-constant systems.

Corollary 3.2. *If* A, B, C *are constant matrices of system* (3.1) *then* $\mathscr{A}_p$ *is closed in* W_1^P *if there is a linear transformation of coordinates in* R^n *yielding a system, equivalent to* (3.1), *of the form:*

$$\dot{x}_1(t) = A_{11}x_1(t) + A_{12}x_2(t) + B_{11}x_1(t-1) + B_{12}x_2(t-1) + C_1u(t),$$

$$\dot{x}_2(t) = A_{22}x_2(t) + B_{22}x_2(t-1),$$

where $\left(x_1(t), x_2(t)\right) \in R^{n_1} \times R^{n_2}$, $n_1 + n_2 = n$ *and* $\operatorname{span}\{A_{11}^iC_1 : i = 0, 1, \ldots, n_1 - 1\} = R^{n_1}$, $\operatorname{im} B_{11} \subset \operatorname{im} C_1$.

The derived conditions may be useful in examining optimal control problems with target sets in the Sobolev space $W_1{}^p[T-1, T; R^n]$.

The more general case with many delays in state variable can also be treated in a similar manner as presented in this paper.

There are some examples where closedness is obtained in a topology different from $W_1{}^p$. Consider the two dimensional system

$$\dot{x}_1(t) = x_2(t-1), \qquad \dot{x}_2(t) = u(t).$$

It may be proved that the attainable subspace of the above system is closed, assuming $(x_1, x_2) \in W_2{}^p(0, T; R^1) \times W_1{}^p(0, T; R^1)$, and it may be checked easily, using criterion 3.1, that the attainable subspace is not closed in $W_1{}^p[0, T; R^2]$. This gives a motivation for the statement of the following problem: Having a system of type (3.1) for which the attainable subspace is not closed in $W_1{}^p[T-1, T; R^n]$, find the right Sobolev space for which the closedness property holds. Determine also when such a choise is possible in the class of Sobolev spaces.

4. Control of time-lag systems to a ball in function space

Let the system equation be of the form

$$\dot{x}(t) = f\big(x(t), x(t-h), u(t), t\big), \; t \in [0, T], \tag{4.1}$$

where $x(t) \in R^n$, $u(t) \in R^r$, the numbers $T > h > 0$ are given. The problem is to find a control $u \in L^\infty[0, T; U]$ such that given an initial state $x(t) = \varphi(t)$, $t \in [-h, 0]$, the final state $x_T(\cdot)$, where $x_T(\theta) = x(T+\theta)$ for $\theta \in [-h, 0]$, satisfies

$$\|x_T(\cdot) - \xi(\cdot)\| < \varepsilon \tag{4.2}$$

and the functional

$$J = \int_0^T f_0\big(x(t), u(t), t\big)\, dt \tag{4.3}$$

achieves its minimum. Here $R^r \supset U$ — a nonempty set, $\varepsilon > 0$ is a given number, $\xi(t)$ is a given element of $W_1{}^2[T-1, T; R^n]$ and $\|t\|$ is a suitable norm in function space. We shall consider the norms in the spaces L^2, C and $W_1{}^2$.

The physical motivation for such a problem statement is that the true final state of the system (4.1) is represented by $x_T(\cdot)$. Therefore if the desired behaviour of a system is required for $t \geqq T$ e.g. small deviations from equilibrium state the right final state should be reached.

The extension of the maximum principle to systems of type (4.1) with constrained trajectory end $x(T) \in R^n$ is due to Kharatishvili [14].

Problems with constraints on the final state $x_T(\cdot)$ were considered in the works of Banks and Kent [15], Jacobs and Kao [16], Banks and Jacobs [17], Kurcyusz [10]. The necessary conditions for optimality were given in the papers cited above but either the nontriviality could not be proved [15] or the assumptions on the system were too strong (unconstrained controls and complete controllability in [16], linear system and quadratic performance index in [17], the system affine with respect to control [10]). In Olbrot [18] it is shown that a nontrivial maximum principle can be stated for problems with constraints of type (4.2) without using any of the above strong assumptions.

The main results in [18] are the following

Theorem 4.1. *Assume the functions* $f_0: R^n \times U \times [0, T] \mapsto R^1$, $f: R^n \times R^n \times U \times [0,T] \mapsto R^n$ *be continuous,* $f_0(\cdot, u, t)$, $f(\cdot, y, u, t)$, $f(x, \cdot, u, t)$ *be of class* C^1 *with respect to the dotted argument. Let* φ *and* ξ *be continuous functions of time* t *and the norm in* (4.2) *be the supremum norm in the space of continuous functions. If* $\hat{x}(\cdot)$, $\hat{u}(\cdot)$ *are respectively the optimal trajectory and optimal control satisfying* (4.1), (4.2) *and minimizing* (4.3) *then there exists a nonzero triple* (p_0, p, λ), *where* $p_0 \leq 0$ *is a constant and* $p: [0, T+h] \mapsto R^n$, $\lambda: [T-h, T] \mapsto R^1$ *are such that*

(i) *p is absolutely continuous on* $[0, T-h]$ *and on* $[T-h, T]$ *satisfying, on both intervals, the equation*

$$\dot{p}(t) = -\frac{\partial}{\partial x(t)} H\big(p(t), \hat{x}(t), \hat{x}(t-h), \hat{u}(t), t\big) - \frac{\partial}{\partial x(t)} H\big(p(t+h), \hat{x}(t+h), \hat{x}(t), \hat{u}(t), t+h\big), \tag{4.4}$$

where[4]

$$H(p, x, y, u, t) = \begin{cases} p_0 f_0(x, u, t) + p^* f(x, y, u, t) & \text{on } [0, T-h], \\ p_0 f_0(x, u, t) + \big[p - 2\lambda(t)\big(x - \xi(t)\big)\big]^* f(x, y, u, t) & \text{on } [T-h, T]. \end{cases} \tag{4.5}$$

The function p satisfies the following boundary conditions

$$p(t) = 0 \quad \text{on } [T, T+h],$$
$$p(T-h-0) = p(T-h+0) + 2\lambda(0)\big(\hat{x}(T-h) - \xi(T-h)\big);$$

(ii) $\lambda(\cdot)$ *is nonincreasing, continuous from the right, constant on subintervals on which*[5] $|x(t)| < \varepsilon$ *and* $\lambda(T) = 0$;

(iii) $H\big(\hat{u}(t)\big) = \max_{u \in U} H(u)$ *for a.a.* $t \in [0, T]$, $\quad$ (4.6)

where $H(u)$ denotes $H\big(p(t), \hat{x}(t), \hat{x}(t-h), u, t\big)$.

Theorem 4.2. *Let the functions* f_0, f, φ *satisfy the assumptions of Theorem* 4.1. *Consider the following norms in* (4.3)

(N 1) $\|x_T\| = \left(|x(T)|^2 + \int_{T-h}^{T} |\dot{x}(t)|^2\, dt\right)^{1/2}$ — $W_1{}^2$ norm,

(N 2) $\|x_T\| = \left(\int_{T-h}^{T} |x(t)|^2\, dt\right)^{1/2}$ — an L^2 norm.

[4] The transposition of a vector (matrix) is denoted by *.

[5] The Euclidean norm in R^n is denoted by $|\cdot|$.

Let the function ξ *be in* $W_1^2[T-h, T; R^n]$ *if the norm* (N 1) *is used. If* $\hat{x}(\cdot)$, $\hat{u}(\cdot)$ *is the optimal pair for the problem defined by* (4.1), (4.2), (4.3) *with any of the above norms then there exists a nonzero triple* (p_0, p, p_a), *where* $p_0, p_a \in R^1$, $p_0 \leqq 0$, $p: [0, T+h] \mapsto R^n$, *such that*

(i) p *satisfies the equation* (4.4) *on* $[0, T]$, *where the Hamiltonian is defined in a somewhat similar way as in* (4.5):

$$H(p, x, y, u, t) = \begin{cases} p_0 f_0(x, u, t) + p^* f(x, y, u, t) & \text{for} \quad t \in [0, T-h], \\ p_0 f_0(x, u, t) + p^* f(x, y, u, t) + p_a F_a(x, y, u, t) & \text{for} \quad t \in [T-h, T] \end{cases} \tag{4.7}$$

and

$$F_a(x, y, u, t) = \begin{cases} 2\big(x - \xi(t)\big)^* \big(f(x, y, u, t) - \dot{\xi}(t)\big) + |f(x, y, u, t) - \dot{\xi}(t)|^2 & \text{for the case (N 1)}, \\ |x - \xi(t)|^2 & \text{for the case (N 2)}. \end{cases}$$

The function p *satisfies the following boundary conditions*

$$p(t) = 0 \quad \text{on} \quad [T, T+h],$$

and in addition, if the norm (N 1) *is taken,*

$$p(T-h-0) = p(T-h+0) + 2p_a\big(x(T-h) - \xi(T-h)\big).$$

Thus the function p *is absolutely continuous except in the case* (N 1) *at the point* $t = T - h$;

(ii) $P_a\big(\|x_T(\cdot) - \xi(\cdot)\| - \varepsilon\big) = 0$, *where* $\|\cdot\|$ *denotes either* (N 1) *or* (N 2) *norm*;

(iii) *condition* (4.6) *holds for* H *defined in* (4.7).

It should be pointed out that the necessary conditions for optimality stated in Theorems 4.1, 4.2 are similar rather to classical results of Kharatishvili [14] than to existing in the literature other results concerning function space constraints retarded systems. The maximum condition (4.6) holds for the set U of admissible control values, not, as e.g. in [10, 15], for the class of admissible control functions. The differential-difference equation for the adjoint variable $p(t)$ can be written down in the same form as in [14]. The presence of either a scalar-valued function λ (the case of sup norm) or an additional constant p_a. The function $p(\cdot)$ is absolutely continuous except at the instance $t = T - h$, where it may have a jump if the norm is of C or W_1^2 space. In the case of L^2 or W_1^2 norm the function space constraints can be treated as the constraints for final values of some coordinates of the system. It is more evident from the proofs of the Theorems [18]. In case of L^2 norm there is no jump of $p(t)$ at $t = T - h$. Moreover, in that case the optimality conditions are simpler.

References

[1] Bazaraa, M. S., J. J. Goode, Necessary optimality criteria in mathematical programming in normed linear spaces. J. Optimization Theory Appl. 11, 235–244 (1973).

[2] Arrow, K. J., L. Hurwicz, H. Uzawa, Studies in linear and nonlinear programming. Stanford University Press 1958.

[3] Varaiya, P. P., Nonlinear programming in Banach spaces, SIAM J. Appl. Math. **15**, 285—293 (1967).
[4] Гирсанов, И. В., Лекции по математической теории экстремальных задач. Изд. Московского Университета 1970.
[5] Findeisen, W., J. Szymanowski, A. Wierzbicki, Computational methods in optimization, part III. Ed. Technical University of Warsaw.
[6] Kurcyusz, S., On the existence and nonexistence of Lagrange multipliers in Banach spaces. J. Optimization Theory Appl. **20**, 81—110 (1976).
[7] Craven, B. D., Nonlinear programming in locally convex spaces. J. Optimization Theory Appl. **10**, 197—210 (1972).
[8] Luenberger, D., Optimization by vector space methods. J. Wiley, 1969.
[9] Rolewicz, S., Functional analysis and control theory [in Polish]. PWN, Warszawa 1975.
[10] Kurcyusz, S., A local maximum principle for operator constraints and its application to systems with time lags, Control and Cybernetics **2**, 99—125 (1973).
[11] Kurcyusz, S., A. W. Olbrot, On the closure in W_1^q of the attainable subspace of linear time lag systems. J. Differ. Equations **24**, 29—50 (1977).
[12] Przyłuski, K. M., Application of the shifted penalty functional to dynamical optimization of delay systems. M. Sc. Thesis, Institute of Automatic Control, Warsaw 1974.
[13] Kalman, R., Canonical structure of linear dynamical systems. Proc. Nat. Acad. Sci. USA **48**, 596—600 (1962).
[14] Харатишвили, Г. Л., Принцип максимума в теории оптимальных процессов с запаздыванием. Докл. Акад. Наук СССР **136**, 39—42 (1961).
[15] Banks, H. T., G. A. Kent, Control of functional differential equations of retarded and neutral type to target sets in function space. SIAM J. Control **10**, 567—593 (1972).
[16] Jacobs, M. Q., T. J. Kao, An optimum settling problem for fime-lag system. J. Math. Anal. Appl. **40**, 687—707 (1972).
[17] Banks, H. T., M. Q. Jacobs, An attainable set approach to optimal control of functial differential equations with function space boundary conditions. J. Diff. Equations **13**, 127 to 149 (1973).
[18] Olbrot, A. W., Optimal control of a nonlinear retarded system with a ball in function space as a target set (to be published).

Address of the author:

Institute of Automatic Control
Technical University of Warsaw
ul. Nowowiejska 15/19
00-665 Warszawa
Poland

Allgemeine Konvergenztheoreme zu Projektionsverfahren für lineare Operatorgleichungen und einige Anwendungen

Siegfried Prößdorf, Berlin

1. Einleitung

In diesem Vortrag formulieren wir einige neue Konvergenztheoreme über Projektionsverfahren (PV) zur näherungsweisen Lösung von linearen Operatorgleichungen der Gestalt

$$Ax + Tx = y;$$

hierbei ist A ein linearer beschränkter und invertierbarer Operator, der zwischen zwei Banach-Räumen X und Y wirkt, und T ist ein linearer kompakter oder (in der Norm) kleiner Operator. Da die Konvergenz eines PV im allgemeinen nicht für eine beliebige rechte Seite $y \in Y$ vorliegt (z. B. wenn X nicht separabel ist und die entsprechenden Projektoren endlichdimensional sind oder wenn die Projektoren nicht beschränkt sind), führen wir zunächst den Begriff der *Konvergenzmannigfaltigkeit* eines PV ein. Das ist nichts anderes als die Menge aller derjenigen Elemente $y \in Y$, für die das PV konvergiert. Wir untersuchen die Beziehungen zwischen den Konvergenzmannigfaltigkeiten der Operatoren A und $A + T$, geben Abschätzungen für die Konvergenzgeschwindigkeit (Fehlerabschätzungen) an und formulieren Kriterien für die Stabilität des PV. Die Ergebnisse werden angewandt auf die Kollokationsmethode für gewöhnliche Differentialgleichungen und singuläre Integralgleichungen, auf die Multhoppsche Methode für singuläre Integrodifferentialgleichungen sowie auf die Reduktionsmethode für singuläre Integralgleichungen und Wiener-Hopf-Gleichungen. Die wichtigsten Ergebnisse dieses Vortrages sind in den Arbeiten [12—19] zu finden.

2. Allgemeine Konvergenztheoreme

Es seien X und Y Banach-Räume. Mit $L(X\ Y,)$ bezeichnen wir die Menge aller linearen beschränkten Operatoren, die von X nach Y wirken, und mit $K(X, Y)$ $(\subset L(X, Y))$ die Teilmenge aller vollstetigen (kompakten) Operatoren. Wir geben zwei Folgen von (i.a. nicht beschränkten) Projektoren $\{P_n\}$, $\{Q_n\}$ mit den Definitionsgebieten $D(P_n) \subset X$, $D(Q_n) \subset Y$ und den abgeschlossenen Bildräumen im $P_n \subset X$, im $Q_n \subset Y$ vor.

Für den Operator $A \in L(X, Y)$ besteht das Projektionsverfahren $\{P_n, Q_n\}$ bekanntlich in folgendem. Man ersetzt die Operatorgleichung

$$Ax = y \qquad (y \in Y) \tag{1}$$

durch die Näherungsgleichung

$$Q_n A P_n x_n = Q_n y, \tag{2}$$

für die eine Lösung $x_n \in$ im P_n gesucht wird. Wenn von einem gewissen n_0 an die Gleichung (2) für *jedes* $y \in Y$ eine und nur eine Lösung x_n besitzt und wenn diese für $n \to \infty$ gegen die Lösung der Gleichung (1) konvergiert, dann sagt man, das Projektionsverfahren $\{P_n, Q_n\}$ sei *auf den Operator A anwendbar*, und schreibt $A \in \Pi\,\{P_n, Q_n\}$.

Im weiteren verlangen wir nur, daß für $n \geqq n_0$ die Beziehung $A(\text{im}\, P_n) \subset D(Q_n)$ gilt und die Operatoren Q_nAP_n: im $P_n \to$ im Q_n umkehrbar stetig sind. Mit $\mathfrak{K}(A; \{P_n, Q_n\})$ bezeichnen wir die Menge aller Elemente $y \in Y$ mit folgenden Eigenschaften: 1) $y \in D(Q_n)$ für $n \geqq n'(y)$; 2) die Folge der Elemente $x_n = (Q_nAP_n)^{-1}\, Q_ny$ konvergiert für $n \to \infty$ in der Norm von X gegen ein Element $x \in X$; 3) es gilt $Ax = y$. Diese Menge heißt *Konvergenzmannigfaltigkeit des Operators A bezüglich des Projektionsverfahrens* $\{P_n, Q_n\}$.

Wenn A invertierbar ist, dann ist offensichtlich die Konvergenzmannigfaltigkeit die maximale lineare Menge, auf der die Operatoren $(Q_nAP_n)^{-1}\, Q_n$ stark gegen A^{-1} konvergieren. Des weiteren gilt $\mathfrak{K}(A; \{P_n, Q_n\}) = Y$ genau dann, wenn $A \in \Pi\{P_n, Q_n\}$ gilt.

Satz 1. *Es sei* $A \in L(X, Y)$ *ein invertierbarer und* $T \in L(X, Y)$ *ein Operator mit folgenden Eigenschaften:*

1) $A + T$ *ist invertierbar;*

2) im $T \subset D(Q_n)$ *und* $Q_nT \in L(X, Y)$ $\quad (n = 1, 2, \ldots)$;

3) $\|A^{-1}T - (Q_nAP_n)^{-1}\, Q_nT\| \to 0$ *für* $n \to \infty$.

Dann sind die Operatoren $Q_n(A + T)\, P_n$: im $P_n \to$ im Q_n *für* $n \geqq n_1$ *umkehrbar stetig und*

$$\mathfrak{K}(A; \{P_n, Q_n\}) \subset \mathfrak{K}(A + T; \{P_n, Q_n\}).$$

Der folgende Satz ist eine Folgerung des Satzes 1.

Satz 2. *Es sei* $A \in L(X, Y)$ *ein invertierbarer Operator und* $Z \subset Y$ *ein in* Y *stetig eingebetteter Banach-Raum, wobei* $Z \subset \mathfrak{K}(A; \{P_n, Q_n\})$ *gelte. Der Operator* $T \in K(X, Z)$ *erfülle die folgenden zwei Bedingungen:*

α) $\dim \ker (A + T) = 0$;

β) $Q_n \mid Z \in L(Z, Y)$.[1]

Dann gilt

$$\mathfrak{K}(A; \{P_n, Q_n\}) = \mathfrak{K}(A + T; \{P_n, Q_n\}).$$

Für $Z = Y$ erhält man aus Satz 2 die bekannte Tatsache, daß im Falle stetiger Projektoren aus $A \in \Pi\{P_n, Q_n\}$ stets $A + T \in \Pi\{P_n, Q_n\}$ folgt, falls $T \in K(X, Y)$ die Bedingung α) erfüllt (vgl. [3]).

Der folgende Satz liefert eine Fehlerabschätzung für eine gewisse Klasse von Operatoren A.

Satz 3. *Es sei* $A = CB$, *wobei* $B \in L(X, Y)$ und $C \in L(Y, Y)$ *invertierbare Operatoren sind, die folgende zwei Bedingungen erfüllen:*

(a) $Q_nBP_nx = BP_nx, \quad \forall x \in D(P_n), \quad n = 1, 2, \ldots;$

(b) $Q_nC^{\pm1}Q_ny = Q_nC^{\pm1}y, \quad \forall y \in D(Q_n)$.

[1] Mit $Q_n \mid Z$ bezeichnen wir die Einschränkung des Operators Q_n auf Z.

Des weiteren seien die Bedingungen des Satzes 1 (*oder des Satzes* 2) *erfüllt.*

Wenn $y \in \mathfrak{K}(A; \{P_n, Q_n\})$ *gilt und* $x \in X$, $x_n \in \operatorname{im} P_n$ *entsprechend die Lösungen der Gleichungen*

$$Ax + Tx = y, \qquad Q_n(A + T)\, x_n = Q_n y$$

sind, dann gilt die Abschätzung

$$c_1 \|Bx - Q_n Bx\|_Y \leqq \|x - x_n\|_X \leqq c_2 \|Bx - Q_n Bx\|_Y$$

und im Falle stetiger Projektoren Q_n *die Abschätzung*

$$\|x - x_n\|_X \leqq c \|Q_n\|\, E_n^Y(Bx).$$

Hierbei sind c_1, c_2, c *positive Konstanten und* $E_n^Y(f) = \inf_{y_n} \|f - y_n\|_Y$ $(y_n \in \operatorname{im} Q_n)$.

Folgerung 1. *Es seien die Bedingungen des Satzes* 3 *erfüllt. Des weiteren gelte* $Bx \in Z$, *wobei* $Z \subset Y$ *Banach-Räume sind, die der folgenden Bedingung genügen:*

(c) $\|y - Q_n y\|_Y \to 0, \quad \forall y \in Z, \quad n \to \infty.$

Dann gilt die Abschätzung

$$\|x - x_n\|_X \leqq c E_n^Z(Bx).$$

Der folgende Satz liefert hinreichende Bedingungen dafür, daß eine Menge Z zur Konvergenzmannigfaltigkeit eines Operators gehört.

Satz 4. *Es sei* $A = CB$, *wobei* $B \in L(X, Y)$ *und* $C \in L(Y, Y)$ *invertierbare Operatoren sind. Des weiteren seien die oben genannten Bedingungen* (b) *und* (c) *sowie auch die beiden folgenden Bedingungen erfüllt:*

(a′) $\operatorname{im} BP_n = \operatorname{im} Q_n \quad (n = 1, 2, \ldots)$;

(d) *es existiert eine Menge* $Z \subset Y$ *mit* $C^{-1}(Z) \subset Z$.

Dann gilt $Z \subset \mathfrak{K}(A; \{P_n, Q_n\})$.

Wir bemerken, daß in dem Spezialfall $X = Y$ und $A = I$ (identischer Operator) die Sätze 1 und 3 sowie die Folgerung 1 wohlbekannte Aussagen über Projektionsverfahren für Riesz-Schaudersche Gleichungen ergeben (vgl. [7], § 15).

Zum Satz 2 läßt sich ein Analogon für Störungen mit kleiner Norm formulieren.

Satz 5. *Es sei* $A \in L(X, Y)$ *ein invertierbarer Operator und* $Z \subset Y$ *ein in* Y *stetig eingebetteter Banach-Raum mit* $Z \subset \mathfrak{K}(A; \{P_n, Q_n\})$. *Des weiteren gelte* $Q_n \mid Z \in L(Z, Y)$, $P_n \in L(X, X)$ *und* $P_n x \to x$ *für alle* $x \in X$. *Dann existiert eine Konstante* $\gamma > 0$ *derart, daß für einen beliebigen Operator* $T \in L(X, Z)$ *mit* $\|T\|_{X \to Z} < \gamma$ *die Aussage des Satzes* 1 *gilt.*

Infolge der Fehler, die in Gleichung (2) bei der Berechnung des Operators $A_n = Q_n A P_n$ sowie auch der rechten Seite $Q_n y$ (Berechnung von Skalarprodukten usw.) auftreten, hat man in Wirklichkeit nicht die Gleichung (2), sondern eine gewisse gestörte Gleichung der Gestalt

$$(A_n + \Gamma_n)\, \tilde{x}_n = Q_n y + \delta_n \tag{3}$$

zu lösen. Dabei wird man natürlich fordern, daß $\tilde{x}_n \in \operatorname{im} P_n$, $\delta_n \in \operatorname{im} Q_n$ und $\Gamma_n : \operatorname{im} P_n \to \operatorname{im} Q_n$ gilt. Das $PV\{P_n, Q_n\}$ erweist sich in folgendem Sinne als stabil (bez. der Definition der Stabilität vgl. [9]).

Satz 6. *Es sei $A \in L(X, Y)$ ein invertierbarer Operator und $Z \subset Y$ ein in Y stetig eingebetteter Banach-Raum mit $Z \subset \mathfrak{K}(A; \{P_n, Q_n\})$. Des weiteren gelte* $\operatorname{im} Q_n \subset Z$, $Q_n \mid Z \in L(Z, Y)$ *und* $y \in Z$.

Dann existieren von n und y unabhängige positive Konstanten p, q, γ derart, daß für $\|\Gamma_n\|_{X \to Z} < \gamma$ folgendes gilt:

1. *Für beliebiges* $\delta_n \in \operatorname{im} Q_n$ $(n \geqq n')$ *ist die Gleichung* (3) *eindeutig lösbar;*

2. $\|\tilde{x}_n - x_n\|_X \leqq p\|y\|_Z \|\Gamma_n\|_{X \to Z} + q\|\delta_n\|_Z$.

Sämtliche Ergebnisse dieses Abschnittes sind in den Arbeiten von S. Prößdorf und B. Silbermann [16, 17] enthalten. Die Sätze 1 und 2 wurden im wesentlichen schon in der Arbeit [13] (vgl. Beweis des Satzes 2.1) bewiesen. Bei den in den Abschnitten 3 bis 5 folgenden konkreten Realisierungen der Resultate dieses Abschnittes sind die Bedingungen des Satzes 6 stets erfüllt. Wir werden deshalb im folgenden von einer Formulierung dieses Satzes in der jeweiligen konkreten Situation Abstand nehmen.

3. Kollokationsmethode für Differentialgleichungen

Als erstes Beispiel der Anwendung der allgemeinen Sätze des vorangegangenen Abschnittes betrachten wir die lineare Differentialgleichung

$$Lu \equiv u^{(m)} + \sum_{k=0}^{m-1} p_k(t)\, u^{(k)} = f(t) \tag{4}$$

mit den linearen homogenen Randbedingungen

$$\varphi_i(u) \equiv \sum_{k=0}^{m-1} [\alpha_{ik} u^{(k)}(a) + \beta_{ik} u^{(k)}(b)] = 0 \qquad (i = 1, \ldots, m). \tag{5}$$

Wir nehmen an, daß die Aufgabe

$$u^{(m)} = 0, \qquad \varphi_i(u) = 0 \qquad (i = 1, \ldots, m)$$

nur die Null-Lösung besitzt. Dann existieren Polynome

$$\psi_j(t) = \sum_{k=0}^{m+j} c_{jk} t^k \qquad (j = 0, 1, \ldots),$$

die den Randbedingungen (5) genügen (z. B. die Lösungen der Aufgabe $u^{(m)} = t^j$, $\varphi_i(u) = 0$ $(i = 1, \ldots, m)$). Bei der Kollokationsmethode sucht man die Näherungslösung der Aufgabe (4), (5) in der Form

$$u_n(t) = \sum_{j=0}^{n} \xi_j \psi_j(t)$$

und bestimmt die Koeffizienten ξ_j so, daß $u_n(t)$ die Gleichung (4) in den Kollokationspunkten $t_{0n}, t_{1n}, \ldots, t_{nn}$ erfüllt:

$$(Lu_n - f)(t_{in}) = 0 \quad (i = 0, \ldots, n); \quad t_{in} = \frac{a+b}{2} + \frac{a-b}{2} \cos \frac{2i+1}{(2n+1)} \pi. \tag{6}$$

Satz 7. *Die Koeffizienten $p_k(t)$ $(k = 0, \ldots, m-1)$ seien stetig und $f(t)$ sei Riemann-integrierbar*[2] *auf $[a, b]$. Wir nehmen an, daß die Aufgabe (4), (5) eine eindeutig bestimmte Lösung $u_0(t)$ besitzt.*

Dann ist das Gleichungssystem (6) für alle hinreichend großen n eindeutig lösbar, und für $n \to \infty$ gilt

$$\|u_n^{(m)} - u_0^{(m)}\|_{L_2} \to 0, \qquad \max_{a \leq t \leq b} |u_n^{(k)}(t) - u_0^{(k)}(t)| \to 0 \quad (k = 0, \ldots, m-1).$$

Im Falle einer stetigen Funktion $f(t)$ gelten die Abschätzungen

$$\|u_n^{(m)} - u_0^{(m)}\|_{L_2} \leqq c e_n(u_0^{(m)}),$$

$$\max_{a \leq t \leq b} |u_n^{(k)}(t) - u_0^{(k)}(t)| \leqq c e_n(u_0^{(m)}) \qquad (k = 0, \ldots, m-1), \tag{7}$$

wobei

$$e_n(z) = \min_{\gamma_j} \max_{a \leq t \leq b} \left| z(t) - \sum_{j=0}^{n} \gamma_j t^j \right|.$$

Satz 7 ergibt sich aus den Ergebnissen des Abschnittes 2 unter folgenden Voraussetzungen: X — Raum aller Funktionen aus dem Sobolewschen Raum $W_2^{(m)}[a, b]$, die die Randbedingungen (5) erfüllen, $Y = L_2(a, b)$, Q_n — Interpolationspolynom der Ordnung n der Funktion φ nach den Knotenpunkten $t_{0n}, t_{1n}, \ldots, t_{nn}$;

$$P_n \varphi = \sum_{j=0}^{n} \xi_j \psi_j(t),$$

wobei ξ_j die Fourierkoeffizienten der Funktion $\varphi \in X$ nach dem System $\{\psi_j(t)\}_{j=0}^{\infty}$ sind. Das Gleichungssystem (6) ist dann der Gleichung $Q_n L P_n u_n = Q_n f$ äquivalent. Wir setzen $Au = Bu = u^{(m)}$, $C = I$ und $Z = R[a, b]$ (Menge der Riemann-integrierbaren Funktionen). Auf Grund einer bekannten Verallgemeinerung des Erdös-Turan-Theorems ist dann die Bedingung (c) aus Folgerung 1 erfüllt. Aus Satz 4 folgt damit $R[a, b] \subset \mathfrak{R}(A; \{P_n, Q_n\})$. Nach Satz 2 ergibt sich mit $Z = C[a, b]$ die Beziehung $\mathfrak{R}(L; \{P_n, Q_n\}) = \mathfrak{R}(A; \{P_n, Q_n\})$ und damit die erste Behauptung des Satzes 7. Wenn $f(t)$ stetig ist, dann kann man auf Grund der Folgerung 1 und des Sobolewschen Einbettungssatzes unmittelbar auf die Abschätzung (7) schließen.

In ähnlicher Weise lassen sich die bekannten Ergebnisse zur Kollokationsmethode für Fredholmsche Integralgleichungen (vgl. [7]) aus den Resultaten des Abschnitts 2 ableiten.

[2] In der Literatur ist bislang nur der Fall einer stetigen Funktion $f(t)$ betrachtet worden (vgl. [7]).

4. Multhoppsche Methode für die Prandtlsche Integrodifferentialgleichung

Wir betrachten die Prandtlsche Integrodifferentialgleichung der Tragflügeltheorie

$$\frac{1}{\pi}\int_{-1}^{1}\frac{\varphi'(\tau)}{\tau - t}\,d\tau + \int_{-1}^{1} K(t, \tau)\,\varphi'(\tau)\,d\tau + a(t)\,\varphi(t) = f(t) \qquad (-1 < t < 1), \tag{8}$$

wobei $K(t, \tau)$ ein schwach singulärer Kern und $a(t)$ eine stetige Funktion auf $[-1, 1]$ ist. Von der gesuchten Funktion $\varphi(t)$ fordern wir, daß sie absolut stetig ist,

$$\varphi'(t) \in L_p([-1, 1]; \varrho) \quad \text{mit} \quad \varrho(t) = (1 - t^2)^{(p-1)/2}, \qquad 1 < p < \infty,$$

gilt und außerdem die Randbedingungen

$$\varphi(-1) = \varphi(1) = 0$$

erfüllt sind. Mittels der Substitutionen

$$t = \cos s\,(0 \leqq s \leqq \pi), \qquad x(s) = \varphi(t), \qquad y(s) = -f(t)\sin s$$

erhält man die zu (8) äquivalente Gleichung

$$(Ax')\,(s) = y(s), \tag{9}$$

wobei

$$(Au)\,(s) = \frac{\sin s}{\pi}\int_0^{\pi}\frac{u(\sigma)\,d\sigma}{\cos\sigma - \cos s} + \int_0^{\pi} T(s, \sigma)\,u(\sigma)\,d\sigma,$$

$$T(s, \sigma) = K_1(\cos s, \cos\sigma)\sin s,$$

$$K_1(t, \tau) = K(t, \tau) \qquad (t < \tau),$$

$$K_1(t, \tau) = K(t, \tau) + a(t) \qquad (t > \tau).$$

Wir fordern, daß der Integraloperator mit dem Kern $T(s, \sigma)$ ein vollstetiger Operator ist, der aus $L_p(0, \pi)$ in den Raum $C[0, \pi]$ wirkt (vgl. z. B. [8]).

Entsprechend der Multhoppschen Methode ([10]) suchen wir die Näherungslösung der Gleichung (9) in der Form

$$x_n(s) = \sum_{k=1}^{n}\frac{c_k}{k}\sin ks, \tag{10}$$

wobei die Koeffizienten c_k aus dem System

$$(Ax_n')\,(s_j) = y(s_j); \qquad j = 1, \ldots, n, \qquad s_j = j\pi/(n + 1),$$

oder, was dasselbe ist, aus dem System

$$\sum_{k=1}^{n} (\alpha_{jk} + \beta_{jk}) c_k = y(s_j) \qquad (j = 1, \ldots, n),$$

$$\alpha_{jk} = \sin k s_j, \qquad \beta_{jk} = \int_0^\pi T(s_j, \sigma) \cos k\sigma \, d\sigma, \tag{11}$$

bestimmt werden.

Im weiteren bezeichnen wir mit $H^\mu[0, \pi]$ $(0 < \mu \leqq 1)$ den Banach-Raum aller Funktionen $u(s)$, $0 \leqq s \leqq \pi$, die eine Hölder-Bedingung mit dem Exponenten μ erfüllen; die Norm ist durch die Gleichung

$$\|u\|_{H^\mu} = \|u\|_C + \sup_{0 \leqq s, s' \leqq \pi} \frac{|u(s) - u(s')|}{|s - s'|^\mu}, \qquad \|u\|_C = \max |u(s)|,$$

erklärt. Mit $C^{r,\mu}[0, \pi]$ bezeichnen wir den Raum aller Funktionen $u(s) \in H^\mu[0, \pi]$, die Ableitungen $u^{(j)}(s) \in H^\mu[0, \pi]$ $(j = 1, \ldots, r)$ besitzen, mit der Normdefinition

$$\|u\|_{C^{r,\mu}} = \sum_{j=0}^{r-1} \|u^{(j)}\|_C + \|u^{(r)}\|_{H^\mu}.$$

Schließlich bedeute $L_p^0(0, \pi)$ den Unterraum der Funktionen $u(s) \in L_p(0, \pi)$ mit $\int_0^\pi u(s) \, ds = 0$.

Satz 8 [17]. *Es seien die folgenden beiden Bedingungen erfüllt:*

1) *Die Gleichung* $Au = 0$ *hat in* $L_p^0(0, \pi)$ *nur die Null-Lösung;*

2) $y(s) \in R[0, \pi]$.

Dann hat das System (11) *für alle hinreichend großen* n *eine eindeutige Lösung, und die Folge der Funktionen* (10) *konvergiert gleichmäßig gegen die Lösung* $x(s)$ *der Gleichung* (9).

Wenn die Funktionen $y(s)$, $a(t)$ *und* $K(t, \tau)$ $(t - \tau)$ *(in beiden Veränderlichen) eine Hölder-Bedingung mit dem Exponenten* μ, $0 < \mu \leqq 1$, *erfüllen, dann gilt für beliebiges* ε, $0 < \varepsilon < \mu$, *die Abschätzung*

$$\|x - x_n\|_{H^1} = O(n^{\varepsilon - \mu}). \tag{12}$$

Wenn die Lösung $x(s)$ *der Gleichung* (9) *zur Klasse* $C^{r,\mu}[0, \pi]$ $(r = 0,1)$ *gehört, dann gilt*

$$\|x - x_n\|_C = O(n^{\varepsilon - \mu - r}). \tag{13}$$

Den Satz 8 erhält man in Anwendung des Satzes 4 (mit $Z = R[0, \pi]$), der Sätze 2, 3 und der Folgerung 1 (mit $Z = C[0, \pi]$) unter den Voraussetzungen $X = L_p^0(0, \pi)$, $Y = L_p(0, \pi)$, $C = I$ sowie einiger Ergebnisse aus der Monographie [11] (Abschnitt 3.4.1). Für $p = 2$ wurde der Satz 8 ohne die Fehlerabschätzungen (12) und (13) in [20] auf andere Weise erhalten (vgl. auch [6]).

5. Kollokations- und Reduktionsmethode für singuläre Integralgleichungen

Als eine weitere wichtige Anwendung der Ergebnisse des Abschnittes 2 ergeben sich Konvergenzaussagen und Fehlerabschätzungen für die Kollokations- sowie die Reduktionsmethode zur nährungsweisen Lösung von singulären Integralgleichungen der Gestalt

$$A\varphi \equiv c(t)\,\varphi(t) + \frac{d(t)}{\pi i}\int_\Gamma \frac{\varphi(\tau)}{\tau - t}\,d\tau + T\varphi = f(t) \tag{14}$$

in den Räumen $L_p(\Gamma)$ $(1 < p < \infty)$ und $H^\mu(\Gamma)$ $(0 < \mu < 1)$. Hierbei bedeutet Γ den Einheitskreis $|t| = 1$, $c(t)$ und $d(t)$ sind Funktionen aus $H^\mu(\Gamma)$, und T ist ein linearer vollstetiger Operator; das Integral in (14) wird im Sinne des Cauchyschen Hauptwertes verstanden. Im weiteren setzen wir noch

$$a(t) = c(t) + d(t), \qquad b(t) = c(t) - d(t).$$

5.1. Kollokationsmethode

Wir betrachten auf dem Einheitskreis die Kollokationsknoten

$$t_j = e^{2\pi i j/(2n+1)}, \qquad j = 0, \pm 1, \ldots, \pm n,$$

und suchen die Näherungslösung der Gleichung (14) in der Form

$$\varphi_n(t) = \sum_{k=-n}^{n} \xi_k t^k, \tag{15}$$

wobei die Koeffizienten ξ_k Lösung des Systems

$$(A\varphi_n)(t_j) = f(t_j) \qquad (j = 0, \pm 1, \ldots, \pm n),$$

oder, was dasselbe ist, des Systems

$$a(t_j)\sum_{k=0}^{n} \xi_k t_j^k + b(t_j)\sum_{k=-n}^{-1} \xi_k t_j^k + \sum_{k=-n}^{n} \xi_k c_{jk} = f(t_j) \qquad (j = 0, \pm 1, \ldots, \pm n) \tag{16}$$

mit $c_{jk} = (Tt^k)(t_j)$ sind.

Satz 9 ([15]). *Es sei* $a(t), b(t) \in H^\mu(\Gamma)$ $(0 < \mu < 1)$, $T\colon L_p(\Gamma) \to C(\Gamma)$ $(1/\mu < p < \infty)$ *ein vollstetiger Operator, und außerdem seien die nachfolgenden drei Bedingungen erfüllt:*

1) $a(t) \neq 0, \quad b(t) \neq 0 \quad (t \in \Gamma)$;

2) $\operatorname{ind} a(t) = \operatorname{ind} b(t)$;

3) *die Gleichung* $A\varphi = 0$ *hat in* $L_p(\Gamma)$ *nur die Null-Lösung.*

Dann hat das System (16) *für* $n \geq n_0$ *und für eine beliebige Funktion* $f(t) \in R(\Gamma)$[3] *eine eindeutige Lösung, und es gilt* $\varphi_n \xrightarrow[L_p]{} \varphi$.

[3] $R(\Gamma)$ bezeichnet die Klasse aller Riemann-integrierbaren 2π-periodischen Funktionen.

Wenn $a(t)$, $b(t) \in C^{r,\mu}(\Gamma)$, $f(t) \in C^{r,\nu}(\Gamma)$ $(r \geqq 0, 0 \leqq \nu < 1)$ *und* $T: L_p(\Gamma) \to C^{r,\nu}(\Gamma)$ *gilt, dann gilt die Abschätzung*

$$\|\varphi - \varphi_n\|_{L_p} = O(n^{-r-\lambda}), \qquad \lambda = \min(\mu, \nu). \tag{17}$$

Im Falle $a(t)$, $b(t)$, $f(t) \in C^{r,\mu}(\Gamma)$ *und* $T: H^\nu(\Gamma) \to C^{r,\mu}(\Gamma)$ $(r \geqq 0, 0 < \nu < \mu)$ *gilt die Abschätzung*

$$\|\varphi - \varphi_n\|_{H^\nu} = O\left(\frac{\ln n}{n^{r+\mu-\nu}}\right). \tag{18}$$

Die Konvergenz der Kollokationsmethode im Raum $H^\nu(\Gamma)$ wurde mit einer etwas schwächeren Fehlerabschätzung als (18) ($\ln^3 n$ anstelle $\ln n$) in [21, 23] bewiesen (vgl. auch [2, 5]).

Ein analoges Ergebnis läßt sich auch im nichtelliptischen Fall gewinnen, nämlich dann, wenn die Bedingung 1 aus Satz 9 in einzelnen Punkten des Einheitskreises verletzt ist. Der Einfachheit halber nehmen wir hier an, daß die Funktionen $a(t)$, $b(t)$ folgende Darstellung gestatten:

$$a(t) = (t^{-1} - \alpha_1^{-1})^{m_1} A(t), \qquad b(t) = (t - \alpha_2)^{m_2} B(t).$$

Hierbei seien $m_1, m_2 \geqq 0$ ganze Zahlen, $\alpha_1, \alpha_2 \in \Gamma$, und es gelte $A(t) \neq 0$, $B(t) \neq 0$ $(t \in \Gamma)$. Wir setzen $\varkappa = \max(m_1, m_2)$.

Satz 10 ([15]). *Es sei* $A(t)$, $B(t) \in C^{\varkappa,\mu}(0 < \mu < 1)$, $T: L_p(\Gamma) \to C^{\varkappa,\mu}(\Gamma)$ *ein vollstetiger Operator, und außerdem seien die folgenden beiden Bedingungen erfüllt:*

1) $\operatorname{ind} A(t) = \operatorname{ind} B(t)$;

2) $\dim \ker A = 0$.

Dann hat das System (16) *für* $n \geqq n_0$ *und für eine beliebige Funktion* $f(t) \in C^{\varkappa,\mu}(\Gamma)$ *eine eindeutige Lösung, und es gilt*

$$\|\varphi - \varphi_n\|_{L_p} = O(n^{-\mu}).$$

Im Falle $A(t)$, $B(t)$, $f(t) \in C^{\varkappa+r,\mu}(\Gamma)$ $(r \geqq 0)$ *und* $T: L_p(\Gamma) \to C^{\varkappa+r,\mu}(\Gamma)$ *gilt die Abschätzung*

$$\|\varphi - \varphi_n\|_{L_p} = O(n^{-\mu-r}).$$

5.2. Reduktionsmethode

Bei dieser Methode sucht man die Näherungslösung der Gleichung (14) wieder in der Form (15), wobei jetzt die Koeffizienten ξ_k aus folgenden Systemen zu bestimmen sind:

$$\sum_{k=0}^{n} a_{j-k}\xi_k + \sum_{k=-n}^{-1} b_{j-k}\xi_k + \sum_{k=-n}^{n} c_{jk}\xi_k = f_j \qquad (j = 0, \pm 1, \ldots, \pm n). \tag{19}$$

Hierbei bedeuten a_j, b_j, f_j, c_{jk} entsprechend den j-ten Fourierkoeffizienten der Funktionen $a(t)$, $b(t)$, $f(t)$, Tt^k.

Im weiteren bezeichnen wir mit $W_p^{(l),\mu}$ die Menge aller Funktionen aus dem Sobolew-Raum $W_p^{(l)}(\Gamma)$, deren l-te Ableitung eine Hölder-Bedingung mit dem Exponenten μ erfüllt, wenn ihr Zuwachs in der L_p-Norm abgeschätzt wird (vgl. [1]).

Satz 11 ([15]). *Es sei $a(t)$, $b(t) \in H^\mu(\Gamma)$ $(0 < \mu < 1)$, T ein vollstetiger Operator in $L_p(\Gamma)$ $(1 < p < \infty)$, und außerdem seien die nachfolgenden drei Bedingungen erfüllt:*

1) $a(t) \neq 0, \qquad b(t) \neq 0 \ (t \in \Gamma)$;

2) $\operatorname{ind} a(t) = \operatorname{ind} b(t) = 0$;

3) *die Gleichung $A\varphi = 0$ hat in $L_p(\Gamma)$ nur die Null-Lösung.*

Dann hat das System (19) *für $n \geqq n_0$ und für eine beliebige Funktion $f(t) \in L_p(\Gamma)$ eine eindeutige Lösung, und es gilt $\varphi_n \underset{L_p}{\rightarrow} \varphi$.*

Wenn $a(t)$, $b(t) \in C^{r,\mu}(\Gamma)$, $f(t) \in W_p^{(r),\nu}$ $(r \geqq 0, 0 \leqq \nu < 1)$ und $T: L_p(\Gamma) \to W_p^{(r),\nu}$ gilt, dann gilt die Abschätzung (17).

Im Falle $a(t)$, $b(t)$, $f(t) \in C^{r,\mu}(\Gamma)$ und $T: H^\nu(\Gamma) \to C^{r,\mu}(\Gamma)$ $(r \geqq 0, 0 < \nu < \mu)$ gilt die Abschätzung (18).

Für den Fall des Raumes $L_p(\Gamma)$ wurde Satz 11 ohne die Fehlerabschätzung (17) bereits in der Monographie [3] bewiesen. Eine etwas gröbere Abschätzung als (18) für den Fehler $\|\varphi - \varphi_n\|_{H^\nu}$ wurde in [22, 23] gegeben (dort steht $\ln^4 n$ statt $\ln n$).

Ähnlich wie im Satz 10 kann man ein entsprechendes Ergebnis auch im nichtelliptischen Fall angeben. Des weiteren lassen sich sämtliche Ergebnisse dieses Abschnittes auf den Systemfall übertragen (s. [15, 18—19, 4, 12—14]).

Zum Satz 11 analoge Resultate erhält man für die Reduktionsmethode bzw. das Galerkinsche Verfahren zur Lösung von diskreten Wiener-Hopf-Gleichungen und Wiener-Hopfschen Integralgleichungen (vgl. [4, 12—14, 18—19]).

Literatur

[1] Butzer, P. L., and R. J. Nessel, Fourier Analysis and Approximation, Vol. I. Basel und Stuttgart 1971.

[2] Габдулхаев, Б. Г., Приближенное решение сингулярных интегральных уравнений методом механических квадратур. Доклады АН СССР **179**, 260—263 (1968).

[3] Gochberg, I. Z., und I. A. Feldman, Faltungsgleichungen und Projektionsverfahren zu ihrer Lösung. Berlin 1974.

[4] Gochberg, I. Z., und S. Prößdorf, Ein Projektionsverfahren zur Lösung entarteter Systeme von diskreten Wiener-Hopf-Gleichungen. Math. Nachr. **65**, 19—45 (1975).

[5] Иванов, В. В., Теория приближенных методов и ее применение к численному решению сингулярных интегральных уравнений. Киев 1968.

[6] Каландия, А. И., Об одном прямом методе решения уравнения теории крыла и его применении в теории упругости. Матем. сборник **42**, 249—272 (1957).

[7] Krasnoselski, M. A., G. M. Wainikko u. a., Näherungsverfahren zur Lösung von Operatorgleichungen. Berlin 1973.

[8] Красносельский, М. А., П. П. Забрейко и др., Интегральные операторы в пространствах суммируемых функций. М., 1966.

[9] Michlin, S. G., Numerische Realisierung von Variationsmethoden. Berlin 1969.

[10] Multhopp, H., Die Berechnung der Auftriebsverteilung von Tragflügeln. Luftfahrt-Forschung **15**, N°. 4, 153—169 (1938).

[11] Prößdorf, S., Einige Klassen singulärer Gleichungen. Berlin 1974.
[12] —, Systeme einiger singulärer Gleichungen vom nicht normalen Typ und Projektionsverfahren zu ihrer Lösung. Studia math. **53**, 125—152 (1975).
[13] Prößdorf, S., und B. Silbermann, Ein Projektionsverfahren zur Lösung abstrakter singulärer Gleichungen vom nicht normalen Typ und einige seiner Anwendungen. Math. Nachr. **61**, 133—155 (1974).
[14] —, —, Verallgemeinerte Projektionsverfahren zur Lösung singulärer Gleichungen vom nicht normalen Typ. Math. Nachr. **68**, 7—28 (1975).
[15] —, —, (З. Пресдорф и Б. Зильберманн), О сходимости методов редукции и коллокации для систем сингулярных интегральных уравнений. Доклады АН СССР **226**, 516—519 (1976).
[16] —, —, Общие теоремы сходимости проекционных методов для операторных уравнений в банаховых пространствах. Доклады АН СССР **230**, 527—529(1976).
[17] —, —, Einige allgemeine Sätze zur Theorie der Projektionsverfahren für lineare Operatorgleichungen in Banachräumen. Math. Nachr. **75**, 61—72 (1976).
[18] —, —, Projektionsverfahren zur Lösung von Systemen singulärer Gleichungen vom nicht normalen Typ. Rev. Roumaine Math. Pures Appl. (im Druck).
[19] —, —, Zur Kollokations- und Reduktionsmethode für Systeme singulärer Integralgleichungen. Vortrag auf dem VII. IKM in Weimar 1975 (erscheint in den Tagungsberichten voraussichtlich 1976).
[20] Schleiff, M., Über Näherungsverfahren zur Lösung einer singulären linearen Integrodifferentialgleichung. Z. Angew. Math. Mech. **48**, 477—483 (1968).
[21] Золотаревский, В. А., О сходимости коллокационного метода для систем сингулярных интегральных уравнений. Матем. исслед., Кишинёв, IX, вып. 1 (31), 56—69 (1974).
[22] —, Решение сингулярных интегральных уравнений методом редукции. Матем. исслед., Кишинёв, IX, вып. 2 (32), 38—52 (1974).
[23] —, О приближенном решении сингулярных интегральных уравнений. Матем. исслед., Кишинёв, IX, вып. 3 (33), 82—94 (1974).

Anschrift des Verfassers:

Akademie der Wissenschaften der DDR
Zentralinstitut für Mathematik und Mechanik
108 Berlin
Mohrenstr. 39
DDR

Über Störungen nichtlinearer Operatorengleichungen

Thomas Riedrich, Dresden

Bei der Untersuchung der Lösungen linearer und nichtlinearer Operatorengleichungen ist die Frage ihres Verhaltens gegenüber „kleinen“ Störungen der in diesen Gleichungen auftretenden Operatoren von Interesse. Während für lineare Operatorengleichungen bereits seit längerer Zeit ausgearbeitete Störungstheorien mit weitreichenden Ergebnissen existieren, sind für nichtlineare Probleme erst in den letzten Jahren größere Anstrengungen in dieser Richtung unternommen worden. Es gibt eine wachsende Zahl von Arbeiten, in denen entsprechende „Stabilitäts-“ oder „Störungs-“ bzw. „stetige Abhängigkeits“-Probleme behandelt werden, und deren Inhalt erstreckt sich von rein topologischen Resultaten bis zu numerischen Aspekten. Zu nennen sind vor allem die Ergebnisse von Daneš [2]; Hahn [5, 6]; Hale [7]; Kasriel u. Nashed [8]; Cain u. Nashed [1]; Kayser [9]; Markin [12].

Analoge Probleme für die Stabilität der Lösungen von Extremalproblemen werden behandelt in Reeken [13, 14, 15] und Esser [3].

In der vorliegenden Arbeit diskutieren wir einfache Störungssätze für den Banachschen und den Schauderschen Fixpunktsatz und betrachten anschließend das Problem der Stabilität von Eigenwerten und Eigenvektoren im Rahmen der Theorie kompakter Vektorfelder in topologischen Vektorräumen. Der Begriff des Abbildungsgrades wird nicht verwendet.

Im folgenden wird stets vorausgesetzt, daß die betrachteten topologischen Vektorräume reell und separiert sind; die Terminologie der topologischen Begriffe richtet sich vor allem nach Bourbaki (vgl. G. Köthe [11]). Eine stetige Abbildung heißt kompakt, wenn ihre Bildmenge eine kompakte Abschließung besitzt. Die Abschließung einer Teilmenge A eines topologischen Raumes wird mit $\bar{A}$, ihr Rand mit ∂A und ihr Inneres $\bar{A} \setminus \partial A$ mit $\operatorname{int} A$ bezeichnet. Eine Nullumgebung U eines topologischen Vektorraumes heiße *einfachberandet* (Bezeichnung nach M. Landsberg; „shrinkable neighborhood“ bei Klee [10]), wenn aus $x \in \bar{U}$, $0 \leqq t < 1$, die Relation $(tx) \in \operatorname{int} U$ folgt. Das Minkowski- Funktional einer einfachberandeten Nullumgebung ist im ganzen Raum stetig. Jeder topologische Vektorraum besitzt ein Fundamentalsystem aus einfachberandeten Nullumgebungen.

1. Störungsaussagen zu den klassischen Fixpunktsätzen

Satz 1 (*Störungssatz zum Banachschen Fixpunktsatz*). *Es sei (X, ϱ) ein vollständiger metrischer Raum mit dem Träger $X \neq \emptyset$ und der Metrik ϱ, sowie $T\colon X \to X$ eine kontrahierende Selbstabbildung von X (es gibt ein reelles q mit $0 \leqq q < 1$ und $\varrho(Tx, Ty) \leqq q\varrho(x, y)$*

für alle $(x, y) \in X \times X$). *Es sei* x_0 *der (vorhandene und einzige) Fixpunkt von* T. *Ist* $S\colon X \to X$ *eine weitere Selbstabbildung von* X, *die einen Fixpunkt* x_1 *besitzt (es gilt* $Sx_1 = x_1$), *so besteht die Ungleichung*

$$\varrho(x_1, x_0) \leqq (1 - q)^{-1}\, \varrho(Sx_1, Tx_1).$$

Beweis. Aus den Eigenschaften der Elemente x_0 und x_1 und der Voraussetzung, daß T kontrahierend ist, folgt mittels der Dreiecksungleichung die Ungleichungskette

$$\begin{aligned}\varrho(x_1, x_0) &= \varrho(Sx_1, Tx_0) \leqq \varrho(Sx_1, Tx_1) + \varrho(Tx_1, Tx_0)\\ &\leqq \varrho(Sx_1, Tx_1) + q\varrho(x_1, x_0),\end{aligned}$$

woraus sich wegen $q < 1$ die Behauptung sofort ergibt.

Eine geringfügige Modifizierung der zugehörigen Störungsaussage ist erforderlich, wenn man sogenannte verallgemeinert-kontrahierende Abbildungen (vgl. z. B. [19]) betrachtet.

Definition 1. Es sei (X, ϱ) ein metrischer Raum mit dem Träger X und der Metrik ϱ. Eine Abbildung $T\colon X \to X$ heißt *verallgemeinert-kontrahierend*, wenn eine Funktion $q(\alpha, \beta)$ $(0 \leqq \alpha < \beta < +\infty)$ mit $0 \leqq q(\alpha, \beta) < 1$ $(0 \leqq \alpha < \beta < +\infty)$ existiert, so daß die Ungleichung

$$\varrho(Tx, Ty) \leqq q(\alpha, \beta)\, \varrho(x, y)$$

für alle Paare $(x, y) \in X \times X$ gilt, die der Bedingung $\alpha \leqq \varrho(x, y) \leqq \beta$ genügen.

Man kann zeigen, daß jede verallgemeinert-kontrahierende Abbildung eines vollständigen metrischen Raumes genau einen Fixpunkt besitzt.

Satz 2 (*Störungssatz für verallgemeinert-kontrahierende Abbildungen*). *Es sei* (X, ϱ) *ein vollständiger metrischer Raum mit endlichem Durchmesser* (d. h., *es existiert ein* $\beta_0 > 0$ *mit* $\varrho(x, y) \leqq \beta_0$ *für alle* $x, y \in X$). *Weiter sei* $T\colon X \to X$ *eine verallgemeinert-kontrahierende Abbildung, und es sei* x_0 *der (vorhandene und einzige) Fixpunkt von* T. *Es sei* (S_n) *eine Folge von Selbstabbildungen von* X, *die gleichmäßig auf* X *gegen* T *konvergieren, d. h.*

$$\lim_{n\to\infty} \left(\sup_{x\in X} \varrho(S_n x, Tx)\right) = 0.$$

Jede der Abbildungen S_n *besitze (mindestens) einen Fixpunkt* $x_n \in X$ $(n = 1, 2, \ldots)$. *Dann gilt die Limesbeziehung*

$$x_0 = \lim_{n\to\infty} x_n.$$

Beweis. Es sei $\varepsilon > 0$ mit $\varepsilon < \beta_0$ beliebig gegeben. Angenommen, es existiert eine Teilfolge (x_{n_j}) der Folge (x_n) der Fixpunkte der Abbildungen S_n mit $\varepsilon \leqq \varrho(x_{n_j}, x_0)$ für alle $j = 1, 2, \ldots$, dann gilt wegen der Vor. über den Durchmesser von X und der Vor. über die Abbildung T die Ungleichung (s. Def. 1)

$$\varrho(Tx_{n_j}, Tx_0) \leqq q(\varepsilon, \beta_0)\, \varrho(x_{n_j}, x_0) \qquad (j = 1, 2, \ldots)$$

(hierbei bezeichne $q(.,.)$ eine der Def. 1 entsprechend gewählte Funktion mit Werten im Intervall $[0, 1]$). Mittels der Dreiecksungleichung folgt (wie im Beweis von Satz 1) die Ungleichungskette

$$\begin{aligned}\varrho(x_{n_j}, x_0) &= \varrho(S_{n_j}x_{n_j}, Tx_0) \leqq \varrho(S_{n_j}x_{n_j}, Tx_{n_j}) + \varrho(Tx_{n_j}, Tx_0)\\ &\leqq \varrho(S_{n_j}x_{n_j}, Tx_{n_j}) + q(\varepsilon, \beta_0)\, \varrho(x_{n_j}, x_0)\end{aligned}$$

und weiter

$$\begin{aligned}0 < \varepsilon \leqq \varrho(x_{n_j}, x_0) &\leqq \big(1 - q(\varepsilon, \beta_0)\big)^{-1} \varrho(S_{n_j}x_{n_j}, Tx_{n_j})\\ &\leqq \big(1 - q(\varepsilon, \beta_0)\big)^{-1} \sup_{x \in X} \varrho(S_{n_j}x, Tx)\end{aligned}$$

Für $j \to \infty$ geht (bei festem $\varepsilon > 0$) die äußerste rechte Seite obiger Ungleichungskette gegen Null, was wegen der Ungleichung $\varepsilon \leqq \varrho(x_{n_j}, x_0)$ $(j = 1, 2, \ldots)$ einen Widerspruch ergibt.

Auf S. Hahn (s. [6]) geht die Beweisidee für den folgenden Störungssatz für die Fixpunkte kompakter Abbildungen zurück. In unserer Formulierung dieses Satzes ist neu, daß die der kompakten Ausgangsabbildung benachbarte Abbildung (Störung), von der gezeigt werden soll, daß ihre Fixpunkte dem (einzigen) Fixpunkt der Ausgangsabbildung benachbart sind, nicht notwendig stetig zu sein braucht.

Satz 3. *Es sei E ein topologischer Vektorraum und G eine abgeschlossene Teilmenge von E. Es sei $T: G \to E$ eine kompakte Abbildung, die genau einen Fixpunkt $x_0 \in G$ besitzt. Dann gibt es zu jeder Nullumgebung U von E eine Nullumgebung V von E, so daß jede Abbildung $\tilde{T}: G \to E$ mit $(\tilde{T}x - Tx) \in V (x \in G)$, nur Fixpunkte haben kann, die in $(x_0 + U)$ liegen.*

Beweis. Es sei U eine (o. B. d. A.) offene Nullumgebung von E. In der Theorie der kompakten Vektorfelder wird gezeigt (s. z. B. V. Klee [10], S. 289), daß der Wertebereich eines auf einer abgeschlossenen Menge erklärten kompakten Vektorfeldes stets abgeschlossen ist. Demzufolge ist die Menge $M = \{z \in E \mid z = Tx - x; x \in G \setminus (x_0 + U)\}$ abgeschlossen und enthält nach Vor. nicht das Nullelement von E. Es existiert somit eine o. B. d. A. symmetrische Nullumgebung V von E mit $V \cap W = \emptyset$. Es sei $\tilde{T}: G \to E$ eine Abbildung mit $(\tilde{T}x - Tx) \in V (x \in G)$. Wäre $\tilde{T}x = x$ für ein $x \in G \setminus (x_0 + U)$, so wäre $-(\tilde{T}x - Tx) = Tx - x$. Das Element $-(\tilde{T}x - Tx)$ gehört nach Wahl von $\tilde{T}$ zur Menge $-V$, also auch zu V (V ist symmetrisch). Das Element $(Tx - x)$ gehört nach Konstruktion von V nicht zu V (als Element von M). Dies widerspricht der Gleichheit der Elemente $-(\tilde{T}x - Tx)$ und $(Tx - x)$. Also gibt es außerhalb von $(x_0 + U)$ keine Fixpunkte von $\tilde{T}$.

2. Störungen nichtlinearer Eigenwertprobleme

Mittels des folgenden von S. Hahn (s. [7]) zuerst bewiesenen Hilfssatzes lassen sich zahlreiche Varianten von Störungsaussagen für Eigenwerte kompakter Abbildungen aufstellen.

Hilfssatz 1. *Es sei E ein topologischer Vektorraum und X eine abgeschlossene Teilmenge von E mit $o \notin X$. Es seien $T: X \to E$ eine kompakte Abbildung von X in E und Λ eine* *abgeschlossene Teilmenge der Menge der reellen Zahlen (versehen mit der üblichen Topologie) mit* $0 \notin \Lambda$. *Dann ist die Menge* $M = \{z \in E \mid z = Tx - \lambda x;\, x \in X;\, y \in \Lambda\}$ *abgeschlossen.*

Beweis (s. [7]). Die Zahl $\sigma = \inf \{|\lambda| \mid \lambda \in \Lambda\}$ ist nach Voraussetzung positiv. Es sei $(z_j)_{j\in J}$ eine Moore-Smith-Folge von Elementen aus M, die gegen ein Element z aus E konvergiert. Zu jedem $j \in J$ gibt es ein $x_j \in X$ und ein $\lambda_j \in \Lambda$ mit $z_j = Tx_j - \lambda_j x_j$. Die Elemente $y_j = Tx_j$ liegen zufolge der Voraussetzungen über T in einer kompakten Teilmenge von E. Wir können daher o. B. d. A. annehmen, daß die Moore-Smith-Folge $(y_j)_{j\in J}$ gegen ein Element y aus E konvergiert (Abkürzung eines Standard-Beweisverfahrens s. z. B. [10], S. 289). Damit konvergiert auch die Folge $(\lambda_j x_j)$, und zwar gegen das Element $w = y - z$. Wegen $o \notin X$ gibt es eine abgeschlossene einfachberandete Nullumgebung V von E mit $V \cap X = \emptyset$. Das Minkowski-Funktional $p(.)$ von V ist stetig, daher konvergiert die Moore-Smith-Folge der reellen Zahlen $p(\lambda_j x_j) = \lambda_j p(x_j)$ gegen die Zahl $p(w)$. Es existieren daher eine positive Zahl m und ein Index $j_0 \in J$ mit $|\lambda_j p(x_j)| \leqq m$ $(j \geqq j_0)$. Wegen $x_j \in X$ $(j \in J)$ gilt $1 \leqq p(x_j)$ $(j \in J)$. Wegen $\lambda_j \in \Lambda$ $(j \in J)$ ergeben sich somit für die Zahlen $|\lambda_j|$ $(j \geqq j_0)$ die Schranken

$$\sigma \leqq |\lambda_j| \leqq m \qquad (j \geqq j_0).$$

Daher können wir weiter annehmen, daß die Moore-Smith-Folge $(\lambda_j)_{j\in J}$ gegen eine reelle Zahl $\lambda \neq 0$ konvergiert, die wegen der Abgeschlossenheit von Λ ein Element von Λ ist. Dann konvergiert aber auch die Moore-Smith-Folge $(x_j)_{j\geqq j_0}$ gegen das Element $x = \frac{1}{\lambda} w$, das wegen der Abgeschlossenheit von X zu X gehört. Die Abbildung T ist (als kompakte Abbildung) stetig, somit konvergiert $(Tx_j)_{j\in J}$ gegen das Element Tx. Andererseits konvergiert $(Tx_j)_{j\in J}$ gegen das Element y. Aus den Gleichheiten $y = Tx$; $x = \frac{1}{\lambda} w$; $w = y - z$ folgt die Beziehung $z = Tx - \lambda x$, also gilt $z \in M$; M ist abgeschlossen.

Mittels des Hilfssatzes 1 erhält man einen im wesentlichen auf S. Hahn zurückgehenden Störungssatz für die Eigenwerte und die Eigenvektoren kompakter Abbildungen (s. auch [7]).

Satz 4. *Es sei E ein topologischer Vektorraum und G eine abgeschlossene Teilmenge von E mit $o \notin G$. Es sei $T: G \to E$ eine kompakte Abbildung von G in E. Die Gleichung $Tx = \lambda x$ ($x \in G$; λ reell) besitze genau eine Lösung (x_0, λ_0), und es gelte $\lambda_0 \neq 0$. Es sei U_0 eine symmetrische abgeschlossene, einfachberandete Nullumgebung von E mit $U_0 \cap G = \emptyset$ sowie $R_0 > 0$ eine positive reelle Zahl derart, daß $T(G) \subset R_0 U_0$ gilt und $\sigma > 0$ eine positive reelle Zahl derart, daß $\sigma < |\lambda_0|$ ist.*

Dann gibt es zu jedem $\varepsilon > 0$ und zu jeder Nullumgebung U von E eine Nullumgebung $V = V(\varepsilon; U)$ von E derart, daß für jede Abbildung $\tilde{T}: G \to E$ mit $(\tilde{T}x - Tx) \in V$ $(x \in G)$ und $\tilde{T}(G) \subset R_0 U_0$ aus dem Bestehen der Gleichung $\tilde{T}\tilde{x} = \tilde{\lambda}\tilde{x}$ für ein $\tilde{x} \in G$ und ein reelles $\tilde{\lambda}$ mit $|\tilde{\lambda}| \geqq \sigma$ die Relationen

$$(\tilde{x} - x_0) \in U \qquad \text{und} \qquad |\tilde{\lambda} - \lambda_0| \leqq \varepsilon$$

folgen.

Beweis. Es seien die Zahl $\varepsilon > 0$ mit o. B. d. A. $\varepsilon \leqq 1$ und die Nullumgebung U beliebig gegeben. Mit $p(.)$ bezeichnen wir das Minkowski-Funktional von U_0, es ist nicht-negativ homogen und stetig (s. [10]). Es sei $\tilde{U}_0$ eine symmetrische abgeschlossene einfachberandete Nullumgebung mit $\tilde{U}_0 + \tilde{U}_0 \subset U_0$. Infolge der Stetigkeit von T gibt es in der Nullumgebung $U_1 = U \cap (\varepsilon\tilde{U}_0)$ eine symmetrische abgeschlossene einfachberandete Nullumgebung V_0 mit der Eigenschaft, daß aus $x \in V_0$ die Relation $(Tx - Tx_0) \in \varepsilon\tilde{U}_0$ folgt. Wir setzen $\Lambda = (-\infty, -\sigma] \cup [\sigma, +\infty)$. Nach Hilfssatz 1 ist dann die Menge $M = \{z \in E \mid z = Tx - \lambda x; x \in G \setminus (x_0 + \operatorname{int} V_0); \lambda \in \Lambda\}$ abgeschlossen. Zufolge der Unität von (x_0, λ_0) gilt $o \notin M$. Es existiert daher eine in V_0 enthaltene symmetrische abgeschlossene Nullumgebung V_1 mit $V_1 \cap M = \emptyset$.

Es sei $\tilde{T}: G \to E$ eine Abbildung mit $\tilde{T}(G) \subset R_0 U_0$ und $(\tilde{T}x - Tx) \in V_1$ $(x \in G)$. Angenommen, es gibt ein $\bar{x} \in G \setminus (x_0 + \operatorname{int} V_0)$ und ein $\bar{\lambda} \in \Lambda$ mit $\tilde{T}\bar{x} = \bar{\lambda}\bar{x}$. Dann folgt die Gleichung

$$T\bar{x} - \bar{\lambda}\bar{x} = T\bar{x} - \tilde{T}\bar{x},$$

deren linke Seite in M, also nicht in V_1 liegt und deren rechte Seite nach Vor. zu V_1 gehört. Dieser Widerspruch zeigt, daß es ein solches $\bar{x}$ nicht geben kann. Es gelte nun $\tilde{T}\tilde{x} = \tilde{\lambda}\tilde{x}$ für ein $\tilde{\lambda}$ mit $|\tilde{\lambda}| \geqq \sigma$ und ein $\tilde{x} \in G$. Nach dem eben bewiesenen Sachverhalt gilt dann $\tilde{x} \in (x_0 + \operatorname{int} V_0)$ und somit $(\tilde{x} - x_0) \in U$. Wegen $p(x) \geqq 1$ für $x \in G$ und $\tilde{T}(G) \subset R_0 U_0$ folgt die Ungleichungskette

$$|\tilde{\lambda}| \leqq |\tilde{\lambda}|\, p(\tilde{x}) = p(\tilde{\lambda}\tilde{x}) = p(\tilde{T}\tilde{x}) \leqq R_0.$$

Weiter besteht die Zerlegung

$$(\tilde{\lambda} - \lambda_0)\, x_0 = \tilde{\lambda}(x_0 - \tilde{x}) + (\tilde{T}\tilde{x} - T\tilde{x}) + (T\tilde{x} - Tx_0).$$

Das Element $\tilde{\lambda}(x_0 - \tilde{x})$ liegt in $R_0 V_0$; das Element $(\tilde{T}\tilde{x} - \tilde{T}\tilde{x})$ liegt in V_1 und damit in V_0 und das Element $(T\tilde{x} - Tx_0)$ liegt in $\varepsilon\tilde{U}_0$. Daher liegt das Element $(\tilde{\lambda} - \tilde{\lambda}_0)\, x_0$ in der Menge $R_0 V_0 + V_0 + \varepsilon\tilde{U}_0$, die ihrerseits in $\varepsilon\tilde{U}_0 + \varepsilon\tilde{U}_0 = \varepsilon(\tilde{U}_0 + \tilde{U}_0)$ enthalten ist. Wegen $\tilde{U}_0 + \tilde{U}_0 \subset U_0$ ergibt sich die Beziehung $(\tilde{\lambda} - \lambda_0)\, x_0 \in \varepsilon\, U_0$, woraus schließlich die Abschätzungen (man beachte, daß $p(x_0) \geqq 1$ gilt)

$$|\tilde{\lambda} - \lambda_0| \leqq |\tilde{\lambda} - \lambda_0|\, p(x_0) = p\big((\tilde{\lambda} - \lambda_0)\, x_0\big) \leqq \varepsilon$$

folgen, womit alles bewiesen ist.

Allgemeinere Störungssätze mit abgeschwächten Voraussetzungen über die Art der Approximation der Ausgangsabbildung durch die gestörte Abbildung, bei denen die Existenz von Lösungen der gestörten Abbildung mit bewiesen wird, erhält man mittels einer Variante der Leray-Schauder-Theorie, deren Grundlagen auf A. Granas [4] und V. Klee [10] zurückgehen. Entscheidendes Hilfsmittel ist ein Homotopie-Erweiterungssatz für kompakte Vektorfelder in Produkträumen, der vom Verf. in [18] bewiesen wurde. Mittels der in [18] für den Fall von Halbeigenwerten ausgeführten Methode lassen sich zahlreiche Störungssätze zu bekannten Eigenwertaussagen beweisen, s. u. a. auch [17]. In der vorliegenden Arbeit soll, um das Wesentliche des Vorgehens besser hervortreten zu lassen, nur eine der grundlegenden Eigenwertaussagen für kompakte Abbildungen im Hinblick auf Störung näher untersucht werden. Zunächst werden die erforderlichen Hilfsmittel — als Zitat — zusammengestellt (s. [18]).

In Verallgemeinerung und Modifizierung einer üblichen Terminologie (vgl. Klee [10], S. 286/287) verwenden wir die folgenden Bezeichnungen.

Es seien E ein topologischer Vektorraum, X eine Teilmenge von E und Y ein topologischer Raum; dann nennen wir jede Abbildung $f: A \to E$ einer Teilmenge A des Produkts $X \times Y$ in E, zu der eine kompakte Abbildung $F: A \to E$ existiert, mit

$$f(x, y) = x - F(x, y) \qquad \big((x, y) \in A\big)$$

ein *kompaktes Vektorfeld auf A mit Werten in E*. Die Menge aller kompakten Vektorfelder auf A mit Werten in E bezeichnen wir mit $D(A, X, Y, E)$, und die Teilmenge aller der $f \in D(A, X, Y, E)$ mit $o \notin f(A)$ bezeichnen wir mit $D_0(A, X, Y, E)$.

Zwei Elemente f und g aus $D_0(A, X, Y, E)$ sollen *homotop* in $D_0(A, X, Y, E)$ heißen, wenn es eine kompakte Abbildung

$$H: A \times [0, 1] \to E$$

gibt mit $x - H(x, y, t) \neq o$ für alle $(x, y, t) \in (A \times [0, 1])$ und die Gleichungen

$$x - H(x, y, 0) = f(x, y)$$

$$x - H(x, y, 1) = g(x, y)$$

für alle $(x, y) \in A$ gelten. Dabei wird die durch die Vorschrift $h(x, y, t) = x - H(x, y, t)$ $\big((x, y) \in A; t \in [0, 1]\big)$ erklärte Abbildung $h: A \times [0, 1] \to E$ eine f und g verbindende *Homotopie* genannt. Häufig bedient man sich des folgenden sehr einfachen Homotopiekriteriums.

Bemerkung 1. *Es sei E ein topologischer Vektorraum und X eine Teilmenge von E. Es sei weiter Y ein topologischer Raum und A eine Teilmenge des Produktraumes $X \times Y$. Ferner sei f ein Element von $D_0(A, X, Y, E)$, zu dem eine sternförmige Nullumgebung V von E existiert mit $f(A) \cap V = \emptyset$. Dann ist jedes $g \in D_0(A, X, Y, E)$ mit $\big(f(x, y) - g(x, y)\big) \in V$ $\big((x, y) \in A\big)$ zu f in $D_0(A, X, Y, E)$ homotop.*

Es gilt nun der folgende Homotopieerweiterungssatz (vgl. [18], S. 182).

Satz 5. *Es sei E ein vollständiger metrisierbarer lokalkonvexer topologischer Vektorraum, X eine Teilmenge von E und Y ein metrisierbarer topologischer Raum; ferner sei A eine Teilmenge von $X \times Y$ und A_0 eine abgeschlossene Teilmenge von A. Es seien weiter f_0 und g_0 Elemente von $D_0(A_0, X, Y, E)$, die zueinander in $D_0(A, X, Y, E)$ homotop sind. Besitzt f_0 eine in $D_0(A, X, Y, E)$ liegende Fortsetzung f, dann besitzt auch g_0 eine in $D_0(A, X, Y, E)$ liegende Fortsetzung g, die zu f in $D_0(A, X, Y, E)$ homotop ist.*

Mittels des obigen Homotopieerweiterungssatzes können wir den folgenden lokalen Störungssatz beweisen, der eine Verallgemeinerung des in [17] angegebenen Störungssatzes in der Hinsicht beinhaltet, daß die gestörte Abbildung nur in einer Umgebung des Eigenvektors der Ausgangsabbildung „in der Nähe" der Ausgangsabbildung zu liegen braucht.

Satz 6. *Es sei E ein vollständiger metrisierbarer lokalkonvexer Raum und U eine abgeschlossene Nullumgebung in E. Es sei $T: \partial U \to E$ eine kompakte Abbildung, zu welcher ein $y \in (E \setminus U)$ mit der Eigenschaft existiert, daß kein Element $x \in \partial U$ auf der Verbin-*

dungsstrecke von y und Tx liegt. Die Gleichung $Tx = \lambda x$ ($x \in \partial U$, $\lambda > 1$) besitze genau eine Lösung (x_0, λ_0) mit $x_0 \in \partial U$ und $\lambda_0 > 1$. Dann gibt es zu jedem $\varepsilon > 0$ und zu jeder Nullumgebung W von E eine Nullumgebung $V = V(\varepsilon; W)$ von E, so daß jede kompakte Abbildung $\tilde{T}: \partial U \to E$ mit

$$(Tx - \tilde{T}x) \in V\big(x \in (x_0 + W) \cap \partial U\big)$$

mindestens einen Eigenvektor $\tilde{x}_0 \in \partial U$ mit einem Eigenwert $\tilde{\lambda}_0 > 1$ besitzt, so daß die Relationen

$$|\tilde{\lambda}_0 - \lambda_0| \leqq \varepsilon \qquad \text{und} \qquad (x_0 - \tilde{x}_0) \in W$$

gelten.

Beweis. Es seien $\varepsilon > 0$ und W eine o. B. d. A. abgeschlossene Nullumgebung von E. Wir setzen $t_0 = \lambda_0^{-1}$ und nehmen o. B. d. A. an, daß $\varepsilon < \min\big(1, 2(1 - t_0)\, t_0^{-2}\big)$ gilt. Auf Grund unserer Voraussetzungen über λ_0 gilt dann $0 < t_0 < 1$, und für die positive Zahl $\varepsilon' = \frac{t_0^2}{2}\varepsilon$ bestehen die Ungleichungen $\varepsilon' < t_0$ und $\varepsilon' < 1 - t_0$. Im folgenden betrachten wir den Teilraum $G = \{x \in \partial U \mid (x - x_0) \in W\} \times [t_0 - \varepsilon', t_0 + \varepsilon']$ des metrisierbaren Raumes $\partial U \times [0, 1]$. Wir definieren die Abbildung $h: \partial U \times [0, 1] \to E$ durch die Vorschrift $h(x, t) = (1 - t)\, x + t(x - Tx)$ $(= x - tTx)$ ($x \in \partial U$; $t \in [0, 1]$). Man überzeugt sich leicht davon, daß die Gleichung $h(x, t) = 0$ genau dann besteht, wenn $x = x_0$ und $t = t_0$ gilt.

Es sei A_0 der Rand von G, gebildet im Raum $\partial U \times [0, 1]$, und es bezeichne h_0 die Einschränkung der Abbildung h auf A_0. A_0 ist eine abgeschlossene Teilmenge von G und ersichtlich ist h_0 ein kompaktes Vektorfeld auf A_0 mit Werten in E. Da der Punkt (x_0, t_0) im Raum $\partial U \times [0, 1]$ ein innerer Punkt von G ist, gilt die Relation $h_0 \in D_0(A_0, \partial U, [0, 1], E)$. Wir zeigen, daß h_0 keine Fortsetzung besitzt, die in $D_0(G, \partial U, [0, 1], E)$ liegt. Angenommen, es gäbe eine solche Fortsetzung $h' \in D_0(G, \partial U, [0, 1], E)$, so wäre die durch die Vorschrift

$$j(x, t) = \begin{cases} h(x, t) & \text{für} \quad (x, t) \in (\partial U \times [0, 1]) \setminus G \\ h'(x, t) & \text{für} \quad (x, t) \in G \end{cases}$$

erklärte Abbildung $j: \partial U \times [0, 1] \to E$ eine Homotopie, die die Abbildungen $I_{\partial U} - T$ und $I_{\partial U}$ in $D_0(\partial U, E)$ verbindet. Dies ist auf Grund der Voraussetzungen über die Abbildung T nicht möglich, ein diesbezüglicher Beweis findet sich in [17].

Mittels einer einfachen Überlegung, die im Beweis der Tatsache besteht, daß die Bildmenge einer abgeschlossenen Menge bezüglich einer Abbildung, die ein kompaktes Vektorfeld ist, wieder abgeschlossen ist (s. Klee [10], S. 289), ergibt sich die Relation $0 \notin \overline{h_0(A_0)}$. Es gibt also eine absolutkonvexe Nullumgebung V mit $V \cap h_0(A_0) = \emptyset$.

Es sei jetzt $\tilde{T}: \partial U \to E$ eine kompakte Abbildung mit $(Tx - \tilde{T}x) \in V$ $\big(x \in (x_0 + W) \cap \partial U\big)$. Wir setzen $K(x, t) = (1 - t)\, x + t(x - \tilde{T}x)$ ($x \in \partial U$; $t \in [0, 1]$) und bezeichnen mit K_0 die Einschränkung der durch die letztere Gleichheit definierten Abbildung $K: \partial U \times [0, 1] \to E$ auf A_0. Wegen des Bestehens der Gleichung $h_0(x, t) - K_0(x, t) = t(\tilde{T}x - Tx)$ $\big((x, t) \in A_0\big)$ gilt die Relation $\big(h_0(x, t) - K_0(x, t)\big) \in V$ $\big((x, t) \in A_0\big)$. Nach der Bemerkung 1 sind daher h_0 und K_0 in $D_0(A_0, \partial U, [0, 1], E)$ homotop. Da h_0 keine Fortsetzung besitzt, die in D_0 $(G, \partial U, [0, 1], E)$ liegt, besitzt nach dem

Homotopie-Erweiterungssatz auch K_0 keine solche Fortsetzung. Weil die Einschränkung von K auf G in $D(G, \partial U, [0, 1], E)$ liegt, muß somit die Abbildung K in G eine Nullstelle besitzen, m. a. W., es gibt ein $(\bar{x}, \bar{t}) \in G$ mit $K(\bar{x}, \bar{t}) = 0$. Nach Definition von K bedeutet dies das Bestehen der Gleichung $\bar{T}\bar{x} = \bar{\lambda}_0\bar{x}$ mit $\bar{\lambda}_0 = \frac{1}{\bar{t}}$. Nach der Konstruktion von G gilt daher $(\bar{x} - x_0) \in W$ und $0 < t_0 - \varepsilon' \leqq t \leqq t_0 + \varepsilon' < 1$. Die Ungleichungskette

$$|\bar{\lambda}_0 - \lambda_0| = \left|\frac{1}{t_0} - \frac{1}{\bar{t}}\right| = \frac{1}{t_0\bar{t}}\,|t_0 - \bar{t}| \leqq \frac{\varepsilon'}{t_0\bar{t}} \leqq \frac{\varepsilon'}{t_0(t_0 - \varepsilon')} \leqq \frac{\varepsilon'}{t_0 \cdot \frac{t_0}{2}} = \varepsilon$$

zeigt, daß das Paar $(\bar{x}, \bar{\lambda}_0)$ alle geforderten Bedingungen erfüllt.

Literatur

[1] Cain, G. L., M. Z. Nashed, Fixed points and stability for a sum of two operators in locally convex spaces. Pacific J. Math. **39**, 581–592 (1971).

[2] Daneš, J., On densifying and related mappings and their application in nonlinear functional analysis. *In:* Theory of Nonlinear Operators. Proceedings of a summer-school 1972 Neuendorf, GDR, 15–56 (1974).

[3] Esser, H., Über die Stetigkeit des Extremalwertes nichtkonvexer Optimierungsaufgaben mit einer Anwendung auf die Diskretisierung von Kontrollproblemen. Z. Angew. Math. Mech. **52**, 535–542 (1972).

[4] Granas, A., The theory of compact vector fields and some of its applications to topology of functional spaces. (I) Rozprawy Mat. XXX, Warszawa 1962.

[5] Hahn, S., Über Näherungslösungen von nichtlinearen Operatorengleichungen in topologischen Vektorräumen. Wiss. Z. Techn. Univ. Dresden **22**, 37–42 (1973).

[6] Hahn, S., Über die Stabilität von Lösungen nichtlinearer Operatorengleichungen in nicht notwendig lokalkonvexen topologischen Vektorräumen. Comment. Math. Univ. Carolinae **17**, 421–440 (1976).

[7] Hale, J. K., Continuous dependence of fixed points of condensing maps. J. Math. Analysis Appl. **46**, 388–394 (1974).

[8] Kasriel, R. H., M. Z. Nashed, Stability of solutions of some classes of nonlinear operator equations. Proc. Amer. Math. Soc. **17**, 1036–1042 (1966).

[9] Kayser, G., Über die Existenz von Lösungen nichtlinearer Operatorengleichungen mit nicht notwendig kompakten Operatoren. Dissertation 1973. TU Dresden (unveröffentlicht).

[10] Klee, V., Leray-Schauder-theory without local convexity. Math. Ann. **141**, 286–296 (1960).

[11] Köthe, G., Topologische lineare Räume I. Springer-Verlag, Berlin–Göttingen–Heidelberg 1960.

[12] Markin, J. T., Continuous dependence of fixed point sets. Proc. Amer. Math. Soc. **38**, 545 to 547 (1973).

[13] Reeken, M., Stability of critical points under small perturbations. Part I: Topological theory. Manuscripta math. **7**, 387–411 (1972).

[14] Reeken, M., Stability of critical points under small perturbations. Part II: Analytic theory. Manuscripta math. 8, 69–92 (1973).

[15] Reeken, M., Stability of critical values and isolated critical continua. Manuscripta math. 12, 163–193 (1974).

[16] Riedrich, T., Über die Stabilität positiver Eigenwerte finiter Abbildungen. Math. Nachr. **41**, 301–307 (1969).

[17] Riedrich, T., Störungen nichtlinearer Operatorengleichungen. Wissenschaftliche Schriftenreihe der TH Karl-Marx-Stadt, 5. TMP, Heft 2, 345—350 (1975).
[18] Riedrich, T., Über die Stabilität positiver Halbeigenwerte kompakter Abbildungen. Beitr. Numer. Math. 4, 179—190 (1975).
[19] SMB Funktionalanalysis. Sammelband herausgegeben von S. G. Krej'n (russisch) Moskau 1972.

Anschrift des Verfassers:

Technische Universität Dresden
Sektion Mathematik
8027 Dresden
Mommsenstr. 13
DDR

Sur la stabilité des inégalités paraboliques de type monotone

Moïse Sibony, Tours

Abstract. Our aim shall be to determine numerically the solution $u(t)$ of a nonlinear parabolic inequality of the form

$$(\mathrm{P})\quad \begin{cases} \displaystyle\int_0^T \left(\frac{du}{dt} + Au, v - u\right) dt \geqq \int_0^T (f, v - u)\, dt, & \forall v \in K \\ u(0) = u_0 \end{cases}$$

u taking its values, for almost all t, in a given convex closed subset K of a real Banach space V. The operator $A(t)$ is monotone nonlinear from V into its dual V' and $f(t)$ is taking its values in V'. We associate to the problem (P) the problem

$$(\mathrm{P}_h, k, \theta)\quad \begin{cases} \left(\dfrac{u_{hk}^{r+1} - u_*{}^r}{k}, v_h - u_{hk}^{r+1}\right) + \left(A_{hk}\left(\theta u_{hk}^{r+1} + (1-\theta)\, u_*{}^r\right), v_h - u_{hk}^{r+1}\right) \\ \geqq (f_{hk}^{r+1}, v_h - u_{hk}^{r+1})\ \forall\ v_h \in K_h \end{cases}$$

(where $u_*{}^r$ is the estimation of u_{hk}^r at the step r) for $r = 0, 1, \ldots, m-1$ with $m = \dfrac{T}{k}$ and $\theta \in [0, 1]$, where K_h denotes a closed convex subset of a finite dimensional space V_h; h is the discrete parameter of the space variable and k is the discrete parameter of the time variable $t \in [0, T]$.

In order to assure the convergence of all the various approximations process, we seek to determine *an optimal number* $m(h, k, r; \theta)$ of iterations to make the global error converge to zero.

Résumé. On se propose de calculer numériquement la solution $u(t)$ d'une inéquation parabolique non linéaire de la forme

$$\left.\begin{aligned} &\int_0^T \left(\frac{du}{dt} + Au, v - u\right) dt \geqq \int_0^T (f, v - u)\, dt \qquad \forall v \in K \\ &u(0) = u_0 \end{aligned}\right\} \tag{1}$$

u étant, pour presque tout t, à valeurs dans un ensemble convexe fermé donné K, d'un espace de Banach réel V. $A(t)$ est pour tout t, un opérateur de V dans son dual V'; $f(t)$ est donné, à valeurs dans V', et u_0 est donné dans K.

On suppose que l'opérateur $A(t)$ est non linéaire, monotone et tel que $\|A(t)\, u(t)\|_{V'} \leqq M\|u(t)\|_V^{p-1}$ pour tout t, où M est une constante positive, p un réel supérieur ou égal à 2.

Après avoir assuré l'existence et l'unicité de la solution $u(t)$ du problème (1), nous sommes amenés à étudier le problème discrétisé

$$\left.\begin{aligned}&\left(\frac{u_{hk}^{r+1} - u_{hk}^{r}}{k}, v_h - u_{hk}^{r+1}\right) + \left(A_{hk}\left(\theta u_{hk}^{r+1} + (1-\theta)\, u_{hk}^{r}\right), v_h - u_{hk}^{r+1}\right)\\ &\geqq \left(f_{hk}^{r+1}, v_h - u_{hk}^{r+1}\right) \qquad \forall v_h \in K_h\end{aligned}\right\} \tag{2}$$

pour $r = 0, 1, \ldots, m-1$ avec $m = \dfrac{T}{k}$ et $\theta \in \,]0, 1]$ où K_h est un convexe fermé d'un espace V_h de dimension finie; h est le paramètre de discrétisation de la variable d'espace et k est le paramètre de discrétisation de la variable de temps $t \in [0, T]$.

Comme le problème (2) est résolu numériquement à l'aide d'une méthode itérative, il apparaît à chaque étape $r + 1$, une erreur due au remplacement de u_{hk}^{r} par sa valeur approchée $u_*{}^r$, et une erreur due au calcul de u_{hk}^{r+1} par des méthodes itératives effectuées jusqu'au rang m. Plus précisément, à l'étape $(r + 1)$ on résoud le problème

$$\left.\begin{aligned}&\left(\frac{u_{hk}^{r+1} - u_*{}^{r}}{k}, v_h - u_{hk}^{r+1}\right) + \left(A_{hk}\left(\theta u_{hk}^{r+1} + (1-\theta)\, u_*{}^{r}\right), v_h - u_{hk}^{r+1}\right)\\ &\geqq \left(f_{hk}^{r+1}, v_h - u_{hk}^{r+1}\right) \qquad \forall v_h \in K_h\end{aligned}\right\} \tag{3}$$

à l'aide d'une méthode itérative du type

$$u_{hk,\underline{m+1}}^{r+1} = P_{K_h}^{S_h}\left(u_{hk,\underline{m}}^{r+1} - \varrho S_h^{-1}\left(B_{hk} u_{hk,\underline{m}}^{r+1} - g_*\right)\right) \tag{4}$$

avec

$$B_{hk} u_{hk}^{r+1} = u_{hk}^{r+1} + k A_{hk}\left(\theta u_{hk}^{r+1} + (1-\theta)\, u_*{}^{r}\right), \quad g_* = u_*{}^{r} + k f_{hk}^{r+1}.$$

On se propose alors de chercher un nombre optimal $m(h, k, r, \theta)$ d'itérations à effectuer à l'étape r, pour assurer la convergence de ces divers procédés d'approximation du problème. Le plan est le suivant:

1. Le problème continu;
2. Schéma de discrétisation — Convergence;
3. Résolution numérique — Méthode itérative — Estimation de l'erreur;
4. Exemples.

1. Formulation du problème continu: Existence et unicite de la solution

Soient H un Hilbert réel séparable muni du produit scalaire $(.,.)$ et de la norme $|.|$, et V un espace de Banach réel réflexif séparable de norme $\|.\|$. On suppose que V est inclus dans H, avec injection continue et image dense, et on identifie H à son dual. On suppose

(H1) $|u| \leqq c_0\|u\| \;\forall\, u \in v$ où c_0 est une constante positive.

Soit K un convexe fermé de V contenant 0. Pour tout $t \in [0, T]$, soit $A(t)$ un opérateur de V dans V' vérifiant les conditions:

$$(\mathrm{H2}) \begin{cases} \text{i) } A(t) \text{ est monotone hémicontinu de } V \text{ dans } V'. \\ \text{ii) } \|A(t)\,u\|_{V'} \leqq M\|u\|_V{}^{p-1} \text{ où } M \text{ est une constante positive,} \\ \quad \big(A(t)\,u, u\big) \geqq \alpha\|u\|_V{}^p \quad \alpha > 0 \quad \text{et} \quad p \geqq 2. \\ \text{iii) } t \mapsto \big(A(t)\,u, v\big) \text{ est mesurable pour tout } t, \text{ et tous } u \text{ et } v \in V. \end{cases}$$

Soit $K_1 = \{v(.) \in L^p(0, T; V)$ tels que $v(t) \in K$ $p.p$ $dt\}$. K_1 est un convexe fermé de L^p. On se donne deux réels p, q tels que $\frac{1}{p} + \frac{1}{q} = 1$, $p \geqq 2$. Nous avons alors le

Théorème 1.1. *Pour f donné dans $L^q(0, T; V')$ et u_0 dans K, il existe $u \in K_1$ tel que $u(0) = u_0$ et tel que*

$$\int_0^T \left(\frac{dv}{dt} + Au,\ v - u\right) dt \geqq \int_0^T (f, v - u)\, dt \tag{1.1}$$

pour tout $v \in K_1$, tel que $\frac{dv}{dt} \in L^q(0, T; V')$ et $v(0) = u_0$. De plus on a $u(t) \in C([0, T]; H)$.

Démonstration. L'opérateur A applique $L^p(0, T; V)$ dans $L^q(0, T; V')$. En notant $\langle .,. \rangle$ la dualité entre $L^p(0, T; V)$ et $L^q(0, T; V')$ on a:

$$\langle Au, v\rangle = \int_0^T \big(A(t)\,u(t), v(t)\big)\, dt.$$

A est monotone et hémicontinu car $\langle A(t)\,\theta u(t) + (1 - \theta)\,v(t), u(t) - v(t)\rangle$ est continue en θ pour tout $t \in [0, T]$ et majorée par une fonction intégrable, d'après (H2), donc, d'après le théorème de Lebesgue, $\langle A\big(\theta u + (1 - \theta)\,v\big), u - v\rangle$ est continue en 0. Ensuite, on utilise un résultat bien connu de H. Brézis [1].

De plus, on a le résultat de régularité suivant dû à H. Brézis [1], [2].

Théorème 1.2. *On suppose A indépendant de t et K tel que $(Id + \lambda A)^{-1} K \subset K, \forall \lambda > 0$. On se donne $f \in L^2(0, T; H)$ et $u_0 \in K$, alors il existe une fonction unique $u \in L^p(0, T; V)$ telle que $\frac{du}{dt} \in L^p(0, T; V')$, $u(0) = u_0$ et*

$$\left(\frac{du}{dt} + Au,\ v - u\right) \geqq (f, v - u)\ p.p\ dt, \qquad \forall v \in K. \tag{1.2}$$

2. Schéma de discrétisation

Notations

* Soit $h \in R^n - \{0\}$ un paramètre destiné à tendre vers 0. On considère un espace V_h de dimension finie et des applications linéaires continues: p_h (prolongement), injective, de V_h dans V et r_h (restriction) de H dans V_h. Soient $u_h, v_h \in V_h$. On munit V_h de la norme

$\|u_h\|_h = \|p_h u_h\|$ et du produit scalaire $(u_h, v_h)_h = (p_h u_h, p_h v_h)$. On définit alors s_h de V' dans V_h par la relation $(s_h f, v_h)_h = (f, p_h v_h)$ et on a $|s_h f|_h \leqq |f|_H$. De plus, on a les relations suivantes entre les deux normes de V_h

$$|u_h|_h \leqq c_0 \|u\|_h \quad \text{et} \quad \|u_h\|_h \leqq C(h)\, |u_h|_h \quad \text{avec} \quad C(h) \xrightarrow[h\to 0]{} +\infty. \tag{2.1}$$

On suppose que

$$\text{(H3)} \quad \|p_h r_h\| \leqq C \quad \text{et} \quad p_h r_h u \to u \quad \text{dans } V \text{ quand } h \to 0.$$

On en déduit que

$$\|r_h\| = \operatorname*{Sup}_{u\in H} \frac{\|r_h u\|_h}{|u|} \leqq C \quad \text{et} \quad |r_h| = \operatorname*{Sup}_{u\in H} \frac{|r_h u|_h}{|u|} \leqq C_0 C = C_1.$$

* Posons $K_h = p_h^{-1}(K)$; K_h est un convexe fermé de V_h. On suppose que

$$\text{(H4)} \quad D = \bigcup_h p_h K_h \text{ est dense dans } K \text{ avec } p_h K_h \subseteqq p_{h'} K_{h'}, \text{ si } h' \leqq h.$$

* On définit l'operateur A_h: $V_h \to V_h$ par

$$(A_h u_h, v_h)_h = (A p_h u_h, p_h v_h).$$

D'après (H2) on a les relations

$$(A_h u_h, u_h)_h \geqq \alpha \|u_h\|^p, \tag{2.2}$$

$$(A_h u_h, u_h)_h \leqq M \|u_h\|_h^{p-1} \|v_h\|_h. \tag{2.3}$$

Pour la discrétisation en temps nous utilisons le paramètre $k = \dfrac{T}{m}$ avec $m \in N$. Soit, pour $r \in N$, $W_r(t)$ la fonction indicatrice de l'intervalle $[rk, (r+1)\,k[$.

On définit $E_k(r_1 k, r_2 k; V_h)$ comme l'ensemble des fonctions de $[0, T]$ dans V_h de la forme

$$u_{h,k}(t) = \sum_{r=r_1}^{r_2-1} u_{hk}(rk)\, W_r(t). \tag{2.4}$$

On définit les opérateurs $A_{hk}(t)$ de V_h dans V_h par

$$A_{hk}(t)\, u_h = \frac{1}{k} \sum_{r=0}^{m-1} \left(\int_{rk}^{(r+1)k} A_h(s) \;\; u_h ds \right) W_r(t). \tag{2.5}$$

Pour f donné dans $L^q(0, T; V')$ on définit

$$f_{h,k}(t) = \frac{1}{k} \sum_{r=0}^{m-1} \left(\int_{rk}^{(r+1)k} S_h f(s) ds \right) W_r(t). \tag{2.6}$$

On définit aussi les symboles de dérivation

$$\nabla_k\varphi(t) = \frac{\varphi(t+k)-\varphi(k)}{k} \quad \text{et} \quad \nabla_k\varphi(t) = \frac{\varphi(t)-\varphi(t-k)}{k}. \tag{2.7}$$

* Problème discrétisé. On cherche $u_{h,k} \in E_k(0, T+k; V_h)$ tel que

$$\left.\begin{aligned} &\left(\frac{u_{hk}(t+k)-u_{hk}(t)}{k} + A_{hk}(t)\big(\theta u_{hk}(t+k) + (1-\theta)\,u_{hk}(t)\big), v_h - u_{hk}(t+k)\right) \\ &\geqq \big(f_{hk}(t), v_h - u_h(t+k)\big) \qquad \forall v_h \in K_h = p_h^{-1}(K), \qquad \forall t \in [0,T] \end{aligned}\right\} \tag{2.8}$$

pour $\theta \in]0,1]$ et $u_{hk}(0) = r_h u_0$, u_0 donné.

Posons $u^{r+1} = u_{hk}\big((r+1)\,k\big)$; $u^r = u_{hk}(rk)$;

$$f^{r+1} = f_{hk}(rk); \qquad A_{r+1}u_h = \frac{1}{k}\int_{rk}^{(r+1)k} A_h(t)\,u_h dt;$$

$$z^{r+1} = \theta u^{r+1} + (1-\theta)\,u^r; \qquad v = v_h \quad \text{et} \quad \nabla_{r+1}u^r = \frac{u^{r+1}-u^r}{k}.$$

Le problème (2.8) équivaut alors à trouver pour $r = 0, 1, \ldots, m-1$; u^{r+1} solution de l'inégalité

$$(\nabla_{r+1}u^r, v - u^{r+1}) + (A_{r+1}z^{r+1}, v - u^{r+1}) \geqq (f^{r+1}, v - u^{r+1}) \quad \forall v \in K_h \tag{2.9}$$

pour $r = 0, 1, \ldots, m-1$ avec $u_{hk}(0) = r_h u_0$.

Le problème (2.9) est de la forme

$$(\tilde{A}u^{r+1}, v - u^{r+1}) \geqq (g^r, v - u^{r+1}) \qquad \forall v \in K_h, \tag{2.10}$$

où $\tilde{A}u^{r+1} = u^{r+1} + A_{r+1}\big(\theta u^{r+1} + (1-\theta)\,u^r\big)$

$$g^r = u^r + \int_{rk}^{(r+1)k} S_h f(t)\,dt.$$

$\tilde{A}$ est hémicontinu d'après (H2) et le théorème de Lebesgue. De plus, $\tilde{A}$ est fortement monotone (i.e.):

$$(\tilde{A}u - \tilde{A}u, v - u) \geqq |u - v|_h^2$$

D'où l'existence et l'unicité de la solution du problème discrétisé.

Remarque 2.1. Comme on a posé $u_{hk}^0 = r_h u_0$, ceci n'est possible que si $r_h u_0 \in K_h$. Sinon on doit prendre $u_{hk}^0 = u_h^0$, où u_h^0 est une suite de vecteurs de K_h telle que $p_h u_h^0 \xrightarrow[h\to 0]{} u$ dans V, ce qui est toujours possible d'après (H4).

Dorénavant, pour abréger l'exposé, nous traiterons le cas où $p \geqq 2$ et $\theta = 1$ (méthode implicite standard). Nous donnerons ensuite tous les autres résultats sans démonstration.

On suppose, comme dans A. Raviart [4]

$$\text{(H5)} \begin{cases} \text{L'application } t \mapsto \big(A_h(t)\, u, v\big) \text{ est de classe } C^1\text{; il existe} \\ \text{une constante } P > 0 \text{ indépendante de } t \text{ et } h \text{ telle que} \\ \left(\dfrac{A_h(t+k) - A_h(t)}{k}\, u, v\right) \leqq P\|u\|^{p-1}\,\|v\| \qquad \forall\, u, v. \end{cases}$$

Nous avons alors le

Théorème 2.1. (cas où $p \geqq 2$ et $\theta = 1$). On suppose $f(t)$ à variation bornée de $[0, T]$ dans H (donc dans L^∞) et

$$\frac{1}{k}\int_0^k |Ap_h r_h u_0|\, dt \leqq c^{te} \quad \text{et} \tag{2.11}$$

$kC^p(h)$ borné (hypothèse nécessaire seulement dans le cas où A dépend de t). (2.12)

On a alors

$$|\nabla_k u_{hk}(t)| \leqq c^{te}, \text{ pour tout } t, h, k. \tag{2.13}$$

Dans le cas où $A(t)$ ne dépend pas de t, cette constante majorante s'écrit

$$c^t/ = \mathfrak{v} + \sqrt{\mathfrak{v}^2 + (|f|_{L^\infty} + \operatorname{Sup}_h |Ap_h r_h u_0|^2}, \tag{2.14}$$

où $\mathfrak{v}$ = variation de f de $[0, T]$ dans H.

Si de plus A est continu de V V', alors la condition $\operatorname{Sup}_h |Ap_h r_h u_0|$ borné est vérifiée si $Au_0 \in H$.

Démonstration. 1°) Cas où $A(t) = A$ indépendant de t. Faisons $v = u^r$ dans (2.9) et $v = u^{r+1}$ dans (2.9) à l'étape $r - 1$ et ajoutons ces deux inéquations. On obtient alors

$$(\nabla_{r+1} - \nabla_r, \nabla_{r+1}) \leqq (f^{r+1} - f^r, \nabla_{r+1}) + (Az^r - Az^{r+1}, u^{r+1} - u^r) \tag{2.15}$$

pour $\theta = 1$ et grâce à la monotonie de A, on a:

$$(\nabla_{r+1} - \nabla_r, \nabla_{r+1}) \leqq (f^{r+1} - f^r, \nabla_{r+1}). \tag{2.16}$$

On a donc

$$\begin{aligned} &\frac{1}{2}(|\nabla_{r+1}|^2 - |\nabla_r|^2) + \frac{1}{2}|\nabla_{r+1} - \nabla_r|^2 \\ &\leqq \int_{rk}^{(r+1)k} \left|\frac{S_h f(t) - S_h f(t-k)}{k}\right| \left|\frac{u^{r+1} - u^r}{k}\right| \\ &\leqq \int_{rk}^{(r+1)k} \left|\frac{f(t) - f(t-k)}{k}\right| \operatorname{Sup}_r |\nabla_{r+1}|. \end{aligned} \tag{2.17}$$

En sommant ces inégalités, il vient:

$$|\nabla_k u_{hk}(rk)|^2 \leqq |\nabla_k u_{hk}(0)|^2 + 2\left(\int\limits_k^{(r+1)k} \left|\frac{f(t) - f(t-k)}{k}\right| dt\right) \operatorname*{Sup}_r |\nabla_{r+1}|.$$

En posant $\mathfrak{v}$ = variation de f de $[0, T]$ dans H on a

$$\left(\operatorname*{Sup}_r |\nabla_k u_{hk}(r+k)|\right)^2 \leqq |\nabla_k u_{hk}(0)|^2 + 2\mathfrak{v} \operatorname*{Sup}_r |\nabla_k u_{hk}(rk)|.$$

D'après les variations de $X^2 - 2\mathfrak{v}X - D$ on a

$$\operatorname{Sup} |\nabla_k u_{hk}(rk)| \leqq \mathfrak{v} + \sqrt{\mathfrak{v}^2 + |\nabla_k u_{hk}(0)|^2}. \tag{2.18}$$

Pour majorer $|\nabla_k u_{hk}(0)|$ on utilise (2.8) pour $r = 1$ et la monotonie de A:

$$\left|\frac{u_{hk}(k) - u_{hk}(0)}{k}\right|^2 \leqq \left|\frac{1}{k}\int\limits_0^k S_h(ft)\, dt\right| |\nabla_k u_{hk}(0)| + \left|\frac{1}{k}\int\limits_0^k A p_h r_h u_0\right| |\nabla_k u_{hk}(0)|.$$

Donc, d'après les hypothèses sur f et u_0, on a démontré le théorème dans le cas où $A(t) = A$.

2⁰) Cas où A dépend de t. On suppose alors (H 5). Dans le problème discrétisé (2.8) on fait $v = 0$ et en utilisant la monotonie de A on obtient:

$$(u^{r+1} - u^r, u^{r+1}) \leqq k(f^{r+1}, u^{r+1}),$$

ce qui donne

$$|u^{r+1}|^2 - |u^r|^2 \leqq k|f^{r+1}|^2 + k|u^{r+1}|^2,$$

d'où

$$|u^r|^2 \leqq \left[C_1{}^2|u_0|^2 + \int\limits_0^T |f(t)|^2\, dt\right] e^T = E.$$

D'après (2.15) on va avoir en plus à majorer le terme

$$\frac{1}{k}(A_{r+1}u^r - A_r u^r, u^{r+1} - u^r). \tag{2.19}$$

D'après l'hypothèse (H 5) on majore (2.19) par

$$PkC^p(h)\, |\nabla_{r+1}|\, |u^r| \leqq kC^p(h)\, P\sqrt{E} \operatorname{Sup} |\nabla_{r+1}|. \tag{2.20}$$

On a alors, avec $kC^p(h)$ borné la même fin de démonstration que dans la première partie où $\mathfrak{v}$ est remplacé par $(\mathfrak{v} + kC^p(h)\, P\sqrt{E})$.

Théorème 2.2 (*Cas où $p \geqq 2$ et $\theta = 1$*). *En supposant f dans $L^2(0, T; H)$, on a les majorations*

$$|u_{hk}(t)|_h^2 \leqq \left[\int_0^T |f(t)|^2\,dt + C_1^2|u_0|^2\right] e^T \tag{2.21}$$

et

$$\int_0^T \||u_{hk}(t+k)\||_h^p\,dt \leqq \frac{1}{2\alpha}\left[\int_0^T |f(t)|^2\,dt + C_1^2\,|u_0|^2\right] e^T \tag{2.22}$$

Démonstration. Faisons $v = 0$ dans (2.8). On a

$$(u^{r+1} - u^r, u^{r+1}) + k(Au^{r+1}, u^{r+1}) \leqq k(f^{r+1}, u^{r+1}),$$

d'où

$$\frac{1}{2}(|u^{r+1}|^2 - |u^r|^2) + \frac{1}{2}|u^{r+1} - u^r|^2 + k\alpha\|u^{r+1}\|^p \leqq \frac{k}{2}|f^{r+1}|^2 + \frac{k}{2}|u^{r+1}|^2$$

et, par sommation

$$|u^{r+1}|^2 + 2\alpha\int_0^{(r+1)k} \|u_{hk}(t+k)\|_k^p \leqq \int_0^T |f(t)^2\,dt + C_1^2|u_0|^2 + \int_0^{(r+1)k} |u_{hk}(t+k)|^2\,dt.$$

D'où le théorème.

Théorème 2.3 (*Convergence faible: cas où $p \geqq 2$ et $\theta = 1$*). *On suppose que les théorèmes 2.1 et 2.2 sont vérifiés. On peut alors extraire de $p_h u_{hk}(t)$ une sous-suite, encore notée $p_h u_{hk}(t)$, telle que*

i) $p_h u_{hk}(t) \rightharpoonup u(t)$ *dans* $L^\infty(0, T; H)$ **faible quand* $(h, k) \to 0$

ii) $p_h u_{hk}(t+k) \rightharpoonup u(t)$ *dans* $L^p(0, T; V)$ *faible avec*

iii) $u(t) \in L^p(0, T; V)$, $\dfrac{du}{dt} \in L^q(0, T; V')$, $u(0) = u_0$ *et*

$$\int_0^T \left(\frac{du}{dt} + Au, v - u\right) ds \geqq \int_0^t (f, v - u)\,ds \qquad \forall t \in [0, T], \qquad \forall v \in L^p(0, T; K).$$

Auparavant, nous démontrons les quatre lemmes suivants:

Lemme 2.1. *Soit H un Hilbert et soit $p_h u_{hk}(t)$ une suite de $L^\infty(0, T; H)$ en escalier et constante sur les intervalles $[rk, (r+1)k[$ (pour $r = 0, 1, \ldots, m-1$ avec $m = \dfrac{T}{k}$ telle que $|p_h u_{hk}(t)|_{L^\infty(0,T;H)} \leqq C$ où C est une constante indépendante de h, k et t, et $p_h u_{hk}(t) \rightharpoonup u(t)$ dans $L^\infty(0, T; H)$ *faible. Si, en outre $|\nabla_k p_h u_{hk}(t)|_H$ (resp. $\int_0^T |\nabla_k p_h u_{hk}(t)|^2\,dt$) est borné indépendamment de h, k et t, alors $\nabla_k p_h u_{hk}(t) \rightharpoonup \dfrac{du}{dt}$ faiblement dans $L^\infty(0, T; H)$ *faible (resp. $L^2(0, T; H)$ faible) et $u \in C([0, T]; H)$.*

Démonstration. Extrayons de $\nabla_k p_h u_{hk}$ une sous-suite, encore notée $\nabla_k p_h u_{hk}$ qui converge vers η dans L^∞*faible (resp. L^2 faible). Montrons que $\eta = \frac{du}{dt}$. On a pour tout $\varphi \in \mathscr{D}([0, T]; H)$

$$\langle \nabla_k p_h u_{hk}, \varphi \rangle = \frac{1}{k} \int_0^T \big(p_h u_{hk}(t+k), \varphi(t)\big)\, dt - \frac{1}{k} \int_0^T \big(p_h, u_{hk}(t), \varphi(t)\big)\, dt$$

$$= \frac{1}{k} \int_0^T \big(p_h u_{hk}(t), \varphi(t-k)\big)\, dt + \frac{1}{k} \int_T^{T+k} \big(p_h u_{hk}(t), \varphi(t-k)\big)\, dt$$

$$- \frac{1}{k} \int_0^T \big(p_h u_{hk}(t), \varphi(t-k)\big)\, dt - \frac{1}{k} \int_0^T \big(p_h u_{hk}(t), \varphi(t)\big)\, dt.$$

On a: $\int_0^k \big(p_h u_{hk}(t), \varphi(t-k)\big)\, dt = 0$ et $\frac{1}{k} \int_T^{T+k} \big(p_h u_{hk}(t), \varphi(t-k)\big)\, dt \leqq C|\varphi(\xi)| \to 0$ quand $h \to 0$ car $\varphi(T) = 0$. De plus $\frac{\varphi(t-k) - \varphi(t)}{k} \to -\varphi'(t)$ fortement dans L^1, donc:

$$\int_0^T \big(\nabla_k p_h u_{hk}(t), \varphi(t)\big)\, dt \to -\int_0^T \big(u(t), \varphi'(t)\big)\, dt - \int_0^T (\eta, \varphi)\, dt.$$

Donc $\eta = \frac{du}{dt}$. D'où $\frac{du}{dt} \in L^\infty(0, T; H)$ (resp. L^2) et comme $u \in L^\infty(0, T; H)$, on a, après modification sur un ensemble de mesure nulle, $u \in C([0, T]; H)$.

Lemme 2.2. *Soit H un Hilbert. Soit $u_n(t)$ une suite de fonctions de $L^\infty(0, T; H)$ telles que $|u_n(t)| \leqq c^t/\ \forall t \in [0, T]$, $\forall n$. Alors, si $u_n(t) \rightharpoonup u(t)$ dans $L^\infty(0, T; H)$ *faible, on a $u_n(t) \rightharpoonup u(t)$ dans H faible pour presque tout t*

Démonstration. $u_n \rightharpoonup u$ dans L^∞ *faible $\rightharpoonup \forall v \in L^1(0, V; H)$ on a

$$\int_0^T \big(u_n(t), v(t)\big)\, dt \to \int_0^T \big(u(t), v(t)\big)\, dt.$$

Soit $v(s) = \mathfrak{v} 1_{[0,T]} \psi(s)$ pour $v \in H, t \in [0, T]$ et $\psi(s) \in L^1(0, T; R)$. On a $v(s) \in L^1(0, T; H)$, donc $\int_0^t \big(u_n(s), v\big)\, \psi(s)\, ds \to \int_0^t \big(u(s), v\big)\, \psi(s)\, ds$. On montre, d'après le lemme de Fatou que

$$\int_0^t \big[u(s), v) - \underline{\lim}\, \big(u_n(s), v\big)\big]\, \psi(s)\, ds \geqq 0 \qquad \forall \psi \in L^1(0, T; R)$$

D'où $\big(u(t), v\big) = \underline{\lim}\, \big(u_n(t), v\big)$ p.p dt $\forall v \in H$ et on reprend le même raisonnement avec

$\overline{\lim}$. On a $(u(t), v) = \overline{\lim}\,(u_n(t), v)$ p.p dt, $\forall v \in H$. Donc p.p dt, $\forall v \in H$, $(u(t), v) = \lim\,(u_n(t), v)$.

Lemme 2.3. *Avec les hypothèses du lemme* 2.1 (*resp. dans le cas où* $\int_0^T |\nabla_k u_{hk}(t)|^2\,dt \leqq c^{te}$ *avec en plus* $|p_h u_{hk}(t+k) - p_h u_{hk}(t)| \to 0$ *pour tout* t *quand* $(h, k) \to 0$) *on a* $p_h u_{hk}(t) \rightharpoonup u(t)$ *dans* H *faible pour tout* $t \in [0, T]$.

Démonstration. On a déjà $\nabla_k p_h u_{hk}(t) \rightharpoonup \dfrac{du}{dt}(t)$ dans $L^\infty(0, T; H)$ *faible (resp. $L^2(0, T; H)$ faible) et, comme u est continue, on a, pour tout $v \in H$

$$(u(t_1) - u(t_0), v) = \int_{t_0}^{t_1} \left(\frac{du}{dt}, v\right) = \lim \int_{t_0}^{t_1} (\nabla_k p_h u_{hk}(t), v)\,dt.$$

La dernière intégrale s'écrit

$$\frac{1}{k}\int_{t_1}^{t_1+k} (p_h u_{hk}(t), v)\,dt - \frac{1}{k}\int_{t_0}^{t_0+k} (p_h u_{hk}(t), v)\,dt.$$

Soit n tel que $nk \leqq t < (n+1)\,k$ pour $t \in [0, T]$ donné

$$\frac{1}{k}\int_t^{t+k} (p_h u_{hk}(s), v)\,ds = \frac{1}{k}\int_t^{(n+1)k} + \frac{1}{k}\int_{(n+1)k}^{t+k}$$

et $p_h u_{hk}(t) = p_h u_{hk}(nk)$ d'où

$$\frac{1}{k}\int_t^{t+k} (p_h u_{hk}(s), v)\,ds = \frac{1}{k}\,[(n+1)\,k - t]\,(p_h u_{hk}(nk), v)$$

$$+ \frac{1}{k}\,[(t+k) - (n+1)\,k]\,(p_h u_{hk}((n+1)\,k), v)$$

$$= (p_h u_{hk}(t), v) + \varepsilon(h, k)$$

avec $\varepsilon(h, k) \to 0$ quand $(h, k) \to 0$, car on a $|p_h u_{hk}(t+k) - p_h u_{hk}(t)| \to 0$ pour tout t. Donc

$$\int_{t_0}^{t_1} (\nabla_k p_h u_{hk}(t), v)\,dt = (p_h u_{hk}(t_1), v) - (p_h u_h(t_0), v) + \varepsilon(h, k).$$

Supposons $p_h u_{hk}(t_0) \rightharpoonup u(t_0)$ dans H, alors on a

$$(u(t_1), v) - (u(t_0), v) = \lim\,(p_h u_{hk}(t_1), v) - (p_h u_{hk}(t_0), v).$$

On en déduit que pour tout $t_1 \in [0, T]$ on a $p_h u_{hk}(t_1) \rightharpoonup u(t_1)$ faiblement dans H. D'où le lemme.

On démontre aussi le

Lemme 2.4. *Soit $v \in L^p(0, T; V)$, telle que $v(t) \in K$ p.p dt et $v' = \frac{dv}{dt} \in L^q(0, T; V')$, $v(0) = u_0$. Alors il existe $v_{hk} \in L^p(0, T; K_h)$ telle que $p_h v_{hk} \to v$ dans $L^p(0, T; V)$ et $p_h \nabla_k v_{hk} \to v'$ dans $L^q(0, T; V')$ avec $v_{hk}(0) \to u_0$ quand $(h, k) \to 0$.*

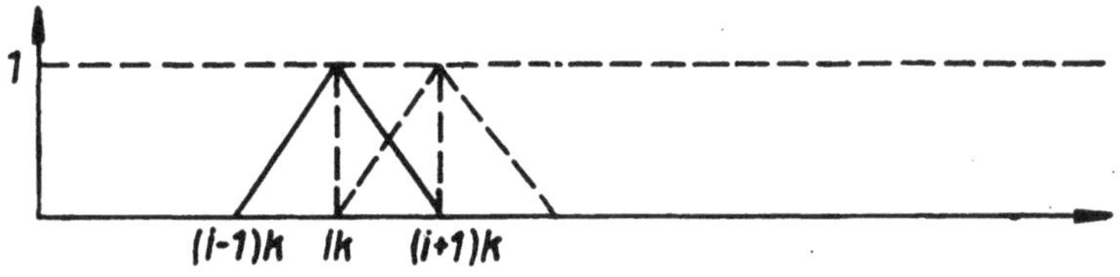

Fig. 1

Démonstration. Soit $w_i(t)$ la fonction indicatrice de l'intervalle $[ik, (i+1)k[$. On définit des fonctions Λ_k^i comme dans la Figure 1 ci-dessus. Pour v vérifiant les hypothèses du lemme, on pose $v^i = \frac{1}{k} \int_{(i-1)k}^{ik} v(t)\, dt$ pour $i = 1, \ldots, m = \frac{T}{k}$. On prolonge v par continuité en dehors de $[0, T]$, sur $[a, T]$, avec $a < -k$, par $v(t) = v(0)$, avec les mêmes propriétés sur $[0, T]$. Pour $i = 0$ on a $v^0 = \frac{1}{k} \int_{-k}^{0} v(t)\, dt = u_0$.

1) Posons $v_k(t) = \sum_{i=0}^{m-1} v^i w_i(t)$ et pour $t \in [0, T]$,

$$\nabla_k v_k(t) = \frac{v_k(t+k) - v_k(t)}{k} = \sum_{i=0}^{m-1} \frac{1}{k^2} \int_{ik}^{(i+1)k} \big(v(s) - v(s-k)\big)\, ds.$$

On montre alors que $\nabla_k v_k(t) \to \frac{dv}{dt}$ dans $L^q(0, T; V')$. Pour cela il suffit de voir que

$$\tau_k v = \sum_{i=0}^{m-1} \left(\frac{1}{k} \int_{(i-1)k}^{(i+1)k} v(u) \Lambda_k^i(u)\, du \right) w_i \to v \text{ dans } L^q(0, T; V'), \text{ pour tout } v \in L^q(0, T; V')$$

quand $k \to 0$.

2) Soit $v_{hk}(t) = \sum_{i=0}^{m-1} v_{hk}^i w_i(t)$ on a:

$$\|p_h \nabla_k v_{hk} - v'\|_{L^q} \leqq \|p_h \nabla_k v_{hk} - \nabla_k v_k\|_{L^q} + \|\nabla_k v_k - v'\|_{L^q}.$$

On a déjà $\|\nabla_k v_k - v'\|_{L^q} \to 0$ quand $k \to 0$. D'après l'hypothèse (H4) on peut choisir les v_{hk}^i, pour $i = 0, 1, \ldots, m = \frac{T}{k}$ de façon que $\|p_h \nabla_k v_{hk} - \nabla_k v_k\|_{L^q} \to 0$ quand $(h, k) \to 0$.

3) Reste à étudier $\|p_h v_{hk}(t) - v(t)\|_{L^q(0,T;V)}$. Posons $\theta_k v = \sum_{i=0}^{m-1} \left(\frac{1}{k} \int_{(i-1)k}^{ik} v(u)\, du \right)$ $\times\, w_i(t)$. Pour $v \in L^p(0, T; V)$, on démontre que $\theta_k v \to v$ dans $L^p(0, T; V)$ quand $k \to 0$. Ensuite $p_h v_{hk} - \theta_k v = \sum_{i=0}^{m-1} (p_h v_{hk}^i - v^i)\, w_i(t)$. D'après le choix des v_{hk}^i fait précédemment, on a $\|p_h v_{hk} - \theta_k v\|_{L^p} \to 0$ quand $(h, k) \to 0$.

Démonstration du théorème 2.3. D'après les majorations de $p_h u_{hk}(t)$ et $\nabla_k p_h u_{hk}(t)$ et d'après les lemmes 2.1, 2.2, 2.3, on peut extraire une sous-suite, encore notée $p_h u_{hk}$, telle que $p_h u_{hk} \rightharpoonup u$ dans $L^\infty(0, T; H)$ *faible avec $u \in C([0, T]; H)$; $p_h u_{hk}(t) \rightharpoonup u(t)$ dans H faible pour tout $t \in [0, T]$ donc en particulier $u(0) = u_0$ avec $p_h u_{hk}(0) \to u_0$ dans V fort; $\nabla_k p_h u_{hk}(t) \rightharpoonup \frac{du}{dt}(t)$ dans H faible p.p dt, et dans $L^\infty(0, T; H)$ *faible. De plus, on peut choisier cette sous-suite de façon que $p_h u_h(t + k) \rightharpoonup \hat{u}(t)$ dans $L^p(0, T; V)$ faible. Comme les convergences ont lieu dans $\mathfrak{D}'([0, T]; H)$, on a $u = \hat{u}$. D'après le lemme 2.4, pour tout $v \in L^p(0, T; K)$, il existe une suite de fonctions $v_{hk}(t) \in L^p(0, T; K_h)$ constantes sur $[rk, (r + 1)\, k[$, telles que $p_h v_{hk}(t) \to v(t)$ dans $L^p(0, T; V)$ fort. A chaque étape r du problème discret, prenons $v_h = v_{hk}(rk)$; on a alors, pour $r = 0, 1, \ldots, m - 1$

$$\int_{rk}^{(r+1)k} \big(\nabla_k p_h u_{hk}(t), p_h v_{hk}(t)\big)\, dt + \int_{rk}^{(r+1)k} \big(A(t)\, p_h u_{hk}(t + k), p_h v_{hk}(t) - p_h u_{hk}(t + k)\big)\, dt$$

$$- \big(u_{hk}((r + 1)) - u_{hk}(rk), u_{hk}((r + 1)k)\big) \geqq \int_{rk}^{(r+1)k} \big(f(t), p_h v_{hk}(t) - p_h u_{hk}(t + k)\big)\, dt. \tag{2.23}$$

Or

$$-(u^{r+1} - u^r, u^{r+1}) = -\frac{|u^{r+1}|^2 - |u^r|^2}{2} - \frac{|u^{r+1} - u^r|^2}{2} \leqq -\frac{1}{2}\, |u^{r+1}|^2 - \frac{1}{2}\, |u^r|^2. \tag{2.24}$$

Soit $t \in [0, T]$ et r tel que $rk \leqq t < (r + 1)\, k$. En sommant de 0 à r les inégalités précédentes on obtient:

$$\int_0^{rk} \big(\nabla_k p_h u_{hk}(s), p_h v_{hk}(s)\big)\, ds$$

$$+ \int_0^{rk} \big(A(s)\, p_h u_{hk}(s + k), p_h v_{hk}(s) - p_h u_{hk}(s + k)\, ds - \frac{|p_h u_{hk}(t)|^2}{2}$$

$$+ \frac{|p_h u_{hk}(0)|^2}{2} \geqq \int_0^{rk} \big(f(s), p_h v_{hk}(s) - p_h u_{hk}(s + k)\big)\, ds. \tag{2.25}$$

Prenons alors la limite supérieure des deux membres de cette inégalité, on a alors, d'après les convergences introduites au début de cette démonstration

$$\int_0^t \left(\frac{du}{dt}, v\right) ds + \overline{\lim} \int_0^{rk} \big(A(s)\, p_h u_{hk}(s+k), p_h v_{hk}(s) - p_h u_{hk}(s+k)\big)\, ds$$

$$-\frac{|u(t)|^2}{2} + \frac{|u(0)|^2}{2} \geqq \int_0^t \big(f(s), v(s) - u(s)\big)\, ds. \tag{2.26}$$

Il reste alors une quantité à simplifier: On sait que $p_h v_{hk}(s)$ converge vers $v(s)$ dans $L^p(0, T; V)$ fort, cette quantité est alors égale à:

$$\overline{\lim} \int_0^{rk} \big(A(s)\, p_h u_{hk}(s+k), v(s) - p_h u_{hk}(s+k)\big)\, ds.$$

Utilisons alors le monotonie de $A(s)$ pour tout $s \in [0, T]$, puis la convergence faible de $p_h u_{hk}(s+k)$, on obtient alors: $\int_0^t (Av, v-u)\, ds$. D'où, d'après la continuité de $u(t)$

$$\int_0^t \left(\frac{du}{dt}, v - u\right) ds + \int_0^t (Av, v-u)\, ds \geqq \int_0^t (f, v-u)\, ds \tag{2.27}$$

pour tout $v(t) \in L^p(0, T; V)$.

En utilisant alors cette relation pour $v = \theta u + (1-\theta)\, w$ où $w \in L^p(0, T; V)$, et en faisant tendre θ vers 1, on a, d'après l'hémicontinuité de $A(t)$ pour tout t, et d'après le théorème de Lebesgue:

$$\int_0^t \left(\frac{du}{dt}, v - u\right) ds + \int_0^t (Au, v-u)\, ds \geqq \int_0^t (f, v-u)\, ds \tag{2.28}$$

et ceci pour $v(t) \in L^p(0, T; V)$.

On a vu, au début, que $u(0) = u_0$; le théorème est ainsi démontré.

Remarque 2.2. Dans le cas où on a unicité de la solution du problème continu, ce n'est plus une sous-suite de $p_h u_{hk}(t)$ qui converge vers la solution $u(t)$ du problème continu, mais la suite $p_h u_{hk}(t)$ elle-même.

Théorème 2.4 (*Cas où* $p \geqq 2$ *et* $\theta = 1$). *Sous les hypothèses du théorème* 2.3, *on a convergence forte de* $p_h u_{hk}(t)$ *vers* $u(t)$, *dans* $L^2(0, T; H)$.

Démonstration. On sait que la solution $u(t)$ du problème continu appartient à $L^p(0, T; V)$. Il existe alors une suite $v_{hk}(t)$, de fonctions en escalier, constantes sur $[rk, (r+1)\, k[$, telles que $p_h u_{hk}(t)$ converge vers $u(t)$ dans $L^p(0, T; V)$ fort (lemme 2.4).

Faisons alors $v_h = v_{hk}(rk)$ dans le problème discrétisé à l'étape r, et faisons la somme des inégalités ainsi obtenues de 0 à $r - 1$, on a alors:

$$\int_0^{rk} \big(\nabla_k p_h u_{hk}(t), p_h v_{hk}(t) - p_h u_{hk}(t+k)\big)\, dt$$
$$+ \int_0^{rk} \big(A(t)\, p_h u_{hk}(t+k), p_h v_{hk}(t) - p_h u_{hk}(t+k)\big)\, dt$$
$$\geqq \int_0^{rk} \big(f(t), p_h v_{hk}(t) - p_h u_{hk}(t+k)\big)\, dt. \tag{2.29}$$

Comme $\nabla_k p_h u_{hk}(t)$ est borné dans $L^\infty(0, T; H)$, $p_h u_{hk}(t+k)$ est borné dans $L^p(0, T; V)$, $A(t)$ est un opérateur borné de V dans V', comme $p_h v_{hk}(t)$ converge vers $u(t)$ dans $L^p(:0, T; V)$ fort, on a:

$$\int_0^{rk} \big(\nabla_k p_h u_{hk}(t), u(t) - p_h u_{hk}(t+k)\big)\, dt$$
$$+ \int_0^{rk} \big(A(t)\, p_h u_{hk}(t+k), u(t) - p_h u_{hk}(t+k)\big)\, dt$$
$$\geqq \int_0^{rk} \big(f(t), u(t) - p_h u_{hk}(t+k)\big)\, dt + \varepsilon_1(h, k) \tag{2.30}$$

avec $\varepsilon_1(h, k) \to 0$ quand $(h, k) \to 0$.

Considérons alors l'inéquation vérifiée par la solution $u(t)$ du problème continu que l'on écrit pour $t = rk$:

$$\int_0^{rk} \left(\frac{du}{dt}, v - u\right) dt + \int_0^{rk} (Au, v - u)\, dt \geqq \int_0^{rk} (f, v - u)\, dt \tag{2.31}$$

pour tout $v \in L^p(0, T; K)$.

On prendra $v(t) = p_h u_{hk}(t+k)$. La solution $u(t)$ est telle que $u(t) \in L^p(0, T; V)$ et $\frac{du}{dt} \in L^q(0, T; V')$, on sait alors (lemme 2.4) que $\nabla_k u(t)$ converge vers $\frac{du}{dt}$ dans $L^q(0, T; V')$ fort. On a alors, comme $p_h u_{hk}(t+k)$ converge vers $u(t)$ faiblement dans $L^p(0, T; V)$:

$$\int_0^{rk} \big(\nabla_k u, p_h u_{hk}(t+k) - u(t)\big)\, dt + \int_0^{rk} \big(Au, p_h u_{hk}(t+k) - u(t)\big)\, dt$$
$$\geqq \int_0^{rk} \big(f, p_h u_{hk}(t+k) - u(t)\big)\, dt + \varepsilon_2(h, k) \tag{2.32}$$

avec $\varepsilon_2(h, k) \to 0$ quand $(h, k) \to 0$.

Faisons alors la somme des deux inégalités (2.30) et (2.32) vérifiées par la problème discret et par le problème continu. On a alors, grâce à la monotonie de l'opérateur $A(t)$ et pour tout t,

$$\int_0^{rk} \big(\nabla_k u - \nabla_k p_h u_{hk}, u(t) - p_h u_{hk}(t+k)\big)\, dt \leqq \varepsilon_1(h, k) + \varepsilon_2(h, k).$$

On sait aussi (lemme 2.1 et théorème 2.3) que u est continue de $[0, T]$ dans H et que $\frac{du}{dt} \in L^\infty(0, T; H)$. Donc, d'après le lemme 2.4 $\nabla_k u$ converge vers $\frac{du}{dt}$ dans $L^2(0, T; H)$, donc aussi dans $L^q(0, T; V')$. Donc $\nabla_k u$ et $\nabla_k p_h u_{hk}$, sous les hypothèses du théorème 2.3, sont bornés dans $L^2(0, T; H)$. Comme $u(t)$ est continue de $[0, T]$ dans H, $u(t + k) - u(t)$ tend vers 0 dans $L^2(0, T; H)$ quand $k \to 0$. On a donc:

$$\int_0^{rk} \big(\nabla_k u - \nabla_k p_h u_{hk}, u(t + k) - p_h u_{hk}(t + k)\big)\, dt \leqq \varepsilon_1(h, k) + \varepsilon_2(h, k) + \varepsilon_3(h, k)$$

avec $\varepsilon_3(h, k) \to 0$ quand $(h, k) \to 0$.

Posons alors $\varepsilon_4(h, k) = \varepsilon_1 + \varepsilon_2 + \varepsilon_3$ et $w_{hk}(t) = u(t) - p_h u_{hk}(t)$; on a alors:

$$\int_0^{rk} \big(w_{hk}(t + k) - w_{hk}(t), w_{hk}(t + k)\big)\, dt \leqq k\varepsilon_4(h, k), \quad \text{soit}$$

$$\int_{rk}^{(r+1)k} |w_{hk}(t)|^2\, dt \leqq 2k\varepsilon_4(h, k) + \int_0^k |w_{hk}(t)|^2\, dt,$$

et on sait que $p_h u_{hk}(0)$ converge vers $u(0)$ dans V fort et que $u(t)$ est uniformément continue de $[0, T]$ dans H. On a donc:

$$\int_0^k |w_{hk}(t)|^2\, dt \leqq k\, 2\varepsilon_5(h, k) \quad \text{avec} \quad \varepsilon_5(h, k) \to 0 \quad \text{quand} \quad (h, k) \to 0.$$

Posons alors $\frac{\varepsilon}{2} = \varepsilon_4(h, k) + \varepsilon_5(h, k)$. On a alors:

$$\int_{rk}^{(r+1)k} |w_{hk}(t)|^2\, dt \leq k\, \varepsilon(h, k).$$

Sommons ces inégalités de 0 à (m − 1); on a alors:

$$\int_0^T u(t) - p_h u_{hk}(t)|^2\, dt \leqq T\varepsilon(h, k) \to 0. \tag{2.33}$$

D'où le théorème.

3. Résolution numérique — Méthode itérative et majoration de l'erreur

Reprenons le problème discrétisé (2.8), (2.9). Soit $u^{r+1} = u_h\big((r + 1)\, k\big)$ la solution exacte du problème discrétisé

$$\left.\begin{aligned} &(u^{r+1} - u^r, v_h - u^{r+1}) + \int_{rk}^{(r+1)k} \big(A_h(t)\, (\theta u^{r+1} + (1 - \theta)\, u^r), v_h - u^{r+1}\big)\, dt \\ &\geqq \int_{rk}^{(r+1)k} \big(S_h f(t), v_h - u^{r+1}\big)\, dt, \qquad \forall v_h \in K_h. \end{aligned}\right\} \tag{3.1}$$

Or, à l'étape (r), u^r est connu approximativement par u_*^r. Soit u_*^{r+1} la solution approchée de

$$\left.\begin{aligned}&(w - u_*^r, v_h - w) + \int_{rk}^{(r+1)k} \big(A_h(t)\,(\theta w + (1-\theta)\,u_*^r,)\, v_h - w\big)\,dt\\ &\geqq \int_{rk}^{(r+1)k} \big(S_h f(t), v_h - w\big)\,dt, \qquad \forall v_h \in K_h.\end{aligned}\right\} \tag{3.2}$$

Soit w^{r+1} la solution exacte de (3.2). Posons $e^{r+1} = u_*^{r+1} - u^{r+1}$. Soit $e_h\colon [0, T] \to V_h$ définie par $e_h(0) = 0$, $e_h(t) = e^r$ pour $rk \leqq t \leqq (r+1)\,k$. On se propose de chercher des conditions suffisantes pour que $\|p_h e_h(t)\| \to 0$ quand $(h, k) \to 0$. Si le problème (3.2) est résolu numériquement à l'aide d'une méthode itérative, on pose alors $u_*^{r+1} = u_m^{r+1}$, m étant le nombre d'itérations auquel on s'est arrêté pour le calcul numérique de w^{r+1}. On a alors

$$e^{r+1} = u_*^{r+1} - u^{r+1} = u_m^{r+1} - w^{r+1} + w^{r+1} - u^{r+1}.$$

On suppose, comme dans G. Geymonat et M. Sibony [3]

$$\text{(H6)}\ \left\{\begin{aligned}&\text{i) } \exists\, L_h > 0 \ \text{ tel que } \ (A_h u_h - A_h v_h, u_h - v_h)_h \geqq L_h |u_h - v_h|_h^2;\\ &\text{ii) } \forall N > 0,\ \ \exists\, \Gamma_h(N) > 0 \ \text{ tel que, si } \ u_h, v_h \in V_h \ \text{ avec } \ |u_h|_h \leqq N,\\ &\quad |v_h|_h \leqq N, \text{ alors } \forall v_h \in V_h \text{ on a } (A_h u_h - A_h v_h, w_h)_h \leqq \Gamma_h(N)\,|u_h - v_h|_h |w_h|_h.\end{aligned}\right.$$

Comme dans [5, 6] on démontre le

Théorème 3.1 (*Cas où $p > 2$ et $\theta \in [0, 1]$*). *Sous les hypothèses* (H6), *la suite* (u_{m+1}^{r+1}) *définie par*

$$u_{m+1}^{r+1} = P_{K_h}[u_m^{r+1} - \varrho(B_{h,k}^r u_m^{r+1} - g_*^r)] \quad \textit{avec} \quad u_0^{r+1} = u_*^r$$

où

$$B_{h,k}^r u = u + \int_{rk}^{(r+1)k} A_h\big(\theta u + (1-\theta)\,u_*^r\big)\,dt$$

$$g_*^r = u_*^r + \int_{rk}^{(r+1)k} S_h f(t)\,dt$$

converge vers w^{r+1} solution de (3.2) quand $m \to \infty$ si le paramètre $\varrho > 0$ vérifie

$$\eta = 1 - 2\varrho(1 + \theta k L_h) + \varrho^2\big(1 + \theta k \Gamma_h(N^*)\big)^2 < 1 \tag{3.4}$$

avec N^* tel que $|\theta w^{r+1} + (1-\theta)\,u_*^r| \leqq \dfrac{N^*}{2}$. *De plus, on a l'estimation de l'erreur:*

$$|u_{m+1}^{r+1} - w^{r+1}|^2 \leqq \eta^{m+1} |u_*^r - w^{r+1}|^2. \tag{3.5}$$

En particulier, l'estimation sur η est satisfaite par

$$\varrho = \frac{1 + k\theta L_h}{\big(1 + k\theta\Gamma_h(N^*)\big)^2} \quad \big(\textit{car } \Gamma_h(N^*) \geqq L_h\big). \tag{3.6}$$

Remarque 3.1: Calcul de la constante N^* pour $\theta = 1$. En faisant $v_h = 0$ dans (3.2), on a pour $\theta = 1$ $(w^{r+1} - u_*{}^r, w^{r+1}) + k\big(A_h(w^{r+1}), w^{r+1}\big) \leqq k(f^{r+1}, w^{r+1})$ d'où $\frac{1}{4}\,|w^{r+1}|^2 - \frac{1}{2}\,|u_*{}^r|^2 + \frac{1}{2}\,|w^{r+1} - u_*{}^r|^2 \leqq k \int\limits_{rk}^{(r+1)k} |f(t)|^2\,dt$. Ce qui donne

$$|w^{r+1}| \leqq 2\left(\frac{1}{2}\,|u_*{}^r|^2 + k \int\limits_{rk}^{(r+1)k} |f(t)|^2\,dt\right)^{1/2} = \frac{N^*}{2}.$$

Posons

$$\lambda = w^{r+1} - u_*{}^r; \qquad x_{r+1} + w^{r+1} - w^r; \quad \eta_{r+1} = w^{r+1} - u_*{}^{r+1}. \tag{3.8}$$

On se propose de donner des conditions assurant la convergence de l'erreur $p_h e_h$ dans $L^\infty(0, V; H)$. Auparavant nous avons le

Lemme 3.1 *(pour $p \geqq 2$ et $\theta = 1$). Sous les hypothèses des théorèmes* 2.2, 2.3 *et* 3.1 *et si* $k\Gamma_h{}^2\left(\frac{N^*}{2}\right) \leqq \frac{2}{3}\,L_h$ *où* N^* *est donné par* (3.7), *il existe deux constantes réelles positives* C *et* D *telles que*

$$|\lambda|^2 = |w^{r+1} - u_*{}^r|^2 \leqq C + D|e^r|^2 \tag{3.9}$$

Démonstration. Posons $v_h = u_*{}^r$ dans (3.2). On a alors

$$|\lambda|^2 + \int\limits_{rk}^{(r+1)k} \big(A_h\big(\theta w^{r+1} + (1-\theta)\,u_*{}^r\big), \lambda\big) \leqq \int\limits_{rk}^{(r+1)k} \big(S_h f(t), \lambda\big)\,dt \tag{3.10}$$

et $\theta w^{r+1} + (1-\theta)\,u_*{}^r = u_*{}^r + \theta\lambda$ ce qui donne

$$|\lambda|^2 + \frac{kL_h}{\theta}\,\|\theta w^{r+1} + (1-\theta)\,u_*{}^r\|_h{}^2 \leqq \int\limits_{rk}^{(r+1)k} \big(S_h f(t), \lambda\big)$$

$$+ \frac{1}{\theta}\int\limits_{rk}^{(r+1)k} \big(A_h(u_*{}^r + \theta\lambda), -\lambda\big) + \frac{1}{\theta}\int\limits_{rk}^{(r+1)k} \big(A_h(u_*{}^r + \theta\lambda)\,w^{r+1}\big). \tag{3.11}$$

On va majorer chacun des 3 termes du second membre de (3.11).

$$\int\limits_{rk}^{(r+1)k} \big(S_h f(t), \lambda\big) \leqq \frac{k}{4\varepsilon}\int\limits_{rk}^{(r+1)k} |f(t)|^2\,dt + \varepsilon|\lambda|^2 \text{ en prenant } \varepsilon = \frac{1}{2\theta}. \tag{3.12}$$

$$\frac{1}{\theta}\int\limits_{rk}^{(r+1)k} \big(A_h(u_*{}^r + \theta\lambda), -\lambda\big) \leqq \frac{k}{\theta}\,\Gamma_h\left(\frac{N^*}{2}\right)\|u_*{}^r + \theta\lambda\|\,|\lambda|.$$

Or, d'après (3.10) on a

$$|\lambda|^2 \leqq \left(\int_{rk}^{(r+1)k} |f(t)|\, dt + k\Gamma_h \| \theta w^{r+1} + (1-\theta)\, u_*^r \| \right) |\lambda|,$$

ce qui donne

$$\frac{1}{\theta} \int_{rk}^{(r+1)k} \left(A_h(u_*^r + \theta\lambda), -\lambda\right) \leqq \frac{k}{2\theta} \int_{rk}^{(r+1)k} |f(t)|^2\, dt$$

$$+ \frac{3}{2} \frac{k^2}{\theta} \Gamma_h^2 \|\theta w^{r+2} + (1-\theta)\, u_*^r\|^2. \tag{3.13}$$

Faisons $v_h = 0$ dans (3.2); en utilisant $w^{r+1} = \lambda + u_*^r$, on a:

$$\frac{1}{\theta} \int_{rk}^{(r+1)k} \left(A_h\left(\theta w^{r+1} + (1-\theta)\, u_*^r\right), w^{r+1}\right) \leqq \frac{1}{\theta} \int_{rk}^{(r+1)k} \left(S_h f(t), w^{r+1}\right)$$

$$- \frac{1}{\theta} (\lambda, u_*^r) - \frac{1}{\theta} |\lambda|^2. \tag{3.14}$$

La relation (3.11) devient alors avec (3.12), (3.13), (3.14):

$$|\lambda|^2 + \frac{k}{\theta} \left[L_h - \frac{3}{2} k\Gamma_h^2 \right] \|\theta w^{r+1} + (1-\theta)\, u_*^r\|^2 \leqq \frac{k\theta}{2} \int_{rk}^{(r+1)k} |f(t)|^2\, dt$$

$$+ \frac{k}{2\theta} \int_{rk}^{(r+1)k} |f(t)|^2\, dt + \frac{1}{\theta} \int_{rk}^{(r+1)k} \left(S_h f(t), w^{r+1}\right) dt - \frac{1}{2\theta} |\lambda|^2 - \frac{1}{\theta} (\lambda, u_*^r). \tag{3.15}$$

Or $w^{r+1} = \lambda + u_*^r \Rightarrow |w^{r+1}|^2 \leqq 2|\lambda|^2 + 2|u_*^r|^2$ et $|u_*^r|^2 \leqq 2|e^r|^2 + 2|u^r|^2$, ce qui donne successivement

$$\frac{1}{\theta} \int_{rk}^{(r+1)k} \left(S_h f(t), w^{r+1}\right) dt \leqq \frac{2k}{\theta} \int_{rk}^{(r+1)k} |f(t)|^2\, dt + \frac{1}{4\theta} |\lambda|^2 + \frac{1}{2\theta} |e^r|^2 + \frac{1}{2\theta} |u^r|^2, \tag{3.16}$$

$$\frac{1}{\theta} (\lambda, u_*^r) \leqq \frac{1}{8\theta} |\lambda|^2 + \frac{4}{\theta} |e^r|^2 + \frac{4}{\theta} |u^r|^2. \tag{3.17}$$

(3.15) devient alors

$$\left(1 + \frac{1}{8\theta}\right) |\lambda|^2 + \frac{k}{\theta} \left[L_h - \frac{3}{2} k\Gamma_h^2 \right] \|\theta w^{r+1} + (1-\theta)\, u_*^r\|^2 \leqq \frac{9}{2\theta} |e^r|^2$$

$$+ \frac{9}{2\theta} |u^r|^2 + \cdots + \left(\frac{k\theta}{2} + \frac{5k}{2\theta}\right) \int_{rk}^{(r+1)k} |f(t)|^2\, dt.$$

D'où, avec $\frac{1}{8\theta} < 1 + \frac{1}{8\theta}$, $\theta \leqq 1$, $k \leqq k_0$ et $k\Gamma_h^2 \left(\frac{N^*}{2}\right) \leqq \frac{2L_h}{3}$

$$|\lambda|^2 \leqq C + D|e^r|^2. \tag{3.18}$$

Théorème 3.2 (*Cas où* $p \geqq 2$ *et* $\theta = 1$). *Sous les hypothèses des théorèmes* 2.2, 3.1. *De plus, on suppose*

$$k\Gamma_h^2 \left(\frac{N^*}{2}\right) \leqq \frac{2}{3} L_h, \tag{3.19}$$

où N^* *est donné par* (3.7),

$$\sup_r \frac{\eta^{m+1}}{k^2} \to 0 \tag{3.20}$$

avec η *donné par* (3.4), (3.6). *Alors* $p_h e_h \to 0$ *dans* $L^\infty(0, T; H)$.

Démonstration. Posons $v_h = w^{r+1}$ dans (3.1) et $v_h = u^{r+1}$ dans (3.2) et ajoutons les deux inégalités. On a alors d'après (3.8):

$$\begin{aligned}(e^{r+1} + \eta_{r+1} - e^r, e^{r+1} + \eta_{r+1}) \leqq \int_{rk}^{(r+1)k} \Big(&A_h(\theta u^{r+1} + (1-\theta)\, u^r) \\ &- A_h\big(\theta w^{r+1} + (1-\theta)\, u_*^{\,r}\big), e^{r+1} + \eta_{r+1}\Big)\end{aligned} \tag{3.21}$$

ce qui donne, d'après la monotonie de A_h:

$$|\eta_{r+1}|^2 + \frac{1}{2}(|e^{r+1}|^2 - |e^r|^2) + \frac{1}{2}|e^{r+1} - e^r|^2 \leqq -(\eta_{r+1}, e^{r+1} - e^r) - (\eta_{r+1}, e^{r+1}).$$

D'où

$$\frac{1}{2}(|e^{r+1}|^2 - |e^r|^2) + \frac{1}{4}|e^{r+1} - e^r|^2 \leqq \frac{1}{k}|\eta_{r+1}|^2 + \frac{k}{4}|e^{r+1}|^2. \tag{3.22}$$

Or, d'après la méthode itérative (3.3), on a $|\eta_{r+1}|^2 \leqq \eta^{m+1}|\lambda|^2$ avec $\lambda = w^{r+1} + u_*^{\,r}$ et on sait que $|\lambda|^2 \leqq C + D|e^r|^2$, d'où

$$\frac{1}{2}(|e^{r+1}|^2 - |e^r|^2) \leqq \frac{k}{4}|e^{r+1}|^2 + C\frac{\eta^{m+1}}{k} + D\frac{\eta^{m+1}}{k}|e^r|^2$$

et, par sommation, avec $e_h(0) = 0$ on a:

$$\frac{1}{2}|e^{r+1}|^2 \leqq \frac{1}{4}\int_0^{(r+1)k} |e_h(t+k)|^2\, dt + D\frac{\eta^{m+1}}{k^2}\int_0^{(r+1)k} |e_h(t)|^2\, dt + CT\frac{\eta^{m+1}}{k^2} \tag{3.23}$$

donc l'erreur totale tend vers 0 dans $L^\infty(0, T; H)$ quand $\frac{1}{k^2}\sup_r \eta^{m+1} \to 0$.

Nous donnons maintenant quelques autres résultats importants sans démonstrations.

Théorème 3.3 (*Cas où* $p \geqq 2$ *et* $\theta = 1$). *Sous les hypothèses des théorèmes* 2.1, 2.2, 3.1 *avec* $A(t)$ *indépendant de t et si* $\operatorname{Sup}_r \frac{\eta^{m+1}}{k^3} \leqq \frac{1}{2}$ *alors l'erreur* $p_h e_h(t) \to 0$ *dans* $L^\infty(0, T; H)$.

Théorème 3.4 (*Cas où* $2 < p < +\infty$ *et* $\theta \in]0, 1]$). 1) *On suppose* (H2), (H3), $f \in L^2(0, T; H)$, $\theta \geqq \theta_0 > 0$ *et* $\frac{k}{\theta} C_h^{2p}$ *borné où* C_h *est donné par* (2.1); *alors*

$$|u_h(t)|^2 + \frac{\theta}{2} \sum_{i=0}^{r-1} |u^{i+1} - u^i|^2 + 2\alpha \int_0^{rk} \|z_h(t)\|^p \, dt \tag{3.24}$$

est borné où $z^{r+1} = \theta u^{r+1} + (1 - \theta) u^r$ *et* $z_h(t) = z^{r+1}$ *pour* $t \in [rk, (r + 1) k[$. *Si, de plus,* $f \in L^\infty(0, T; H)$, *alors on a*:

$$\left|u_{hk}\big((r + 1) k\big) - u_{hk}(rk)\right| \to 0 \quad \text{quand} \quad (h, k) \to 0. \tag{3.25}$$

2) *On peut alors extraire de* $p_h u_{hk}(t)$ *une sous-suite encore notée* $p_h u_{hk}$, *telle que* $p_h u_{hk} \rightharpoonup u$ *dans* $L^\infty(0, T; H)$ ***faible *et* $p_h\big(\theta u_{hk}(t + k) + (1 - \theta) u_{hk}(t)\big) \rightharpoonup u$ *dans* $L^p(0, T; V)$ *faible, et qui vérifie*

$$\int_0^t \left(\frac{dv}{dt} + Av, v - u\right) ds \geqq \int_0^t (f, v - u) \, ds \quad \text{avec} \quad t \in [0, T] \tag{3.26}$$

pour tout $v \in K$ *p.p dt*, $v \in L^p(0, T; V)$, $\frac{dv}{dt} \in L^q(0, T; V')$, $v(0) = u_0$.

3) *Si l'opérateur* A *est tel que*

$$\int_0^t \left(A(s) \, v(s), \frac{dv}{dt}\right) ds \geqq \varphi(v) \qquad \forall v \in L^p(0, T; K) \quad \text{tel que} \quad \frac{dv}{dt} \in L^p(0, T; V), \tag{3.27}$$

où φ *est une fonction bornée inférieurement sur les bornés de* $L^p(0, T; V)$; *alors, pour tout* $f \in L^2(0, T; H)$, *la solution* $u(t)$ *de* (3.26) *vérifie*

$$\int_0^t \left(\frac{du}{dt} + Au, v - u\right) ds \geqq \int_0^t (f, v - u) \, ds \tag{3.28}$$

pour tout $v \in L^p(0, T; V)$, $\frac{dv}{dt} \in L^q(0, T; V')$ *et* $v(0) = u_0$. *De plus,* $u(t) \in C([0, T]; H)$ *et* $\frac{du}{dt} \in L^2(0, T; H)$, *avec* $u(0) = u_0$.

4) *Si on pose* $V = W_0^{1,p}(\Omega)$, $H = L^2(\Omega)$, $V' = W^{-1,q}(\Omega)$ *pour* Ω *ouvert borné de* R^n *et*

$Au = -\sum_{i=1}^{n} \frac{\partial}{\partial x_i}\left(\left|\frac{\partial u}{\partial x_i}\right|^{p-2} \frac{\partial u}{\partial x_i}\right)$ *on a alors*

$$\int_0^t \left(A(s)\, v(s), \frac{dv}{dt}\right) ds = \frac{1}{p}\, \|v(t)\|_V^p - \frac{1}{p}\, \|v(0)\|_V^p$$

et $A(t)$ *vérifie la condition* (3.27).

Théorème 3.5 (*Convergence de l'erreur dans le cas* $2 < p < \infty$, $\theta \in]0, 1]$). *On suppose les conditions du théorème* 3.4. *Si de plus on a*

$$k^2 \Gamma_h^2(N_1) \leqq L_h \frac{\theta_0}{1-\theta_0} \tag{3.29}$$

où $\theta \geqq \theta_0 > 0$ *et* N_1 *donné par les estimations a priori*: $|u_{hk}(rk)| \leqq N_1$ *et* $|u_*^r| \leqq N_1$;

$$\frac{3}{2}\left(\frac{N^*}{2}\right)^{p-2} M^2 k C^p(h) \leqq \alpha \tag{3.30}$$

où N^* *est une constante telle que* $|\theta w^{r+1} + (1-\theta)\, u_*^r| \leqq \frac{N^*}{2}$, $C(h)$ *est donné par* (2.1) *et où* α, M *vérifient les conditions* $(Au, u) \geqq \alpha\|u\|^p$, $\|Au\|_{V'} \leqq M\|u\|_V^{p+1}$;

$$\frac{\eta^{m+1}}{k^2}\, e^{T\frac{(1-\theta)}{2k^2}} \to 0 \quad \textit{avec } \eta \textit{ donné par } (3.4), (3.6), \tag{3.31}$$

alors on a la convergence de l'erreur $p_h e_h \to 0$ *dans* $L^\infty(0, T; H)$.

4. Exemples

Soit Ω un ouvert borné de R^n, de côtés parallèles aux axes et $x = (x_1, \ldots, x_n) \in \Omega$. Posons $V = W_0^{1,p}(\Omega) \subset H = L^2(\Omega) \subset V' = W^{-1,q}(\Omega)$, $\|u\|_V = \left(\sum_{i=1}^{n} |D_i u|_{L^p}^p\right)^{1/p}$, $|u|_H = |u|_{L^2}$. Soit K un convexe fermé de V contenant 0 et $f \in L^2(0, T; L^2(\Omega))$. Posons

$$Au = -\mu \sum_{i=1}^{n} D_i(|D_i u|^{p-2} D_i u) - \lambda \Delta u;\ D_i = \frac{\partial}{\partial x_i},\ \lambda, \mu \geqq 0 \tag{4.1}$$

Nous démontrons alors les inégalités suivantes:

$$|u|_{L^2} \leqq C(\Omega)\, |u|_{L^p} \quad \text{avec} \quad C(\Omega) = \{\mu(\Omega)\}^{\frac{p-2}{2p}} \tag{4.2}$$

où μ est la mesure de Lebesgue

$$\|Au\|_{V'} \leqq \mu\|u\|_V^{p-1} + \lambda C(\Omega)\, n^{\frac{p-2}{2p}} \|u\|_V = M\|u\|_V^{p-1} + M_1\|u\|_V. \tag{4.3}$$

$$(Au, u) \geqq \mu\|u\|_V^p + \lambda\left(\sum_{i=1}^{n} |D_i u|_{L^2}^2\right) \geqq \mu\|u\|_V^p. \tag{4.4}$$

Sous les conditions du théorème 2.3 les hypothéses sur le convexe K du théorème 1.2 ne sont plus nécessaires et peuvent être remplacées par les conditions sur la discrétisation. Nous aurons en même temps l'existence et l'unicité de la solution $u(t)$ du problème (4.5) et la convergence de $p_h u_{hk}(t)$, solution du problème discrétisé, vers $u(t)$.

$$\int_0^T \left(\frac{du}{dt} + Au, v - u\right) \geqq \int_0^T (f, v - u)\, dt \tag{4.5}$$

pour $v \in L^p(0, T; V)$ avec $u(x, 0) = 0\ \forall x \in \Omega,\ u(x, t) = 0\ \forall x \in \Gamma = \partial\Omega,\ \forall t \in [0, T]$.

4.1. Schéma de discrétisation

Posons $h = (h_1, \ldots, h_n)$ et considérons le réseau de points $\mathfrak{R}_h = \{M \in \Omega \mid M = (m_1 h_1, \ldots, m_n h_n),\ m_j \in Z$ et $h_i = h, \forall i\}$.

On construit un espace V_h comme dans [5] et [6], en posant pour $M \in \mathfrak{R}_h$; $\omega_{h,q}^M = \prod_{i=1}^{n}\left[\left(m_i - \frac{q+1}{2}\right)h, \left(m_i + \frac{q+1}{2}\right)h\right]$ et $\mathfrak{R}_h^1 = \{M \in \mathfrak{R}_h, \omega_{h,1}^M, \subset \Omega\}$. V_h est identifié à l'espace des fonctions $u_h(x) = \sum_{M \in \mathfrak{R}_h^1} u_h^M \theta_h^M$ où θ_h^M est la fonction indicatrice de $\omega_{h,0}^M$ et u_h^M est une composante réelle quand M décrit $\mathfrak{R}_h^1$. On pose

$$|u_h|_h = |u_h(x)|_{L^2} = \left\{\sum_M h^n\, |u_h^M|^2\right\}^{1/2} \tag{4.6}$$

et on sait qu'il existe p_h de V_h dans $V = W_0^{1,p}(\Omega)$ tel que

$$\left.\begin{aligned} &|u_h|_h = |p_h u_h|_{L^2};\ \|u_h\|_h = \left(\sum_{i=1}^{n} |\nabla_i u_h|_{L^p}^p\right)^{1/p} = \|p_h u_h\|_V \quad \text{et} \\ &|||u_h|||_h = \left(\sum_{i=1}^{n} |\nabla_i u_h|_{L^2}^2\right)^{1/2} = \|p_h u_h\|_{H_0^1(\Omega)}. \end{aligned}\right\} \tag{4.7}$$

On introduit, comme dans [3], un opérateur S_h défini sur V_h par $S_h = I - k\lambda\theta \sum_{i=1}^{n} \nabla_i^2$ et une norme

$$[u_h]^2 = |u_h|^2 + k\lambda\theta\, |||u_h|||^2. \tag{4.8}$$

Pour $\Omega[= \prod_{i=1}^{n}]a_i, b_i[$ on calcule facilement les constantes d'équivalences des normes ainsi introduites:

$$\left.\begin{aligned}
&|u_h|_h \leqq C_0\|u_h\|_h \quad \text{avec} \quad C_0 = \frac{1}{n\pi}\left[\sum_{i=1}^{n}(b_i-a_i)^2\right]^{1/2}\{n\mu(\Omega)\}^{\frac{p-2}{2p}},\\
&\|u_h\|_h \leqq C(h)\,|u_h|_h \quad \text{avec} \quad C(h) = C_h = \frac{2n^{1/p}}{h},\\
&|||u_h|||_h \leqq C(\Omega)\,n^{\frac{p-2}{2p}}\,\|u_h\|_h \quad \text{avec} \quad C(\Omega) = \{\mu(\Omega)\}^{\frac{p-2}{2p}},\\
&\|u_h\|_h \leqq h^{\frac{2n}{p-1}}\,|||u_h|||_h\\
&|||u_h|||_h^2 \leqq \beta(h)\,|u_h|_h^2 \quad \text{avec} \quad \beta(h) = \left\{C(\Omega)\,C(h)\,n^{\frac{p-2}{2p}}\right\}^2,\\
&|u_h|_h \leqq d|||u_h|||_h \quad \text{avec} \quad d = C_0/\{\mu(\Omega)\,n\}^{\frac{p-2}{2p}}.
\end{aligned}\right\} \qquad (4.9)$$

Posons

$$A_h u_h = -\mu\sum_{i=1}^{n}\nabla_i(|\nabla_i u_h|^{p-2}\,\nabla_i u_h) - \lambda\sum_{i=1}^{n}\nabla_i^2 u_h, \qquad K_h = p_h^{-1}(K) \qquad (4.10)$$

(cf. [3]) et considérons alors deux cas:

1) *Méthode implicite standard* $\theta = 1$

Dans ce cas, la condition $\|Au\|_V \leqq M\|u\|_{V'}^{p-1}$, qui n'est pas vérifié par l'opérateur A, n'est pas utilisée dans les démonstrations de la convergence. En supposant $f \in L^2(0, T; H)$, on a alors la convergence forte de la suite $p_h u_{hk}$, solution du problème discrétisé, vers la solution $u(t)$, unique et régulière de (4.5).

2) *Cas où* $\theta \in\,]0, 1]$

On peut alors remplacer dans toutes les démonstrations la condition $\|Au\|_{V'} \leqq M\|u\|_V^{p-1}$ par l'hypothèse plus faible

$$\|Au\|_{V'} \leqq M\|u\|_V^{p-1} + M_1\|u\|_V \qquad (4.11)$$

vérifiée par notre opérateur. On obtient les mêmes majorations et la convergence de l'erreur; seules les constantes intervenant dans les majorations sont modifiées. On montre alors que si

$$f \in L^\infty(0, T; H), \quad \theta \geqq \theta_0 > 0, \quad k \leqq \frac{\theta}{2} \quad \text{et} \quad k\frac{C_h^{2p}}{\theta} \quad \text{borné} \qquad (4.12)$$

on a la convergence de $p_h u_{hk}$ vers la solution $u(t)$ (théorème 3.4).

4.2. Itérations sur le problème discrétisé

Posons:

$$u_{hk,m+1}^{r+1} = P_{K_h}^{S_h}\{u_{hk,m}^{r+1} - \varrho S_h^{-1}(B_{hk}u_{hk,m}^{r+1} - g_*)\} \qquad (4.13)^1$$

[1] Ici l'introduction de l'opérateur S_h accellère la convergence. Evidemment on peut prendre $S_h = Id$. *Cf.* [3].

où $B_{hk}u = u + kA_h(\theta u + (1-\theta)\, u_*^r)$, $g_* = u_*^r + k \int_{rk}^{(r+1)k} S_h f(t)\, dt$, $u_{hk}^0 = u_h^0$, $r = 0, 1, \ldots, m-1$ $\left(m = \dfrac{T}{k}\right)$, $P_{K_h}^{S_h}$ étant l'opérateur de projection sur K_h suivant la norme [.] donnée par (4.8). Les itérations (4.13) convergent vers u_{hk}^{r+1} quand $m \to \infty$ pour

$$\varrho = 1 \Big/ \left(1 + \frac{\theta k \beta(h)\, C(N)}{1 + \theta k \lambda \beta(h)}\right)^2 \tag{4.14}$$

où $\beta(h)$ est donné par (4.9), où $C(N) = \dfrac{\mu(p-1)\, N^{p-2}}{h^{n(p-2)/2}} + \lambda$, $N = N_0 \left(\dfrac{\beta(h)}{1 + k\lambda\theta\beta(h)}\right)^{1/2}$ et N_0 tel que $[\theta w^{r+1} + (1-\theta)\, u_*^r] \leqq \dfrac{N_0}{2}$.

1) *Dans le cas* $\theta = 1$, on a l'estimation de N^* tel que $|\theta w^{r+1} + (1-\theta)\, u_*^r| \leqq \dfrac{N^*}{2}$ donnée par

$$N^* = 4 \left\{\frac{1}{2}\, |u_*^r|^2 + k \int_{rk}^{(r+1)k} |f(t)|^2\, dt\right\}^{1/2} \tag{4.15}$$

et on en déduit N_0 et N.

2) *Dans le cas* $\theta \in \,]0, 1]$ nous avons

$$N^* = 2C_0 \left\{\frac{\theta}{2} \int_{rk}^{(r+1)k} |f(t)|^2\, dt + \frac{1}{2pq^{p-1}} \left[4^p \|u_*^r\|^p + \left(\frac{4M_1}{\mu}\right)^q \|u_*^r\|^q\right]\right\}^{1/p} \tag{4.16}$$

et on en déduit N_0 et N.

On a alors $|u_{hk}^{r+1} - w^{r+1}|^2 \leqq \eta^{m+1} |u_*^r - w^{r+1}|^2$ où η est le coefficient de contraction des itérations (4.13) donné par

$$\eta = (1 + k\lambda\theta\beta(h)\, (1 - \varrho). \tag{4.17}$$

4.3. Estimation asymptotique de l'erreur

1) *Pour* $\theta = 1$ on a $p_h e_h \to 0$ dans $L^\infty(0, T; H)$ pour

$$\operatorname*{Sup}_r \frac{\eta^{m+1}}{k^3} \leqq \frac{1}{2}. \tag{4.18}$$

Si $\lambda = \mu = 1$, $T = 1$, $p = 3$, $\Omega = (]0, 1[)^2$, $h_1 = h_2 = h$.

Si $\dfrac{8k}{h^2} = k\beta(h) \leqq a = c^t/$, alors une condition suffisante sur le nombre d'itérations m à effectuer à l'étape (r) pour avoir $p_h e_h \to 0$ est donnée par

$$m + 1 \geqq (3|\log k| + C_1) \frac{C_2}{k^2} \tag{4.19}$$

où $C_1 = \log 2 + \log(1 + a)$, $C_2 = \left\{\frac{a}{8}\left[1 + \frac{a}{1+a}\left(1 + 4\sqrt{2}\, N^*\right)\right]\right\}^2$.

2) *Pour* $\theta \in]0, 1]$

*Pour le même exemple, avec $\theta \in]0, 1]$ on montre que pour $\theta \geqq \theta_0 > 0$, $k \leqq \frac{\theta}{2}$, $f \in L^\infty(0, T; H)$ et $\frac{k}{h^6} \leqq a = \text{c}^{\text{t}}/$, une condition suffisante pour avoir $\operatorname{Sup}_r \frac{\eta^{m+1}}{k^2} e^{T/2k^2} < \varepsilon$ est donnée par

$$\left.\begin{aligned} m + 1 \geqq &\left\{|\log \varepsilon| + \frac{T(1-\theta)}{2k^2} + 2|\log k| + \log(1 + 8\theta a h^4)\right\} \\ &\times \left\{1 + \frac{8\theta a h^2}{1 + 8\theta a h^4}\left(h^2 + 4\sqrt{2}\, N^*\right)\right\}^2 \end{aligned}\right\} \tag{4.20}$$

où N^* est donné par (4.16). On aura alors $p_h e_h \to 0$ dans $L^\infty(0, T; H)$.

*Si le convexe K est borné dans V, dans ce cas les majorations de la solution du problème discrétisé ne nécessitent plus l'hypothèse $\frac{k}{h^6}$ borné. L'hypothèse $f \in L^\infty(0, T; H)$ et $\frac{k}{h^3} \to 0$ permet d'avoir la relation $|u_{hk}(t + k) - u_{hk}(t)| \to 0$.

Si $\frac{k}{h^3} \leqq b = \text{c}^{\text{t}}/$, on a la convergence de l'erreur pour

$$\begin{aligned} m + 1 \geqq &\frac{1}{h}\left\{h + \frac{8\theta b}{1 + 8\theta b}\left(h^2 + 4\sqrt{2}\, N^*\right)\right\}^2 \times \\ &\times \left\{|\log \varepsilon| + \frac{T(1-\theta)}{2k^2} + 2|\log h| + \log(1 + 8\theta b h)\right\}. \end{aligned} \tag{4.21}$$

3) De nombreux résultats numériques confirmant les estimations a priori sont présentés dans G. Geymonat et M. Sibony [3] dans le cas où $K = V$.

Bibliographie

[1] Brezis, H., Problèmes unilatéraux. J. Math. Pures Appl. **51**, 1—168 (1972).

[2] Brezis, H., Opérateurs maximaux monotones et semi-groupes de contractions dans les espaces de Hilbert. Notas de Matematica, North-Holland Mathematics Studies (1973).

[3] Geymonat, G., et M. Sibony, Approximation de certaines équations paraboliques non linéaires. Calcolo **13**, fasc. 2 (1976).

[4] Raviart, P. A., Sur l'approximation de certaines équations d'évolution linéaires et non linéaires. J. Math. Pures Appl. **46**, 109—183 (1967).

[5] Sibony, M., Sur l'approximation d'équations et inéquations aux dérivées partielles non linéaires de type monotone. J. Math. Analysis Appl. 34, 502—564 (1971).
[6] Sibony, M., Méthodes itératives pour les équations et inéquations aux dérivées partielles non linéaires de type monotone. Calcolo 7, 65—183 (1970).
[7] Saramito, B., These de 3e cycle Mathematiques, Université de Paris VII, 1975.

Adresse de l'auteur:

Université de Tours
Faculté des Sciences
Laboratoire de Mathématiques
Parc de Grandmont
37200 Tours
France

Über die Invarianz der Rotation bei Approximation der Vektorfelder

Геннади Вайникко, Тарту

Konvergenzuntersuchungen zu Näherungsverfahren für nichtlineare Gleichungen nutzen die Invarianzsätze für die Rotation (oder für den Abbildungsgrad) aus, wenn ein Vektorfeld mittels endlichdimensionaler Vektorfelder approximiert wird. In [1, 2] sind Invarianzbedingungen für kompakte Vektorfelder angegeben worden. In der vorliegenden Arbeit verallgemeinern und verstärken wir diese Ergebnisse für eine Klasse von Vektorfeldern, die kompakte und limeskompakte Vektorfelder enthält.

1. Rotationsbegriff für eine Klasse von Vektorfeldern

Es sei E ein reeller Banachraum und $\Omega \subset E$ eine beschränkte, offene Menge in E mit dem Rand $\partial\Omega$ und mit der Abschließung $\bar{\Omega}$. Wir betrachten einen (nicht unbedingt stetigen) Operator $T\colon \bar{\Omega} \to E$. Eine Menge $\Sigma \subset E$ heißt (vgl. mit [3]) *fundamental* für T, wenn die folgenden Bedingungen (1)—(3) erfüllt sind:

Σ ist konvex, kompakt (damit abgeschlossen) und enthält alle Fixpunkte von T; (1)

T ist stetig auf $\bar{\Omega} \cap \Sigma$ und $T(\bar{\Omega} \cap \Sigma) \subseteq \Sigma$; (2)

$x_0 \in \bar{\Omega},\ x_0 \notin \Sigma \Rightarrow x_0 \notin \operatorname{co}\{Tx_0, \Sigma\}$ (3)

(co — konvexe Hülle). Die leere Menge $\emptyset$ ist fundamental für T, falls T keinen Fixpunkt in $\bar{\Omega}$ hat.

Die Klasse von Operatoren, die eine Fundamentalmenge besitzen, ist ziemlich umfassend. Beispielsweise ist die abgeschlossene, konvexe Hülle $\overline{\operatorname{co}}\,\{T\bar{\Omega}\}$ des Wertebereiches $T\bar{\Omega}$ eines vollstetigen (stetigen und kompakten) Operators $T\colon \bar{\Omega} \to E$ fundamental für T; für einen stetigen limeskompakten Operator $T\colon \bar{\Omega} \to E$ ist der Limeswertebereich fundamental (über diese Begriffe und Behauptung siehe [4, 5]). Andererseits gibt es Operatoren, die nicht kompakt oder limeskompakt sind, jedoch eine Fundamentalmenge besitzen. Ein Beispiel stellt der Verschiebungsoperator T in einem Hilbertraum H mit einer orthonormalen Basis $\{e_k\}_{k=-\infty}^{\infty}$ dar:

$$Tx = \sum_{k=-\infty}^{\infty} \xi_k e_{k+1} \qquad \text{für} \qquad x = \sum_{k=-\infty}^{\infty} \xi_k e_k \in \bar{\Omega} = \{x \in H : \|x\| \leqq 1\}$$

— der Operator T ist nicht limeskompakt, hat jedoch eine Fundamentalmenge, nämlich $\Sigma = \{0\}$.

Für einen Operator $T: \bar{\Omega} \to E$, der eine Fundamentalmenge Σ besitzt und keinen Fixpunkt auf dem Rand $\partial\Omega$ hat, kann man leicht den Rotationsbegriff einführen (vgl. mit [3, 5]). Sei $\bar{\Omega} \cap \Sigma$ nicht leer. Die Einschränkung

$$T_\Sigma : \bar{\Omega} \cap \Sigma \to \Sigma \subset E$$

des Operators T auf die Menge $\bar{\Omega} \cap \Sigma$ ist stetig (siehe (2)). Betrachten wir eine stetige Erweiterung[1]

$$\hat{T}: \bar{\Omega} \to \Sigma \subset E$$

von T_Σ auf $\bar{\Omega}$. Der Operator $\hat{T}$ ist vollstetig, weil Σ kompakt ist (siehe (1)); er hat keinen Fixpunkt auf dem Rand $\partial\Omega$, weil die Fixpunkte zu Σ gehören und $\hat{T}x = Tx$ für $x \in \bar{\Omega} \cap \Sigma$ gilt. Somit ist die Rotation $\gamma(I - \hat{T}; \partial\Omega)$ des Vektorfeldes $x - \hat{T}x$ auf dem Rand $\partial\Omega$ definiert. Wir setzen

$$\gamma(I - T; \partial\Omega) = \gamma(I - \hat{T}; \partial\Omega); \tag{4}$$

ist $\bar{\Omega} \cap \Sigma$ leer, so setzen wir $\gamma(I - T; \partial\Omega) = 0$.

Es ist leicht nachzuprüfen, daß die Rotation $\gamma(I - T; \partial\Omega)$ von der Erweiterungsweise von T_Σ oder von der Wahl der Fundamentalmenge Σ nicht abhängt. Es seien Σ_1 und Σ_2 zwei Fundamentalmengen für T. Dann ist $\Sigma := \Sigma_1 \cap \Sigma_2$ ebenso fundamental für T. Zunächst sei $\bar{\Omega} \cap \Sigma$ nicht leer. Mit $\hat{T}_1$, $\hat{T}_2$ und $\hat{T}$ bezeichnen wir die Operatoren, die wie oben konstruiert sind und den Fundamentalmengen Σ_1, Σ_2 bzw. Σ entsprechen.

Wir behaupten, daß die vollstetigen Vektorfelder $x - \hat{T}_1x$ und $x - \hat{T}x$ linear homotop auf $\partial\Omega$ sind. Nehmen wir das Gegenteil an: Für ein $x \in \partial\Omega$ und ein $\lambda \in [0, 1]$ gilt

$$x = \lambda\hat{T}x + (1 - \lambda)\,\hat{T}_1x.$$

Weil $\hat{T}x \in \Sigma \subseteqq \Sigma_1$, $\hat{T}_1x \in \Sigma_1$, gilt $x \in \Sigma_1$, $\hat{T}_1x = \hat{T}x$ und

$$x = \lambda\hat{T}x + (1 - \lambda)\,Tx \in \operatorname{co}\{Tx, \Sigma\};$$

auf Grund von (3) gilt $x \in \Sigma$, $\hat{T}x = Tx$ und

$$x = \lambda Tx + (1 - \lambda)\,Tx = Tx,$$

was unserer Voraussetzung widerspricht, daß T keinen Fixpunkt auf $\partial\Omega$ hat. Analog zeigt man, daß die vollstetigen Vektorfelder $x - \hat{T}_2x$ und $x - \hat{T}x$ auf $\partial\Omega$ linear homotop sind. Folglich gilt

$$\gamma(I - \hat{T}_1; \partial\Omega) = \gamma(I - \hat{T}; \partial\Omega) = \gamma(I - \hat{T}_2; \partial\Omega).$$

Es bleibt der Fall zu betrachten, daß $\bar{\Omega} \cap \Sigma$ leer ist. Dann hat T in $\bar{\Omega}$ keinen Fixpunkt; ist $\bar{\Omega} \cap \Sigma_1$ oder $\bar{\Omega} \cap \Sigma_2$ nicht leer, so hat ebenso $\hat{T}_1$ bzw. $\hat{T}_2$ keinen Fixpunkt, und jedenfalls erhalten wir $\gamma(I - T; \partial\Omega) = 0$.

Damit ist die Korrektheit der Definition (4) bewiesen. Ebenso leicht ist es, zu sehen, daß für einen vollstetigen oder einen stetigen limeskompakten Operator die Definition (4)

---

[1] Eine solche Erweiterung existiert nach dem Satz von Dugundji [6], weil Σ konvex ist. Der Satz von Dugundji lautet: Ist $f: M_0 \subseteqq M \to X$ ein stetiger Operator von einer abgeschlossenen Menge M_0 eines metrischen Raumes M in einen lokalkonvexen topologischen Raum X, so existiert ein stetiger Operator $g: M \to X$, so daß $f(x) = g(x)$ für $x \in M_0$ und $g(M) \subseteqq \operatorname{co}\{(M_0)\}$ gilt.

den bekannten Definitionen der Rotation äquivalent ist. Auch kann man beweisen, daß die eingeführte Rotation die üblichen Eigenschaften der Rotation besitzt.

Leider ist die Definition (4) für unsere Ziele nicht konstruktiv, und wir führen noch eine Konstruktion ein (dabei folgen wir der Arbeit [5]). Der Operator $T : \bar{\Omega} \to E$ besitze eine Fundamentalmenge Σ und habe keinen Fixpunkt auf dem Rand $\partial\Omega$; die Menge $\bar{\Omega} \cap \Sigma$ sei nicht leer. Auf der kompakten Menge $\bar{\Omega} \cap \Sigma$ ist T gleichmäßig stetig, und es gibt ein $\varepsilon > 0$, so daß

$$\sup_{x, x' \in \bar{\Omega} \cap \Sigma, \|x - x'\| \leq \varepsilon} \|Tx - Tx'\| < \alpha \tag{5}$$

mit

$$\alpha := \inf_{x \in \partial\Omega \cap \Sigma} \|x - Tx\| \tag{6}$$

gilt (es ist leicht zu sehen, daß $\alpha > 0$; ist $\partial\Omega \cap \Sigma$ leer, so können wir $\alpha > 0$ beliebig fixieren). Wir wählen für die kompakte Menge $\bar{\Omega} \cap \Sigma$ ein ε-Netz $\{x^1, \ldots, x^r\} \subset \bar{\Omega} \cap \Sigma$. Durch die Formeln

$$Sx = \sum_{k=1}^{r} \mu^k(x)\, Tx^k \Big/ \sum_{k=1}^{r} \mu^k(x), \tag{7}$$

$$\mu^k(x) = \begin{cases} \varepsilon + \eta - \|x - x^k\|, & \text{falls } \|x - x^k\| \leqq \varepsilon, \\ \eta, & \text{falls } \|x - x^k\| > \varepsilon \end{cases}$$

definieren wir einen Operator $S : \bar{\Omega} \to \Sigma \subset E$. Der Operator S ist für jedes $\eta > 0$ auf $\bar{\Omega}$ erklärt, stetig und endlichdimensional. Die Zahl $\eta > 0$ wählen wir genügend klein, so daß

$$\sup_{x \in \bar{\Omega} \cap \Sigma} \|Tx - Sx\| < \alpha. \tag{8}$$

Das ist möglich: Für $\eta = 0$ folgt (8) aus (5) und S hängt von $\eta \geqq 0$ gleichmäßig stetig bezüglich $x \in \bar{\Omega} \cap \Sigma$ ab.

Wir wollen beweisen, daß die vollstetigen Vektorfelder $x - \hat{T}x$ und $x - Sx$ linear homotop auf $\partial\Omega$ sind. Nämlich ergibt sich aus den Bedingungen

$$x \in \partial\Omega, \qquad x = \lambda \hat{T}x + (1 - \lambda) Sx, \qquad 0 \leq \lambda \leq 1,$$

daß $x \in \Sigma$. Folglich gilt $\hat{T}x = Tx$ und

$$\|x - Tx\| = (1 - \lambda) \|Sx - Tx\| < \alpha,$$

was (6) widerspricht. Die Definition (4) können wir nun mit der folgenden Gleichheit ergänzen:

$$\gamma(I - T; \partial\Omega) = \gamma(I - \hat{T}; \partial\Omega) = \gamma(I - S; \partial\Omega). \tag{9}$$

2. Invarianz der Rotation

Es seien E und E_n $(n \in N = \{1, 2, \ldots\})$ reelle Banachräume und $P = (p_n)_{n \in N}$ ein System von Operatoren $p_n : E \to E_n$, so daß für beliebige $x, x' \in E$ und $a, a' \in R$

$$\|p_n(ax + a'x') - (ap_nx + a'p_nx')\| \to 0 \qquad (n \in N),$$

$$\|p_nx\| \to \|x\| \qquad (n \in N).$$

Eine Folge $(x_n)_{n \in N' \subseteqq N}$ mit $x_n \in E_n$ *P-konvergiert* gegen $x \in E$, wenn $\|x_n - p_n x\| \to 0$ $(n \in N')$; wir bezeichnen das mit $x_n \xrightarrow{P} x$ $(n \in N')$. Eine Folge $(x_n)_{n \in N'}$ mit $x_n \in E_n$ ist *P-kompakt*, wenn für jede unendliche Menge $N'' \subset N'$ eine unendliche Menge $N''' \subset N''$ existiert, so daß die Teilfolge $(x_n)_{n \in N'''}$ (gegen ein gewisses $x \in E$) P-konvergiert. Ausführlicher über diese Begriffe siehe z. B. [7].

Betrachten wir Operatoren $T : \bar{\Omega} \to E$ und $T_n : \bar{\Omega}_n \to E_n$, wobei $\Omega \subset E$ und $\Omega_n \subset E_n$ beschränkte, offene Mengen mit den Rändern $\partial\Omega$ bzw. $\partial\Omega_n$ und den Abschließungen $\bar{\Omega}$ bzw. $\bar{\Omega}_n$ sind.

$$\begin{array}{ccc} E \supset \bar{\Omega} & \xrightarrow{T} & E \\ \downarrow{\scriptstyle p_n} & & \downarrow{\scriptstyle p_n} \\ E_n \supset \bar{\Omega}_n & \xrightarrow{T_n} & E_n \end{array}$$

Satz 1. *Die folgenden Bedingungen seien erfüllt:*

(i) *Der Operator $T : \bar{\Omega} \to E$ hat eine Fundamentalmenge Σ und hat keinen Fixpunkt auf dem Rand $\partial\Omega$;*

(ii) *Die Mengen Ω und Ω_n sind konsistent:*

$$x \in \bar{\Omega} \cap \Sigma \Rightarrow \exists x_n \in \bar{\Omega}_n, \quad x_n \xrightarrow{P} x \ (n \in N);$$

$$x_n \in \bar{\Omega}_n, \quad x_n \xrightarrow{P} x \ (n \in N' \subseteqq N) \Rightarrow x \in \bar{\Omega};$$

$$x_n \in \partial\Omega_n, \quad x_n \xrightarrow{P} x \ (n \in N' \subseteqq N) \Rightarrow x \in \partial\Omega;$$

(iii) *Die Operatoren $T_n : \bar{\Omega}_n \to E_n$ sind vollstetig, gleichmäßig beschränkt* $\left(\sup_{x_n \in \Omega_n} \|T_n x_n\| \leqq \text{const } (n \in N)\right)$ *und mit T konsistent:*

$$x_n \in \bar{\Omega}_n, \quad x_n \xrightarrow{P} x, \quad T_n x_n \xrightarrow{P} x' \ (n \in N' \subseteqq N) \Rightarrow x' = Tx;$$

$$x_n \in \bar{\Omega}_n, \quad x_n \xrightarrow{P} x \in \Sigma \Rightarrow T_n x_n \xrightarrow{P} Tx;$$

(iv) *Für $0 < \lambda \leqq 1$ ist die folgende Regularitätsbedingung erfüllt:*

$$x_n \in \bar{\Omega}_n, \quad (x_n - \lambda T_n x_n)_{n \in N' \subseteqq N} \ P\text{-kompakt} \Rightarrow (x_n)_{n \in N'} \ P\text{-kompakt}.$$

Dann hat T_n für fast alle n keinen Fixpunkt auf dem Rand $\partial\Omega_n$, und

$$\gamma(I_n - T_n; \partial\Omega_n) = \gamma(I - T; \partial\Omega) \qquad (n \geqq n_0) \tag{10}$$

(I bzw. I_n ist der Einheitsoperator in E bzw. E_n).

Beweis. I. Zunächst betrachten wir den Fall, daß $\bar{\Omega} \cap \Sigma$ leer ist. Dann hat T in $\bar{\Omega}$ keinen Fixpunkt. Wir behaupten, daß T_n für fast alle n keinen Fixpunkt in $\bar{\Omega}_n$ hat. Ist nämlich $T_n x_n = x_n \in \bar{\Omega}_n$ $(n \in N' \subseteqq N)$, so ist die Folge (x_n) wegen (iv) P-kompakt, eine Teilfolge P-konvergiert: $x_n \xrightarrow{P} x \in \bar{\Omega}$ $(n \in N'' \subseteqq N')$. Dann gilt $T_n x_n \xrightarrow{P} x$ $(n \in N'')$

und dank (iii) $Tx = x$, was unmöglich ist. Damit gilt für fast alle n

$$\gamma(I_n - T_n; \partial\Omega_n) = 0 = \gamma(I - T; \partial\Omega),$$

und im Falle $\bar{\Omega} \cap \Sigma = \emptyset$ ist der Satz bewiesen.

II. Nun sei $\bar{\Omega} \cap \Sigma$ nichtleer. Betrachten wir noch einmal das ε-Netz $\{x^1, \ldots, x^r\} \subset \bar{\Omega} \cap \Sigma$ der kompakten Menge $\bar{\Omega} \cap \Sigma$ (siehe 1). Wählen wir $\{x_n^1, \ldots, x_n^r\} \subset \bar{\Omega}_n$ so, daß $x_n^k \xrightarrow{P} x^k$ $(n \in N)$, $k = 1, \ldots, r$. Analog dem Operator $S : \bar{\Omega} \to \Sigma \subset E$ (siehe (7)) definieren wir endlichdimensionale Operatoren $S_n : \bar{\Omega}_n \to E_n$,

$$S_n x_n = \sum_{k=1}^{r} \mu_n^k(x_n)\, T_n x_n^k \Big/ \sum_{k=1}^{r} \mu_n^k(x_n),$$

$$\mu_n^k(x_n) = \begin{cases} \varepsilon + \eta - \|x_n - x_n^k\|, & \text{falls} \quad \|x_n - x_n^k\| \leqq \varepsilon, \\ \eta, & \text{falls} \quad \|x_n - x_n^k\| > \varepsilon. \end{cases}$$

Es ist leicht zu sehen, daß

$$x_n \in \bar{\Omega}_n, \quad x_n \xrightarrow{P} x \Rightarrow s_n x_n \xrightarrow{P} Sx;$$

$$x_n \in \bar{\Omega}_n \quad (n \in N' \subseteqq N) \Rightarrow (S_n x_n)_{n \in N'} \quad P\text{-kompakt}.$$

Aus dem Hauptergebnis der Arbeit [2] folgt dann, daß S_n für fast alle n auf $\partial\Omega_n$ keinen Fixpunkt hat und

$$\gamma(I_n - S_n; \partial\Omega_n) = \gamma(I - S; \partial\Omega) \qquad (n \geqq n_0). \tag{11}$$

In unserem Fall kann man (11) leicht auch unmittelbar beweisen.

III. Wir behaupten, daß T_n für fast alle n auf $\partial\Omega_n$ keinen Fixpunkt hat und

$$\gamma(I_n - T_n; \partial\Omega_n) = \gamma(I_n - S_n; \partial\Omega_n) \qquad (n \geqq n_0). \tag{12}$$

Nämlich zeigen wir, daß die vollstetigen Vektorfelder $x_n - T_n x_n$ und $x_n - S_n x_n$ auf $\partial\Omega_n$ für fast alle n linear homotop sind. Nehmen wir das Gegenteil an: Für gewisse $x_n \in \partial\Omega_n$ und $\lambda_n \in [0, 1]$ gilt

$$x_n = \lambda_n T_n x_n + (1 - \lambda_n) S_n x_n \qquad (n \in N' \subseteqq N). \tag{13}$$

O. B. d. A. nehmen wir an, daß $\lambda_n \to \lambda \in [0, 1]$ $(n \in N')$.

Ist $\lambda = 0$, so ist die Folge (x_n) P-kompakt, $x_n \xrightarrow{P} x \in \partial\Omega$ $(n \in N'' \subset N')$, und der Grenzübergang in (13) gibt $x = Sx$. Damit gilt $x \in \Sigma$ und

$$\|x - Tx\| = \|Sx - Tx\| < \alpha \qquad \text{(siehe (8))},$$

was (6) widerspricht.

Nun sei $\lambda > 0$. Schreiben wir (13) um:

$$x_n - \lambda T_n x_n = (\lambda_n - \lambda)\, T_n x_n + (1 - \lambda_n)\, S_n x_n \qquad (n \in N'). \tag{14}$$

Die Folge auf der rechten Seite ist P-kompakt. Dank (iv) ist die Folge (x_n) ebenso

P-kompakt; aus (14) sehen wir, daß auch die Folge $(T_n x_n)$ P-kompakt ist. Es konvergiere $x_n \xrightarrow{P} x$, $T_n x_n \xrightarrow{P} x'$ $(n \in N'' \subseteq N')$. Dann gilt $x \in \partial\Omega$, $x' = Tx$, und der Grenzübergang in (13) gibt

$$x = \lambda Tx + (1 - \lambda) Sx \in \text{co} \{Tx, \Sigma\}.$$

Dank (3) erhalten wir $x \in \Sigma$ und

$$\|x - Tx\| = (1 - \lambda) \|Sx - Tx\| < \alpha,$$

was wieder (6) widerspricht. Damit ist (12) bewiesen.

Aus (12), (11) und (9) schließen wir (10).

Der Satz 1 ist bewiesen.

Satz 2. *Die Bedingungen des Satzes 1 seien erfüllt, und es sei* $\gamma(I - T; \partial\Omega) \neq 0$. *Dann hat* T_n *für fast alle* n *mindestens einen Fixpunkt* x_n^* *in* $\bar{\Omega}_n$. *Jede Fixpunktfolge* (x_n^*) *ist* P*-kompakt und ihre Häufungspunkte* x^* *sind Fixpunkte von* T. *Insbesondere gilt* $x_n^* \xrightarrow{P} x^*$, *falls* T *in* $\bar{\Omega}$ *nur einen Fixpunkt hat.*

Beweis: Diese Behauptungen folgen unmittelbar aus (10), (iv) und (iii).

3. Bemerkungen

Die Bedingungen der Sätze 1 und 2 kann man etwas abschwächen.

1. Die Operatoren T_n müssen nicht unbedingt kompakt sein. Es genügt, wenn aus der linearen Homotopie des Feldes $x_n - T_n x_n$ mit einem endlichdimensionalen Felde $x_n - S_n x_n$ die Gleichheit entsprechender Rotationen folgt. So ist es beispielsweise für Operatoren T_n, die bezüglich eines „guten" Nichtkompaktheitsmaßes verdichtend sind (genauere Formulierungen siehe in [4]).

2. Der Operator T muß nicht unbedingt eine fundamentale Menge Σ besitzen. Nehmen wir an, daß Σ den Bedingungen (1)—(3) außer der Kompaktheit genügt. Ist die Bedingung (5) mit einem $\varepsilon > 0$ erfüllt, so ist die Ungleichung $\chi(\bar{\Omega} \cap \Sigma) < \varepsilon$ hinreichend für die Gültigkeit der Sätze 1 und 2 (χ ist das Nichtkompaktheitsmaß von Hausdorff). Die Rotation $\gamma(I - T; \partial\Omega)$ können wir dabei mit $\gamma(I - T; \partial\Omega) = \gamma(I - S, \partial\Omega)$ definieren (siehe (7) und (9)).

3. Die Regularitätsbedingung (iv) haben wir ebenso nicht im ganzen Umfang benutzt. Die folgende Formulierung ist genügend:

$$x_n \in \bar{\Omega}_n, \quad 0 < \lambda \leqq 1, \quad d(x_n - \lambda T_n x_n, \text{co} \{0, p_n \Sigma\}) \to 0$$

$$(n \in N') \Rightarrow (x_n) \ P\text{-kompakt} \qquad (d - \text{Abstand}).$$

4. Ist (10) irgendwie bewiesen, so ist für die Gültigkeit des Satzes 2 die folgende Abschwächung der Bedingung (iv) hinreichend:

$$x_n \in \bar{\Omega}_n, \quad x_n - T_n x_n = 0 \qquad (n \in N') \Rightarrow (x_n) \ P\text{-kompakt}.$$

5. Vielleicht scheint es unerwartet, aber der Satz 1 verliert seine Gültigkeit, wenn man anstatt (iv) die Regularität nur für $\lambda = 1$ voraussetzt. Wir bringen ein entsprechendes

Beispiel. Es sei $E = E_n = H$ ein Hilbertraum mit einer orthonormalen Basis $\{e_k\}_{k=1}^{\infty}$; $p_n = I$;

$$\Omega = \Omega_n = \{x \in H : \|x\| < 1\};$$

$$Tx = 2 \sum_{k=1}^{\infty} \xi_k e_{k+1}, \quad T_n x = \frac{2}{n} e_1 + 2 \sum_{k=1}^{n} \xi_k e_{k+1} \quad \text{für} \quad x = \sum_{k=1}^{\infty} \xi_k e_k \in H.$$

Die Menge $\Sigma = \{0\}$ ist Fundamentalmenge für $T : \bar{\Omega} \to H$. Weil $\hat{T}x = 0$ für jedes $x \in \bar{\Omega}$ ist, gilt $\gamma(I - T; \partial\Omega) = 1$. Andererseits folgt aus $T_n x_n = x_n$, daß $x_n = \frac{1}{n} \sum_{k=1}^{n} 2^k e_k \notin \bar{\Omega}_n$, somit gilt

$$\gamma(I_n - T_n; \partial\Omega_n) = 0 \neq 1 = \gamma(I - T; \partial\Omega) \qquad (n \in N).$$

Doch sind die Bedingungen (i)—(iii) und die Regularitätsbedingung für $\lambda = 1$ erfüllt. Für $\lambda = \frac{1}{2}$ ist die Regularitätsbedingung nicht erfüllt.

In diesem Beispiel ist T limeskompakt und $\{0\}$ sein Limeswertebereich. Eine Modifizierung des Beispiels ist möglich, so daß T vollstetig wird:

$$Tx = \sum_{k=1}^{\infty} \frac{\xi_k}{k} e_{k+1}, \quad T_n x_n = \frac{1}{n} e_1 + \sum_{k=1}^{n} \frac{\xi_k}{k} e_{k+1} + 2 \sum_{k=n+1}^{k_n} \xi_k e_{k+1}$$

(k_n genügend groß, so daß aus $T_n x = x$ die Ungleichung $\|x\| > 1$ folgt).

Literatur

[1] Вайникко, Г. М., Компактная аппроксимация операторов и приближенное решение операторных уравнений. Докл. АН СССР **189**, 237—240 (1969).

[2] Вайникко, Г., О приближении неподвижных точек вполне непрерывных операторов. Уч. зап. Тартуск. ун-та **342**, 225—236 (1974).

[3] Забрейко, П. П., М. А. Красносельский, В. В. Стрыгин, О принципах инвариантности вращения. Изв. ВУЗ, Математика 1972, № 5, 51—57 (1972).

[4] Садовский, Б. Н., Предельно компактные и уплотняющие операторы. Успехи матем. наук **27**, 81—146 (1972).

[5] Потапов, А. С., К теории вращения предельно компактных векторных полей. Comment. Math. Univ. Carolinae **15**, 693—710 (1974).

[6] Dugundji, J., An extention of Tietze's theorem. Pacific J. Math. **1**, 353—367 (1951).

[7] Vainikko, G., Über Konvergenzbegriffe für lineare Operatoren in der Numerischen Mathematik. Math. Nachr. (erscheint 1976).

Anschrift des Verfassers:

Тартуский гос. университет
ул. Лийви 2—205
202 400 Тарту
СССР

Maximum principle for control processes governed by mildly nonlinear parabolic equations

Lothar v. Wolfersdorf, Freiberg

Abstract. In the report maximum principles of Pontryagin type for various classes of control processes governed by semilinear parabolic equations of second order with Hölder continuous coefficients are derived by a modification of a method of L. Bittner. Some optimal control problems for systems of parabolic differential and integrodifferential equations of second order and equations of higher order are considered, too.

As is well-known, questions of optimal performance and identification for various technical and biological processes lead to mathematical problems of optimal control for nonlinear initial-boundary value problems of parabolic differential equations [1—6]. In the sequel we report on a method for deriving maximum principles of Pontryagin type for some classes of such control problems by semilinear parabolic equations of second order with Hölder continuous coefficients. Suitable classical integral representations of the solutions of corresponding linear problems are employed to reduce the considered nonlinear initial-boundary value problems to Hammerstein integral equations with weakly singular kernels and to define generalized solutions of the initial-boundary value problems as continuous or bounded measurable solutions of the integral equations. To the obtained optimal control problems for these integral equations we apply a modification of a method of L. Bittner [7] for deriving maximum principles for optimal control problems in Banach spaces, which was given by the author in [8, 9]. If the coefficients of the differential equations are sufficiently smooth, as usual, the H-functions in maximum principle can be expressed by (in similar manner defined) generalized solutions of an adjoint initial-value problem. We deal with boundary and distributed control for differential and integro-differential equations and systems of second order and also for differential equations of higher order, where the controls are bounded measurable or piecewise continuous functions with values in prescribed subsets of Euclidean spaces. There is not supposed that the solutions to the nonlinear initial-boundary value problems exist and are uniquely determined for every admissible control. The report represents a survey of the author's papers [11—13].

We briefly mention some related papers on maximum principles for parabolic differential equations stated under various other assumptions. Maximum principles for control problems of analogous classes of semilinear parabolic equations and systems of second order (with piecewise continuous control functions) are derived by A. I. Egorov [14, 15] and N. N. Golub [16, 17] with a generalization of a method of L. I. Rozonoer for control problems of ordinary differential equations. In N. N. Golub [17, 18] and

M. M. Denn [19] also some more general classes of equations are considered. Further, A. I. Egorov [20] deals with the basic problem of boundary control with bounded measurable control functions for the linear heat transfer equation in one space variable. General methods for deriving maximum principles for control processes of semi-linear partial differential equations with bounded measurable controls are developed by L. Cesari [21] and V. I. Plotnikov [22]. Maximum principles for linear parabolic equations considered as abstract evolution equations of first order in Banach spaces are derived by Ju. V. Egorov [23], A. Friedman [24, 25] and J. L. Lions [26]. Finally, P. K. C. Wang [2], W. L. Brogan [27] and others treat control problems for parabolic equations by means of Bellman's method of dynamic programming.

1. Optimal control for equations of second order with boundary conditions of Neumann type [11, 13]

Let B be a bounded domain of R^n, $n \geqq 2$ with boundary $S \in C^{1+\lambda}$, $\lambda > 0$. We consider the *control process*

$$(Lw)(x, t) = \Phi\big(x, t, w(x, t), u_0(x, t)\big) \quad \text{in} \quad D = B \times (0, T), \tag{1}$$

$$w(x, 0) = 0 \quad \text{in } B, \tag{2}$$

$$\frac{\partial w}{\partial \nu}(x, t) + \beta(x, t)\, w(x, t) = \Psi\big(x, t, w(x, t), u_1(x, t), u_2(t)\big) \quad \text{in} \quad Q = S \times (0, T), \tag{3}$$

where

$$(Lw)(x, t) \equiv \sum_{i,j=1}^{n} a_{ij}(x, t) \frac{\partial^2 w}{\partial x_i\, \partial x_j} + \sum_{i=1}^{n} b_i(x, t) \frac{\partial w}{\partial x_i} + c(x, t)\, w - \frac{\partial w}{\partial t} \tag{4}$$

$$(a_{ij} = a_{ji})$$

is a uniformly parabolic operator of second order in D with Hölder continuous coefficients and $\partial w/\partial \nu$ means the outer conormal derivative of w in $(x, t) \in Q$ (cf. [28, 29]). The given functions Φ, β, Ψ are continuous in their arguments and possess continuous derivatives with respect to w. The control functions u_0, u_1, u_2 are bounded measurable (or piecewise continuous) scalar or vector functions in D, Q, and $(0, T)$, respectively, which satisfy the restrictions

$$u_0(x, t) \in U_0 \text{ a.e. in } D, \quad u_1(x, t) \in U_1 \text{ a.e. in } Q, \quad u_2(t) \in U_2 \text{ a.e. in } (0, T), \tag{5}$$

where U_i, $i = 0, 1, 2$ are arbitrary non-void fixed subsets of Euclidean spaces or (closed) domains depending continuously on (x, t) or t, respectively. (The case of additional distributed controls depending only on x or t, respectively, can be treated, too. See the next section.)

We define generalized solutions $w(x, t)$ to (1)–(3) as functions $w(x, t)$, $(x, t) \in D$, which together with a function $w_1(x, t)$, $(x, t) \in Q$ (the trace of $w(x, t)$ on Q) represent in $\bar{D}$ and $\bar{Q}$, respectively, continuous solutions to the following *system of Hammerstein integral equations*:

$$\{w(x,t), w_1(x,t)\} = -\int_0^t \int_B G(x,t,\xi,\tau)\, \Phi\big(\xi,\tau,w(\xi,\tau),u_0(\xi,\tau)\big)\, d\xi\, d\tau \tag{6}$$

$$+ \int_0^t \int_S G(x,t,\xi,\tau)\, \Psi\big(\xi,\tau,w_1(\xi,\tau),u_1(\xi,\tau),u_2(\tau)\big)\, dS_\xi\, d\tau$$

in D, Q, respectively.

Here $G(x, t, \xi, \tau)$ denotes the Green's function of the corresponding linear initial-boundary value problem (1)—(3) with $\Phi = \Psi = 0$.

Aim of control is to minimize the sum J of the functionals

$$J_0 = \int_0^T \int_B F_0\big(x,t,w(x,t),u_0(x,t)\big)\, dx\, dt, \tag{7a}$$

$$J_1 = \int_0^T \int_S F_1\big(x,t,w(x,t),u_1(x,t),u_2(t)\big)\, dS_x\, dt,\quad J_2 = \int_B F_2\big(x,w(x,T)\big)\, dx, \tag{7b}$$

and (for $\Phi = \Phi(x, t)$ not depending on w and u_0)

$$J_3 = \int_0^T F_3\big(t, w(y_1,t),\ldots,w(y_s,t)\big)\, dt,\quad J_4 = F_4\big(w(y_1,T),\ldots,w(y_s,T)\big), \tag{7c}$$

where $y_l \in B$, $l = 1, \ldots, s$ and F_i, $i = 0, \ldots, 4$ are continuous functions, which possess continuous (in cases $i = 2, 3, 4$ Lipschitz continuous) derivatives with respect to the w arguments. The function $w(x, t)$ in J_0 and J_1 is given by the state functions $w(x, t)$ and $w_1(x, t)$, respectively, whereas $w(x, T)$ in J_2, $w(y_l, t)$ and $w(y_l, T)$, $l = 1, \ldots, s$ in J_3 and J_4 are defined by the corresponding expressions from the right-hand side of (6).

The control problem (5)—(7) is a special case of the *abstract problem* [8—10]

$$\operatorname{Min} J(\xi,\omega) \quad \text{subject to } \xi = T(\xi,\omega),\ (\xi,\omega) \in X \times W, \tag{8}$$

where $\xi = \{w, w_1\}$, $X = C(\bar{D}) \times C(\bar{Q})$, $\omega = \{u_0, u_1, u_2\}$, the set W is the subset of the corresponding bounded measurable (or piecewise continuous) vector functions defined by the constraints (5), and the operator T is given by (6). As auxiliary space we use the space $Y = L_{p_1}(D) \times L_{p_2}(Q)$ with $1 < p_1 < 1 + 2/n$, $1 < p_2 < 1 + 1/n$. (If $\Phi = \Phi(x,t)$ the operator T is defined only by the second equation of (6) with $X = C(\bar{Q})$ and $Y = L_p(Q)$, $1 < p < 1 + 1/n$).

There exist the Fréchet derivatives T_ξ and J_ξ in X and the inverse operator $[I - T_\xi]^{-1}$ for a supposed pair of optimal solutions (ξ^0, ω^0), $\xi^0 = \{w^0, w_1^0\}$, $\omega^0 = \{u_0^0, u_1^0, u_2^0\}$ to (8), and they can be extended to linear bounded operators or functionals in Y, respectively. Moreover, the relation

$$\|T(\xi + h,\omega) - T(\xi,\omega) - T_\xi(\xi,\omega)\, h\|_Y = o(\|h\|_Y) \quad \text{as} \quad \|h\|_X \to 0$$

for any $\xi \in X$ holds, and some continuity relations in ξ for the norms of the differences

$$T_\xi(\xi,\omega) - T_\xi(\xi^0,\omega), \qquad J_\xi(\xi,\omega) - J_\xi(\xi^0,\omega)$$

in the spaces X and Y are valid uniformly in ω. Let $\{\omega_k\}$, $k = 1, 2, \ldots$ be a sequence of Pontryagin's variations of the optimal control triplet $\omega^0 = \{u_0^0, u_1^0, u_2^0\}$ with respect to u_0, u_1 and u_2, respectively, e.g. $\omega^k = \{u_0^k, u_1^0, u_2^0\}$ with

$$u_0^k(x, t) = \begin{cases} u_0 \in U_0 & \text{in a neighbourhood } V_k \text{ of a fixed point } (\bar{x}, \bar{t}) \in D \\ u_0^0(x, t) & \text{else in } D \end{cases}$$

and $\gamma_k = \operatorname{mes} V_k > 0$, and analogously for u_1^0 and u_2^0. Then for $\gamma_k \to 0$ we have

$$\|T(\xi^0, \omega_k) - T(\xi^0, \omega^0)\|_X \to 0, \quad \|T_\xi(\xi^0, \omega_k) - T_\xi(\xi^0, \omega^0)\|_{X,Y} \to 0,$$

$$\frac{1}{\gamma_k} \|J_\xi(\xi^0, \omega_k) - J_\xi(\xi^0, \omega^0)\|_X \text{ is bounded,}$$

and there exist the limits

$$\frac{1}{\gamma_k} [T(\xi^0, \omega_k) - T(\xi^0, \omega^0)] \to \delta T \text{ in } Y, \quad \frac{1}{\gamma_k} [J(\xi^0, \omega_k) - J(\xi^0, \omega^0)] \to \delta J.$$

In the proofs of these relations we make essential use of the well-known estimations

$$\begin{aligned} |G(x, t, \xi, \tau)| &\leqq C\,|t - \tau|^{-n/2} \exp\{-c\,|x - \xi|^2/|t - \tau|\} \quad (C, c > 0) \\ &\leqq \text{const}\,|t - \tau|^{-\mu}\,|x - \xi|^{2\mu - n}, \qquad \mu = n/(n + 1) \end{aligned} \tag{9}$$

for the Green's function (cf. [28, 29]).

There follows the inequality

$$\delta J + \zeta(\delta T) \geqq 0, \tag{10}$$

where ζ is the uniquely determined solution to the adjoint variational equation

$$\zeta = T_\xi^*(\xi^0, \omega^0)\,\zeta + J_\xi(\xi^0, \omega^0) \quad \text{in} \quad Y^* = L_{q_1}(D) \times L_{q_2}(Q) \tag{11}$$

with T_ξ^* the operator adjoint to T_ξ in Y and q_i the exponents conjugate to p_i, $i = 1, 2$. (10) with (11) yields the maximum conditions for $\omega^0 = \{u_0^0, u_1^0, u_2^0\}$:

$$\operatorname*{Max}_{u_i \in U_i} H_i(x, t, u_i) = H_i\big(x, t, u_i^0(x, t)\big) \quad \text{for a.e.} \quad (x, t) \in D, Q \text{ or } t \in (0,T) \tag{12}$$

$$(i = 0, 1, 2)$$

with the *H-functions*

$$\begin{aligned} H_0(x, t, u_0) = &-F_0\big(x, t, w^0(x, t), u_0\big) \\ &+ \Phi\big(x, t, w^0(x, t), u_0\big) \cdot z(x, t), \; (x, t) \in D, \end{aligned} \tag{13a}$$

$$\begin{aligned} H_1(x, t, u_1) = &-F_1\big(x, t, w^0(x, t), u_1, u_2^0(t)\big) \\ &- \Psi\big(x, t, w^0(x, t), u_1, u_2^0(t)\big)\, z(x, t), \quad (x,t) \in Q, \end{aligned} \tag{13b}$$

$$H_2(t, u_2) = -\int_S F_1\big(x, t, w^0(x, t), u_1{}^0(x, t), u_2\big)\, dS_x$$

$$-\int_S \Psi\big(x, t, w^0(x, t), u_1{}^0(x, t), u_2\big) \cdot z(x, t)\, dS_x, \qquad t \in (0, T). \tag{13c}$$

The function $z(x, t)$, $(x, t) \in D \cup Q$, is given by

$$z(x, t) = \int_B G(\xi, T, x, t)\, F_{2w}\big(\xi, w^0(\xi, T)\big)\, d\xi$$

$$+ \int_t^T \int_B G(\xi, \tau, x, t)\, \zeta(\xi, \tau)\, d\xi\, d\tau + \int_t^T \int_S G(\xi, \tau, x, t)\, \zeta_1(\xi, \tau)\, dS_\xi\, d\tau$$

$$+ \sum_{l=1}^{s} \Bigg[\int_t^T G(y_l, \tau, x, t)\, F_{3w_l}\big(\tau, w^0(y_1, \tau), \ldots\big)\, d\tau$$

$$+ G(y_l, T, x, t)\, F_{4w_l}\big(w^0(y_1, T), \ldots\big)\Bigg], \tag{14}$$

where $\zeta(x, t) \in L_{q_1}(D)$, $\zeta_1(x, t) \in L_{q_2}(Q)$ represent the solutions of the following linear integral equation system

$$\zeta(x, t) = F_{0w}\big(x, t, w^0(x, t), u_0{}^0(x, t)\big) - \Phi_w\big(x, t, w^0(x, t), u_0{}^0(x, t)\big) \cdot z(x, t) \quad \text{in } D,$$

$$\zeta_1(x, t) = F_{1w}\big(x, t, w^0(x, t), u_1{}^0(x, t), u_2{}^0(t)\big) + \Psi_w\big(x, t, w^0(x, t), u_1{}^0(x, t), u_2{}^0(t)\big) \cdot z(x, t)$$

$$\text{in } Q. \tag{15}$$

If the coefficients $a_{ij}(x, t)$ and $b_i(x, t)$ possess Hölder continuous derivatives of second and first order, respectively, the function $z(x, t)$ can be interpreted as (in analogous way as above defined) generalized solution to the *adjoint initial-boundary value problem*

$$(L^*z)(x, t) = -F_{0w}\big(x, t, w^0(x, t), u_0{}^0(x, t)\big) + \Phi_w\big(x, t, w^0(x, t), u_0{}^0(x, t)\big) \cdot z(x, t)$$

$$- \sum_{l=1}^{s} F_{3w_l}\big(t, w^0(y_1, t), \ldots\big) \cdot \delta(x - y_l) \quad \text{in } D, \tag{16}$$

$$z(x, T) = F_{2w}\big(x, w^0(x, T)\big) + \sum_{l=1}^{s} F_{4w_l}\big(w^0(y_1, T), \ldots\big) \cdot \delta(x - y_l) \quad \text{in } B, \tag{17}$$

$$\frac{\partial z}{\partial \nu}(x, t) + \gamma(x, t)\, z(x, t) = F_{1w}\big(x, t, w^0(x, t), u_1{}^0(x, t), u_2{}^0(t)\big)$$

$$+ \Psi_w\big(x, t, w^0(x, t), u_1{}^0(x, t), u_2{}^0(t)\big) \cdot z(x, t) \text{ in } Q, \tag{18}$$

where L^* is the (formally) adjoint differential operator to L, and $\delta(x - y_l)$ the delta distribution with support y_l, and

$$\gamma(x, t) = \beta(x, t) + \sum_{i=1}^{n} \left(\sum_{j=1}^{n} \frac{\partial a_{ij}(x, t)}{\partial x_j} - b_i(x, t)\right) \cos(N_x, x_i)$$

with N_x the outer normal in $x \in S$.

Remarks. 1. The functions Φ, Ψ, F_i, $i = 0, \ldots, 3$ and their derivates with respect to the w arguments need only be bounded and measurable with respect to (x, t) and continuous or Lipschitz continuous, respectively, with respect to the other variables uniformly in (x, t).

2. It is also possible to prescribe the inhomogeneous initial condition

$$w(x, 0) = \varphi(x) \quad \text{in } B \tag{2'}$$

with given bounded measurable function $\varphi(x)$ on B instead of (2). In this case the expression

$$\int_B G(x, t, \xi, 0)\, \varphi(\xi)\, d\xi$$

has to be added in the right sides of (6), and the whole analysis has to be done in the corresponding space X of bounded measurable functions in D and Q.

3. If $n \leqq 3$ the functional J_3 is also possible for general Φ depending on w and u_0.

4. Analogous assertions as above are valid in case $n = 1$, where the functional J_4 is allowed for general Φ, too.

5. Still another kind of functionals will be considered:

$$\hat{J} = \int_0^T \hat{F}\big(t, y_1(t), \ldots, y_r(t)\big) dt, \quad y_k(t) = \int_B A_k(x, t)\, w(x, t)\, dx \quad (k = 1, \ldots, r), \tag{19}$$

where $A_k(x, t)$, $k = 1, \ldots, r$ are bounded measurable functions in D and the function $\hat{F}$ is bounded and measurable with respect to t and continuously differentiable with respect to the y_k uniformly in t. This leads to the additional term

$$\pm \sum_{k=1}^{r} A_k(x, t)\, \hat{F}_{y_k}\big(t, y_1^0(t), \ldots, y_r^0(t)\big)$$

in formula (15) for $\varphi(x, t)$ and the right side of the adjoint equation (16), respectively.

6. We wish to point out also, that functionals of the more general form

$$\tilde{J} = \int_0^T \int_S \tilde{F}\big(x, t, w(x, t), \partial w/\partial \nu(x, t), u_1(x, t), u_2(t)\big) dS_x\, dt$$

can be reduced to functionals of the form J_1 using the boundary condition (3) in its classical meaning.

Finally, we remark that instead of the Green's function also a *fundamental solution* $\Gamma(x, t, \xi, \tau)$ of the suitably extended operator L can be employed to our problem with $\Phi = \Phi(x, t)$ and homogeneous initial condition (2). Then, generalized solutions $w(x, t)$ to (1)—(3) are defined as functions $w(x, t)$, $(x, t) \in D$, which together with functions $w_1(x, t)$ and $\varrho(x, t)$, $(x, t) \in Q$, represent in $\bar{D}$ and $\bar{Q}$, respectively, continuous solutions to the following *system of integral equations*:

$$\{w(x, t), w_1(x, t)\} = -\int_0^t \int_B \Gamma(x, t, \xi, \tau)\, \Phi(\xi, \tau)\, d\xi\, d\tau + 2 \int_0^t \int_S \Gamma(x, t, \xi, \tau) \big[\varrho(\xi, \tau)$$

$$+ \Psi\big(\xi, \tau, w_1(\xi, \tau), u_1(\xi, \tau), u_2(\tau)\big)\big]\, dS_\xi\, d\tau \quad \text{in } D \text{ and } Q, \text{ respectively,} \tag{20}$$

$$\varrho(x,t) = \int_0^t \int_B \left[\frac{\partial \Gamma(x,t,\xi,\tau)}{\partial \nu(x,t)} + \beta(x,t)\,\Gamma(x,t,\xi,\tau)\right] \Phi(\xi,\tau)\,d\xi\,d\tau$$

$$- 2\int_0^t \int_S \left[\frac{\partial \Gamma(x,t,\xi,\tau)}{\partial \nu(x,t)} + \beta(x,t)\,\Gamma(x,t,\xi,\tau)\right] \big[\varrho(\xi,\tau)$$

$$+ \Psi\big(\xi,\tau,w_1(\xi,\tau),u_1(\xi,\tau),u_2(\tau)\big)\big]\,dS_\xi\,d\tau \quad \text{in } Q. \tag{21}$$

The parameters p_i, $i = 1, 2, 3$ in $Y = L_{p_1}(D) \times L_{p_2}(Q) \times L_{p_3}(Q)$ must then be chosen as $1 < p_1 < 1 + 2/n$, $1 < p_2 < 1 + 1/n$, $1 < p_3 < 1 + \beta/(n + 1 - \beta)$ with $\beta = \min(\lambda, \alpha)$, where α is the common Hölder exponent of the coefficients of L (cf. [28]). By means of fundamental solutions also analogous problems for the equation (1) in non cylindrical domains can be handled, for instance in case $n = 1$ in domains bounded by Gevrey curves (cf. [30]).

The above control problem for the one-dimensional heat transfer operator is dealt with separately in [10] together with some minor modifications of the problem. Some complements to this case are given by F. Unger [31, 32]. Especially in [31] a boundary control problem for an infinite cylinder is considered.

2. Distributed control for equations of second order with boundary conditions of Dirichlet type [11, 13]

We retain the notations and assumptions of the preceding section and deal with the following *control process*:

$$(Lw)(x,t) = \Phi\left(x,t,w(x,t),\frac{\partial w}{\partial x}(x,t),u_0(x,t),u_1(x),u_2(t)\right) \text{ in } D = B \times (0,T), \tag{22}$$

$$w(x,0) = 0 \quad \text{in } B, \tag{23}$$

$$w(x,t) = 0 \quad \text{in } Q = S \times (0,T), \tag{24}$$

where $\partial w/\partial x = \{\partial w/\partial x_1, \dots, \partial w/\partial x_n\}$ and the distributed control functions u_0, u_1, u_2 are now defined in D, B and $(0, T)$, respectively. (The case of homogeneous boundary conditions of second or third kind on Q instead of (24) can be treated in the same way.) And Φ shall be continuously differentiable with respect to $\partial w/\partial x_i$, $i = 1, \dots, n$, too. Generalized solutions $w(x, t)$ to (22)—(24) are functions, which together with a n-dimensional vector function $w_1(x, t)$ (the gradient $\partial w/\partial x$ of $w(x, t)$ in D) represent in $\bar{D}$ continuous solutions to the following *system of integral equations*:

$$\{w(x,t), w_1(x,t)\} = -\int_0^t \int_B \left\{G(x,t,\xi,\tau), \frac{\partial G(x,t,\xi,\tau)}{\partial x}\right\} \Phi\big(\xi,\tau,w(\xi,\tau),$$

$$w_1(\xi,\tau),u_0(\xi,\tau),u_1(\xi),u_2(\tau)\big)\,d\xi\,d\tau \quad \text{in } D. \tag{25}$$

Here $G(x, t, \xi, \tau)$ means the Green's function for the Dirichlet problem of the operator L constructed by W. Pogorzelski (cf. [29]).

The functional J to be minimized is the sum of the functionals

$$J_0 = \int_0^T \int_B F_0\left(x, t, w(x, t), \frac{\partial w}{\partial x}(x, t), u_0(x, t), u_1(x), u_2(t)\right) dx\, dt, \tag{26a}$$

$$J_1 = \int_0^T \int_S F_1\left(x, t, \frac{\partial w}{\partial \nu}(x, t)\right) dS_x\, dt, \quad J_2 = \int_B F_2(x, w(x, T))\, dx, \tag{26b}$$

where F_i, $i = 0, 1, 2$ are continuous functions, which possess continuous (in cases $i = 1, 2$ Lipschitz continuous) derivatives with respect to the w arguments.

We use the auxiliary space $Y = L_{p_1}(D) \times L_{p_2}^{(n)}(D)$ with $1 < p_1 < 1 + 2/n$ and $1 < p_2 < 1 + 1/(n + 1)$. Because of the estimates (9) and

$$\left|\frac{\partial}{\partial x_i} G(x, t, \xi, \tau)\right| \leqq C|t - \tau|^{-(n+1)/2} \exp\{-c|x - \xi|^2/|t - \tau|\} \qquad (C, c > 0)$$

$$\leqq \text{const}\, |t - \tau|^{-\mu}\, |x - \xi|^{2\mu - n - 1}, \qquad \mu = (n + 1)/(n + 2) \tag{27}$$

the needed assumptions follow. We get maximum conditions with the *H-functions*

$$H_0(x, t, u_0) = -F_0\left(x, t, w^0(x, t), \frac{\partial w^0}{\partial x}(x, t), u_0, u_1{}^0(x), u_2{}^0(t)\right)$$

$$+ \Phi\left(x, t, w^0(x, t), \frac{\partial w^0}{\partial x}(x, t), u_0, u_1{}^0(x), u_2{}^0(t)\right) z(x, t), \quad (x, t) \in D, \tag{28a}$$

$$H_1(x, u_1) = -\int_0^T F_0\left(x, t, w^0(x, t), \frac{\partial w^0}{\partial x}(x, t), u_0{}^0(x, t), u_1, u_2{}^0(t)\right) dt$$

$$+ \int_0^T \Phi\left(x, t, w^0(x, t), \frac{\partial w^0}{\partial x}(x, t), u_0{}^0(x, t), u_1, u_2{}^0(t)\right) z(x, t)\, dt, \quad x \in B, \tag{28b}$$

$$H_2(t, u_2) = -\int_B F_0\left(x, t, w^0(x, t), \frac{\partial w^0}{\partial x}(x, t), u_0{}^0(x, t), u_1{}^0(x), u_2\right) dx$$

$$+ \int_B \Phi\left(x, t, w^0(x, t), \frac{\partial w^0}{\partial x}(x, t), u_0{}^0(x, t), u_1{}^0(x), u_2\right) z(x, t)\, dx, \quad t \in (0, T). \tag{28c}$$

The function $z(x, t)$, $(x, t) \in D \cup B$ is given by

$$z(x, t) = \int_B G(\xi, T, x, t)\, F_{2w}\big(\xi, w^0(\xi, T)\big)\, d\xi + \int_t^T \int_B G(\xi, \tau, x, t)\, \zeta(\xi, \tau)\, d\xi\, d\tau$$

$$+ \sum_{k=1}^{n} \int_t^T \int_B \frac{\partial G}{\partial \xi_k}(\xi, \tau, x, t)\, \zeta_k(\xi, \tau)\, d\xi\, d\tau, \tag{29}$$

where $\zeta(x, t) \in L_{q_1}(D)$, $\zeta_k(x, t) \in L_{q_2}(D)$, $k = 1, \ldots, n$ represent the solutions of the following integral equation system

$$\zeta(x, t) = F_{0w}\left(x, t, w^0(x, t), \frac{\partial w^0}{\partial x}(x, t), u_0{}^0(x, t), u_1{}^0(x), u_2{}^0(t)\right)$$
$$- \Phi_w\big(x, t, w^0(x, t), \ldots\big)\, z(x, t) \quad \text{in } D, \tag{30}$$
$$\zeta_k(x, t) = F_{0\partial w/\partial x_k}\big(x, t, w^0(x, t), \ldots\big)$$
$$- \Phi_{\partial w/\partial x_k}\big(x, t, w^0(x, t), \ldots\big)\, z(x, t) \quad \text{in } D, \qquad k = 1, \ldots, n.$$

If the functions Φ and F_0 do not depend on $\partial w/\partial x$ and the coefficients of the operator L fulfil additional smoothness assumptions, as above, the function $z(x, t)$ can be interpreted as generalized solution to the *adjoint problem*

$$(L^*z)(x, t) = -F_{0w}\big(x, t, w^0(x, t), u_0{}^0(x, t), u_1{}^0(x), u_2{}^0(t)\big)$$
$$+ \Phi_w\big(x, t, w^0(x, t), \ldots\big)\, z(x, t) \quad \text{in } D, \tag{31}$$

$$z(x, T) = F_{2w}\big(x, w^0(x, T)\big) \quad \text{in } B, \tag{32}$$

$$z(x, t) = -F_{1\partial w/\partial \nu}\left(x, t, \frac{\partial w^0}{\partial \nu}(x, t)\right) \quad \text{in } Q. \tag{33}$$

Remarks. 1. The remarks 1, 2, 5 of the preceding section apply also here, where for inhomogeneous initial condition the functions Φ and F_0 may not depend on $\partial w/\partial x$ and the functional J_1 is not allowed. (For including functionals J_1 see later in section 4.)

2. For $n \leq 3$ and Φ not depending on $\partial w/\partial x$ or for $n \leqq 2$, respectively, also functionals

$$J_3 = \int_0^T F_3\big(t, w(y_1, t), \ldots, w(y_s, t)\big)\, dt \quad \text{with} \quad y_l \in B, \quad l = 1, \ldots, s$$

are possible, which leads to an additional term in (29) and (31) as in (14) and (16), respectively. Further for $n = 1$ also functionals

$$J_3' = \int_0^T F_3'\left(t, w(y_1, t), \ldots, w(y_s, t), \frac{\partial w}{\partial x}(y_1, t), \ldots, \frac{\partial w}{\partial x}(y_s, t)\right) dt, \tag{34}$$

and if Φ does not depend on $\partial w/\partial x$

$$J_4 = F_4(w(y_1, T), \ldots, w(y_s, T))$$

are allowed. The first functional yields an additional term in (29) analogously as J_3 and the last one an additional term in (29) and (32) as in (14) and (17).

Finally, in case $n = 1$ the following *problem of point control* can be handled:

$$(Lw)\,(x, t) = \Phi(x, t, w(x, t)) + \sum_{k=1}^{r} g_k(t, u_k(t))\, \delta(x - x_k) \quad \text{in } (0, 1) \times (0, T), \tag{35}$$

$$w(x, 0) = 0 \quad \text{in } (0, 1), \tag{36}$$

and

$$w(0, t) = w(1, t) = 0 \quad \text{in } (0, T) \tag{37}$$

(or other linear homogeneous boundary conditions), where $0 < x_k < 1$, $k = 1, \ldots, r$, $g_k(t, u_k)$ are continuous functions and the values $u_k(t)$ of the bounded measurable controls u_k are restricted to some fixed subsets of the real axis or to some intervals depending continuously on t. By definition this problem is equivalent to the *integral equation*

$$\begin{aligned} w(x, t) = &-\int_0^t \int_0^1 G(x, t, \xi, \tau)\, \Phi(\xi, \tau, w(\xi, \tau))\, d\xi\, d\tau \\ &- \sum_{k=1}^{r} \int_0^t G(x, t, x_k, \tau)\, g_k(\tau, u_k(\tau))\, d\tau \quad \text{in } (0, 1) \times (0, T) \end{aligned} \tag{38}$$

with the corresponding Green's function $G(x, t, \xi, \tau)$. The functionals are analogous to above:

$$J_0 = \int_0^T \int_0^1 F_0(x, t, w(x, t), u_1(t), \ldots, u_r(t))\, dx\, dt, \tag{39a}$$

$$J_1 = \int_0^T F_1\left(t, \frac{\partial w}{\partial x}(0, t), \frac{\partial w}{\partial x}(1, t)\right) dt, \qquad J_2 = \int_0^1 F_2\,(x, w(x, T))\, dx, \tag{39b}$$

$$J_3 = \int_0^T F_3(t, w(y_1, t), \ldots, w(y_s, t))\, dt, \qquad J_4 = F_4(w(y_1, T), \ldots, w(y_s, T)),$$

$$\text{where } \; 0 < y_l < 1, \qquad l = 1, \ldots, s. \tag{39c}$$

Then we get maximum principles for optimal controls u_k^0, $k = 1, \ldots, r$ with the *H-functions*

$$H_k(t, u_k) = -\int_0^1 F_0(x, t, w^0(x, t), u_1^0(t), \ldots, u_k, \ldots, u_r^0(t))\, dt + g_k(t, u_k) \cdot z(x_k, t), \tag{40}$$

where $z(x, t)$ is the solution to the corresponding adjoint problem.

3. Optimal control for equations of higher order and systems of second order [11, 13]

Let B be a bounded domain of R^n, $n \geqq 2$, with boundary $S \in C^{2m+\lambda}$, $\lambda > 0$ for some $m = 1, 2, \ldots$ now. In $D = B \times (0, T)$ the differential operator

$$(Lw)(x, t) = \sum_{|\nu| \leqq 2m} a_\nu(x, t) \left(\frac{\partial w}{\partial x}\right)^\nu - \frac{\partial w}{\partial t}, \qquad \nu = (\nu_1, \ldots, \nu_n)$$

and on $Q = S \times (0, T)$ the boundary operators

$$B_j\left(t, x, \frac{\partial}{\partial x}\right) = \sum_{|\nu| \leqq r_j} b_{j\nu}(x, t) \left(\frac{\partial}{\partial x}\right)^\nu, \quad j = 1, \ldots, m \quad (0 \leqq r_j \leqq 2m - 1)$$

are given, which are supposed to satisfy the conditions of R. Arima [33]. (See also S. D. Ivasišen, S. D. Eidel'man [34] and S. D. Ivasišen [35] for a treatment of corresponding parabolic systems.) Then the following *problem of distributed control* can be dealt with as in the preceding section:

$$(Lw)(x, t) = \Phi\left(x, t, w(x, t), \frac{\partial w}{\partial x}, \ldots, \frac{\partial^{2m-1} w}{\partial x^{2m-1}}, u_0(x, t), u_1(x), u_2(t)\right) \text{ in } D, \quad (41)$$

$$w(x, 0) = 0 \quad \text{in } B, \quad (42)$$

$$B_j\left(t, x, \frac{\partial w}{\partial x}\right) = 0 \quad \text{in } Q, \qquad j = 1, \ldots, m \quad (43)$$

with

$$J = \int_0^T \int_B F\left(x, t, w(x, t), \frac{\partial w}{\partial x}, \ldots, \frac{\partial^{2m-1} w}{\partial x^{2m-1}}, u_0(x, t), u_1(x), u_2(t)\right) dx\, dt$$
$$+ \int_B G\big(x, w(x, T)\big)\, dx = \text{Min!} \quad (44)$$

For restricted values of n also functionals of the form J_3 and J_4 of the preceding section can be included.

We only consider two types of *parabolic systems of second order*, which contain the differential operators

$$(L_k w)(x, t) = \sum_{i,j=1}^n a_{ij}^k(x, t) \frac{\partial^2 w}{\partial x_i \partial x_j} + \sum_{i=1}^n b_i^k(x, t) \frac{\partial w}{\partial x_i}$$
$$+ c^k(x, t)\, w - \frac{\partial w}{\partial t} \qquad (k = 1, \ldots, m). \quad (45)$$

The coefficients of these operators shall fulfil analogous assumptions as the coefficients of the operator L in the preceding sections. We employ the abbreviations

$$W(x, t) = \big(w_1(x, t), \ldots, w_m(x, t)\big) \quad \text{in } D$$

and for fixed $r \in \{0, \ldots, m\}$

$$\hat{W}(x, t) = \big(\hat{w}_1(x, t), \ldots, \hat{w}_r(x, t)\big) \text{ on } Q, \quad \widetilde{W}(x, t) = \big(\tilde{w}_{r+1}(x, t), \ldots, \tilde{w}_m(x, t)\big) \text{ in } D,$$

where $\hat{w}_k(x, t)$, $k = 1, \ldots, r$ are the boundary values of $w_k(x, t)$ on $Q = S \times (0, T)$ and $\tilde{w}_k(x, t) = \partial w_k/\partial x$, $k = r + 1, \ldots, m$ is the gradient of $w_k(x, t)$ in D.

Problem I:

$$(L_k w_k)(x, t) = \Phi_k\big(x, t, W(x, t), \widetilde{W}(x, t), u_0(x, t)\big) \text{ in } D, \quad k = 1, \ldots, m, \tag{46}$$

$$w_k(x, 0) = 0 \quad \text{in } B, \qquad k = 1, \ldots, m; \tag{47}$$

$$\frac{\partial w_k}{\partial \nu_k}(x, t) + \beta_k(x, t)\, w_k(x, t) = \Psi_k\big(x, t, \hat{W}(x, t), u_1(x, t), u_2(t)\big)$$

$$\text{in } Q, \quad k = 1, \ldots, r, \tag{48}$$

$$w_k(x, t) = 0 \quad \text{in } Q, \qquad k = r + 1, \ldots, m, \tag{49}$$

where $0 \leqq r \leqq m$. The functionals have similar form as in the preceding sections, especially

$$J_0 = \int_0^T \int_B F_0\big(x, t, W(x, t), \widetilde{W}(x, t), u_0(x, t)\big)\, dx\, dt \tag{50a}$$

and

$$J_1 = \int_0^T \int_S F_1\big(x, t, \hat{W}(x, t), u_1(x, t), u_2(t)\big)\, dS_x\, dt. \tag{50b}$$

Problem II:

$$(L_k w_k)(x, t) = \Phi_k\big(x, t, W(x, t), \widetilde{W}(x, t), Y(t), u(x, t), v(t)\big)$$

$$\text{in } D, \qquad k = 1, \ldots, m, \tag{51}$$

$$w_k(x, 0) = 0 \quad \text{in } B, \qquad w_k(x, t) = 0 \quad \text{in } Q, \qquad k = 1, \ldots, m, \tag{52}$$

where $\widetilde{W}(x, t) = \partial W/\partial x$ in D, $u(x, t)$ and $v(t)$ are the controls, and $Y(t) = \big(y_1(t), \ldots, y_l(t)\big)$, $t \in [0, T]$ is the solution to the initial value problem

$$\frac{dy_j}{dt} = \Psi_j\big(t, Y(t), v(t)\big) \quad \text{in } (0, T), \qquad y_j(0) = y_j^0, \qquad j = 1, \ldots, l. \tag{53}$$

The functionals J_0, J_1, J_2 are now given by

$$J_0 = \int_0^T \int_B F_0\big(x, t, W(x, t), \widetilde{W}(x, t), Y(t), u(x, t), v(t)\big)\, dx\, dt, \tag{54a}$$

$$J_1 = \int_0^T \int_S F_1\left(x, t, \frac{\partial w_1}{\partial \nu_1}(x, t), \ldots, \frac{\partial w_m}{\partial \nu_m}(x, t)\right) dS_x\, dt, \tag{54b}$$

$$J_2 = \int_B F_2\big(x, W(x, T), Y(T)\big)\, dx. \tag{54c}$$

Taking W, $\tilde{W}$, $\hat{W}$ in problem I and W, $\tilde{W}$, Y in problem II as state functions and supposing analogous smoothness assumptions on the given functions as in the preceding sections we get analogous assertions as there for the problems I and II.

4. Optimal control for integro-differential equations of second order [12, 13]

We consider some problems of distributed and boundary control for parabolic integro-differential equations, too. The basic notations and assumptions of the preceding sections remain unchanged.

Problem III a:

$$(L_k w_k)(x, t) = \Phi_k\big(x, t, W(x, t), \tilde{W}(x, t), p(x, t), q(x, t)\big) \quad \text{in } D, \quad k = 1, \ldots, m, \tag{55}$$

$$w_k(x, 0) = 0 \quad \text{in } B, \qquad k = 1, \ldots, m, \tag{56}$$

$$w_k(x, t) = 0 \quad \text{in } Q, \qquad k = 1, \ldots, r, \tag{57}$$

$$\frac{\partial w_k}{\partial \nu_k}(x, t) + \beta_k(x, t)\, w_k(x, t) = 0 \quad \text{in } Q, \qquad k = r + 1, \ldots, m, \tag{58}$$

where $1 \leqq r \leqq m$ and $p(x, t) = \big(p_1(x, t), \ldots, p_s(x, t)\big)$, $q(x, t) = \big(q_1(x, t), \ldots, q_l(x, t)\big)$ are defined by

$$p_i(x, t) = \int_0^T \int_S K_i(x, t, \xi, \tau)\, \Psi_i\big(\xi, \tau, u_0(\xi, \tau), u_1(\tau)\big)\, dS_\xi\, d\tau \quad \text{in } D, \tag{59a}$$

$$q_j(x, t) = \int_0^T \int_B L_j(x, t, \xi, \tau)\, X_j\big(\xi, \tau, W(\xi, \tau), \tilde{W}(\xi, \tau), p(\xi, \tau), q(\xi, \tau)\big)\, d\xi\, d\tau \quad \text{in } D. \tag{60a}$$

Problem III b has the same form as Problem III a with functions Φ_k in (55) not depending on $q(x, t)$ and $p(x, t)$, $q(x, t)$ are now defined by

$$p_i(x, t) = \int_0^T \int_S K_i\big(x, t, \xi, \tau)\, \Psi_i(\xi, \tau, q(\xi, \tau), u_0(\xi, \tau), u_1(\tau)\big)\, dS_\xi\, d\tau \quad \text{in } D, \tag{59b}$$

$$q_j(x, t) = \int_0^T \int_B L_j(x, t, \xi, \tau)\, X_j\big(\xi, \tau, W(\xi, \tau), \tilde{W}(\xi, \tau), p(\xi, \tau)\big)\, d\xi\, d\tau \quad \text{in } Q. \tag{60b}$$

The functions Ψ_i in (59a, b) are also allowed to depend on $\hat{W}(x, t)$, too.

Problem IV is given by (55)—(58) with

$$p_i(x, t) = \int_0^T \int_B K_i(x, t, \xi, \tau)\, \Psi_i\big(\xi, \tau, W(\xi, \tau), \tilde{W}(\xi, \tau), p(\xi, \tau), u_0(\xi, \tau), u_1(\tau), u_2(\xi)\big)\, d\xi\, d\tau \quad \text{in } D \tag{61}$$

and

$$\frac{\partial q_j(x,t)}{\partial t} = X_j\big(x, t, W(x,t), \widetilde{W}(x,t), p(x,t), q(x,t), u_1(t)\big),$$

$$q_j(x, 0) = q_j^0(x) \text{ in } D. \tag{62}$$

Additional differential equations of the form (62) are also possible in problems III a, b.

The functionals to be minimized are analogous to the ones in the preceding sections.

We make the following *additional assumptions*:

1. The kernels $K_i(x, t, \xi, \tau)$, $i = 1, \dots, s$ and $L_j(x, t, \xi, \tau)$, $j = 1, \dots, l$ are summable in their domains $D \times Q$ and $D \times D$, respectively, summable with respect to (ξ, τ) uniformly in (x, t) and summable to power $p > 1$ with respect to (x, t) uniformly in (ξ, τ), i.e. in problems III a, b e.g.

$$\sup_{(x,t)\in D} \int_0^T\int_S |K_i(x, t, \xi, \tau)|\, dS_\xi\, d\tau < \infty, \quad \sup_{(\xi,\tau)\in Q} \int_0^T\int_B |K_i(x, t, \xi, \tau)|^p\, dx\, dt < \infty, \tag{63}$$

and finally continuous in the mean with power 1 with respect to (x, t) and with power $p > 1$ with respect to (ξ, τ), i.e. in problems III a, b e.g.

$$\lim_{(x,t)\to(x_0,t_0)} \int_0^T\int_S |K_i(x, t, \xi, \tau) - K_i(x_0, t_0, \xi, \tau)|\, dS_\xi\, d\tau = 0 \quad \forall (x_0, t_0) \in \bar{D} \tag{64}$$

and

$$\lim_{(\xi,\tau)\to(\xi_0,\tau_0)} \int_0^T\int_B |K_i(x, t, \xi, \tau) - K_i(x, t, \xi_0, \tau_0)|^p\, dx\, dt = 0 \quad \forall (\xi_0, \tau_0) \in Q. \tag{65}$$

(The last and the corresponding conditions need be fulfilled only for almost every (ξ_0, τ_0)).

2. The functions X_j, $j = 1, \dots, l$ are continuous with respect to their arguments and continuously differentiable with respect to the components of $W, \widetilde{W}, p, q$. The functions $q_j^0(x)$, $j = 1, \dots, l$ are bounded and measurable.

Then it is possible to introduce $p(x, t)$ and $q(x, t)$ as *additional state functions* besides $W(x, t)$ and $\widetilde{W}(x, t)$. Under the assumption that no bifurcation is present for the optimal controls and states we get maximum principles for the problems III a, b and IV as above. In particular this is always the case, if the kernels $K_i(x, t, \xi, \tau)$ and $L_j(x, t, \xi, \tau)$ are Volterra kernels with the above mentioned properties.

In some applications the assumption (64) or the corresponding assumption in problem IV is not fulfilled. Therefore it is sometimes not possible to consider $p(x, t)$ as additional state function, but it has to be put directly in the right sides of the differential equations (55). We only consider the problem III a, b without $q(x, t)$ here. Then our method works with state functions $W(x, t)$ and $\widetilde{W}(x, t)$, if the conditions (63) and (65) are fulfilled and the derivatives Φ_{kw_j}, $\Phi_{k\tilde{w}_j}$ and Φ_{kp_i} are Lipschitz continuous with respect to p now. Possible kinds of functionals are

$$J_0 = \int_0^T\int_B F_0\big(x, t, W(x,t), \widetilde{W}(x,t), p(x,t)\big)\, dx\, dt, \tag{66}$$

$$J_1 = \int_0^T \int_S F_1\left(x, t, \frac{\partial w_1}{\partial v_1}, \ldots, \frac{\partial w_r}{\partial v_r}, w_{r+1}, \ldots, w_m, p^1(x, t)\right) dS_x\, dt, \tag{67}$$

$$J_2 = \int_B F_2\big(x, W(x, T), p^2(x)\big)\, dx, \tag{68}$$

where the vector functions $p^1(x, t)$, $p^2(x)$ are defined analogously to $p(x, t)$ and also the derivatives of the functions F_0, F_1, F_2 with respect to the w- and p-arguments satisfy corresponding Lipschitz conditions. (For details see [12]). Also functionals of types J_3 and J_4 of the preceding sections are allowed (if there is no dependence of Φ_k on $\tilde{W}$) for $n \leqq 3$ and $n = 1$, respectively.

As an example for the last case we mention the following boundary control problem (problems of such type especially occur in control of biochemical systems, cf. [6]):

$$(L_1 w_1)(x, t) = \Phi_1\big(x, t, w_1(x, t), w_0(x, t)\big) \quad \text{in } D, \tag{69a}$$

$$(L_2 w_0)(x, t) = \Phi_2\big(x, t, w_1(x, t), w_0(x, t)\big) \quad \text{in } D \tag{69b}$$

with

$$w_1(x, 0) = w_0(x, 0) = 0 \quad \text{in } B, \tag{70}$$

$$w_1(x, t) = 0, \qquad w_0(x, t) = \Psi\big(x, t, u_0(x, t), u_1(t)\big) \quad \text{in } Q, \tag{71}$$

where Ψ is a continuous function and the coefficients of L_2 are assumed to be sufficiently smooth. Then we can put

$$w_0(x, t) = w_2(x, t) + p(x, t)$$

with

$$p(x, t) = -\int_0^t \int_S \frac{\partial G_2(x, t, \xi, \tau)}{\partial v_2(\xi, \tau)}\, \Psi\big(\xi, \tau, u_0(\xi, \tau), u_1(\tau)\big)\, dS_\xi\, d\tau, \tag{72}$$

where $G_2(x, t, \xi, \tau)$ denotes the Green's function for the Dirichlet problem of the operator L_2, and obtain

$$(L_k w_k)(x, t) = \Phi_k\big(x, t, w_1(x, t), w_2(x, t) + p(x, t)\big) \quad \text{in } D, \quad k = 1, 2; \tag{69'}$$

with homogeneous initial and boundary conditions for w_k, $k = 1, 2$. The kernel of the integral expression (72) fulfils (63) and (65) for $1 < p < 1 + 1/(n + 1)$, but not (64). Possible functionals are for example

$$J_1 = \int_0^T \int_S F_1\left(x, t, \frac{\partial w_1}{\partial v_1}(x, t)\right) dS_x\, dt, \tag{73}$$

$$J_2 = \int_B F_2\big(x, w_1(x, T), w_0(x, T)\big)\, dx, \tag{74}$$

and with $y_l \in B$, $l = 1, \ldots, s$

$$J_3 = \int_0^T F_3\big(t, w_1(y_1, t), \ldots, w_1(y_s, t), w_0(y_1, t), \ldots, w_0(y_s, t)\big)\, dt \qquad (\text{for } n \leqq 3), \tag{75}$$

$$J_4 = F_4\big(w_1(y_1, T), \ldots, w_1(y_s, T), w_0(y_1, T), \ldots, w_0(y_s, T)\big) \qquad (\text{for } n = 1). \tag{76}$$

In analogous way initial control problems of the following form can be handled as special cases of problem IV: (69a, b) with

$$w_1(x, 0) = 0, \qquad w_0(x, 0) = \Psi\big(x, u(x)\big) \quad \text{in } B, \tag{77}$$

$$w_1(x, t) = w_0(x, t) = 0 \quad \text{in } Q \tag{78}$$

(or homogeneous boundary conditions of third kind).

Finally, we briefly discuss the case of *inhomogeneous initial and boundary values.* In this case difficulties arise by considering functionals of type J_1, if we do not assume strong compatibility conditions on these values. We deal with the following problem (cf. section 2):

$$(Lw)\,(x, t) = \Phi\big(x, t, w(x, t), u_0(x, t), u_1(x), u_2(t)\big) \quad \text{in } D, \tag{79}$$

$$w(x, 0) = \varphi(x) \quad \text{in } B, \qquad w(x, t) = \psi(x, t) \quad \text{in } Q, \tag{80}$$

where $\varphi(x)$, $\psi(x, t)$ are given bounded measurable functions on B and Q, respectively. The functional is a quadratic one of the form

$$J_1 = \int_0^T \int_S a(x, t) \left[\frac{\partial w}{\partial \nu}(x, t) - c(x, t)\right]^2 dS_x\, dt \tag{81}$$

with given bounded measurable functions $a(x, t)$, $c(x, t)$ on Q. We suppose the existence of a bounded measurable function $q(x, t)$ in D with

$$(Lq)\,(x, t) = 0 \text{ in } D, \quad q(x, t) = \psi(x, t) \text{ in } Q, \quad q(x, 0) \text{ continuous and bounded in } B \tag{82}$$

and $\partial q/\partial \nu$ exists and is continuous and bounded on $Q \cap (t \geqq \delta > 0)\ \forall\ \delta > 0$ with

$$\left|\frac{\partial q}{\partial \nu}(x, t)\right| = O(t^{-1/2-\varepsilon}), \quad \varepsilon > 0 \text{ arbitrary}, \quad \text{as } t \to 0 \text{ on } Q. \tag{83}$$

Then for the function

$$p(x, t) = q(x, t) + \int_B G(x, t, \xi, 0)\, [\varphi(\xi) - q(\xi, 0)]\, d\xi \tag{84}$$

with the corresponding Green's function $G(x, t, \xi, 0)$ the analogous estimation to (83) holds. Therefore, if we make the assumption

$$a(x, t) = O(t^{1/2+\varepsilon_0}) \quad \text{for some } \varepsilon_0 > 0, \tag{85}$$

we can put

$$w(x, t) = \overline{w}(x, t) + p(x, t)$$

to obtain a related control problem for $\overline{w}(x, t)$ with homogeneous initial and boundary values. In the same way the above mentioned boundary and initial control problems for differential equations of the form (69a, b) with inhomogeneous initial and boundary conditions for $w_1(x, t)$ and functionals (73) can be dealt with.

As is well known, for the *heat operator* $Lw = w_{xx} - w_t$ in $(0, 1) \times (0, T)$ with the conditions

$$w(x, 0) = \gamma \quad \text{in } (0, 1), \qquad w(0, t) = \alpha \quad \text{and} \quad w(1, t) = \beta \quad \text{in } (0, T) \tag{86}$$

$(\alpha, \beta, \gamma$ constants)

the function $p(x, t)$ in (84) is given by the formula

$$p(x, t) = (\beta - \alpha)\, x + \alpha + (\beta - \alpha) \frac{1}{\pi} \sum_{k=1}^{\infty} \frac{1}{k} \sin 2k\pi x\, e^{-4k^2\pi^2 t}$$
$$+ (2\gamma - \alpha - \beta) \frac{2}{\pi} \sum_{k=0}^{\infty} \frac{1}{2k+1} \sin (2k+1)\, \pi x\, e^{-(2k+1)^2\pi^2 t} \tag{87}$$

with the derivative

$$\frac{\partial p}{\partial x}(x, t) = (2\gamma - \alpha - \beta)\, \vartheta_2(x, e^{-4\pi^2 t}) + (\beta - \alpha)\, \vartheta_3(x, e^{-4\pi^2 t})$$
$$= O(t^{-1/2}) \text{ as } t \to 0, \tag{88}$$

where ϑ_2 and ϑ_3 denote the Jacobi's theta functions of second and third kind, respectively.

References

[1] Бутковский, А. Г., Теория оптимального управления системами с распределёнными параметрами. Изд-во Наука, Москва 1965.

[2] Wang, P. K. C., Control of distributed parameter systems. *In:* Advances in Control Systems, ed. C. T. Leondes, vol. 1, 75—172. Academic Press, New York 1964.

[3] Лионс, Ж. Л., (Lions, J. L.), Об оптимальном управлении распределёнными системами. Успехи Мат. Наук 28, 4 (172), 15—46 (1973).

[4] Ostrowski, G. M., und J. W. Wolin, Methoden zur Optimierung chemischer Reaktoren. Akademie-Verlag, Berlin 1973 (Translation from the Russian).

[5] Рудик, А. П., Ядерные реакторы и принцип максимума Понтрягина. Атомиздат, Москва 1971.

[6] Kernevez, J. P., and D. Thomas, Numerical analysis and control of some biochemical systems. Appl. Mathem. Optimization 1, 222—285 (1975).

[7] Bittner, L., On optimal control of processes governed by abstract functional, integral and hyperbolic differential equations. Math. Operationsforschung Statistik 6, 107—134 (1975).

[8] v. Wolfersdorf, L., Optimale Steuerung bei Hammersteinschen Integralgleichungen mit schwach singulären Kernen. Math. Operationsforschung Statistik 6, 609—626 (1975).

[9] v. Wolfersdorf, L., Optimal control of a class of processes described by general integral equations of Hammerstein type. Math. Nachr. **71,** 115—141 (1976).

[10] v. Wolfersdorf, L., Optimale Steuerung einer Klasse nichtlinearer Aufheizungsprozesse. Z. Angew. Math. Mech. **55,** 353—362 (1975).

[11] v. Wolfersdorf, L., Optimal control for processes governed by mildly nonlinear differential equations of parabolic type I, II. Z. Angew. Math. Mech. **56,** 531—538 (1976); **57,** 11—17 (1977).

[12] v. Wolfersdorf, L., Optimal control of processes described by mildly nonlinear parabolic integro-differential equations of second order (in preparation).

[13] v. Wolfersdorf, L., On optimal control and identification of processes governed by parabolic differential equations of second order. Z. Angew. Math. Mech. **57,** 315—322 (1977).

[14] Егоров, А. И., Об оптимальных процессах в системах, содержащих обьекты с распределёнными параметрами I, II. Автом. Телемех. **26**, 6, 977—994 (1965); **26**, 7, 1188 до 1196 (1965).

[15] Егоров, А. И., Необходимые условия оптимальности для систем с распределёнными параметрами, Матем. Сборник **69** (111), 3, 371—421 (1966).

[16] Голубь, Н. Н., Экстремальное управление линейными и нелинейными процессами в системах с бесконечным числом степеней свободы. Автом. Телемех. **32**, 2, 24—35 (1971).

[17] Голубь, Н. Н., Оптимальное управление линейными и нелинейными системами с распределёнными параметрами. Автом. Телемех. **30**, 9, 16—28 (1969).

[18] Голубь, Н. Н., Оптимальное управление нелинейными системами с распределёнными параметрами. Автом. Телемех. **30**, 10, 35—44 (1969).

[19] Denn, M. M., Optimal boundary control for a non-linear distributed system. Intern. J. Control **4,** 167—178 (1966).

[20] Егоров, А. И., Об условиях оптимальности в одной задаче управления процессом теплопередачи. Ж. Вычисл. Матем. Матем. Физ. **12**, 3, 791—799 (1972).

[21] Cesari, L., Optimization with partial differential equations in Dieudonné-Rashevsky form and conjugate problems. Arch. Rational Mech. Anal. **33,** 339—357 (1969).

[22] Плотников, В. И., Необходимые и достаточные условия оптимальности и условия единственности оптимизирующих функции для управляемых систем общего вида. Изв. Акад. Наук СССР, Сер. Матем. **36**, 652—679 (1972).

[23] Егоров, Ю. В., Необходимые условия оптимальности управления в Банаховых пространствах. Матем. Сборник **64** (106), 1, 79—101 (1964).

[24] Friedman, A., Optimal control for parabolic equations. J. Math. Anal. Appl. **18,** 479—491 (1967).

[25] Friedman, A., Optimal control in Banach spaces. J. Math. Anal. Appl. **19,** 35—55 (1967).

[26] Lions, J. L., Contrôle optimal de systèmes gouvernés par des équations aux dérivées partielles. Dunod, Gauthier-Villars, Paris 1968.

[27] Brogan, W. L., Dynamic programming and a distributed parameter maximum principle. Trans. ASME, Ser. D, **90,** 2, 152—156 (1968).

[28] Friedman, A., Partial differential equations of parabolic type. Prentice-Hall, Inc., Englewood Cliffs 1964.

[29] Ладыженская, О. А., Солонников, В. А., Уральцева, Н. Н., Линейные и квазилинейные уравнения параболического типа. Москва 1967.

[30] Камынин, Л. И., Метод тепловых потенциалов для параболического уравнения с разрывными коэффициентами. Сибирск. Матем. Журнал **4**, 5, 1071—1105 (1963).

[31] Unger, F., Zum Maximumprinzip bei Randsteuerproblemen der eindimensionalen Wärmeleitungsgleichung. Z. Angew. Math. Mech. **56,** 25—32 (1976).

[32] Unger, F., Zum Maximumprinzip bei der eindimensionalen Wärmeleitungsgleichung mit Rand- und verteilter Steuerung. Z. Angew. Math. Mech. **57**, 123—124 (1976).
[33] Arima, R., On general boundary value problem for parabolic equations. J. Math. Kyoto Univ. **4**, 207—243 (1964).
[34] Ивасишен, С. Д., Эйдельман, С. Д., Оценки матрици Грина однородной параболической граничной задачи. Доклады Акад. наук СССР **172**, 1262—1265 (1967).
[35] Ивасишен, С. Д., Матрица Грина неоднородной параболической граничной задачи. Доклады Акад. наук СССР **187**, 730—733 (1969).

Address of the author:

Bergakademie Freiberg
Sektion Mathematik
92 Freiberg
Bernhard-v.-Cotta-Str. 2
DDR

Short communications

Control of real large-scale systems by interaction balance techniques

Mieczysław Brdyś, Warszawa

Abstract. In the paper the coordination of a real large system is discussed. It is assumed that the available model of the system is not precise. Two versions of a two-level method (Interaction Balance Method with Feedback — IBMF) are discussed. All results which are presented in this paper are attained by members of the academic staff of the Institute of Automatic Control, Technical University of Warsaw.

1. Introduction

The interaction balance method of coordination finds the optimal controls for the system upon the base of the mathematical models of all subsystems. Such model-optimal controls applied to a real system will yield the truly optimal value of the system performance generally only in the case when the mathematical models are an ideal description of the reality. In practice however the models are always less or more precise, but never the ideal image of reality. There are numerous reasons of such situation. First, when a mathematical model is being built some simplifying assumptions are always made. It is well known, too, that many of the model coefficients are taken from experiments. At last, the real process is influenced by various disturbances, which cannot be exactly foreseen. In the following sections several possibilities of coupling the interaction balance method using mathematical models with the real system will be analysed. The aim is to find some suboptimal controls.

2. Problem formulation

Assume that there are given: the mathematical models of the subsystems

$$y^i = f^i(c^i, u^i), \qquad i = 1, \ldots, N, \tag{1}$$

the system structure equations

$$u^i = M^i y = \sum_{j=1}^{N} M_j{}^i y^j, \qquad i = 1, \ldots, N, \tag{2}$$

and the system performance

$$Q(c, u) = \sum_{i=1}^{N} Q^i(c^i, u^i). \tag{3}$$

We assume that $c^i \in C^i$, $u^i \in U^i$, $y^i \in Y^i$, where C^i, U^i, Y^i are Hilbert spaces and $f^i: C^i \times U^i \to Y^i$, $Q^i: C^i \times U^i \to R$, $M_j^i: Y^j \to U^i$ (M_j^i — linear continuous). For compactness of notation we denote $c = (c^1, \ldots, c^N)$, similiarly with u and y. Also for convenience we will write, when needed, the equations (1) and (2) in the compact form

$$y = f(c, u)$$

and

$$u = My.$$

The variables c^i, y^i, u^i are, respectively, control, output and interaction (input) of the i-th subsystem. The interaction u^i represents an action of the other subsystems on the i-th subsystem.

The mapping M describes the structure of the system (i.e. the way in which the subsystems are connected).

The real subsystems are assumed to be described by operator equations which are different from (1) (and of course unknown):

$$y^i = f_*^i(c^i, u^i), \qquad i = 1, \ldots, N, \tag{4}$$

where

$$f_*^i : C^i \times U^i \to Y^i.$$

We assume that the structure equations in the real system are the same as in the mathematical description (eq. (2)). Controls c^i, interactions u^i and outputs y^i should also satisfy the constraints

$$H_0^i(c^i, u^i, y^i) \in S^i \subset Z^i, \qquad i = 1, \ldots, N, \tag{5}$$

where $H_0^i : C^i \times U^i \times Y^i \to Z^i$, ($Z^i$ — Hilbert space), and S^i is a closed convex cone in Z^i.

The set of all (c^i, u^i, y^i) satisfying (5) we denote by CUY^i. We assume that for every control c, where $c^i \in P_{C^i}(CUY^i)$ (P_{C^i} — projection on C^i), $i = 1, \ldots, N$, there exists a unique u, such that

$$u = Mf(c, u).$$

If we substitute equations (11) into (5) we will obtain constraints for (c^i, u^i) in the model, i.e.

$$H^i(c^i, u^i) = H_0^i\big(c^i, u^i, f^i(c^i, u^i)\big) \in S^i. \tag{6}$$

The set of (c^i, u^i) satisfying (5) we denote by CU^i. Similiarly the set of (c^i, u^i) satisfying

$$H_*^i(c^i, u^i) = H_0^i\big(c^i, u^i, f_*^i(c^i, u^i)\big) \in S^i,$$

we denote by CU_*^i. Of course in a general case $CU^i \neq CU_*^i$, but if the constraints H_0^i in (5) do not depend on y^i then $CU_*^i = CU^i$.

Def. 1. We say that $\hat{c} = (\hat{c}^1, \ldots, \hat{c}^N)$ is a *model-optimal control* if $(\hat{c}^i, \hat{u}^i) \in CU^i$, where $\hat{u} = Mf(\hat{c}, \hat{u})$, and $Q(\hat{c}, \hat{u}) \leq Q(c, u)$ for every $(c, u) \in CU = CU^1 \times \cdots \times CU^N$ such that $u = Mf(c, u)$.

Def. 2. We say that $\hat{c}_* = (\hat{c}_*^1, \ldots, \hat{c}_*^N)$ is a *real-optimal control* if $(\hat{c}_*, \hat{u}_*) \in CU_* = CU_*^1 \times \cdots \times CU_*^N$, where $\hat{u}_* = Mf_*(\hat{c}_*, \hat{u}_*)$ and

$$Q(\hat{c}_*, \hat{u}_*) \leqq Q(c, u),$$

for every $(c, u) \in CU$ such that $u = Mf(x, u)$.

Since the operators f_*^i are unknown the real-optimal control cannot be found unless experimental optimization on the real system is applied. In the case of a complex system with many controls the extremal regulation can hardly be supposed to be successful. On the other hand the model-optimal control can be found through application of a suitable minimization procedure. But if this model-optimal control $\hat{c}$ is applied to the real process, then it will be nonoptimal in general. It may also very easily happen that the pair $(\hat{c}, \hat{u})$ does belong to CU_* even in the case, when $CU = CU_*$. In [2, 3] it had been suggested to look for another control $\bar{c}$ through the specific coordination method. To explain this idea we must remind that thanks to the special structure of the system a model-optimal control $\hat{c}$ can be found through the so called interaction balance method [5]. This is a two-level method: on the lover (infimal) level the infimal decision units solve independently their local problems:

I) given $p = (p^1, \ldots, p^N)$, $p^i \in U^i$, find

$$\min_{(c^i, u^i)} \left[Q^i(c^i, u^i) + \langle p^i, u^i\rangle - \sum_{j=1}^{N} \langle p^i, M_j^i f^i(c^i, u^i)\rangle\right], \tag{7}$$

where

$$(c^i, u^i) \in CU^i.$$

The set of solutions of all infimal problems we denote by $\overline{CU}(p)$. Of course $\overline{CU}(p) \subset CU$ for every p. On the upper (supremal) level the coordinator task is:

II M) find $\hat{p} \in U$ such that $\overline{CU}(\hat{p})$ is nonempty and

$$\bar{u}(\hat{p}) - Mf(\bar{c}(\hat{p}), \bar{u}(\hat{p})) = 0, \tag{8}$$

for a certain $(\bar{c}(\hat{p}), \bar{u}(\hat{p})) \in \overline{CU}(\hat{p})$.

If there exists a solution $\hat{p}$ of the supremal problem, then every c such that (c, u) solves local problems and $u - Mf(c, u) = 0$ is model-optimal, and we say that the model of the system is *coordinable by interaction balance method*.

3. Presentation of Interaction Balance Method with Feedback (IBMF)

Now the Interaction Balance Method with Feedback (see Fig. 1) can be proposed. Let the infimal decision problems be the same as above (i.e. of the form I). But the coordinator task on the supremal level is modified:

II R) find $\tilde{p} \in U$ such that $\overline{CU}(\tilde{p})$ is nonempty and

$$\bar{u}(\tilde{p}) - u^*(\bar{c}(\tilde{p})) = 0, \tag{9}$$

for a certain $(\bar{c}(\tilde{p}), \bar{u}(\tilde{p})) \in \overline{CU}(\tilde{p})$, where $u_*(\bar{c}(\tilde{p})) = Mf_*(\bar{c}(\tilde{p})), u_*(\bar{c}(\tilde{p}))$, i.e., $u_*(c)$ is the value of interaction in the real system, when control c is applied.

According to the previously made uniqueness assumption $u_*(\cdot)$ is a single-point valued mapping. We assume of course that the value of interactions in the real system can be measured. It is evident that a point $c = c(p)$ is in general case neither model-optimal nor real-optimal. This point may also not belong to the set CU_*. However, in the case, when $CU_* = CU$ (i.e., when the constraints H_0^i do not depend on y^i) we know that $(\bar{c}, u_*(\bar{c})) \in CU_*$ which is very important. Note that if the models are ideal, i.e., $f = f_*$, then every $(\bar{c}, u_*(\bar{c}))$ is real optimal.

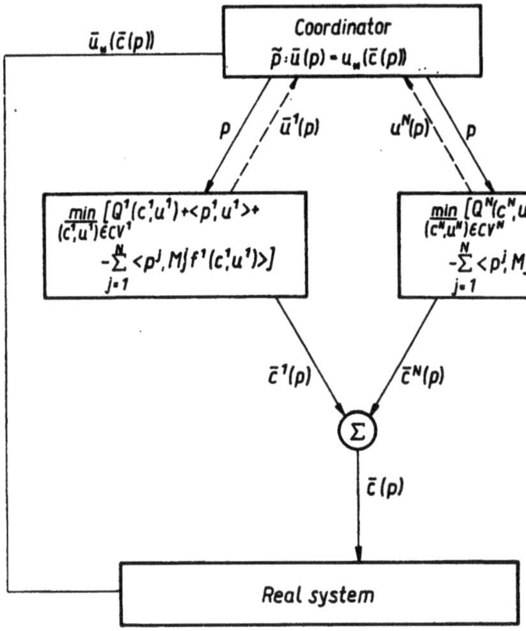

Fig. 1. Information and computing structure for IBMF

Conditions under which the solution $\tilde{p}$ of the IBMF supremal problem exists and a suboptimality analysis of IBMF are given in [4]. One from the existence theorems has the following form:

Theorem 1. *Assume that $CU = CU^1 \times \cdots \times CU^N \subset C \times U$ is a bounded, closed and convex set, the operators f^i and f_*^i are weakly continuous on CU^i, $i = 1, \ldots, N$, Q is weakly lower semicontinuous and upper bounded on CU and*

i) *there exists $K_1 > 0$, $K_2 \in R$, such that for every $h \in U$, $\|h\| \leqq K_1$ exists $(c, u) \in CU$ such that $u = Mf(c, u) + h$,*

ii) *for every $(c, u) \in CU$ we have*

$$\|Mf_*(c, u) - Mf(c, u)\| \leqq K_2 \qquad \text{where} \qquad 0 < K_2 \leqq K_1,$$

iii) *for every $p \in U$, $\|p\| \leqq r$ where* $r \geqq \dfrac{K_0' - K_0''}{K_1 - K_2}\left(K_0' = \sup\limits_{CU} Q(c, u), K_0'' = \min\limits_{CU} Q(c, u)\right)$,

the infimal problems I) *have unique solutions,*

iv) *the mappings* $\bar{c}(.)$, $\bar{u}(.)$ *are single-point valued and weakly continuous on the set* $\mathscr{U} = \{s \in U : \|s\| \leqq K_2\}$.

Then there exists a solution $\tilde{p}$ *of the IBMF supremal problem* II R *and* $\|\tilde{p}\| \leqq r$.

We will indicate now a certain coordination strategy for the coordinator of IBMF. Assume, that for every $p \in U$ the infimal problem I) has a unique solution. Then the task of the coordinator will be the following: find the solution $\tilde{p}$ of the operator equation

$$F_*(p) = \bar{u}(p) - u_*(\bar{c}(p)) = 0. \tag{10}$$

Since very little more can be said about F_* in the general case without making explicit assumptions on $\bar{c}(.)$ and $\bar{u}(.)$ it is difficult to propose any general strategy for the solution of (10). If, however, we assume that the sets CU^i have a special form, i.e., $Z^i = R^{k_i}$, $S^i = \{z^i \in R^{k_i} : z^i \leqq 0\}$ and $H^i(c^i, u^i) = (h_1{}^i(c^i, u^i), \ldots, h^i_{k_i}(c^i, u^i))$, such a strategy could be given. First we assume further what follows:

a) there exists a bounded operator F^{-1}, where $F : U \to U$,

$$F = I - Mf_u(c, u_*(c)),$$

b) f^i, $u_*(.)\, Qh_l{}^i$ $i = 1, \ldots, N$, $l = 1, \ldots, k_i$, are twice Fréchet differentiable,

c) $h_l{}^i$ are convex for every $i = 1, \ldots, N$, $l = 1, \ldots, k_i$,

d) there exist a number $\nu > 0$ and convex neighbourhoods $\omega(x^0)$, $\Omega(p^0)$ of the points $x^0 = (c^0, u^0)$ and p^0, where x^0 and p^0 are starting points, such that

$$\forall (x \in \omega(x^0)) \forall (p \in \Omega(p^0))\, Q_{xx}(x) + \langle p, G(x)\rangle_{xx} \geqq \nu I,$$

where $x = (c, u)$, $G(x) = u - Mf(c, u)$,

e) for each $p \in \Omega(p^0)$ denote $I^0(p) = \{(i, l) : h_l{}^i(\bar{c}^i(p), \bar{u}^i(p)) = 0\}$ and define an operator $\bar{H} : \Omega(p^0) \times C \times U \to R^{I_0(p)}$ as follows:

$$\bar{H}(p, c, u) = (h^{i_1}_{l_1}(c, u), \ldots, h^{i_n}_{l_n}(c, u)), \tag{11}$$

where $(i, k) \in I^0(p)$ for all $k = 1, \ldots, n$. Assume that $\forall (p \in \Omega(p^0))$ the derivative $\bar{H}_x$ of the operator $\bar{H}$ with respect to x taken at the point $\bar{x}(p) = (\bar{c}(p), \bar{u}(p))$ is a surjection.

f) for any $p_1, p_2 \in \Omega(p^0)$ the interval $[p_1, p_2] = \{p : p = tp_2 + (1 - t)\, p_1, 0 \leqq t \leqq 1\}$ can be divided into a countable number of sub-intervals so that within each of them the set $I^0(p)$ is constant,

g) $\exists (\delta > 0) \forall (p \in \Omega(p^0)) \forall (v \in R(G_x{}^*(\bar{x}(p^0))))$

$$d(v, R(H_x{}^*(p, x)\big|_{(p, \bar{x}(p))})) \geqq \delta \|v\|$$

where $d : X \times 2^X \to R^1$, $X = C \times U$, is a distance between the point and set involved.

Roughly speaking this assumption has the purpose to give a sufficient freedom in manipulation with $\bar{x}(p)$ through changes in p.

To solve the operator equation (10) the following iteration formula is proposed:

$$p^{k+1} = p^k + \varepsilon E_0 D(\bar{x}(p^k)), \tag{12}$$

where ε is a suitably choosen positive number, $D(x) = u - u_*(c)$, $E_0 = (D_0 A G_0^*)^{-1}$ where $D_0 = D_x(\bar{x}(p^0))$, $G_0 = G_x(\bar{x}(p^0))$, A is such a self-conjugate positive semidefinite operator that $N(A) \cap R(G_0^*) = \{0\}$. $N(.)$, $R(.)$ denote the null-space and range of A and G_0^*. We have the following formulae:

$$D_0 = \left[-\left(I - Mf_u(c^0, u_*(c^0))\right)^{-1} Mf_c(c^0, u_*(c^0)), I\right]$$

and

$$G_0 = [-Mf_c(c_0, u^0), I - Mf_u(c^0, u^0)], \qquad \text{where} \qquad (c^0, u_0) = \bar{x}(p^0).$$

Convergence properties of the above algorithm are formulated in the following theorem:

Theorem 2. *Suppose that assumptions* a), ..., g) *are satisfied and*

(i) $\beta = \|E_{00}(D_{00} - D_0) A G_0^*\| < 1$,

where $E_{00} = (D_{00} A G_0^*)^{-1}$, $\quad D_{00} = \left(I - Mf_u(c^0, u_*(c^0))\right)^{-1} G_0$,

(ii) $\exists\, (m_A > 0, M_A > 0)\, \left(\forall x \in R(G_0^*)\right) \quad m_A\|x\|^2 \leqq \langle x, Ax\rangle \leqq M_A\|x\|^2$,

(iii) $\exists\, (M_W > 0)\, \forall\, \left(p \in \Omega(p^0)\right) \forall\, (x) \quad \langle x, W(p)\, x\rangle \leqq M_W\|x\|^2$,

where

$$W(p) = Q_{xx}(\bar{x}(p)) + \langle p, G(x)\rangle_{xx}|_{x=\bar{x}(p)} + \langle \mu, H(p, x)\rangle|_{x=\bar{x}(p)},$$

where μ is the vector of the Lagrange multipliers for the infimal level problems,

(iv) $\forall\left((p_1, p_2) \in \Omega(p^0)\right) \|D_x(\bar{x}(p_1))\, B(p_2)\, G_x^*(\bar{x}(p_1))$

$- D_x(\bar{x}(p_2))\, B(p_2)\, G_x^*(\bar{x}(p_2))\|_0 \leqq L\|p_1 - p_2\|_0$,

where

$$B(p) = \begin{cases} W^{-1} & \text{if} \quad I^0(p) = \emptyset, \\ W^{-1} - W^{-1}\bar{H}_x^*(\bar{H}_x W^{-1}\bar{H}_x^*)^{-1}\bar{H}_x W^{-1} & \text{if} \quad I^0(p) \neq \emptyset. \end{cases}$$

Denote

$$\xi(\varepsilon) = \frac{1 - \alpha(\varepsilon)}{\varepsilon}, \; b_0 = \|E_0\|_0, \; \eta_0 = \|E_0 D(\bar{x}(p^0))\|_0,$$

$$h_0 = b_0 L \eta_0, \; r^* = \frac{\xi - \sqrt{\xi^2 - 2h_0}}{h_0}\eta_0, \; r^{**} = \frac{\xi + \sqrt{\xi^2 - 2h_0}}{h_0}\eta_0.$$

Assume further that $\sqrt{2h_0} < \xi$, $\bar{S}(p^0, r^*) = \{p : \|p - p^0\|_0 \leqq r^*\} \subset \Omega(p^0)$, *where*

$$\|u\|_0 = \langle G_0^* u, A G_0^* u\rangle.$$

Then

1) *Equation* (10) *has a solution $\bar{p}$ on $\bar{S}(p^0, r^*)$ which is a limit of a sequence* (12).

2) *The following inequality is satisfied*:

$$\|p^{k+1} - \bar{p}\|_0 \leqq \frac{\varepsilon}{1 - q_0}\|E_0 D(\bar{x}(p^k))\|_0,$$

where

$$q_0 = \alpha(\varepsilon) + \varepsilon b_0 L r^* < 1.$$

The proof of Theorem 2 is given in [6]. Note that the convergence conditions obtained above are identical to those of Kantorovich for the Newton method [7].

4. Presentation of IBMF with feasible control generation (IBMF with FCG)

As it was mentioned above, the solution $\hat{c}$ of the IBMF supremal problem can be non feasible for the real system, when $CU \neq CU_*$. Elements of a sequence $\{\bar{c}_k(p)\}_k$ which is generated by IBMF are also nonfeasible in general. To eliminate this fault we must do a certain modification in feedback. Computing and information structure in which the feedback is modified is given on Fig. 2. There are two new elements in this structure

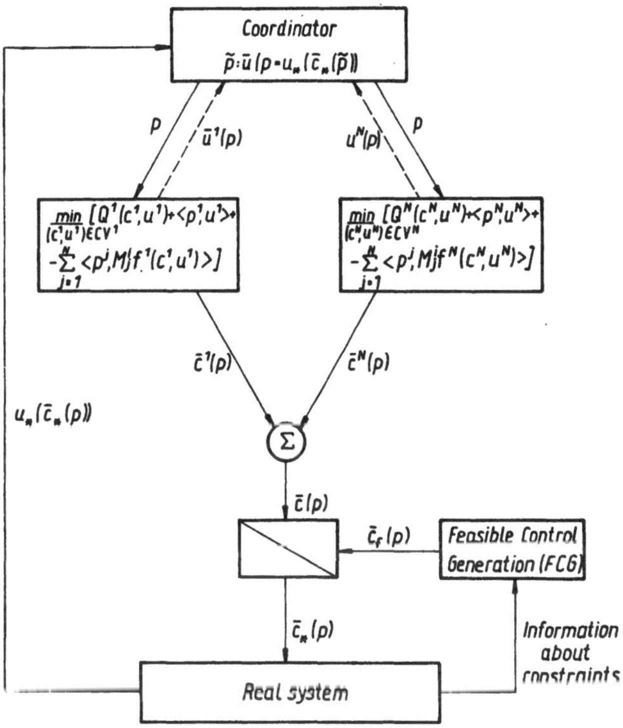

Fig. 2. Information and computing structure for IBMF with FCG

with compare to the structure from Fig. 1. First, the so called Feasible Control Generation block generates on the base of information about the actual state of the constraints in the real system a feasible control $\bar{c}_f(p)$ when there is a need for it. Second, it decides which control $\bar{c}(p)$ or $\bar{c}_f(p)$ will be applied to the real system. More precisely,

$$\bar{c}^*(p) = \begin{cases} \bar{c}(p) & \text{if} \quad \bar{c}(p) \in C_*, \\ \bar{c}_f(p) & \text{in the opposite case,} \end{cases} \tag{13}$$

where C^* is a feasible control set for the real system.

To make this structure applicable we must first show how to construct the FCG block. Two methods of doing it are proposed. To present the first of them we must assume more about the mathematical model of the system and its real mathematical description. Assume what follows:

$$\exists\big(\bar{\alpha} = (\bar{\alpha}^1, \ldots, \bar{\alpha}^N) \in A^1 \times \cdots \times A^N\big) \ \forall\ (c, u) \in C \times U \quad f_*(c, u) = \tilde{f}(c, u, \bar{\alpha}),$$

where

$$\tilde{f}: C \times U \times A \to Y, \qquad \tilde{f}(c, u, \alpha) = \big(\tilde{f}^1(c^1, u^1, \alpha^1), \ldots, \tilde{f}^N(c^N, u^N, \alpha^N)\big),$$

$$A = A^1 \times \cdots \times A^N,$$

and

$$\exists\big(\bar{\beta} = (\bar{\beta}^1, \ldots, \bar{\beta}^N) \in B^1 \times \cdots \times B^N\big) \ \forall\ \big((c, u) \in C \times U\big) \ \forall\ \big(y \in f_*(C \times U)\big)$$

$$H_*(c, u) = \tilde{H}(c, u, \bar{\beta}),$$

where

$$\tilde{H}: C \times U \times B \to Z, \qquad \tilde{H}(c, u, \beta) = \big(\tilde{H}^1(c^1, u^1, \beta^1), \ldots, \tilde{H}^N(c^N, u^N, \beta^N)\big),$$

$$B = B^1 \times \cdots \times B^N.$$

Assume further that $\tilde{H}, \tilde{f}, A, B$ are well known and $\exists\,(\hat{\alpha} \in A)\ \exists\ (\hat{\beta} \in B)$ such that $\tilde{f}(c, u, \hat{\alpha}) = f(c, u)$ and $\tilde{H}(c, u, \hat{\beta}) = H(c, u)$ for all $(c, u) \in C \times U$.

Roughly speaking it is assumed that the real mathematical model of the system is known with accuracy to parameters α, β. We only know that the real values $\bar{\alpha}, \bar{\beta}$ of the parameters α, β belong to certain sets A and B. Denote as $C_f^{(A,B)}$ the set of all feasible controls for all values of α, β when mappings $\tilde{H}$ and $\tilde{f}$ are considered. It is an obvious fact that $C_f^{(A,B)} \subset C^*$. It was proved in [1] that the class of problems for which $C_f^{(A,B)} \neq \emptyset$ is quite wide.

The algorithms for control which belong to $C_f^{(A,B)}$ generating were also given in [1]. Hence, it is realistic to construct the FCG block as follows (see Fig. 3):

$$\bar{c}_f(p) \triangleq P_{C_f^{(A,B)}}\big(\bar{c}(p)\big). \tag{14}$$

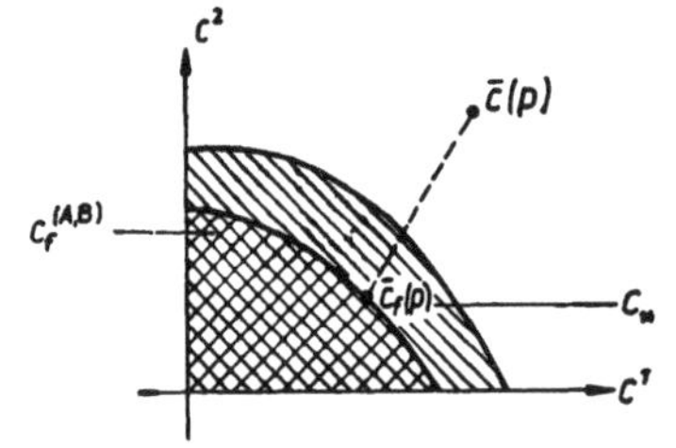

Fig. 3. First method of feasible control generation

Under certain assumptions $C_f^{(A,B)}$ is a convex, closed set and formula (14) takes sense. The important property of such a realization of the FCG block ist that only the model is used here.

Suppose now that there is given $\bar{c}(p)$ and a certain set $\tilde{C}_*(p) = \{c_1, \ldots, c_k\}$ such that $c_i \in C$ for every $i = 1, \ldots, k$. Since C_* is upon easy assumptions convex then conv $\tilde{C}_*(p) \subset C_*$. The second method of FCG block realization is given by the formula (see Fig. 4)

$$\bar{c}_f(p) \triangleq P_{\mathrm{conv}\,\tilde{C}_*(p)}\big(\bar{c}(p)\big). \tag{15}$$

If it is sensible, we can join $\bar{c}_f(p)$ to the set $\tilde{C}_*(p)$. It is naturally to expect that for a certain sequence $\{p^k\}$ which is generated by the coordinator we will be given the sets $\tilde{C}_*(p^k)$ such that $\lim_{k\to\infty} \operatorname{conv} \tilde{C}_*(p^k) = C_*$. The problem of generating a sequence $\{p^k\}$ such that convergence is guaranteed is still open.

If the system has a "fast dynamic" we can slightly modify formula (15) as follows:

$$\bar{c}_f(p) \triangleq \bar{c}(p) + \hat{\varrho}\big[P_{\operatorname{conv}\tilde{C}_*(p)}\big(\bar{c}(p)\big) - \bar{c}(p)\big],$$

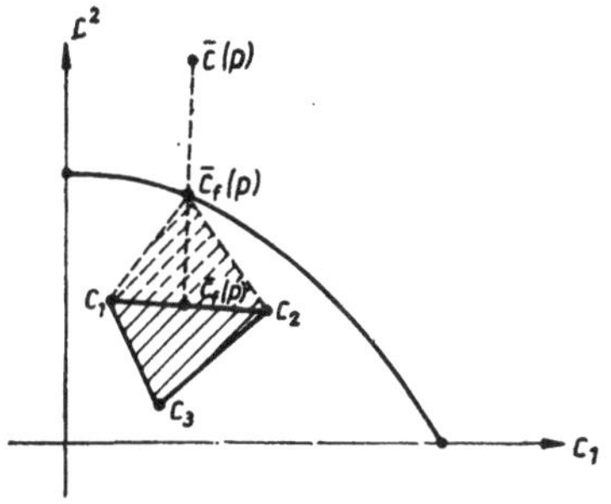

Fig. 4. Second method of feasible control generation

where $\hat{\varrho}$ is a smallest real nonnegative number for which $\bar{c}_f(p) \in C_*$. The new set $\tilde{C}_*(p)$ estimates the set C_* better than it did before. Another technique of real large-scale system control, the so-called parametric method, is given in [8].

References

[1] Brdyś, M., Finding a feasible control for real process under uncertainty. Presented on the 7-th IFIP Conference, Nice, September 8—13, 1975.

[2] Findeisen, W., Multilevel control systems (in Polish). PWN, Warsaw 1974.

[3] Findeisen, W., Control and coordination in multilevel systems. Technical Report No 5/1974, Institute of Automatic Control, Technical University of Warsaw.

[4] Malinowski, K., Multilevel methods of mathematical programming and complex system optimization (in Polish). Ph. D. Dissertation, Institute of Automatic Control, Techn. University of Warsaw 1974.

[5] Mesarović, M., D. Macko, Y. Takahara, Theory of hierarchical, multilevel, systems. Academic Press, New York 1970.

[6] Ruszczyński, A., Convergence conditions for the interaction balance algorithm based on approximate mathematical models (in preparation).

[7] Kantorovich, L., G. Akilov, Functional analysis in normed spaces. Fizmatgiz, Moscow 1959, translated into English by Pergamon Press, Oxford 1964.

[8] Woźniak, A., Parametric method of coordination using feedback from the real process. Technical Report No 3/1975, Institute of Automatic Control, Technical University of Warsaw.

Address of the author:

Institute of Automatic Control
Technical University of Warsaw
ul. Nowowiejska 15/19
00-665 Warszawa
Poland

Some remarks on branching theory

Slavomír Burýšek, Praha

Introduction. In connection with branching phenomena two important problems are studied. The first one is the existence of bifurcation points and the second is the existence and properties of branches of nontrivial solutions in a neighbourhood of a bifurcation point. In section 1 we recall some results concerning the first problem and show a generalization of results contained in [17] which is closely related to some results in [7]. In section 2 we deal with the second problem for a special class of nonlinear equations and show asymptotic formulae for nontrivial small solutions. These results complete in some way results obtained in [5, 24].

1. The existence of bifurcation points

Let X be a real Banach space, X^* its dual space and let λ be a real parameter. Suppose $S, T: X \to X^*$ are homogeneous operators of the degree $\varkappa > 0$. Denote $\Lambda_{S,T}$ the set of all normalized eigenvalues of the couple (S, T) (i.e., $\lambda \in \Lambda_{S,T}$ if and only if there is an element $x \in X$, $\|x\| = 1$ such that $Sx - \lambda Tx = 0$), $\mathscr{S}_{S,T}$ the approximate spectrum of the couple (S, T) (i.e., the set of all λ such that there is a sequence $\{x_n\} \subset X$, $\|x_n\| = 1$ satisfying $\|Sx_n - \lambda Tx_n\| \to 0$) and $W_{S,T}$ the numerical range of the couple (S, T) (i.e., the closure of the set $\left\{\frac{(Sx, x)}{(Tx, x)}: \|x\| = 1\right\}$). Assume, further, that $Tx = 0$ if and only if $x = 0$. It is easy to show that every eigenvalue $\lambda_0 \in \Lambda_{S,T}$ with an eigenvector σ_0 is a bifurcation point of the equation

$$Sx - \lambda Tx = 0 \tag{1.1}$$

with respect to zero. Recall that if $c(t) = (\lambda(t), x(t))$, $t \in I$ is a simple continuous curve in $R^1 \times X$ (i.e., c is a continuous mapping from an interval I into the Banach space $R^1 \times X$ such that $c(t) \neq c(s)$ for any $t, s \in I$, $t \neq s$) and if $F: R^1 \times X \to X$ is a continuous operator satisfying $F(\lambda(t), x(t)) = 0$ for any $t \in I$ then $(\lambda_0, x_0) = (\lambda(t_0), x(t_0))$, $t_0 \in I$ is called a bifurcation point of the equation

$$F(\lambda, x) = 0 \tag{1.2}$$

with respect to the curve c if and only if for any neighbourhood $U(\lambda_0, x_0)$ of the point (λ_0, x_0) there is a point $(\lambda, x) \in U(\lambda_0, x_0)$ satisfying the equation (1.2) such that (λ, x)

$\neq (\lambda(t), x(t))$ for any $t \in I$. In case of the equation (1.1) we have $c(t) = (t, 0)$, $t \in [0, +\infty)$ so that we can speak about the bifurcation with respect to zero or about the bifurcation from the zero solution.

Suppose that $F(\lambda, x) = \lambda x - A(x)$, $x \in X$ where $A: X \to X$, $A(0) = 0$ is a nonlinear Fréchet differentiable operator in a neighbourhood of the point zero. It follows from the well-known implicite function theorem that the bifurcation points with respect to zero of the equation (1.2) can lie only in the spectrum $\sigma(A'(0))$ of the linear operator $A'(0)$ and if, in addition, the range of the operator $\lambda I - A'(0)$ is closed then each bifurcation point of the equation (1.2) with respect to zero is contained in the point spectrum of the operator $A'(0)$. If A is completely continuous in a neighbourhood of the point zero then the well-known Kasnoselskij's theorem says that each eigenvalue of odd multiplicity of the linear operator $A'(0)$ is a bifurcation point. If, in addition, A is a uniformly differentiable potential operator on a Hilbert space with a weakly continuous potential then another Krasnoselskij's theorem says that every eigenvalue of $A'(0)$ is a bifurcation point of the equation (1.2) (see [15]). Recently generalizations of the last theorem were proved as, for example, the following.

Theorem 1.1 [2]. *Let $L: X \to X$ be a selfadjoint continuous linear operator on a Hilbert space X and let $T: R^1 \times X \to X$ be a potential operator with respect to $x \in X$ twice continuously differentiable and such that $T(0, \lambda) = T_x'(0, \lambda) = 0$ for any $\lambda \in R^1$. Suppose further that $I - \lambda_0 L$ is a Fredholm operator which is not invertible in X. Then λ_0 is a bifurcation point with respect to zero of the equation*

$$x - \lambda Lx + T(x, \lambda) = 0. \tag{1.3}$$

An important generalization of Krasnoselskij's theorem is shown in [7]. Roughly speaking, if S, T, R are odd potential operators on a reflexive Banach space with S, T homogeneous and R of "higher order" then, under further suitable hypotheses, each normalized eigenvalue of the couple (S, T) is a bifurcation point of the equation

$$Sx - \lambda Tx = R(x). \tag{1.4}$$

Suppose further that X is a reflexive Banach space and let $K, P: X \to X^*$ be nonlinear operators satisfying $K(0) = P(0) = 0$. Consider the equation

$$K(x) - \lambda P(x) = 0. \tag{1.5}$$

In [17] results were obtained concerning bifurcation points with respect to zero of the equation (1.5) where K has the Fréchet derivative $K'(0)$ and P has the weak Fréchet derivative $P'(0)$. Now we extend some of these results assuming, instead of the differentiability, that K and P are near to some homogeneous operators S and T.

Definition 1.2. Let $S, T: X \to X^*$ be homogeneous operators of the degree $\varkappa > 0$. We shall say that the operator K *is weakly near to S at the point zero* if $\lim_{\|x\| \to 0} \frac{(K(x) - S(x), h)}{\|x\|^\varkappa} = 0$ for any $h \in X$. We shall say that P *is near to T* if $\lim_{\|x\| \to 0} \frac{\|P(x) - T(x)\|}{\|x\|^\varkappa} = 0$.

Denote $B_{K,P}$ the set of all bifurcation points of the equation (1.5) with respect to zero. Then the following theorem holds.

Theorem 1.3. *Let S, T be strongly continuous homogeneous operators of the degree $\varkappa > 0$. Let K be weakly near to S and P be near to T. Suppose further that there is a constant $\gamma > 0$ such that*

$$\|P(x)\| \geqq \gamma \|x\|^{\varkappa} \quad \text{for any} \quad x \in X.$$

Then $B_{K,P} \subset \mathcal{S}_{S,T}$.

Further information about the set $B_{K,P}$ is contained in the following proposition.

Proposition 1.4. *Suppose S and T are strongly continuous homogenous operators of the degree $\varkappa > 0$. Let K be near to S and P be near to T. Assume further that there is a constant $\gamma_0 > 0$ such that*

$$\big(P(x), x\big) \geqq \gamma_0 \|x\|^{\varkappa+1} \quad \text{for any} \quad x \in X.$$

Then $B_{K,P} \subset \Lambda_{S,T} \subset \mathcal{S}_{S,T} \subset W_{S,T}$.

Now we state a theorem on the existence of a bifurcation point of the equation (1.5).

Theorem 1.5. *Suppose that the operators K and P satisfy the following conditions.*

(i) *P is a potential and monotone operator near to a continuous homogeneous operator T at the point zero and P is continuous on every twodimensional subspace in X.*

(ii) *There is a constant $\gamma_0 > 0$ such that $\big(P(x), x\big) \geqq \gamma_0 \|x\|^{\varkappa+1}$ for any $x \in X$.*

(iii) *K is a completely continuous potential operator near to a continuous homogeneous operator S of the degree $\varkappa$ at the point zero.*

Then the equation (1.5) has a bifurcation point with respect to zero if either

(a) *K is positive (i.e., $\big(K(x), x\big) > 0$ for any nonzero point $x \in X$) or*

(b) *the couple (S, T) has a positive eigenvalue $\lambda_0 \in \Lambda_{S,T}$.*

Remark 1.6. If, in addition to the assumptions of Theorem 1.5, the operator K satisfies $(K(x), x) \geqq \delta \|x\|^{\varkappa+1}$ for any $x \in X$ and some $\delta > 0$ then the equation (1.5) has a nonzero bifurcation point. Sufficient conditions for the assumption (b) of Theorem 1.5 to be satisfied can be derived from results of [3].

2. Small solutions in a neighbourhood of a bifurcation point

In recent years many authors deal with the problem of small solutions. Let X be a Banach space and let $G: R^1 \times X \to X$ be an operator satisfying $G(\lambda_0, 0) = 0$ for some $\lambda_0 \in R^1$. Consider the equation

$$G(\lambda, x) = 0. \tag{$*$}$$

A nontrivial continuous function f from a neighbourhood $U(\lambda_0)$ of the point λ_0 into X we shall call a small solution of the equation $(*)$ if $G\big(\lambda, f(\lambda)\big) = 0$ for any $\lambda \in U(\lambda_0)$ and

$f(\lambda_0) = 0$. Sometimes it is useful to consider small solutions of (*) restricted to $U(\lambda_0) \cap M$ where $M \subset R^1$ is a given set containing the point λ_0. In this case we shall speak about small solutions with respect to M.

Most of the authors consider the equation (*) with an operator $G(\lambda, x) = \lambda x - Lx + F(\lambda, x)$ where $L\colon X \to X$ is a linear operator and $F\colon R^1 \times X \to X$ is a nonlinear operator of "higher order". The important case is when λ_0 is a bifurcation point of the equation (*). In this case λ_0 is an eigenvalue of the linear operator L and the well-known Ljapunov-Schmidt method is used to reduce the problem of finding small solution to a branching equation in an appropriate finite dimensional space. Then the branching equation is studied by various methods. For example, the method of topological degree theory in [4, 5], variational methods in [21] and iterations in [13] are used while the method of implicite functions in [24, 25] and monotone operator theory in [9] are employed. In the analytical case the method of Někrasov-Nazarov and the abstract power series theory are applied in [8, 18, 26]. Very important is the method of Newton's diagram [23]. Some results concerning a global behaviour of small solutions are contained in [6] and new results in perturbed bifurcation theory are in [12]. Stability of bifurcating solutions is studied in [4, 22].

In order to solve the difficult problem of branching near to multiple eigenvalues another information is needed about the nonlinear term. For example, in [21] the operator F is supposed in the form $F(x) = T(x) + R(x)$ where T is a homogeneous polynomial operator and R is a higher order operator. The exact number of small solutions has been found in some cases of multiplicity and the degree of T. But, in general, the number and the form of small solutions is an open question. In spite of it there are many of interesting applications of the branching theory [11, 14, 16, 19].

In the rest of this section we shall try to give an asymptotic formula for a small solution of the following equation

$$(\lambda J - L)\, x - T(\lambda, x) - R(\lambda, x) = 0\,, \tag{2.1}$$

where $J, L\colon X \to X$ is a couple of commuting continuous linear operators with an eigenvalue λ_0 and $T, R\colon R^1 \times X \to X$ are real analytic operators in a neighbourhood of the point $(\lambda_0, 0)$ satisfying the following assumptions.

(a) T is a homogeneous polynomial operator of the degree $k > 1$ in the variable x for any λ in a neighbourhood $U(\lambda_0)$ of λ_0.

(b) $R(\lambda, 0) = R_x'(\lambda, 0) = 0$ for any $\lambda \in U(\lambda_0)$ and $\lim\limits_{\|x\| \to 0} \dfrac{\|R(\lambda, x)\|}{\|x\|^k} = 0$ uniformly with respect to $\lambda \in U(\lambda_0)$.

(c) $X = N(\lambda_0 J - L) \oplus R(\lambda_0 J - L)$, where $N(\lambda_0 J - L)$ is the nullspace, $R(\lambda_0 J - L)$ is the range of $\lambda_0 J - L$ and $R(\lambda_0 J - L)$ is closed.

Under the above assumptions there are projectors $P\colon X \to N(\lambda_0 J - L)$ and $Q\colon X \to R(\lambda_0 J - L)$. Since J and L commute the operators $\lambda J - L$ and $\lambda_0 J - L$ also commute and thus $N(\lambda_0 J - L)$ and $R(\lambda_0 J - L)$ are invariant subspaces for the operators J and $\lambda J - L$. It follows that P and Q commute with $\lambda J - L$. Let $x \in X$, $x = u + v$, where $u = Px$ and $v = Qx$. Applying P and Q to the equation (2.1) we obtain the system

$$(\lambda J - L)\, u = PT(\lambda, u + v) + PR(\lambda, u + v)\,, \tag{2.2}$$

$$(\lambda J - L)\, v = QT(\lambda, u + v) + QR(\lambda, u + v)\,. \tag{2.3}$$

Using the implicite function theorem we can show that for each (λ, u) in a neighbourhood of $(\lambda_0, 0)$ there is unique solution $v = f(\lambda, u)$ of the equation (2.3) such that f is real analytic and $f(\lambda_0, 0) = 0$. Substituting $f(\lambda, u)$ into (2.2) we obtain the branching equation for the equation (2.1) in the following form

$$(\lambda - \lambda_0)\, Ju = PT\big(\lambda, u + f(\lambda, u)\big) + PR\big(\lambda, u + f(\lambda, u)\big). \tag{2.4}$$

Now it is obvious that λ_0 is a bifurcation point of the equation (2.1) if and only if λ_0 is a bifurcation point of the branching equation (2.4).

Remark 2.1. If $N(\lambda_0 J - L)$ has a finite dimension n then, according to [24], Theorem 5.1, the assumption (c) holds if and only if $\lambda_0 J - L$ is a Fredholm operator of the index zero. The branching equation (2.4) turns to a system of n scalar equations with n unknowns. In the space $N(\lambda_0 J - L)$ we define an inner product to obtain a Hilbert space with an equivalent norm.

Recall that an equivalent branching equation we obtain using Schmidt's lemma or Nikolskij's theorem (see [23]). The result of Petryshyn in [20] implies that a generalization of the reduction process can be shown. Indeed, it suffices to assume that $\lambda_0 J - L$ is a continuous "A-proper" operator. Another extension to a closed unbounded operator is obtained in [10].

In order to find small solutions of the equation (2.1) we can use the following lemma.

Lemma 2.2. *Suppose that the branching equation* (2.4) *has a small solution* $u(\lambda)$ *in a neighbourhood of* λ_0. *Then the equation* (2.1) *has a small solution* $x(\lambda)$ *of the form*

$$x(\lambda) = u(\lambda) + f\big(\lambda, u(\lambda)\big).$$

Let us pay attention to the branching equation (2.4). Suppose at first that (2.4) has the following form

$$(\lambda - \lambda_0)\, Ju = S(\lambda, u), \qquad u \in N(\lambda_0 J - L), \tag{2.4'}$$

where S is a continuous operator in a neighbourhood of $(\lambda_0, 0)$ and homogeneous of the degree $\varkappa > 1$ in the variable u for any λ.

Proposition 2.3. *Suppose that for any* λ *in a neighbourhood* $U(\lambda_0)$ *of* λ_0 *there is an eigenvalue* $\mu_0(\lambda)$, $\mu_0(\lambda_0) \neq 0$, *of the operator* S *with a normalized eigenvector* $u_0(\lambda)$ (*i.e.*, $S\big(\lambda, u_0(\lambda)\big) = \mu_0(\lambda)\, u_0(\lambda)$) *such that* u_0 *is a continuous function in* $U(\lambda_0)$ *and* $Ju_0(\lambda_0) \neq 0$. *Then there is a* $\delta > 0$ *such that the following assertions hold.*

(i) *If* $\varkappa$ *is even then there is a small solution* $u_1(\lambda)$ *of the equation* (2.4′) *with respect to the interval* $(\lambda_0 - \delta, \lambda_0 + \delta)$.

(ii) *If* $\varkappa$ *is odd and* $\mu_0(\lambda_0) > 0$, $\big(\mu_0(\lambda_0) < 0\big)$ *then there are two small solutions* $u_{1,2}$ *of* (2.4′) *with respect to* $[\lambda_0, \lambda_0 + \delta)$ (*with respect to* $(\lambda_0 - \delta, \lambda_0]$).

(iii) *These small solutions have the form*

$$u_{1,2}(\lambda) = \pm \left[\frac{\lambda - \lambda_0}{\mu_0(\lambda)}\right]^{\frac{1}{\varkappa - 1}} u_0(\lambda). \tag{2.5}$$

Corollary. *Let S be continuous and let $S(\lambda, .)$ be a completely continuous J-symmetric homogeneous polynomial operator (i.e. J^*S is a potential operator). Suppose that $N(\lambda_0 J - L)$ is a Hilbert space. Then the assertions of Proposition 2.3 hold.*

Finally we show sufficient conditions for the existence of small solutions of the equation (2.1).

Theorem 2.4. *Let the assumptions* (a), (b), (c) *be satisfied and let λ_0 be a simple eigenvalue of the couple (J, L) with a normalized eigenvector u_0. Suppose, further, that $(PT(\lambda_0, u_0), u_0) \neq 0$ and $(Ju_0, u_0) \neq 0$. Then the following assertions hold.*

(i) *If k (the degree of T) is even then there is a unique small solution $x(\lambda)$ of the equation (2.1) in a neighbourhood $U(\lambda_0)$ such that*

$$x(\lambda) = \left[\frac{\lambda - \lambda_0}{a}\right]^{\frac{1}{k-1}} u_0 + v\left[(\lambda - \lambda_0)^{\frac{1}{k-1}}\right],$$

where $a = (PT(\lambda_0, u_0), u_0)/(Ju, u_0)$ and v is an analytic function in a neighbourhood of zero such that $v(0) = 0$.

(ii) *If k is odd then there are two small solutions $x_{1,2}(\lambda) = \pm x(\lambda)$ of (2.1) with respect to $[\lambda_0, \lambda_0 + \delta)$ if $a > 0$ and with respect to $(\lambda_0 - \delta, \lambda_0]$ if $a < 0$, where $\delta > 0$.*

Let us remark that the existence of a bifurcation point λ_0 does not imply the existence of small solutions in a neighbourhood of λ_0. For example, let $S, T: X \to X$ be differentiable potential and homogeneous operators of the degree $k > 0$. Let $Tx = 0$ if and only if $x = 0$. Then there are no small solutions of the equation $Sx - \lambda Tx = 0$.

References

[1] Bazley, N. W., Pimbley, G. H., A region of no secondary bifurcation for nonlinear Hammerstein equations. Z. Angew. Math. Phys. **25**, 743–752 (1974).

[2] Berger, M. S., Applications of global analysis to specific nonlinear problems. Rocky Mountain J. Math. **3**, 319–354 (1973).

[3] Burýšek, S., Eigenvalue problem, bifurcations and equations with analytic operators in Banach spaces. *In:* Theory of Nonlinear Operators, Proceedings of a summer-school held in October 1972, 1–14. Akademie-Verlag, Berlin 1974.

[4] Crandall, M. G., Rabinowitz, P. H., Bifurcation, perturbation of simple eigenvalue, linearized stability. Arch. Rational Mech. Anal. **52**, 161–180 (1973).

[5] Dancer, E. N., Bifurcation theory in real Banach spaces. Proc. London Math. Soc. **23**, 699–734 (1971).

[6] Dancer, E. N., Global solution branches for positive mappings. Arch. Rational Mech. Anal. **52**, 181–192 (1973).

[7] Fučík, S., Nečas, J., Souček, J., Souček, V., Krasnoselskii's main bifurcation theorem. Arch. Rational Mech. Anal. **54**, 328–339 (1974).

[8] Громов, А. В., Об аналитическом решении некоторых нелинейных уравнений. Сибир. Мат. Ж. **15**, 680–682 (1974).

[9] Gustavson, K., Sather, D., Large nonlinearities and monotonicity. Arch. Rational Mech. Anal. **48**, 109–122 (1972).

[10] —, Large nonlinearities and closed linear operators. Arch. Rational Mech. Anal. **52**, 10–19 (1973).

[11] Joseph, D. D., Repeated supercritical branching of solutions arising in the variational theory of turbulence. Arch. Rational Mech. Anal. **53**, 101—130 (1974).

[12] Keener, J. P., Perturbed bifurcation theory at multiple eigenvalues. Arch. Rational Mech. Anal. **56**, 348—366 (1974).

[13] Keller, H. B., Langford, W. F., Iterations, perturbations and multiplicities for nonlinear bifurcation problems. Arch. Rational Mech. Anal. **48**, 83—108 (1972).

[14] Knightly, G. H., Sather, D., Nonlinear buckled states of rectangular plates. Arch. Rational Mech. Anal. **54**, 356—372 (1974).

[15] Красносельский, М. А., Топологические методы в теории нелинейных интегральных уравнений. Москва 1956.

[16] Langenbach, A., Bifurkation und differenzierbare Zweige von Eigenfunktionen in der Theorie der Kármánschen Platten. Math. Nachr. **61**, 181—199 (1974).

[17] Langenbach, A., Lösungsbifurkation bei Gleichungen mit schwach differenzierbaren Operatoren. Math. Nachr. **64**, 105—117 (1974).

[18] Меламед, В. Б., Об ветвлении решений нелинейных уравнений с несколькими комплексными параметрами. Сибир. Мат. Ж. **12**, 805—811 (1971).

[19] Pazy, A., Rabinowitz, P. H., On a branching process in neutron transport theory. Arch. Rational Mech. Anal. **51**, 153—164 (1973).

[20] Petryshyn, W. V., Fredholm alternatives for nonlinear A-proper mappings with applications to nonlinear elliptic boundary value problems. J. Functional Anal. **18**, 288—317 (1975).

[21] Sather, D., Nonlinear gradient operators and the method of Ljapunov-Schmidt. Arch. Rational Mech. Anal. **43**, 222—244 (1971).

[22] Sattinger, D. H., Stability of bifurcation solutions by Leray-Schauder degree. Arch. Rational Mech. Anal. **43**, 154—165 (1971).

[23] Вайнберг, М. М., Треногин, В. А., Теория ветвления решений нелинейных уравнений. Москва 1969.

[24] Westreich, D., Banach space bifurcation theory. Trans. Amer. Math. Soc. **171**, 135—156 (1972).

[25] —, Bifurcation at eigenvalues of odd multiplicity, Proc. Amer. Math. Soc. **41**, 609—614 (1973).

[26] Забрейко, П. П., Кац, В. П., О методе Некрасова-Назарова решения нелинейных операторных уравнений. Сибир. Мат. Ж. **12**, 1026—1040 (1971).

Address of the author:

Technical University of Prague (CVUT)
Katedra matematiky 202
Karlovo náměstí 13
121 35 Praha 2
Czechoslovakia

Superlinear convergence of conjugate gradient method in Hilbert space

Zenon Fortuna, Warszawa

Abstract. An example of a quadratic convex functional is presented, for which the conjugate gradient method gives only linear convergence. For the case when the Hessian has the form $A = A_1 + A_2$, where A_1 is an algebraic operator and A_2 is a compact operator, the superlinear convergence is proved.

Let us consider a quadratic functional in a real Hilbert space U:

$$Q(u) = \frac{1}{2} \langle u - \hat{u}, Au - \hat{u} \rangle + \hat{Q},$$

where $u, \hat{u} \in U$, $\hat{u}$ is the optimal point, $\hat{Q}$ is the optimal value, the linear Hessian A is assumed to be bounded:

$$\operatorname{sp} A \subset [m, M]; \qquad m, M > 0.$$

It is a well known fact that when we start from an arbitrarily chosen starting point $u_0 \in U$, the steepest descent method (SDM) generates a sequence $\bar{Q}_0, \bar{Q}_1, \bar{Q}_2, \ldots, \bar{Q}_n, \ldots$ of values of the functional Q, converging to the optimal value $\hat{Q}$. Due to the Kantorovič inequality we can estimate the ratio of convergence $\bar{q}_n = \dfrac{\bar{Q}_{n+1} - \hat{Q}}{\bar{Q}_n - \hat{Q}}$ in the following way:

$$\bar{q}_n \leqq \left(\frac{M - m}{M + m}\right)^2, \qquad n = 0, 1, 2, \ldots$$

Because of $\left(\dfrac{M - m}{M + m}\right)^2 < 1$ this inequality means that the sequence $\bar{Q}_0, \bar{Q}_1, \ldots$ generated by SDM converges at least linearly (like some geometric progression). However, it is possible to find such a Hessian A and such a starting point u_0, that we obtain exact linear convergence, namely

$$\bar{q}_n = \left(\frac{M - m}{M + m}\right)^2, \qquad n = 0, 1, 2, \ldots$$

Because of this fact we say that SDM assures only linear convergence. The conjugate gradient method (see algorithm (a)), when applied to a functional Q with the starting point u_0, generates a sequence $Q_0, Q_1, Q_2, \ldots$ which converges to $\hat{Q}$, too, and one can prove that $Q_2 \leqq \bar{Q}_2, Q_3 \leqq \bar{Q}_3, \ldots, Q_n \leqq \bar{Q}_n, \ldots$ $(Q_0 = \bar{Q}_0, Q_1 = \bar{Q}_1)$. Except the trivial cases we obtain $Q_2 < \bar{Q}_2, Q_3 < \bar{Q}_3, \ldots$; so we say, that CGM is better than SDM.

Because of the fact that $q_0 = \bar{q}_0$, it is impossible to prove a better general evaluation of the ratio of convergence q_n, which would be valid for all n, than the evaluation for SDM. In [1] we find, however, that CGM has an asymptotically better evaluation:

$$q_n \leqq \left(\frac{\sqrt{M} - \sqrt{m}}{\sqrt{M} + \sqrt{m}}\right)^2 + \varepsilon_n, \qquad n = 0, 1, 2, \ldots; \qquad \varepsilon_n > 0, \quad \varepsilon_n \to 0.$$

This shows that we can reduce our estimate of the convergence factor from $\left(\frac{M - m}{M + m}\right)^2$ to at least $\left(\frac{\sqrt{M} - \sqrt{m}}{\sqrt{M} + \sqrt{m}}\right)^2$, but this is still only linear convergence.

In this paper we show that in some cases CGM gives only linear convergence, but for a wide class of quadratic functionals the convergence is superlinear.

First we shall construct an example of a quadratic functional for which CGM gives only linear convergence.

Example. Let $(e_i)_0^\infty$ be an orthonormal basis in U. We define a linear operator A:

$$Ae_k \triangleq b_{k-1}e_{k-1} + a_k e_k + b_k e_{k+1}; \qquad b_{-1} = 0, \; b_i = \frac{1 - m}{4}, \; a_i = \frac{1 + m}{2}, \qquad i = 0, 1, \ldots$$

This is a linear, symmetric operator which has a representation by a Jacobi matrix J,

$$J = \begin{bmatrix} a_0 & b_0 & 0 & 0 & \ldots \\ b_0 & a_1 & b_1 & 0 & \ldots \\ 0 & b_1 & a_2 & b_2 & \ldots \\ \ldots & \ldots & \ldots & \ldots & \ldots \end{bmatrix},$$

and one can prove that this operator is bounded. Its closure $\bar{A}$ is a linear bounded selfadjoint operator, and $\operatorname{sp} \bar{A} \subset [m, 1]$.

We construct $Q(u) = \frac{1}{2} \langle u - u^1, \bar{A}u - \hat{u} \rangle$, and let us start with the first gradient $\bar{A}(u_0 - \hat{u}) = e_0$. Now, using the theory of moments we can prove (see [2]) that

$$\frac{Q_{n+1}}{Q_n} \geqq \left(\frac{1 - m}{4}\right)^2, \qquad n = 0, 1, \ldots$$

So, generally in Hilbert space CGM has only linear convergence.

Now, let us consider the following dynamic problem of optimization:

$$\begin{aligned} \dot{x}(t) &= Dx(t) + Ex(t - h) + Fu(t), \\ x(t) &= \varphi(t) \quad \text{for} \quad t \in [t_0 - h, t_0]. \end{aligned} \tag{1}$$

We want to find a control vector-function $\hat{u}(t)$ which minimizes the functional

$$Q(u) = \frac{1}{2} \int_{t_0}^{t_1} [\langle x(t), Nx(t)\rangle + \langle u(t), Ru(t)\rangle] \, dt. \tag{2}$$

Let us assume that the matrices $R > 0$, $N \geqq 0$, D, E, F are constant. Under these assumptions exists a unique solution $\hat{u}(t)$.

In [3] we find the solution of equation (1) in the form:

$$x(t) = x_0(t) + \int_{t_0}^{t_1} K(t, \tau) \, Fu(\tau) \, d\tau,$$

where $x_0(t)$ is the trajectory generated by $u(t) \equiv 0$, and $K(t, \tau)$ is the continuous kernel of the corresponding integral operator. Let us denote it by shorter notation:

$$x = x_0 + Lu, \tag{3}$$

where L is the above defined compact linear operator (a Volterra operator with continuous kernel) in the space of piecewise continuous control functions U.

After substituting (3) into (2), we obtain

$$Q(u) = \frac{1}{2} \langle u - \hat{u}, A(u - \hat{u})\rangle + Q,$$

where

$$A = R + L^*NL, \quad \hat{u} = -A^{-1}L^*Nx_0, \quad \hat{Q} = Q(\hat{u}).$$

Now we prove the following

Theorem. *If the Hessian A can be split into two parts: $A = A_1 + A_2$ such that A_1 is an algebraic operator (i.e. there exist coefficients $a_1, a_2, \ldots, a_N$ such that $I + a_1A_1 + a_2A_1^2 + \cdots + a_NA_1^N = 0$) and A_2 is a compact operator, then, for any starting point u_0, the sequence $Q_0, Q_1, \ldots, Q_n, \ldots$ generated by CGM converges to the optimal value $\hat{Q}$ superlinearly, namely $\sqrt[n]{Q_n - \hat{Q}} \to 0$.*

It is easy to see that for problem (1), (2) the above assumptions are satisfied (because the matrix R satisfies its characteric equation), so if the Theorem is true, we would have superlinear convergence in these quadratic dynamic cases.

Proof. The algorithm of CGM is the following:

(a)
0. Choose u_0 — starting point; $i = 0$.
1. Compute g_i — gradient of Q in point u_i.
2. Find $s_i = g_i + \alpha_{i-1}$; $\alpha_{-1} = 0$; for $i > 0$, α_{i-1} is such, that $\langle s_i, As_{i-1}\rangle = 0$.
3. Find $u_{i+1} = u_i + \eta_i s_i$ where η_i minimizes the function $f(\eta) = Q(u_i + \eta_{s_i})$.
4. $i : i + 1$, go to 1.

Let us assume (for convenience) that $\hat{Q} = 0$ and $\operatorname{sp} A \subset [\alpha, 1]$.

Lemma 1. $\lim\limits_{k\to\infty} \dfrac{Q_{k+N}}{Q_k} = 0.$

Proof. It is a well known fact (see [1]) that if we have already generated the directions $s_0, s_1, \ldots, s_{k-1}$, then

$$Q_k = \frac{1}{2} \min_{z\in S_{k-1}} \langle g_0 - z, A^{-1}(g_0 - z)\rangle,$$

where

$$S_{k-1} = \operatorname{lin}\{As_0, As_1, \ldots, As_{k-1}\}.$$

We say that Q_k is a suboptimal value of Q on the subspace S_{k-1}. From algorithm (a) it is easy to see that $S_{k-1} = \operatorname{lin}\{Ag_0, A^2g_0, \ldots, A^kg_0\}$, so

$$\begin{aligned} Q_k &= \frac{1}{2} \min_{l_1,\ldots,l_k} \langle g_0 + l_1Ag_0 + \cdots + l_kA^kg_0, A^{-1}(g_0 + \cdots)\rangle \\ &= \frac{1}{2} \langle g_0 + c_1^{(k)}Ag_0 + \cdots + c_k^{(k)}A^kg_0\ A^{-1}(g_0, + \cdots)\rangle = \frac{1}{2} \langle g_k, A^{-1}g_k\rangle, \end{aligned}$$

where g_k is the k-th gradient generated by algorithm (a).

When we change the optimal coefficients $c_i^{(k)}$ in any way, the value of Q_k will increase, so

$$\begin{aligned} \frac{Q_{k+N}}{Q_k} &= \frac{\langle g_0 + c_1^{(k+N)}Ag_0 + \cdots + c_{k+N}A^{(k+N)}g_0, A^{-1}(g_0 + \cdots)\rangle}{\langle g_0 + c_1^{(k)}Ag_0 + \cdots + c_k^{(k)}A^kg_0, A^{-1}(g_0 + \cdots)\rangle} \\ &\leqq \frac{\langle (I + a_1A + a_2A^2 + \cdots + a_NA^N)(g_0 + c_1^{(k)}Ag_0 + \cdots + c_k^{(k)}A^kg_0), A^{-1}(\ldots)\rangle}{\langle g_0 + c_1^{(k)}Ag_0 + \cdots + c_k^{(k)}A^kg_0, A^{-1}(\ldots)\rangle} \\ &= \frac{\langle Cg_k, A^{-1}Cg_k\rangle}{\langle g_k, A^{-1}g_k\rangle}, \end{aligned}$$

where

$$\begin{aligned} C &= I + a_1A + \cdots + a_NA^N = I + a_1(A_1 + A_2) + a_2(A_1 + A_2)^2 + \cdots \\ &\quad + a_N(A_1 + A_2)^N = [I + a_1A_1 + \cdots + a_NA_1^N] + [a_1A_2 + a_2A_1A_2 + a_2A_2A_1 + \cdots]. \end{aligned}$$

The first bracket is equal to 0 because of the assumption on the operator A_1. The second bracket contains a linear, selfadjoint and compact operator (because A_2 is compact). So C is a compact operator. Because of $\operatorname{sp} A \subset [m, 1]$ we have

$$\frac{1}{\langle g_k, A^{-1}g_k\rangle} \leqq \frac{1}{\langle g_k, g_k\rangle}$$

and

$$\frac{Q_{k+N}}{Q_k} \leqq \langle C\bar{g}_k, A^{-1}C\bar{g}_k\rangle, \qquad \bar{g}_k = \frac{g_k}{\|g_k\|}.$$

The sequence $(\bar{g}_k)_0^\infty$ is weakly convergent to 0 (because $\langle \bar{g}_i, \bar{g}_j \rangle = \delta_{ij}$). Thus $C\bar{g}_k \to 0$, and $\frac{Q_{k+N}}{Q_k} \to 0$.

Lemma 2. *If* $\frac{Q_{k+N}}{Q_k} \to 0$, *then* $\sqrt[n]{Q_n} \to 0$.

Proof. Let $n = rN + s$, $0 \leqq s < N$. Then

$$\sqrt[n]{Q_n} = \sqrt[n]{Q_0} \cdot \sqrt[n]{\frac{Q_N}{Q_0} \cdot \frac{Q_{2N}}{Q_N} \cdots \frac{Q_{rN}}{Q_{(r-1)N}} \cdot \frac{Q_n}{Q_{rN}}}.$$

For sufficiently large n we have $\sqrt[n]{Q_0} \leqq 2$. By assumption $\frac{Q_{rN}}{Q_{(r-1)N}} \xrightarrow[r\to\infty]{} 0$, so for any $\varepsilon \in (0, 1)$ there exists an integer r_0 such that for all $k \geqq r_0$,

$$\frac{Q_{kN}}{Q_{(k-1)N}} \leqq \varepsilon^N.$$

Because

$$\frac{Q_n}{Q_{rN}} \leqq 1, \qquad \frac{Q_{iN}}{Q_{(i-1)N}} \leqq 1, \qquad i = 1, 2, \ldots, r_0,$$

then

$$\sqrt[n]{Q_n} \leqq 2 \cdot \sqrt[n]{\frac{Q_{(r_0+1)N}}{Q_{r_0N}} \cdot \frac{Q_{(r_0+2)N}}{Q_{(r_0+1)N}} \cdots \frac{Q_{rN}}{Q_{(r-1)N}}} \leqq 2 \cdot \sqrt[n]{\varepsilon^{N(r-r_0)}} = \frac{2}{\sqrt[n]{\varepsilon^{Nr_0+s}}} \cdot \varepsilon.$$

For sufficiently large n

$$\sqrt[n]{\varepsilon^{Nr_0+s}} \geqq \sqrt[n]{\varepsilon^{N(r_0+1)}} \geqq \frac{1}{2}.$$

Finally, there exists such a n_0 that for every $n \geqq n_0$

$$\sqrt[n]{Q_n} \leqq 4\varepsilon.$$

This means

$$\sqrt[n]{Q_n} \to 0.$$

From Lemmas 1 and 2 follows the proof of the Theorem. Q.E.D.

Remark 1. The assumption that the part A_1 is an algebraic operator is important, as the following example shows.

Let be $Q(u) = \frac{1}{2} \int_1^3 u(t)\, tu(t)\, dt$, where $u \in C_{[1,3]}$. The gradient of this functional has the form $g(u)(t) = A(u - \hat{u})(t) = tu(t)$, so the Hessian $A = t$, and $\operatorname{sp} A \subset [1, 3]$.

Let us start with an algorithm (a), taking $u_0(t) = \frac{1}{t}$, $t \in [1, 3]$. One can prove, using the theory of moments (see [2]), that now

$$Q_n \geqq \frac{1}{3} \cdot \left(\frac{1}{36}\right)^n, \qquad n = 0, 1, 2, \ldots$$

So, the sequence $Q_0, Q_1, \ldots$ converges to 0 bounded from below by a geometrical progression, and the convergence is only linear.

Remark 2. When we can split the Hessian $A = A_1 + A_2$ so that the part A_2 is compact but A_1 is not algebraic, we can simply modify the standard CGM to assure the strongly superlinear convergence. Namely, instead of the standard 2-nd step we propose to compute the conjugate directions in the following way:

$$S_i = A_1^{-1} g_i + \alpha_{i-1} S_{i-1}, \qquad \alpha_{-1} = 0.$$

Now, the actual Hessian has the form $\tilde{A} = I + \tilde{A}_2$ where $\tilde{A}_2$ is compact, and by Lemma 1 we obtain $\frac{Q_{n+1}}{Q_n} \to 0$ (strongly superlinear convergence).

References

[1] Daniel, J. W., The approximate minimization of functionals. Prentice-Hall 1971.

[2] Fortuna, Z. F., Doctor's dissertation. Technical University of Warsaw, Institute of Automatic Control 1974.

[3] Габасов, Р., Кириллова, Ф. М., Качественная теория оптимальных процессов. Москва 1971.

Address of the author:

Institute of Automatic Control
Technical University of Warsaw
ul. Nowowiejska 15/19
00-665 Warszawa
Poland

Zur Polyoptimierung in verallgemeinerten topologischen Vektorräumen

Siegfried Gähler, Berlin

Abstract. The paper deals with polyoptimization problems in generalized topological vector spaces in the general as well as in the convex case and presents among others an extension and strengthening of the theorem of Dubovicki-Milutin [1, 2].

1. Verallgemeinerte Vektortopologien

R sei ein linearer Raum der Dimension $\geqq 1$, $\boldsymbol{V}(R) = \boldsymbol{V}$ die Gesamtheit aller T_1- (und damit T_2-) Vektortopologien über R und $T^{\boldsymbol{V}}$ die feinste Topologie aus $\boldsymbol{V}$.

Eine Topologie T über R heißt *translationsinvariant*, falls für jeden Punkt $p \in R$ und jede T-offene Menge $U \subseteqq R$ die Menge $p + U$ T-offen ist. Die Gesamtheit aller translationsinvarianten T_1-Topologien über R, die jeweils eine Basis $\mathfrak{U}$ äquilibrierter absorbierender Nullumgebungen mit $\alpha U \in \mathfrak{U}$ für alle $\alpha > 0$ und $U \in \mathfrak{U}$ besitzen, bezeichnen wir mit $\boldsymbol{V}'(R)$ oder kurz mit $\boldsymbol{V}'$. Die $T \in \boldsymbol{V}'$ sind verallgemeinerte Vektortopologien über R, d. h., es gilt $\boldsymbol{V} \subseteqq \boldsymbol{V}'$. Die $T \in \boldsymbol{V}$ zeichnen sich gegenüber den $T \in \boldsymbol{V}'$ durch die zusätzliche Forderung, daß zu jedem $U \in \mathfrak{U}$ ein $V \in \mathfrak{U}$ mit $V + V \subseteqq U$ existieren soll, aus. Für sie ist die Voraussetzung, daß $\alpha U \in \mathfrak{U}$ für jedes $\alpha > 0$ und $U \in \mathfrak{U}$ sein soll, eine Folge der übrigen Forderungen. Die verallgemeinerten Vektortopologien $T \in \boldsymbol{V}'$ wurden in [3] und [4] näher untersucht. Mittels [4], Satz 7, folgt, daß jede bezüglich $T \in \boldsymbol{V}'$ offene Menge algebraisch offen ist, und nach [4], Satz 9, besteht, sofern R von endlicher Dimension ist, zwischen beliebigem $T' \in \boldsymbol{V}'$ und der natürlichen Vektortopologie T von R die Beziehung $T' \subseteqq T$. Nach [4], Satz 13, existiert im Fall $\dim R \geqq 2$ zu jedem $T \in \boldsymbol{V}$ eine Topologie $T' \in \boldsymbol{V}'$, die echt gröber als T ist und nicht zu $\boldsymbol{V}$ gehört. In $\boldsymbol{V}'$ gibt es — siehe [4], Satz 11 — eine feinste Topologie $T^{\boldsymbol{V}'}$. Nach [4], Satz 12, besitzt $T^{\boldsymbol{V}'}$ die Gesamtheit $\mathfrak{U}$ aller Teilmengen U von R, derart daß für jeden endlichdimensionalen linearen Teilraum R' von R stets $U \cap R'$ bezüglich der natürlichen Topologie von R' eine äquilibrierte Nullumgebung in R' ist, als Basis, und es gilt $T^{\boldsymbol{V}'} \in \boldsymbol{V}$ genau dann, wenn die Dimension von R endlich ist.

Die Gesamtheit aller translationsinvarianten T_1-Topologien über R, die jeweils eine Basis $\mathfrak{U}$ algebraisch offener Nullumgebungen mit $\alpha U \in \mathfrak{U}$ für alle $\alpha > 0$ und $U \in \mathfrak{U}$ haben, bezeichnen wir mit $\boldsymbol{V}^*(R)$ oder kurz mit $\boldsymbol{V}^*$. Es gilt $\boldsymbol{V}' \subseteqq \boldsymbol{V}^*$. Jede Nullumgebung einer Topologie $T \in \boldsymbol{V}^*$ ist absorbierend. Nach [5], Satz 4 und Satz 5, gibt es in $\boldsymbol{V}^*$ eine feinste — aus der Gesamtheit aller algebraisch offenen Mengen aus R bestehende —

Topologie T^{V^*} und gilt genau dann $T^{V^*} \in V'$ (und damit $T^{V^*} \in V$), wenn $\dim R = 1$ ist.

Jedes lineare Funktional φ über R ist bezüglich T^V und damit bezüglich $T^{V'}$ sowie T^{V^*} stetig. Ein von 0 verschiedenes lineares Funktional φ über R ist nach [5], Satz 3, zu beliebigem $T \in V'$ genau dann T-stetig, wenn sein Nullraum N bezüglich T nicht in R dicht ist.

$\mathfrak{K}$ sei ein Kegel in R mit 0 als Scheitel. Zu beliebigem $T \in V^*$ ist nach [5], Satz 6, das T-Innere von $\mathfrak{K}$ ein Kegel in R mit 0 als Scheitel, der im Falle der Konvexität von $\mathfrak{K}$ konvex ist. Mit $\mathfrak{K}^*$ bezeichnen wir den zu $\mathfrak{K}$ T^{V^*}- (also algebraisch, also $T^{V'}$-, also T^V-) dualen Kegel, d. h. die Gesamtheit aller linearen Funktionale φ über R mit $\varphi(p) \geqq 0$ für alle $p \in \mathfrak{K}$. Zu beliebigem $T \in V^*$ gilt für jedes von 0 verschiedene $\varphi \in \mathfrak{K}^*$ und für jeden T-inneren Punkt p von $\mathfrak{K}$ die Beziehung $\varphi(p) > 0$. Enthält $\mathfrak{K}$ bezüglich einer Topologie $T \in V'$ innere Punkte, dann ist jedes $\varphi \in \mathfrak{K}^*$ T-stetig.

Satz 1 (*vgl.* [1] *bzw.* [6], *Lemma* 5.11). *$\mathfrak{K}_1, \ldots, \mathfrak{K}_n$ ($n \geqq 2$) seien nichtleere konvexe Kegel in R mit 0 als Scheitel, und $\mathfrak{K}_1, \ldots, \mathfrak{K}_{n-1}$ seien T^{V^*}-offen. Dann gilt*

$$\bigcap_{i=1}^{n} \mathfrak{K}_i = \emptyset$$

genau dann, wenn $\varphi_i \in \mathfrak{K}_i'^*$, $i = 1, \ldots, n$, die nicht alle gleich 0 sind, mit

$$\sum_{i=1}^{n} \varphi_i = 0$$

existieren.

Beweis. [5], Satz 7.

2. Eine Ausdehnung und Verschärfung des Satzes von Dubovickij-Miljutin

R sei mit einer Topologie $T \in V'$ ausgestattet und m eine Mächtigkeit $\geqq \aleph_0$. $\mathfrak{Q}$ bezeichne die Gesamtheit aller Teilmengen Q_1 von R von der Art, daß zu beliebigem $q \in Q_1$ eine T^{V^*}-Umgebung U_{q,Q_1} von q mit folgender Eigenschaft existiert: Zu jedem linearen Teilraum R' von R der Dimension $< m$, zu jedem Punkt $r \in Q_1 \cap U_{q,Q_1}$, für den in R' eine $T|R'$-Nullumgebung U mit $r + U \in Q_1$ existiert, sowie zu jedem Punkt $r' \in Q_1 \cap U_{q,Q_1}$ und jedem $\alpha \in (0, 1]$ gibt es eine reelle Zahl $\varepsilon > 0$ und in R' eine $T|R'$-Nullumgebung V mit

$$q + [0, \varepsilon] (\alpha r + (1 - \alpha) r' - q + V) \subseteqq Q_1 .$$

Zu jedem $r \in Q_1 \cap U_{q,Q_1}$ existiert dann, wie sich unter der Einschränkung $R' = \{0\}$ und $r = r'$ ergibt, insbesondere ein $\varepsilon > 0$ mit $q + [0, \varepsilon] (r - q) \subseteqq Q_1$. Die T-offenen Mengen aus R, die konvexen Mengen aus R sowie der Durchschnitt endlich vieler zu $\mathfrak{Q}$ gehörender Mengen gehören nach [5], Satz 9, zu $\mathfrak{Q}$.

Q_1 sei eine Menge des Systems $\mathfrak{Q}$ und q beliebig aus Q_1. $\mathfrak{K} = [0, \infty) (Q_1 \cap U_{q,Q_1} - q)$ ist ein von der speziellen Wahl von U_{q,Q_1} unabhängiger konvexer Kegel in R mit 0 als

Scheitel. Zu beliebigem $r \in \mathfrak{K}$ sei $\mathfrak{N}_r$ die Gesamtheit aller Mengen $N \subseteqq R$, die von $\{0\}$ verschieden sind und zu denen für jede Teilmenge M von $\mathfrak{K}$ der Mächtigkeit $< m$ ein $\varepsilon > 0$ und in dem durch M erzeugten linearen Teilraum lM von R eine $T|lM$-Nullumgebung V so existieren, daß $[0, \varepsilon]\,(r + V) \subseteqq N$ gilt. $\mathfrak{N}_r$ ist nach [5], Satz 11, ein Filter in R mit folgenden Eigenschaften:

1. Für beliebiges $\alpha > 0$ und $N \in \mathfrak{N}_r$ gilt $\alpha N \in \mathfrak{N}_r$.
2. Für jedes $\alpha > 0$ gilt $\mathfrak{N}_{\alpha r} = \mathfrak{N}_r$.

Mit $\widetilde{\mathfrak{N}_r}$ bezeichnen wir die Gesamtheit aller Mengen $[0, \varepsilon]\,(r + V)$, wobei ε eine reelle Zahl > 0 und V eine T-Nullumgebung ist. Es gilt $\widetilde{\mathfrak{N}_r} \subseteqq \mathfrak{N}_r$.

Q_2 und Q_3 seien weitere Teilmengen von R, und q liege in $Q = \bigcap_{i=1}^{3} Q_i$. S sei ein linearer Raum der Dimension $\geqq 1$, K eine nichtleere Teilmenge von S und f eine Abbildung von Q_1 in S. Wir sagen, daß f in q (bezüglich Q, T und K) ein *lokales Extremum* besitzt, wenn eine T-Umgebung U von q mit

$$f[Q \cap U] - f(q) \subseteqq K \tag{$*$}$$

existiert. Wenn an Stelle von $(*)$ die Beziehung $f[Q] - f(q) \subseteqq K$ gilt, sagen wir, daß f in q ein *Extremum* besitzt. Jede Extremstelle von f ist bezüglich T eine lokale Extremstelle von f.

Ein $r \in \mathfrak{K}$ heißt *ausgezeichnete Richtung in* q, falls ein $N \in \mathfrak{N}_r$ mit $q + N \subseteqq Q_1$ und $\big(f[(q + N) \setminus \{q\}] - f(q)\big) \cap K = \emptyset$ existiert. Ein $r \in \mathfrak{K}$ wird *zulässige Richtung in* q genannt, falls es ein $N \in \mathfrak{N}_r$ mit $q + N \subseteqq Q_1 \cap Q_2$ gibt. Ein $r \in \mathfrak{K}$ wird als *Tangentialrichtung in* q bezeichnet, falls $(q + N) \cap Q_3$ für jedes $N \in \mathfrak{N}_r$ einen von q verschiedenen Punkt enthält. $\mathfrak{K}_1$ sei die Gesamtheit aller ausgezeichneten Richtungen, $\mathfrak{K}_2$ die Gesamtheit aller zulässigen Richtungen und $\mathfrak{K}_3$ die Gesamtheit aller Tangentialrichtungen in q. $\mathfrak{K}_1$, $\mathfrak{K}_2$ sowie $\mathfrak{K}_3$ sind Kegel in R mit 0 als Scheitel. Ist $m \geqq \aleph_1$ und q aus dem T-Inneren von Q_1, dann gehört nach [5], Satz 18, jeder Punkt $r \in \mathfrak{K}$, zu dem eine Abbildung p eines Intervalles $(0, \alpha]$ in die Menge Q_3 mit $q \notin p\,[(0, \alpha]]$ und

$$T\text{-}\lim_{\tau \to 0} \frac{p(\tau) - q - \tau r}{\tau} = 0$$

existiert, zu $\mathfrak{K}_3$.

Eine Ausdehnung und Verschärfung des Satzes von Dubovickij-Miljutin ([1]; siehe auch [6], Theorem 6.1; siehe ferner [5], Satz 19, sowie die anschließenden Ausführungen) bringt der nachstehende

Satz 2. *Für ein lokales Extremum von f in q sind die beiden folgenden Bedingungen notwendig:*

1. *Es ist* $\bigcap_{i=1}^{3} \mathfrak{K}_i = \emptyset$.

2. *Für* $i = 1, 2, 3$ *seien* $\mathfrak{K}_i'$ *nichtleere konvexe Kegel in* R *mit* 0 *als Scheitel und mit* $\mathfrak{K}_i' \subseteqq \mathfrak{K}_i$, *wobei die* $\mathfrak{K}_i'$, $i = 1, 2$, $T^{V^*(R)}$-*offen sein mögen. Dann existieren* $\varphi_i \in K_i'^*$, $i = 1, 2, 3$, *die nicht alle gleich* 0 *sind, mit*

$$\sum_{i=1}^{3} \varphi_i = 0 \qquad (\textit{Euler-Lagrangesche Gleichung}).$$

Beweis. [5], Beweis von Satz 19[1].

Unter Berücksichtigung von Satz 1 ergibt sich, daß das φ_1 in Bedingung 2 von Satz 2 im Fall $\mathfrak{K}_2' \cap \mathfrak{K}_3' \neq \emptyset$ nicht gleich 0 sein kann.

Bei verschiedener Wahl von K ergeben sich verschiedene Arten lokaler Extremstellen, von denen die drei im folgenden angeführten Fälle besonders interessieren. Es sei dazu S mit einer Topologie aus $\boldsymbol{V}^*(S)$ ausgestattet und in S ein konvexer Kegel $\boldsymbol{K}$ mit $\boldsymbol{K} \cap (-\boldsymbol{K}) = \{0\}$ und $\underline{\boldsymbol{K}} \neq \emptyset$ gegeben. Es sei

1. $K = \boldsymbol{K}$,
2. $K = S \setminus (-\boldsymbol{K} \setminus \{0\})$

bzw.

3. $K = S \setminus -\underline{\boldsymbol{K}}$.

Eine lokale Extremstelle von f der ersten Art ist eine lokale Extremstelle von f der zweiten Art und eine solche ist eine lokale Extremstelle von f der dritten Art.

$\mathfrak{K}_0$ sei die Gesamtheit aller $r \in \mathfrak{K}$, für die jeweils ein $N \in \mathfrak{N}_r$ mit $q + N \subseteqq Q_1$ existiert, und $\mathfrak{K}_0'$ das $T^{V^*(R)}$-Innere von $\mathfrak{K}_0$. $D_q(f)$ bezeichne die Gesamtheit aller linearen Abbildungen $\Phi\colon R \to S$ von der Art, daß zu jedem $r \in \mathfrak{K}_0'$ mit $\Phi(r) \in -\underline{\boldsymbol{K}}$ ein $N \in \mathfrak{N}_r$ mit $q + N \subseteqq Q_1$ und mit $f[(q + N) \setminus \{q\}] - f(q) \subseteqq -\underline{\boldsymbol{K}}$ existiert. Die Nullabbildung gehört zu $D_q(f)$. $\alpha > 0$ und $\Phi \in D_q(f)$ ziehen $\alpha\Phi \in D_q(f)$ nach sich.

Korollar. *In jedem der für K zugelassenen drei Fälle ist für ein lokales Extremum von f in q die folgende Bedingung notwendig:*

Es sei Φ beliebig aus $D_q(f)$ und $\mathfrak{K}_2'$ sowie $\mathfrak{K}_3'$ wie in Bedingung 2 von Satz 2 gewählt. Dann existieren lineare Funktionale $\psi \in -\boldsymbol{K}^*$ und $\varphi_i \in \mathfrak{K}_i'^*$, $i = 2, 3$, die nicht alle gleich 0 sind, mit

$$\psi \circ \Phi(p) + \sum_{i=2}^{3} \varphi_i(p) \leqq 0$$

für alle $p \in \mathfrak{K}_0'$. Ist $U_{q,Q_1} = R$ möglich und $\mathfrak{K}_0' \neq 0$, so ist die angegebene Ungleichung für alle $p \in Q_1 - q$ richtig.

Beweis. [5], Satz 16 und Beweis von Satz 22[1].

3. Der konvexe Fall

Q_1, Q_2 und Q_3 seien nun konvex. Ferner sei f konvex, d. h., es gelte für beliebiges $p_1, p_2 \in Q_1$ und beliebiges $\alpha \in [0, 1]$ stets

$$\alpha f(p_1) + (1 - \alpha) f(p_2) - f\big(\alpha p_1 + (1 - \alpha) p_2\big) \in \boldsymbol{K}.$$

Satz 3 (vgl. [6], Theorem 15.1). *In jedem der für K zugelassenen drei Fälle ist jede lokale Extremstelle von f zugleich eine Extremstelle von f.*

Beweis. q sei eine lokale Extremstelle von f und U eine T-Umgebung von q mit $f[Q \cap U] - f(q) \subseteqq K$. p sei ein beliebiger Punkt aus Q. Da $U - q$ absorbierend ist, gibt es

---

[1] In [5] wird an Stelle von $\mathfrak{K}_2$ der Kegel aller $r \in \mathfrak{K}$ verwendet, für die jeweils ein $N \in \mathfrak{N}_v$ mit $q + N \subseteqq Q_2$ existiert.

ein $\alpha \in (0, 1)$ mit $p_\alpha = q + \alpha(p - q) \in Q \cap U$, woraus $f(p_\alpha) - f(q) \in K$ folgt. Im Fall 1 erhält man somit auf Grund der Konvexität von f die Beziehung $f(p) - f(q) \in K$. In den Fällen 2 und 3 führen wir den Nachweis, daß $f(p) - f(q) \in K$ ist, indirekt, nehmen also an, es wäre $f(p) - f(q) \in S \setminus K$. Dann ergibt sich auf Grund der Konvexität von f die Beziehung $f(p_\alpha) - f(q) \in S \setminus K$, die jedoch nicht richtig ist. Der Satz ist damit bewiesen.

Wir setzen im weiteren zusätzlich zu den obigen Annahmen voraus, daß $m = \aleph_0$ gilt, $\underline{Q_1} \cap \underline{Q_2} \cap Q_3$ mehr als einen Punkt enthält und f stetig ist.

Satz 4. *Es liege der Fall* 3 (*d. h.* $K = S \setminus -\underline{\boldsymbol{K}}$) *vor. Dann gilt*

$$\mathfrak{K}_1 = \left\{r \in \mathfrak{K};\ \exists N \in \widetilde{\mathfrak{N}_r}: q + N \subseteqq Q_1,\ \left(f[(q + N) \setminus \{q\}] - f(q)\right) \cap K = \emptyset\right\}$$

und

$$\mathfrak{K}_2 = \{r \in \mathfrak{K};\ \exists N \in \widetilde{\mathfrak{N}_r}: q + N \subseteqq Q_1 \cap Q_2\}.$$

Die Kegel $\mathfrak{K}_2$ und $\mathfrak{K}_3$ sind nicht leer. Die Kegel $\mathfrak{K}_1$ und $\mathfrak{K}_2$ sind T-offen. Die Kegel $\mathfrak{K}_1$, $\mathfrak{K}_2$ und $\mathfrak{K}_3$ sind konvex.

Beweis. i. r sei beliebig aus $\mathfrak{K}_1$. Es gilt $r \neq 0$, da andernfalls ein von 0 verschiedener Punkt $r^* \in R$ mit $\{q + r^*, q - r^*\} \subseteqq Q_1$ und $\{f(q + r^*) - f(q), f(q - r^*) - f(q)\} \subseteqq -\underline{\boldsymbol{K}}$ existieren müßte, was im Widerspruch zur Voraussetzung $\boldsymbol{K} \cap (-\boldsymbol{K}) = \{0\}$ auf

$$0 = f(q) - f(q) \in \frac{1}{2}\left(f(q + r^*) - f(q)\right) + \frac{1}{2}\left(f(q - r^*) - f(q)\right) - \boldsymbol{K} \subseteqq -\underline{\boldsymbol{K}}$$

führen würde. p^* sei beliebig aus $\underline{Q_1}$. Da ein $\varepsilon > 0$ und in dem durch $\{r, p^* - q\}$ erzeugten linearen Teilraum $l\{r, p^* - q\}$ eine $T \mid l\{r, p^* - q\}$-Nullumgebung V mit $q + [0, \varepsilon]\,(r + V) \subseteqq Q_1$ existieren, gibt es einen Punkt $p \in Q_1$ mit $q + \varepsilon r = \beta p^* + (1 - \beta)\,p$ für geeignetes $\beta \in (0, 1)$, womit $q + \varepsilon r \in \underline{Q_1}$ folgt. ε kann so gewählt werden, daß $f(q + \varepsilon r) - f(q) \in -\underline{\boldsymbol{K}}$ ist. U sei eine T-Nullumgebung in R mit $q + \varepsilon(r + U) \subseteqq Q_1$ und $f[q + \varepsilon(r + U)] - f(q) \subseteqq -\underline{\boldsymbol{K}}$, $N = [0, \varepsilon]\,(r + U)$ gehört zu $\widetilde{\mathfrak{N}_r}$. Es gilt $q + N \subseteqq Q_1$ und, da für jedes $\varepsilon' \in (0, 1]$

$$\begin{aligned} f[q + \varepsilon'\varepsilon(r + U)] - f(q) &\subseteqq -\boldsymbol{K} + \varepsilon' f[q + \varepsilon(r + U)] + (1 - \varepsilon')\,f(q) - f(q) \\ &= -\boldsymbol{K} + \varepsilon'\left(f[q + \varepsilon(r + U)] - f(q)\right) \subseteqq -\underline{\boldsymbol{K}} \end{aligned}$$

ist, $\left(f[(q + N) \setminus \{q\}] - f(q)\right) \cap K = \emptyset$. Daher folgt $\mathfrak{K}_1 \subseteqq \{r \in \mathfrak{K};\ \exists N \in \widetilde{\mathfrak{N}_r}: q + N \subseteqq Q_1$, $\left(f[(q + N) \setminus \{q\}] - f(q)\right) \cap K = \emptyset\}$. Die Enthaltenseinsbeziehung $\supseteqq$ ist evident, womit die Gleichheit der beiden Mengen bewiesen ist.

ii. r sei nun beliebig aus $\mathfrak{K}_2$. p^* sei ein Punkt aus $\underline{Q_1} \cap \underline{Q_2}$. Da ein $\varepsilon > 0$ und in $l\{r, p^* - q\}$ eine $T \mid l\{r, p^* - q\}$-Nullumgebung V mit $q + [0, \varepsilon]\,(r + V) \subseteqq Q_1 \cap Q_2$ existieren, gibt es einen Punkt $p \in Q_1 \cap Q_2$ mit $q + \varepsilon r = \beta p^* + (1 - \beta)\,p$ für geeignetes $\beta \in (0, 1)$. $U = \frac{1}{\varepsilon}\left(\beta(\underline{Q_1} \cap \underline{Q_2}) + (1 - \beta)\,p - q\right) - r$ ist eine T-Nullumgebung. Es gilt $q + \varepsilon(r + U) \subseteqq Q_1 \cap Q_2$ und folglich $q + [0, \varepsilon]\,(r + U) \subseteqq Q_1 \cap Q_2$. Damit ergibt sich

$\mathfrak{K}_2 \subseteqq \{r \in \mathfrak{K};\ \exists\ N \in \widetilde{\mathfrak{N}_r}: q + N \subseteqq Q_1 \cap Q_2\}$. Da die Enthaltenseinsbeziehung $\supseteqq$ evident ist, folgt die Gleichheit der beiden Mengen.

iii. Zu beliebigem von q verschiedenem $p^* \in \underline{Q_1} \cap \underline{Q_2} \cap Q_3$ gilt offensichtlich $p^* - q \in \mathfrak{K}_2 \cap \mathfrak{K}_3$, die Kegel $\mathfrak{K}_2$ und $\mathfrak{K}_3$ sind mithin nicht leer.

iv. Unter Verwendung der im Satz angegebenen Darstellung für $\mathfrak{K}_1$ und $\mathfrak{K}_2$ folgt, daß die Kegel $\mathfrak{K}_1$ und $\mathfrak{K}_2$ T-offen sind.

v. r_1 und r_2 seien beliebige Punkte aus $\mathfrak{K}_1$ und α eine reelle Zahl aus $(0, 1)$. Mittels der im Satz gegebenen Darstellung für $\mathfrak{K}_1$ folgt die Existenz eines $\varepsilon > 0$ und einer T-Nullumgebung V, so daß für $N_i = [0, \varepsilon]\,(r_i + V)$, $i = 1, 2$, die Beziehung $q + N_i \subseteqq Q_1$ und $f[(q + N_i) \setminus \{q\}] - f(q) \subseteqq -\underline{\boldsymbol{K}}$ gilt. Wegen $[0, \varepsilon]\,(\alpha r_1 + (1 - \alpha)\, r_2 + V) \subseteqq \alpha N_1 + (1 - \alpha)\, N_2$ gehört $N = \alpha N_1 + (1 - \alpha)\, N_2$ zu $\mathfrak{N}_{\alpha r_1 + (1-\alpha) r_2}$. Auf Grund der Konvexität von Q_1 ergibt sich $q + N = \alpha(q + N_1) + (1 - \alpha)\,(q + N_2) \subseteqq Q_1$. Jeder Punkt $p \in (q + N) \setminus \{q\}$ hat die Gestalt $p = \alpha(q + p_1) + (1 - \alpha)\,(q + p_2)$ mit $p_1 \in N_1$, $p_2 \in N_2$ und $\alpha p_1 + (1 - \alpha)\, p_2 \neq 0$. Es gilt

$$f(p) - f(q) \in -\boldsymbol{K} + \alpha\big(f(q + p_1) - f(q)\big) + (1 - \alpha)\,\big(f(q + p_2) - f(q)\big) \subseteqq -\underline{\boldsymbol{K}},$$

woraus $\big(f[(q + N) \setminus \{q\}] - f(q)\big) \cap \boldsymbol{K} = \emptyset$ folgt. Es ist damit $\alpha r_1 + (1 - \alpha)\, r_2 \in \mathfrak{K}_1$, also die Konvexität von $\mathfrak{K}_1$, bewiesen.

vi. r_1 und r_2 seien beliebige Punkte aus $\mathfrak{K}_2$ und α eine reelle Zahl aus $(0, 1)$. Mittels der im Satz gegebenen Darstellung für $\mathfrak{K}_2$ folgt die Existenz eines $\varepsilon > 0$ und einer T-Nullumgebung V, so daß für $N_i = [0, \varepsilon]\,(r_i + V)$, $i = 1, 2$, die Beziehung $q + N_i \subseteqq Q_1 \cap Q_2$ gilt. $N = \alpha N_1 + (1 - \alpha)\, N_2$ gehört zu $\mathfrak{N}_{\alpha r_1 + (1-\alpha) r_2}$. Auf Grund der Konvexität von Q_1 und Q_2 ergibt sich $q + N = \alpha(q + N_1) + (1 - \alpha)\,(q + N_2) \subseteqq Q_1 \cap Q_2$, womit die Konvexität von $\mathfrak{K}_2$ bewiesen ist.

vii. Zum Nachweis der Konvexität von $\mathfrak{K}_3$ seien $r_1, r_2 \in \mathfrak{K}_3$, $\alpha \in (0, 1)$ und $N \in \mathfrak{N}_{\alpha r_1 + (1-\alpha) r_2}$ beliebig gegeben. Da für $i = 1, 2$ zu beliebigem $N_i \in \mathfrak{N}_{r_i}$ in $(q + N_i) \cap Q_3$ ein von q verschiedener Punkt liegt, gibt es eine Teilmenge M^* von $\mathfrak{K}$ der Mächtigkeit $< m$, so daß für jedes $\varepsilon > 0$ und jede $T|lM^*$-Nullumgebung V in $\big(q + [0, \varepsilon]\,(r_i + V)\big) \cap Q_3$ ein von q verschiedener Punkt liegt. Es existiert ein $\varepsilon^* > 0$ und eine äquilibrierte $T|lM^*$-Nullumgebung V^* mit $[0, \varepsilon^*]\,\big(\alpha r_1 + (1 - \alpha)\, r_2 + V^*\big) \subseteqq N$. p_1 sei ein von q verschiedener Punkt aus $\big(q + [0, \varepsilon^*]\,(r_1 + V^*)\big) \cap Q_3$. Dieser hat die Gestalt $p_1 = q + \varepsilon_1(r_1 + q_1)$ mit $\varepsilon_1 \in (0, \varepsilon^*]$, $q_1 \in V^*$ und $r_1 + q_1 \neq 0$. $W = \dfrac{1}{1 - \alpha}\,(V^* - \alpha q_1)$ ist eine $T|lM^*$-Nullumgebung. p_2 sei ein von q verschiedener Punkt aus $\big(q + [0, \varepsilon^*]\,(r_2 + W)\big) \cap Q_3$. Dieser hat die Gestalt $p_2 = q + \varepsilon_2(r_2 + q_2)$ mit $\varepsilon_2 \in (0, \varepsilon^*]$, $q_2 \in W$ und $r_2 + q_2 \neq 0$ und kann so gewählt werden, daß $\alpha(r_1 + q_1) \neq -(1 - \alpha)\,(r_2 + q_2)$ ist. Mit

$$\beta = \frac{\alpha\varepsilon_2}{\alpha\varepsilon_2 + (1-\alpha)\,\varepsilon_1} \quad \text{und} \quad \bar{\varepsilon} = \frac{\varepsilon_1\varepsilon_2}{\alpha\varepsilon_2 + (1-\alpha)\,\varepsilon_1}$$

ergibt sich

$$\begin{aligned}\beta p_1 + (1 - \beta)\, p_2 &= \frac{1}{\alpha\varepsilon_2 + (1 - \alpha)\,\varepsilon_1}\Big(\alpha\varepsilon_2\big(q + \varepsilon_1(r_1 + q_1)\big) + (1 - \alpha)\,\varepsilon_1\big(q + \varepsilon_2(r_2 + q_2)\big)\Big)\\ &= q + \bar{\varepsilon}\big(\alpha r_1 + (1 - \alpha)\, r_2 + \alpha q_1 + (1 - \alpha)\, q_2\big)\\ &\in q + \bar{\varepsilon}\big(\alpha r_1 + (1 - \alpha)\, r_2 + V^*\big) \subseteqq q + N.\end{aligned}$$

Da auf Grund von $p_1, p_2 \in Q_3$ und der Konvexität von Q_3 ferner $\beta p_1 + (1 - \beta) p_2 \in Q_3$ ist, gilt $\beta p_1 + (1 - \beta) p_2 \in (q + N) \cap Q_3$. Wegen $\alpha(r_1 + q_1) \neq -(1 - \alpha)(r_2 + q_2)$ ist $\beta p_1 + (1 - \beta) p_2 \neq q$, womit sich $\alpha r_1 + (1 - \alpha) r_2 \in \mathfrak{K}_3$, also die Konvexität von $\mathfrak{K}_3$, ergibt. Der Satz ist damit bewiesen.

Der Kegel $\mathfrak{K}_1$ für den Fall 3 (d. h. $K = S \setminus -\underline{\boldsymbol{K}}$) ist im Kegel $\mathfrak{K}_1$ für den Fall 2 (d. h. $K = S \setminus (-\boldsymbol{K} \setminus \{0\})$) und dieser im Kegel $\mathfrak{K}_1$ für den Fall 1 (d. h. $K = \boldsymbol{K}$) enthalten.

Satz 5 (vgl. [6], Theorem 15.2). *Für ein Extremum von f in q ist in jedem der für K zugelassenen drei Fälle die folgende Bedingung notwendig:*

2'. *Es existieren $\varphi_i \in \mathfrak{K}_i^*$, $i = 1, 2, 3$, wobei φ_1 im Fall $\big(f[Q_1] - f(q)\big) \cap (-\underline{\boldsymbol{K}}) \neq \emptyset$ nicht gleich 0 ist, mit*

$$\sum_{i=1}^{3} \varphi_i = 0.$$

Im Fall 3 (d. h. $K = S \setminus -\underline{\boldsymbol{K}}$) ist sowohl die Bedingung 1 von Satz 2 als auch die Bedingung 2' für ein Extremum von f in q auch hinreichend.

Beweis. 1. Es sei $\big(f[Q_1] - f(q)\big) \cap (-\underline{\boldsymbol{K}}) \neq \emptyset$. Dann existiert auf Grund von $\underline{Q_1} \neq \emptyset$ und der Stetigkeit von f ein Punkt $r \in R$ und eine T-Nullumgebung U mit $q + r + \overline{U} \subseteqq Q_1$ und $f[q + r + U] - f(q) \subseteqq -\underline{\boldsymbol{K}}$. $N = [0, 1]\,(r + U)$ gehört zu $\mathfrak{N}_r$, und mittels der Konvexität von Q_1 und f ergibt sich $q + N \subseteqq Q_1$ und $f[(q + N) \setminus \{q\}] - f(q) \subseteqq -\underline{\boldsymbol{K}}$, woraus folgt, daß $\mathfrak{K}_1$ im Fall 3 den Punkt r enthält, also nicht leer ist. Unter Verwendung der eben bewiesenen Aussage, der Bemerkung vor Satz 5, der Sätze 1, 2 und 4, sowie des Teiles iii des Beweises von Satz 4 zeigt sich, daß in jedem der für K zugelassenen drei Fälle die Bedingung 2' für ein Extremum von f in q notwendig ist.

2. Es liege nun der Fall 3 vor. Den Nachweis, daß die Bedingung 1 von Satz 2 für ein Extremum von f in q hinreichend ist, führen wir indirekt. Wir nehmen an, es gebe einen Punkt $q^* \in Q$ mit $f(q^*) - f(q) \in -\underline{\boldsymbol{K}}$. p^* sei ein von q verschiedener Punkt aus $\underline{Q_1} \cap \underline{Q_2} \cap Q_3$. Dieser kann so gewählt werden, daß $f(p^*) - f(q) \in -\underline{\boldsymbol{K}}$ ist. Es sei $r^* = p^* - q$ und U eine T-Nullumgebung mit $p^* + U \subseteqq Q_1 \cap Q_2$ und $f[p^* + U] - f(q) \subseteqq -\underline{\boldsymbol{K}}$. $N = [0, 1]\,(r^* + U)$ gehört zu $\mathfrak{N}_{r^*}$. Mittels $q + [0, 1]\, r^* \subseteqq Q_1 \cap Q_3$ ergibt sich $r^* \in \mathfrak{K}_3$. Es ist $q + N = q + [0, 1]\,(p^* + U - q) \subseteqq Q_1 \cap Q_2$ und damit $r^* \in \mathfrak{K}_2$. Für jedes $\varepsilon \in (0, 1]$ gilt

$$\begin{aligned} f[q + \varepsilon(p^* + U - q)] - f(q) &\in -\boldsymbol{K} + (1 - \varepsilon) f(q) + \varepsilon f[p^* + U] - f(q) \\ &= -\boldsymbol{K} + \varepsilon\big(f[p^* + U] - f(q)\big) \subseteqq -\underline{\boldsymbol{K}}, \end{aligned}$$

womit $\big(f[(q + N) \setminus \{q\}] - f(q)\big) \cap K = \emptyset$, also $r^* \in \mathfrak{K}_1$ folgt. Es ist daher $r^* \in \bigcap_{i=1}^{3} \boldsymbol{K}_i$, was einen Widerspruch ergibt. Die Bedingung 1 von Satz 2 ist folglich für ein Extremum von f in q hinreichend. Da $\mathfrak{K}_1$ nach Teil 1 des vorliegenden Beweises im Fall $(f[Q_1] - f(q)) \cap (-\underline{\boldsymbol{K}}) \neq \emptyset$ nicht leer ist, ergibt sich unter Berücksichtigung von Satz 1 und Satz 4 weiterhin, daß auch Bedingung 2' für ein Extremum von f in q hinreichend ist. Der Satz ist damit bewiesen.

Wie man mühelos sieht, kann in Satz 5 der Kegel $\mathfrak{K}_3$ ersetzt werden durch den in $\mathfrak{K}_3$ liegenden konvexen Kegel aller $r \in \mathfrak{K}$, für die ein $\varepsilon > 0$ mit $q + \varepsilon r \in Q_1 \cap Q_3$ existiert.

Literatur

[1] Дубовицкии, А. Я., и Милютин, А. А., Задачи на экстремум при наличии ограничении. ДАН СССР **149**, 759—762 (1963).
[2] —, Задачи на экстремум при наличии ограничении. Журн. вычисл. матем. и матем. физ. **5**, 395—453 (1965).
[3] Gähler, S., Modifizierte Topologien und deren Anwendungen. Math. Nachr. **66**, 217—230 (1975).
[4] —, m-Verfeinerungen von Topologien und verallgemeinerte topologische Vektorräume. Math. Nachr. **73**, 225—241 (1976).
[5] —, Eine Verallgemeinerung des Satzes von Dubovickij-Miljutin. Math. Nachr. **77**, 117—138 (1977).
[6] Гирсанов, И. В., Лекции по математической теории экстремальных задач, Москва 1970.
[7] Neustadt, L. W., An abstract variational theory with applications to a broad class of optimization problems I, General theory. J. SIAM Control **4**, 505—527 (1966); II, J. SIAM Control **5**, 90—137 (1967).

Anschrift des Verfassers:

Akademie der Wissenschaften der DDR
Zentralinstitut für Mathematik und Mechanik
108 Berlin
Mohrenstr. 39
DDR

Dual finite element analysis of unilateral boundary value problems

Ivan Hlaváček, Praha

Abstract. Second order elliptic equations with unilateral boudary conditions are formulated by means of two "dual" variational problems over convex sets: (i) for the solution itself, (ii) for its cogradient. Using piecewise linear polynomials on triangulations of the given domain, algorithms are proposed for approximate solutions to both the problems and rates of convergence deduced. Some a posteriori error estimates are also included.

In the analysis of the classical elliptic boundary value problems (with equalities on the boundary) the dual variational approach can be used, yielding not only approximate values of the solution and of its cogradient but also a posteriori error bounds and two-sided energy estimates (cf. [1, 2, 3] and the literature therein). In the present contribution we extend the approach to second order elliptic equations with unilateral boundary conditions.

We distinguish two classes of equations, namely (i) those with a strictly positive "absolute" term and (ii) those without an absolute term. For the class (i) we choose the following model problem

$$-\Delta u + u = f, \qquad x = (x_1, x_2) \in \Omega \subset R^2, \tag{1}$$

$$u \geqq 0, \qquad \frac{\partial u}{\partial \nu} \geqq 0, \qquad u \frac{\partial u}{\partial \nu} = 0, \qquad x \in \partial\Omega \equiv \Gamma,$$

and the class (ii) will be represented by the model problem

$$-\Delta u = f, \qquad x \in \Omega \subset R^2, \tag{2}$$

$$u = 0, \qquad x \in \Gamma_u \subset \Gamma,$$

$$u \geqq 0, \qquad \frac{\partial u}{\partial \nu} \geqq 0, \qquad u \frac{\partial u}{\partial \nu} = 0, \qquad x \in \Gamma_a = \Gamma \dot{-} \Gamma_u,$$

where Γ_u and Γ_a are nonempty subsets of positive length.

The variational formulations in terms of u and of the gradient-vector $\lambda = \operatorname{grad} u$ (minimum of potential and of complementary energy, respectively) are established and justified, using the duality theory.

Restricting the admissible functions to finite elements with continuous, piecewise linear polynomials on triangulations of the given polygonal domain Ω, we come to minimization problems over finite-dimensional convex sets. The algorithms of Gauss-Seidel and of Uzawa, respectively, are proposed for the numerical solution of these problems.

Making use of a one-sided approximation on the boundary by piecewise linear splines, we prove the convergence of the finite element models. Finally, a posteriori error estimates and two-sided energy estimates are given.

1. Equations with an absolute term

As a model problem, let us consider the unilateral boundary value problem (1), where Ω is any bounded domain with Lipschitz boundary (cf. [5]) and $f \in L_2(\Omega)$. It is well known (cf. [4], chpt. 5) that the problem can be recast as follows.

First introduce

$$K = \{v \mid v \in H^1(\Omega),\ v|_\Gamma \geqq 0\},$$

$$\mathscr{L}(v) = \frac{1}{2} \|v\|_1^2 - (f, v)_0,$$

where $H^1(\Omega)$ denotes the Sobolev space of functions and $(u, v)_1$ and $\|v\|_1$ the associate scalar product and norm, respectively. Henceforth we denote

$$(f, g)_0 = \int_\Omega fg\, dx \qquad \forall f, g \in L_2(\Omega).$$

Then the problem to find $u \in K$ such that

$$\mathscr{L}(u) \leqq \mathscr{L}(v) \qquad \forall v \in K \tag{3}$$

represents a variational formulation of the problem (1) (minimum of potential energy). It has a unique solution. The problem (3) will be called primal.

We can show that the problem (1) can also be reformulated in terms of the "extended" gradient vector

$$\lambda = \{\operatorname{grad} u, u\}. \tag{4}$$

To this end, introduce the following sets

$$Q = \{\lambda \mid \lambda \in [L_2(\Omega)]^3, \quad \operatorname{div} \lambda \in L_2(\Omega)\}, \tag{5}$$

where the operator

$$\operatorname{div} \lambda = \sum_{i=1}^{2} \frac{\partial \lambda_i}{\partial x_i}$$

is taken in the sense of distributions:

$$\int_\Omega \lambda \cdot \operatorname{grad} v \, dx = -\int_\Omega v \operatorname{div} \lambda \, dx \qquad \forall v \in C_0^\infty(\Omega), \tag{6}$$

$$Q_f = \{\lambda \mid \lambda \in Q, \ \lambda_3 = \operatorname{div} \lambda + f\}. \tag{7}$$

For $\lambda \in Q$, the functional $\lambda \cdot \nu \in H^{-1/2}(\Gamma)$ can be defined by means of the relation

$$\langle \lambda \cdot \nu, \gamma v \rangle = \int_\Omega (\lambda \cdot \operatorname{grad} v + v \operatorname{div} \lambda) \, dx, \tag{8}$$

where γv denotes the trace of a function $v \in H^1(\Omega)$ on Γ. (Note that any function $w \in H^{1/2}(\Gamma)$ can be identified with a trace of a $v \in H^1(\Omega)$, the mapping $H^{1/2}(\Gamma) \to H^1(\Omega)$ being linear and continuous — cf. [5].)

We define $\lambda \cdot \nu|_\Gamma \geqq 0$ if

$$\langle \lambda \cdot \nu, \gamma v \rangle \geqq 0 \qquad \forall \gamma v \geqq 0. \tag{9}$$

Next introduce the set

$$\mathcal{U} = \{\lambda \mid \lambda \in Q_f, \lambda \cdot \nu|_\Gamma \geqq 0\} \tag{10}$$

and the functional

$$\varphi(\lambda) = \frac{1}{2} \sum_{i=1}^{3} \|\lambda_i\|_0^2.$$

We define the dual problem: to find a $\lambda^0 \in \mathcal{U}$ such that

$$\varphi(\lambda^0) \leqq \varphi(\lambda) \qquad \forall \lambda \in \mathcal{U}. \tag{11}$$

The dual problem has also a unique solution. For the detailed proofs of this and the following results see the forthcoming paper [6]. Using the duality theory and the associated saddle point theorem (see e.g. [4], chpt. 5), one can give the interpretation of the minimizing element λ^0:

$$\lambda_i^0 = \frac{\partial u}{\partial x_i} \qquad (i = 1, 2), \qquad \lambda_3^0 = u,$$

where u is the solution of the primal problem (3), i.e.,

$$-\operatorname{div} \operatorname{grad} u + u = f \quad \text{in } \Omega,$$

$$u|_\Gamma \geqq 0, \qquad \left.\frac{\partial u}{\partial \nu}\right|_\Gamma \geqq 0, \qquad \left\langle \frac{\partial u}{\partial \nu}, u \right\rangle = 0,$$

where the second inequality is taken in the sense of (9) and the last equality in the sense of (8).

In order to obtain *finite-element approximations to the primal problem,* we suppose that Ω has a polygonal boundary. Let us carve Ω into triangles of side less than h. Let V_h be the space of continuous piecewise linear functions on this triangulation. Introduce the set

$$K_h = \{v | v \in V_h, v|_\Gamma \geqq 0\} = V_h \cap K.$$

Instead of the minimization of $\mathscr{L}(v)$ over K, we minimize it over K_h only, i.e. we find $u_h \in K_h$ such that

$$\mathscr{L}(u_h) \leqq \mathscr{L}(v) \qquad \forall v \in K_h. \tag{3'}$$

The function u_h will be called *Ritz approximation.*

Any $v \in V_h$ can be written in the form

$$v(x) = \sum_{j=1}^{M} v_j \varphi_j(x),$$

where $\varphi_j(x)$ create the basis of V_h ($\varphi_j(P_i) = \delta_{ij}$ for any vertex P_i). Let P_{j_k}, $k = 1, 2, \ldots, p$; $(p < M)$, be all vertices on the boundary Γ. Then

$$v \in K_h \Leftrightarrow \{v \in R^M, v_{j_k} \geqq 0, k = 1, \ldots, p\}, \tag{12}$$

$$\mathscr{L}(v) = \frac{1}{2} v^T A v - f^T v,$$

where

$$v = (v_1, \ldots, v_M)^T, \qquad f_j = (f, \varphi_j)_0, \qquad j = 1, \ldots, M,$$

$$A_{ij} = (\varphi_i, \varphi_j)_1, \qquad i, j = 1, \ldots, M.$$

Making use of the *Gauss-Seidel algorithm,* we choose an initial vector $v^0 \in R^M$, satisfying all inequalities in (12). Then the auxiliary vectors $v^{1,j}$, $v^{2,j}$, ... $(j = 1, \ldots, M)$ are calculated step by step, where

$$v^{m,j} = (v_1{}^m, \ldots, v_j{}^m, v_{j+1}^{m-1}, \ldots, v_M{}^{m-1})^T \qquad (m = 1, \ldots),$$

$$v_{j_k}^m = \operatorname{Proj}_{\langle 0,\infty)} \bar{v}_{j_k}^m \quad \text{for} \quad k = 1, \ldots, p,$$

$$v_j{}^m = \bar{v}_j{}^m \text{ for the remaining subscripts,}$$

$$\bar{v}_j{}^m = \frac{1}{A_{jj}} \left(-\sum_{i<j} A_{ji} v_i{}^m - \sum_{i>j} A_{ji} v_i{}^{m-1} + f_j \right), \qquad j = 1, \ldots, M;$$

$$\operatorname{Proj}_{\langle 0,\infty)} w = \begin{cases} w & \text{if } w \geqq 0, \\ 0 & \text{if } w < 0. \end{cases}$$

Introducing the notation

$$v^m = (v_1{}^m, \ldots, v_M{}^m)^T, \qquad \omega = (\omega_1, \ldots, \omega_M)^T,$$

$$u_h = \sum_{j=1}^{M} \omega_j \varphi_j,$$

it holds (cf. e.g. [4], chpt. 4, § 1.4)

$$\lim \|v^m - \omega\|_{R^M} = 0 \quad \text{for} \quad m \to \infty.$$

Next we estimate the distance between the solutions u and u_h of the problems (3) and (3'), respectively. First of all, let us derive an important

Lemma 1 (cf. [7]). *Let $J(v)$ be a functional defined on the closed convex subset $M \subset B$, where B is a reflexive Banach space. Suppose that J is twice continuously differentiable in the Gâteaux sense and its second differential is positive definite and continuous, i.e., positive constants α and c exist such that*

$$J''(u, z, z) \geqq \alpha \|z\|^2,$$

$$J''(u, z, z) \leqq c\|z\|^2$$

holds for any $u \in M$ and any $z \in B$.

Let $M_h \subset M$ be a closed convex subset. Denote the minimizing elements of $J(v)$ over M and M_h by u and u_h, respectively. Then if there exists a $w_h \in M_h$ such that $2u - w_h \in M$, it holds

$$\|u - u_h\| \leqq \sqrt{\frac{c}{\alpha}}\, \|u - w_h\|. \tag{13}$$

Proof. From the Taylor's theorem it follows that such ϑ exists that $0 < \vartheta < 1$,

$$\begin{aligned} J(u_h) &= J(u) + J'(u, u_h - u) + J''\big(u + \vartheta(u_h - u), u_h - u, u_h - u\big) \\ &\geqq J(u) + \alpha\|u_h - u\|^2, \end{aligned} \tag{14}$$

because $J'(u, u_h - u) \geqq 0$.

On the other hand, for any $v \in M_h$ we have

$$J(v) = J(u) + J'(u, v - u) + J''(u + \vartheta_1(v - u), v - u, v - u) \geqq J(u_h). \tag{15}$$

Inserting $v = w_h$ and $v = 2u - w_h$ into the inequality

$$J'(u, v - u) \geqq 0,$$

we obtain $J'(u, w_h - u) = 0$ and therefore

$$\begin{aligned} J(u_h) \leqq J(w_h) &= J(u) + J''\big(u + \vartheta_2(w_h - u), w_h - u, w_h - u\big) \\ &\leqq J(u) + c\|w_h - u\|^2. \end{aligned} \tag{16}$$

From (14) and (16) the inequality (13) follows. Q.E.D.

Applying Lemma 1 to the problems (3), (3') with $J = \mathscr{L}$, $B = H^1(\Omega)$, $M = K$, $M_h = K_h$, $J''(u, y, z) = (z, y)_1$, $\alpha = c = 1$, we come to the conclusion that for any $w_h \in K_h$ such that $2u - w_h \in K$ it holds

$$\|u - u_h\|_1 \leqq \|u - w_h\|_1. \tag{17}$$

Theorem 1. *Assume that* $u \in H^2(\Omega)$ *and* $u \in H^2(\Gamma_m)$ $\forall m = 1, \ldots, G$, *where* Γ_m *denotes any side of the polygonal boundary* Γ.

Then there exists $w_h \in V_h$ *such that*

$$0 \leqq w_h \leqq u \quad \text{on } \Gamma$$

and, if the triangulation is regular[1], *it achieves the optimal order of approximation*:

$$\|u - w_h\|_1 \leqq Ch \left(\|u\|_{H^2(\Omega)} + \sum_{m=1}^{G} \|u\|_{H^2(\Gamma_m)} \right) \tag{18}$$

with C *independent of* h *and* u.

Proof is based on a one-sided approximation of u on Γ by a linear non-negative spline, which is sufficiently "close" to the Lagrange interpolate of u on Γ (see [6]).

Corollary 1. *Under the assumptions of Theorem* 1 *the Ritz approximations* u_h *converge to* u *if* $h \to 0$, *because it holds*

$$\|u - u_h\|_1 = O(h).$$

Next define the *approximations to the dual problem* (11). Let $\mathscr{T}_h$ be regular triangulations of Ω, satisfying moreover the following requirement:

a positive constant C exists, independent of h and such that

$$h/h_{\min} < C,$$

where $h_{\min}$ denotes the minimal side of $\mathscr{T}_h$.

Using the definition (7), the problem (11) can be transformed into an equivalent one:

$$J(q) = \frac{1}{2} \left(\sum_{i=1}^{2} \|q_i\|_0^2 + \|\text{div } q\|_0^2 \right) + (f, \text{div } q)_0 = \min \tag{19}$$

over the following convex set of "reduced" vectors

$$\mathscr{U}_0 = \{q | q \in [L_2(\Omega)]^2, \text{div } q \in L_2(\Omega), q \cdot \nu|_\Gamma \geqq 0\}.$$

Introduce the set

$$\mathscr{U}_{0h} = \mathscr{U}_0 \cap [V_h]^2$$

and define: a vector $q^h \in \mathscr{U}_{0h}$ will be called the Ritz approximation to the dual problem, if

$$J(q^h) \leqq J(q) \qquad \forall q \in \mathscr{U}_{0h}. \tag{19'}$$

Let $\{w^1, \ldots, w^N\}$ create the basis of $[V_h]^2$. Then

$$q \in \mathscr{U}_{0h} \Leftrightarrow \left\{ q = \sum_{j=1}^{N} y_j w^j, By \geqq 0 \right\},$$

---

[1] A triangulation is called regular, if there exists a positive α_0 independent of h and such that the minimal angle of all triangles in $\mathscr{T}_h$ is not less than α_0.

where B is a $(p \times N)$ matrix and the rank of B is equal to p, $p < N$. The problem (19′) is then equivalent with

$$F(y) = \frac{1}{2} y^T A y - y^T b = \min \tag{20}$$

over the set

$$y \in R^N, \qquad By \geqq 0, \tag{21}$$

where A is the corresponding Gram matrix and b a vector.

As the matrix B is not diagonal, we employ the *algorithm of Uzawa* (cf. [4]): Denote

$$R_+ = \{z \mid z \in R^p, z \geqq 0\},$$

and choose an arbitrary $z^1 \in R_+$. Then we solve the problem

$$Ay^1 = b + B^T z^1.$$

Having y^m, z^m, we calculate

$$z^{m+1} = P_{R_+}(z^m - \varrho B y^m), \qquad Ay^{m+1} = b + B^T z^{m+1},$$

where P_{R_+} is the projection of R^p onto R_+, i.e.

$$z_j^{m+1} = \max\{0, z_j - \varrho(By^m)_j\}, \qquad j = 1, 2, \ldots, p$$

and ϱ is a sufficiently small positive number.

It holds (cf. [4], chpt. 4, § 5.1)

$$\|y^m - y\|_{R^N} \to 0 \quad \text{if} \quad m \to \infty,$$

where y is the solution of (20), (21).

To obtain the estimate of the distance between the solutions q and q^m of the problems (19) and (19′), respectively, we may apply Lemma 1 with $J = I$,

$$B = \{q \mid q \in [L_2(\Omega)]^2, \operatorname{div} q \in L_2(\Omega)\},$$

$$\|q\|^2 = \sum_{i=1}^{2} \|q_i\|_0^2 + \|\operatorname{div} q\|_0^2, \qquad M = \mathcal{U}_0, \qquad M_h = \mathcal{U}_{0h}. \tag{22}$$

Then we have again $\alpha = c = 1$. Consequently, for any vector $t^h \in \mathcal{U}_{0h}$ such that $2q - t^h \in \mathcal{U}_0$ it holds

$$\|q - q^h\| \leqq \|q - t^h\|. \tag{23}$$

Theorem 2. *Assume that $q \in [H^2(\Omega)]^2$ and $q \cdot \nu \in H^2(\Gamma_m)$, $\forall m = 1, \ldots, G$, where Γ_m denotes any side of the polygonal boundary Γ.*

Then there exists $t^h \in \mathcal{U}_{0h}$ such that

$$0 \leqq t^h \cdot \nu \leqq q \cdot \nu \quad \text{on } \Gamma, \qquad \|q - t^h\| \leqq Ch \left(\|q\|_{[H^2(\Omega)]^2} + \sum_{m=1}^{G} \|q \cdot \nu\|_{H^2(\Gamma_m)} \right)$$

with C independent of h and q.

Proof is based on one-sided approximations of the "outward flux" $q \cdot \nu$ on Γ by piecewise linear splines $\psi_h^{(m)} \in C(\Gamma_m)$, which are "close" to the linear interpolate of the flux.

Corollary 2. *Under the assumptions of Theorem 2 the Ritz approximations q^h converge to q with $h \to 0$, because it holds*

$$\|q - q^h\| = O(h).$$

If q^h solves (19′) then

$$\lambda^h = \{q_1^h, q_2^h, f + \operatorname{div} q^h\}$$

is an approximation to the dual problem (11). Besides, we may write the following *a posteriori estimates of errors* for the Ritz approximations to both the primal and dual problems:

$$\|u - u_h\|_1^2 \leqq 2(\mathcal{L}(u_h) + \varphi(\lambda^h)) = E$$
$$= \sum_{i=1}^{2} \left\| q_i^h - \frac{\partial u_h}{\partial x_i} \right\|_0^2 + \|f + \operatorname{div} q^h - u_h\|_0^2 + 2 \int_\Gamma u_h q^h \cdot \nu \, ds, \tag{24}$$

$$\sum_{i=1}^{3} \|\lambda_i^0 - \lambda_i^h\|_0^2 \leqq 2(\mathcal{L}(u_h) + \varphi(\lambda^h)) = E. \tag{25}$$

Note that whereas $\mathcal{L}(u_h) + \varphi(\lambda^h)$ may not be numerically suitable, being a difference of great numbers, the expression E is the sum of non-negative numbers.

Moreover, we have lower and upper bounds of the energy:

$$\|u_h\|_1^2 \leqq \|u\|_1^2 \leqq \sum_{i=1}^{2} \|q_i^h\|_0^2 + \|f + \operatorname{div} q^h\|_0^2,$$

$$\|u_h\|_1^2 \leqq (f, u)_0 \leqq \sum_{i=1}^{2} \|q_i^h\|_0^2 + \|f + \operatorname{div} q^h\|_0^2.$$

2. Equations without absolute term

As a model problem, we consider the unilateral boundary value problem (2). We assume that Γ_u is of positive length.

Let the boundary Γ of a bounded domain Ω consist of M closed polygons $\partial\Omega_j$, i.e.

$$\Gamma = \bigcup_{j=1}^{M} \partial\Omega_j, \qquad \partial\Omega_j \cap \partial\Omega_k = \emptyset \quad \text{for} \quad j \neq k$$

and

$$\text{mes}\,(\partial\Omega_j \dot{-} \Gamma_a) > 0, \qquad j = 1, \ldots, M.$$

Let the triangulations $\mathcal{T}_h$ be regular and the endpoints of Γ_a be also the vertices of $\mathcal{T}_h$. We employ again the spaces V_h of piecewise linear polynomials for approximations to the primal problem. As the approximations to the dual problem are concerned, we use the "equilibrium" linear finite element model, described in [3] and a projection defined there instead of the Lagrange linear interpolation.

Results analogous to those of the previous section can be obtained and algorithms of Gauss-Seidel and Uzawa applied to the solution of the discretized variational problems (see [6]).

References

[1] Aubin, J. P., H. G. Burchard, Some aspects of the method of the hypercircle applied to elliptic variational problems. *In:* Numerical Solution of Partial Differential Equations II (SYNSPADE 1970), 1–67. Academic Press, New York—London 1971.

[2] Hlaváček, I., Some equilibrium and mixed models in the finite element method. *In:* Proceedings of the St. Banach Intern. Math. Center, Warsaw (1977).

[3] Haslinger, J., I. Hlaváček, Convergence of a finite element method based on the dual variational formulation. Apl. mat. **21**, 43–65 (1976).

[4] Céa, J., Optimisation, théorie et algorithmes. Dunod, Paris 1971.

[5] Nečas, J., Les méthodes directes en théorie des équations elliptiques. Academia, Prague 1967.

[6] Hlaváček, I. Dual finite element analysis of unilateral boundary value problems. Apl. mat. **22**, 14–51 (1977).

[7] Mosco, U., G. Strang, One-sided approximation and variational inequalities. Bull. Am. Math. Soc. **80**, 308–312 (1974).

Address of the author:

Mathematical Institute of the Czechoslovak Academy of Sciences
Žitná 25
11567 Praha 1
Czechoslovakia

Notwendiges Optimalitätskriterium für eine Steueraufgabe mit nicht eindeutig lösbarer Zustandsgleichung

Rainer Lorentz, Ilmenau

Abstract. A problem of optimal control formulated in Banach-spaces will be considered. Using a generalized lemma of E. Schmidt it is possible to deal with state-equations with a nonlinear operator whose Fréchet-derivate T_x is a Fredholm-operator with dim $N(T_x) > 0$. For such a problem will be given necessary conditions of optimality.

1. Das abstrakte Grundproblem

B_1, B_2, B_3 seien Banach-Räume, U eine gewisse Teilmenge des Raumes B_2, J ein Funktional mit Definitionsbereich $B_1 \times B_2$, T ein Operator: $B_1 \times B_2 \to B_3$. Es schließt sich die folgende Aufgabe an:

$$\text{P:} \qquad J(x, u) \to \min, \quad T(x, u) = \Theta, \quad u \in U.$$

Wir interpretieren x als Phase, u als Steuerung und die Operatorgleichung $T(x, u) = \Theta$ als das mathematische Modell, das einen Prozeß beschreibt.

2. Zwei Folgerungen aus dem Satz von Dubovickij/Miljutin

Durch Anwendung des Satzes von Dubovickij/Miljutin erhalten wir unter gewissen Voraussetzungen für P notwendige Optimalitätsbedingungen.

Satz 1. *Wir setzen voraus, daß*

(A) *das Paar* $(x^0, u^0) \in B_1 \times B_2$ P *löse,*

(B) J *im Punkt* $(x^0, u^0) \in B_1 \times B_2$ *nach* (x, u) *Fréchet-differenzierbar sei,*

(C) T *im Punkt* $(x^0, u^0) \in B_1 \times B_2$ *zur Klasse* C^1 *gehöre,*[1]

[1] Ein Operator A, der aus einem Banach-Raum X in einen Banach-Raum Y abbildet, gehört im Punkt $x^0 \in X$ zur Klasse C^1, falls er auf einer Umgebung von $x^0 \in X$ Fréchet-differenzierbar ist und die Abbildung $x \to A_x(x)$ stetig bezüglich der Normtopologie im Raum $L(X, Y)$ ist.

(D) $T_{(x,u)}(x^0, u^0)$ *den Raum* $B_1 \times B_2$ *auf* B_3 *abbilde,*
(E) U *konvex sei und ein nichtleeres Inneres besitze.*

Dann existieren eine reelle Zahl l_0 *und ein lineares stetiges Funktional* $v \in B_3^*$ *mit den Eigenschaften:*

(1) $l_0 \geqq 0$,
(2) $(l_0, v) \neq \Theta$,
(3) $T_x^*(x^0, u^0)\, v - l_0 J_x(x^0, u^0) = \Theta$,
(4) $\langle u^0 - u, T_u^*(x^0, u^0)\, v - l_0 J_u(x^0, u^0)\rangle \geqq 0$ *für alle* $u \in U$.

Wir nehmen nun an, daß der Raum B_3 sich als direktes Produkt zweier Banach-Räume B_3^1 und B_3^2 schreiben läßt, d. h. $B_3 = B_3^1 \times B_3^2$. Der Operator T im Problem P läßt sich dann als

$$T(x, u) = \big(T^1(x, u), T^2(x, u)\big)$$

schreiben. Dabei ist $T^1\colon B_1 \times B_2 \to B_3^1$ und $T^2\colon B_1 \times B_2 \to B_3^2$. Dafür bekommen wir auf der Grundlage des Satzes 1 folgende notwendige Optimalitätsbedingung:

Satz 2. *Wir setzen voraus, daß*

(A) *das Paar* $(x^0, u^0) \in B_1 \times B_2$ P *löse,*
(B) J *im Punkt* $(x^0, u^0) \in B_1 \times B_2$ *nach* (x, u) *Fréchet-differenzierbar sei,*
(C) T^1, T^2 *im Punkt* $(x^0, u^0) \in B_1 \times B_2$ *in der Klasse* C^1 *liegen,*
(D) *der Operator* $T_x^1(x^0, u^0)$ *den Raum* $B_1 \times B_2$ *auf* B_3^1 *abbilde,*
(E) *die Menge* $T^2_{(x,u)}(x^0, u^0)\,\big(N\big(T^1_{(x,u)}(x^0, u^0)\big)\big)$ *abgeschlossen sei,*[2]
(F) $R\big(T_x^{1*}(x^0, u^0)\big) \supseteqq R\big(T_x^{2*}(x^0, u^0)\big)$,[3]
(G) U *konvex sei und ein nichtleeres Inneres besitze.*

Dann existieren eine reelle Zahl l_0, *ein lineares stetiges Funktional* $v^1 \in B_3^{1*}$ *und ein lineares stetiges Funktional* $v^2 \in B_3^{2*}$ *mit den Eigenschaften:*

(1) $l_0 \geqq 0$,
(2) $(l_0, v^1, v^2) \neq \Theta$,
(3) $T_x^{1*}(x^0, u^0)\, v^1 + T_x^{2*}(x^0, u^0)\, v^2 - l_0 J_x(x^0, u^0) = \Theta$,
(4) $\langle u^0 - u, T_u^{1*}(x^0, u^0)\, v^1 + T_u^{2*}(x^0, u^0)\, v^2 - l_0 J_u(x^0, u^0)\rangle \geqq 0$ *für alle* $u \in U$.

3. Der Fall, daß $T_x(x^0, u^0)$ Fredholm-Operator ist

In diesem Punkt beweisen wir den folgenden

Satz 3. *Wir setzen voraus, daß*

(A) *das Paar* $(x^0, u^0) \in B_1 \times B_2$ P *löse,*

[2] $N(\circ)$ bezeichne den Nullraum des Operators $\circ$.
[3] $R(\circ)$ bezeichne den Wertevorrat des Operators $\circ$.

(B) *J im Punkt $(x^0, u^0) \in B_1 \times B_2$ nach (x, u) Fréchet-differenzierbar sei,*

(C) *T im Punkt $(x^0, u^0) \in B_1 \times B_2$ zur Klasse C^1 gehöre,*

(D) *$T_x(x^0, u^0)$ ein Fredholm-Operator sei,*

(E) *U konvex sei und ein nichtleeres Inneres besitze.*

Dann existieren eine reelle Zahl l_0 und ein lineares stetiges Funktional $v \in B_3^*$ mit den Eigenschaften:

(1) $l_0 \geqq 0$,

(2) $(l_0, v) \neq \Theta$,

(3) $T_x^*(x^0, u^0)\, v - l_0 J_x(x^0, u^0) = \Theta$,

(4) $\langle u^0 - u, T_u^*(x^0, u^0)\, v - l_0 J_u(x^0, u^0)\rangle \geqq 0$ *für alle* $u \in U$.

Beweis. Wir nehmen zunächst an, daß

$$\dim N\big(T_x(x^0, u^0)\big) = n > 0$$

gelte.

$p_1, \ldots, p_n$; $p_i \in B_1$, $i = 1, \ldots, n$; sei eine Basis des Nullraumes des linearen stetigen Operators $T_x(x^0, u^0)$. Nach dem Satz von Hahn/Banach existieren n lineare stetige Funktionale $p_1^*, \ldots, p_n^*$; $p_i^* \in B_1^*$, $i = 1, \ldots, n$; für die

$$\langle p_i, p_j^*\rangle = \delta_{ij}, \qquad i = 1, \ldots, n; \quad j = 1, \ldots, n;$$

gilt.

$q_1^*, \ldots, q_n^*$; $q_i^* \in B_3^*$, $i = 1, \ldots, n$; sei eine Basis des Defektraumes des linearen stetigen Operators $T_x(x^0, u^0)$. Nach dem Satz von Hahn/Banach existieren n Punkte $q_1, \ldots, q_n$, $q_i \in B_3$, $i = 1, \ldots, n$; für die

$$\langle q_i, q_j^*\rangle = \delta_{ij}, \qquad i = 1, \ldots, n; \quad j = 1, \ldots, n;$$

gilt.

Damit definieren wir auf dem Raum B_1 einen linearen stetigen Operator $\tilde{T}_x(x^0, u^0)$, der in den Raum B_3 abbildet:

$$\tilde{T}_x(x^0, u^0) = -T_x(x^0, u^0) + \sum_{i=1}^{n} \langle \cdot, p_i^*\rangle\, q_i.$$

Nach einem verallgemeinerten Schmidtschen Lemma existiert $\tilde{T}_x^{-1}(x^0, u^0)$.

Wir ordnen nun der Aufgabe P eine Aufgabe $\tilde{\text{P}}$ zu:

$$\tilde{\text{P}}: \qquad J(x, u) \to \min,$$

$$T^1(x, u, y) = \tilde{T}_x(x^0, u^0)\, x + R(x, u) - \sum_{i=1}^{n} y_i q_i = \Theta,$$

$$T^2(x, u, y) = \begin{bmatrix} y_1 - \langle x, p_1^*\rangle \\ \vdots \\ y_n - \langle x, p_n^*\rangle \end{bmatrix} = \Theta,$$

$$u \in U, \quad y \in R^n.$$

Dabei ist $R(x, u) = -T(x, u) + T_x(x^0, u^0)\, x$. $x \in B_1$ wird als Phase und das Paar $(u, y) \in B_2 \times R^n$ wird als Steuerung aufgefaßt. Zwischen den Problemen P und $\tilde{\mathrm{P}}$ besteht der folgende Zusammenhang: Falls der Punkt $(x^0, u^0) \in B_1 \times B_2$ Problem P löst, so ist der Punkt $(x^0, u^0, \langle x^0, p_1{}^*\rangle, \ldots, \langle x^0, p_n{}^*\rangle) \in B_1 \times B_2 \times R^n$ Lösung von $\tilde{\mathrm{P}}$.

Leicht läßt sich nachprüfen, daß das Problem $\tilde{\mathrm{P}}$ unter den getroffenen Voraussetzungen in diesem Punkt die Voraussetzungen des Satzes 2 erfüllt, d. h., es existieren eine reelle Zahl l_0, ein lineares stetiges Funktional $w \in B_3{}^*$ und ein Vektor $a \in R^n$ mit den Eigenschaften:

(1) $l_0 \geqq 0$,

(2) $(l_0, w, a) \neq \Theta$,

(3) $\tilde{T}_x{}^*(x^0, u^0)\, w - \sum\limits_{i=1}^{n} a_i p_i{}^* - l_0 J^*(x^0, u^0) = \Theta$,

(4) $\langle u^0 - u, -T_u{}^*(x^0, u^0)\, w - l_0 J_u(x^0, u^0)\rangle \geqq 0$ für alle $u \in U$, $a_i = \langle q_i, w\rangle$, $i = 1, \ldots, n$.

Nun setzen wir $v = -w$, und der Satz ist bewiesen, da aus $(l_0, v) = \Theta$ wegen (4) $(l_0, w, a) = \Theta$ folgen würde.

Sei nun $\dim N\big(T_x(x^0, u^0)\big) = 0$, d. h., es existiere $T_x{}^{-1}(x^0, u^0)$. In diesem Fall liefert der Satz 1 sofort das Resultat. Damit ist der Satz vollständig bewiesen.

4. Zusammenfassung

Ein Problem der optimalen Steuerung wird in Banach-Räumen betrachtet. Unter Verwendung eines verallgemeinerten Schmidtschen Lemmas ist es möglich, Zustandsgleichungen mit nichtlinearen Operatoren zu behandeln, deren Fréchet-Ableitungen T_x Fredholm-Operatoren mit $\dim N(T_x) > 0$ sind. Für ein solches Problem werden notwendige Optimalitätsbedingungen angegeben.

Anschrift des Verfassers:

Technische Hochschule Ilmenau
Sektion Mathematik, Rechentechnik und ökonomische Kybernetik
63 Ilmenau
Ehrenberg, Block G
DDR

Zur Lösbarkeit nichtlinearer parabolischer Gleichungen mit Hölder-Stetigkeit

Peter Mänz, Berlin

Abstract. This paper gives conditions of the solvability for second order quasilinear parabolic equations and some nonlinear eigenvalue problem which can be proved by an extension of the Leray-Schauder degree for Fredholm maps.

1. Der Vortrag enthält Aussagen über die Lösbarkeit quasilinearer parabolischer Gleichungen sowie gewisser nichtlinearer Eigenwertprobleme, die mit Hilfe einer Verallgemeinerung des Abbildungsgrades von Leray-Schauder für Fredholm-Operatoren gewonnen werden können. Die Untersuchungen wurden auf Anregung von Prof. J. G. Borissowitsch (Woronesh, UdSSR) durchgeführt. Ähnliche Ergebnisse für elliptische Gleichungen wurden von W. G. Swjagin erzielt [3].

2. Wir betrachten das Randwertproblem

$$Lu = f, \qquad u|_{G \cup S_T} = 0 \tag{1}$$

für einen quasilinearen Operator

$$L = \sum_{i,j=1}^{n} a_{ij} \frac{\partial^2}{\partial x_i \, \partial x_j} + \sum_{i=1}^{n} a_i \frac{\partial}{\partial x_i} + a - \frac{\partial}{\partial t}.$$

$u(x, t)$, $f(x, t)$ seien Funktionen auf einem abgeschlossenen Zylinder $\bar{G}_T = \bar{G} \times [0, T]$ mit dem Mantel $S_T = \partial G \times [0, T]$. Der Rand des beschränkten Gebietes $G \subset R^n$ gehöre zur Klasse $C_{2+\alpha}$. Die a_{ij}, a_i, a seien Funktionen von x, t, u, ∇u. Der Operator L wird als gleichmäßig parabolisch vorausgesetzt, d. h., die quadratische Form $\sum_{i,j=1}^{n} a_{ij}\xi_i\xi_j$ sei gleichmäßig positiv definit auf $\bar{G}_T \times R^{n+1}$.

Mit $\Lambda_{\varkappa+\alpha, \frac{\varkappa}{2}+\frac{\alpha}{2}}$ bezeichnen wir die Abschließung der Funktionen-Klasse C^∞ in der Norm von $C_{\varkappa+\alpha, \frac{\varkappa}{2}+\frac{\alpha}{2}}$. Eine Funktion $\varphi(x, t, u, p_1, \ldots, p_n)$ gehöre zur Klasse $\Lambda_{(1)+\alpha, 0+\frac{\alpha}{2}}$, wenn $\varphi \in \Lambda_{1+\alpha}$ in den Argumenten $u, p_1, \ldots, p_n$, $\varphi \in \Lambda_{0+\alpha}$ für x und $\varphi \in \Lambda_{0+\frac{\alpha}{2}}$ für t. Wir schreiben $L \in \Lambda_{(1)+\alpha, 0+\frac{\alpha}{2}}$, falls die Koeffizienten von L zu dieser Klasse gehören. Schließlich bezeichnen wir noch mit $\Lambda^0_{2+\alpha, 1+\frac{\alpha}{2}}$ denjenigen Teilraum von $\Lambda_{2+\alpha, 1+\frac{\alpha}{2}}$, der durch die Bedingung $u|_{G \cup S_T} = 0$ gegeben ist.

3. Als erstes zeigen wir die Lösbarkeit des quasilinearen Problems (1) im Raum $\Lambda^0_{2+\alpha,1+\frac{\alpha}{2}}$. Unter den Voraussetzungen von Punkt 2 bildet der Operator L offensichtlich den Raum $\Lambda^0_{2+\alpha,1+\frac{\alpha}{2}}$ in $\Lambda_{0+\alpha,0+\frac{\alpha}{2}}$ ab. Es gelte $\partial G \in \Lambda_{2+\alpha}$.

Satz 1. *Sei* $L \in \Lambda_{(1)+\alpha,0+\frac{\alpha}{2}}(\overline{G}_T \times R^{n+1})$ *ein quasilinearer gleichmäßig parabolischer Operator. Dann hat das Randwertproblem* (1) *für jedes* $f \in \Lambda_{0+\alpha,0+\frac{\alpha}{2}}(\overline{G}_T)$ *eine Lösung* $u \in \Lambda^0_{2+\alpha,1+\frac{\alpha}{2}}(\overline{G}_T)$, *die der Apriori-Abschätzung*

$$\|u\|_{2+\alpha,1+\frac{\alpha}{2}} \leqq C \cdot \|f\|_{0+\alpha,0+\frac{\alpha}{2}} \tag{2}$$

genügt.

Wir beweisen die Behauptung mit Hilfe der Sätze 2 und 3.

Satz 2. *Für einen quasilinearen gleichmäßig parabolischen Operator* $L \in \Lambda_{(1)+\alpha,0+\frac{\alpha}{2}}(\overline{G}_T \times R^{n+1})$ *gilt:*

1. *L besitzt in jedem Punkt* $u_0 \in \Lambda_{2+\alpha,1+\frac{\alpha}{2}}(\overline{G}_T)$ *eine Fréchet-Ableitung* $L' = L'(u_0) \in \Lambda_{0+\alpha,0+\frac{\alpha}{2}}(\overline{G}_T \times R^{n+1})$.
2. *L' hat die gleichen Elliptizitätskonstanten wie L.*
3. $L' : \Lambda^0_{2+\alpha,1+\frac{\alpha}{2}} \to \Lambda_{0+\alpha,0+\frac{\alpha}{2}}$ *ist eine lineartopologische Isomorphie.*

Beweis. 1. und 2. folgen sofort aus den Voraussetzungen über die Koeffizienten von L. 3. ist bewiesen, wenn wir zeigen, daß das lineare Randwertproblem

$$L' u = f, \qquad u|_{G \cup S_T} = 0 \tag{3}$$

für jedes $f \in \Lambda_{0+\alpha,0+\frac{\alpha}{2}}$ genau eine Lösung $u_0 \in \Lambda^0_{2+\alpha,1+\frac{\alpha}{2}}$ besitzt, die (2) genügt. Die Lösung von (3) kann man als Grenzwert einer Folge eindeutig bestimmter Lösungen von linearen Näherungsproblemen

$$L_m' u = f_m, \qquad u|_{G_m \cup S_T^{(m)}} = 0 \tag{4}$$

erhalten. Hierzu braucht man nur G von außen durch Gebiete G_m, deren Rand $\partial G_m \in C^\infty$, zu approximieren, L' und f unter Erhalt der Klasse und der Norm auf G_1 fortzusetzen (s. z. B. [1]) und dann durch Folgen L_m', $f_m \in C^\infty$ in den Normen $C_{(1)+\alpha,0+\frac{\alpha}{2}}$, $C_{0+\alpha,0+\frac{\alpha}{2}}$ anzunähern. Man sieht leicht, daß die Folge der Einschränkungen $u_m|_{\overline{G}_T}$ der eindeutig bestimmten Lösungsfolge (s. z. B. [1]) für (4) in $C_{2+\alpha,1+\frac{\alpha}{2}}(\overline{G}_T)$ kompakt ist und nur einen Häufungspunkt besitzt, der Lösung von (3) ist. Da außerdem alle $u_m|_{\overline{G}_T}$ einer gemeinsamen Apriori-Abschätzung der Form (2) genügen, ist auch für den Grenzwert (2) gesichert.

Ein nichtlinearer Operator heißt *Fredholm-Operator vom Index k*, wenn er stetig differenzierbar ist und seine Ableitung in jedem Punkt ein linearer Fredholm-Operator vom Index k ist. Aus Satz 2 folgt, L ist ein Fredholm-Operator vom Index 0.

Ein Operator $F: X \to Y$ heißt *eigentlich* (*proper*), wenn für jede kompakte Menge $K \subset Y$ das Urbild $F^{-1}(K)$ kompakt ist.

Satz 3. *Der quasilineare gleichmäßig parabolische Operator* $L \in \Lambda_{(1)+\alpha,0+\frac{\alpha}{2}}$ *ist auf jeder beschränkten Menge von* $\Lambda_{2+\alpha,1+\frac{\alpha}{2}}(\overline{G}_T)$ *eigentlich.*

Dieser Satz läßt sich nach einem Schema von Elworthy und Tromba [2] beweisen.

Wir gehen nun zum Beweis von Satz 1 über. Zunächst bemerken wir, daß alle Lösungen von (1) der Apriori-Abschätzung (2) genügen. Setzt man nämlich eine Lösung von (1) in die Koeffizienten von L ein, so erhält man ein lineares Problem. Seine Lösung, die (2) genügt, ist aber gerade die schon bekannte Lösung der quasilinearen Aufgabe.

Im folgenden wenden wir erneut eine Methode aus [2] an. Auf Grund der Sätze 2 und 3 ist L auf der Kugel $B = B(\Theta; 2C) \subset \Lambda^0_{2+\alpha,1+\frac{\alpha}{2}}$ mit dem Mittelpunkt Θ und dem Radius $2C$ ein eigentlicher Fredholm-Operator vom Index 0. Somit ist für ihn der verallgemeinerte Abbildungsgrad $\deg(L, \partial B, f)$ definiert. Da die Homöomorphie L_λ mit den Koeffizienten $a_{ij}^{(\lambda)} = a_{ij}(x, t, \lambda u, \lambda \nabla u)$ auf $\overline{B} \times [0, 1]$ eigentlich ist, haben wir

$$\deg(L_1, \partial B, f) = \deg(L_0, \partial B, f).$$

Genau so wie für L' in Satz 2 prüft man noch nach, daß der lineare Operator $L_0: \Lambda^0_{2+\alpha,1+\frac{\alpha}{2}} \to \Lambda_{0+\alpha,0+\frac{\alpha}{2}}$ eine lineartopologische Isomorphie ist. Folglich ist $\deg(L_0, \partial B, f) = 1$. Damit erhalten wir für $L = L_1$ ebenfalls $\deg(L, \partial B, f) = 1$. (1) hat also sicher eine Lösung.

4. Wir betrachten jetzt folgendes nichtlineare Eigenwertproblem

$$Lu + \lambda K u = 0, \tag{5}$$

wobei L wieder den Voraussetzungen von Punkt 2 genüge und $Ku = k(x, t, u, \nabla u)$ ein nichtlinearer Operator sei.

Satz 4. *Sei* $L \in \Lambda_{(1)+\alpha,0+\frac{\alpha}{2}}$ *ein quasilinearer gleichmäßig parabolischer Operator und* $K \in \Lambda_{0+\beta,0+\frac{\beta}{2}}$ $(\beta > \alpha)$ *beschränkt. Für alle* $u \in \partial B$, *wobei* $B = B(\Theta; r)$ *und* $r \geqq 2C$, *gelte*

$$\sup_{(x,t)\in\overline{G}_T} |k(x, t, u, \nabla u)| \geqq \varepsilon > 0.$$

Dann existiert eine Zahl λ_0, *für die die Gleichung*

$$Lu + \lambda_0 K u = 0$$

eine Lösung $u_0 \in \partial B$ *besitzt.*

Beweis. Da die Werte von K in $\Lambda_{0+\beta,0+\frac{\beta}{2}}$ liegen, ist K als Operator von $\Lambda^0_{2+\alpha,1+\frac{\alpha}{2}}$ in $\Lambda_{0+\alpha,0+\frac{\alpha}{2}}$ kompakt. Außerdem ist $\|\overline{K(\partial B)}\|_{0+\alpha,0+\frac{\alpha}{2}} \geqq \varepsilon$. Die Behauptung des Satzes folgt deshalb unmittelbar aus einer Verallgemeinerung des Satzes von Birkhoff-Kellog-Rothe [3]:

Seien X, Y Banach-Räume, $L: X \to Y$ ein nichtlinearer Fredholm-Operator vom Index 0, $\deg(L, \partial B, \Theta_Y) \neq 0$, *$K: X \to Y$ kompakt und $\Theta_Y \in \overline{K(\partial B)}$. Dann existiert eine Zahl λ derart, daß die Gleichung $Lu + \lambda Ku = 0$ eine Lösung $u \in \partial B$ hat.*

Literatur

[1] Ладыженская, О. А., Солонников, В. А., Уральцева, Н. Н., Линейные и квазилинейные уравнения параболического типа. ,,Наука'', Москва 1967.

[2] Elworthy, K. D., Tromba, A. J., Degree theory on Banach manifolds. *In:* Proc. Sympos. Pure Math., vol. 18, 86–94. Amer. Math. Soc., Providence, R. I. 1970.

[3] Звягин, В. Г., Диссертация. Воронежский Гос. Университет.

Anschrift des Verfassers:

Humboldt-Universität Berlin
Sektion Mathematik
108 Berlin
Unter den Linden 6
DDR

Über die Regularität der schwachen Lösungen gewisser Typen nichtlinearer elliptischer Differentialgleichungen

Manfred Müller, Berlin

Abstract. In this paper, the Dirichlet problem for some nonlinear generalizations of the biharmonic operator is considered. Then regularity theorems for the weak solutions of these boundary value problems are presented which strengthen known results concerning general nonlinear problems. The proofs rest upon the method of dual norms and the utilization of the structure of the differential operators. The paper concludes with some references to possible generalizations.

1.

In der vorliegenden Arbeit wird die Dirichletsche Randwertaufgabe für einige nichtlineare Verallgemeinerungen des biharmonischen Operators betrachtet. Für die schwachen Lösungen dieser Aufgaben lassen sich Regularitätsaussagen gewinnen, die über die bisher bekannten Ergebnisse für allgemeine nichtlineare Probleme (vgl. [2]) hinausgehen. Dazu wird die Methode der dualen Norm (vgl. z. B. [3]) verwendet und die konkrete Form des jeweiligen Differentialoperators ausgenutzt. Auf mögliche Verallgemeinerungen der vorgelegten Resultate wird am Schluß der Arbeit hingewiesen.

2.

Sei $\Omega \subset R_n$ ein beschränktes Gebiet mit dem Rand Γ, der der Einfachheit halber beliebig oft stetig differenzierbar sein soll. Wir betrachten die nichtlineare partielle Differentialgleichung vierter Ordnung

$$Mu = \Delta[B(x, \Delta u)] = \sum_{|\alpha| \leq 2} (-1)^{|\alpha|} D^\alpha f_\alpha \tag{1}$$

unter den Randbedingungen

$$D^\alpha u|_\Gamma = 0, \qquad |\alpha| \leq 1. \tag{2}$$

Wir geben 3 Typen von Funktionen $B(x, \xi)$ an, die sich durch die Wachstumsbedingungen bez. ihrer Argumente unterscheiden. Dadurch charakterisiert der Operator M drei mögliche nichtlineare Verallgemeinerungen des biharmonischen Operators Δ^2. Im folgenden sei p eine Zahl ≥ 2.

Voraussetzungen:

I. $$B_1(x,\xi),\quad B_2(x,\xi),\quad B_3(x,\xi)\in C^2(\bar\Omega\times R_1),$$

II.1. $$|B_1(x,\xi)|+\sum_{i=1}^n\left|\frac{\partial}{\partial x_i}B_1(x,\xi)\right|+\left|\frac{\partial}{\partial\xi}B_1(x,\xi)\right|\cdot|\xi|\leqq C_1|\xi|^{p-1},$$

2. $$|B_2(x,\xi)|+\sum_{i=1}^n\left|\frac{\partial}{\partial x_i}B_2(x,\xi)\right|+\sum_{i,j=1}^n\left|\frac{\partial^2}{\partial x_i\,\partial x_j}B_2(x,\xi)\right|\leqq C_2(1+|\xi|^{p-1}),$$

$$\left|\frac{\partial}{\partial\xi}B_2(x,\xi)\right|+\sum_{i=1}^n\left|\frac{\partial^2}{\partial x_i\,\partial\xi}B_2(x,\xi)\right|\leqq C_2(1+|\xi|^{p-2}),$$

$$\left|\frac{\partial^2}{\partial\xi^2}B_2(x,\xi)\right|\leqq C_2(1+|\xi|^{p-3});$$

3. $$|B_3(x,\xi)|+\sum_{i=1}^n\left|\frac{\partial}{\partial x_i}B_3(x,\xi)\right|+\sum_{i,j=1}^n\left|\frac{\partial^2}{\partial x_i\,\partial x_j}B_3(x,\xi)\right|$$

$$+\left|\frac{\partial}{\partial\xi}B_3(x,\xi)\right|(1+|\xi|)+\sum_{i=1}^n\left|\frac{\partial^2}{\partial x_i\,\partial\xi}B_3(x,\xi)\right|(1+|\xi|)$$

$$+\left|\frac{\partial^2}{\partial\xi^2}B_3(x,\xi)\right|(1+|\xi|)^2\leqq C_3(1+|\xi|)^{p-1};$$

III.1. $$\left|\frac{\partial}{\partial\xi}B_1(x,\xi)\right|\geqq b_1|\xi|^{p-2},$$

2. $$\left|\frac{\partial}{\partial\xi}B_2(x,\xi)\right|\geqq b_2(1+|\xi|^{p-2}),$$

3. $$\left|\frac{\partial}{\partial\xi}B_3(x,\xi)\right|\geqq b_3(1+|\xi|)^{p-2},$$

$$b_1,b_2,b_3>0;$$

IV. $$f_\alpha\in L_{p'}(\Omega),\qquad|\alpha|\leqq 2,\qquad p'=\frac{p}{p-1}.$$

M_iu, $i=1,2,3$, seien die durch die Funktionen $B_i(x,\xi)$, $i=1,2,3$, erzeugten Operatoren.

Einfachste Beispiele für Funktionen $B_i(x,\xi)$, die die Voraussetzungen I.—III. erfüllen, sind:

$$B_1{}^0(x,\xi)=|\xi|^{p-2}\cdot\xi,$$

$$B_2{}^0(x,\xi)=(1+|\xi|^{p-2})\cdot\xi,$$

$$B_3{}^0(x,\xi)=(1+|\xi|)^{p-2}\cdot\xi.$$

Damit sind folgende Spezialfälle der Differentialoperatoren M_i erfaßt:

$$M_1^0 u = |\Delta u|^{p-2} \cdot \Delta u,$$

$$M_2^0 u = (1 + |\Delta u|^{p-2}) \cdot \Delta u,$$

$$M_3^0 u = (1 + |\Delta u|)^{p-2} \cdot \Delta u.$$

3.

Mit $W_q^l(\Omega)$, $q > 1$, $l \geqq 1$ ganzzahlig, bezeichnen wir die Vervollständigung der Menge $C^l(\overline{\Omega})$ aller in $\overline{\Omega}$ l-mal stetig differenzierbaren Funktionen in der Norm

$$\|v\|_{W_q^l(\Omega)} = \left(\int_\Omega \sum_{|\alpha| \leqq l} |D^\alpha v|^q \, dx \right)^{\frac{1}{q}}.$$

$\mathring{W}_q^l(\Omega)$ bezeichnet die Abschließung der Menge $\mathfrak{D}(\Omega)$ aller in Ω beliebig oft stetig differenzierbaren Funktionen mit kompaktem Träger in der $W_q^l(\Omega)$-Norm.

Existenz und Einzigkeit der Lösungen u_i, $i = 1, 2, 3$, der Randwertprobleme

$$M_i u = \sum_{\alpha \leqq 2} (-1)^{|\alpha|} D^\alpha f_\alpha, \tag{1_i}$$

$$D^\alpha u|_\Gamma = 0, \qquad |\alpha| \leqq 1 \tag{2}$$

werden im Raum $\mathring{W}_p^2(\Omega) = V$ untersucht. Jede der Aufgaben (1_i), (2), $i = 1, 2, 3$, kann dann aufgefaßt werden als Operatorengleichung

$$A_i u = f,$$

wobei die Operatoren $A_i \colon V \to V'$ beschränkt, strong monoton, hemistetig und koerzitiv sind. Aus einem allgemeinen Satz über Gleichungen mit derartigen Operatoren [1] folgt dann

Satz 1. *Unter den Voraussetzungen* I, II.i, III.i, IV, i = 1, 2, 3, *besitzt jedes der Randwertprobleme* (1_i), (2) *genau eine schwache Lösung* $u_i \in \mathring{W}_p^2(\Omega)$, $i = 1, 2, 3$.

Diese Lösungen genügen den Integralidentitäten

$$\int_\Omega B_i(x, \Delta u_i) \, \Delta\varphi \, dx = \int_\Omega \sum_{|\alpha| \leqq 2} f_\alpha D^\alpha \varphi \, dx \tag{3}$$

für jedes $\varphi \in \mathring{W}_p^2(\Omega)$.

Es gelten die Abschätzungen

$$\|u_1\|_{W_p^2(\Omega)} \leqq C \sum_{|\alpha| \leqq 2} \|f_\alpha\|_{L_{p'}(\Omega)}, \tag{4}$$

$$\|u_2\|_{W_p^2(\Omega)}, \|u_3\|_{W_p^2(\Omega)} \leqq C \left(1 + \sum_{|\alpha| \leqq 2} \|f_\alpha\|_{L_{p'}(\Omega)} \right). \tag{5}$$

4.

Wenn $P_0 \in \Gamma$ ein beliebiger, fixierter Punkt ist, so lassen sich wegen der vorausgesetzten beliebig oftmaligen Differenzierbarkeit von Γ in einer gewissen Kugel $K_r(P_0)$ mit dem Mittelpunkt in P_0 Funktionen $z_i = z_i(x_1, \ldots, x_n)$, $i = 1, \ldots, n$, oder $z = Tx$ derart einführen, daß folgende Eigenschaften erfüllt sind (vgl. [3]):

a) Das Bild von $\Gamma \cap K_r(P_0)$ liegt in der Hyperebene $z_n = 0$.

b) $\Omega \cap K_r(P_0)$ wird in den Halbraum $z_n > 0$ abgebildet.

c) Die Abhängigkeit zwischen den Koordinaten x und z ist ein-eindeutig. Die Funktionen $z_i(x_1, \ldots, x_n)$, $i = 1, \ldots, n$, und $x_j(z_1, \ldots, z_n)$, $j = 1, \ldots, n$, sind beliebig oft stetig differenzierbar, die Funktionaldeterminante $\dfrac{D(z)}{D(x)}$ ist in $\Omega \cap K_r(P_0)$ verschieden von Null.

Für hinreichend kleines ϱ gehört dann der Zylinder

$$Z = \left\{z \,\middle|\, \sum_{i=1}^{n-1} z_i^2 < \varrho^2,\ 0 < z_n < \varrho\right\}$$

zum Bild von $\Omega \cap K_r(P_0)$.

Sei $w(x)$ eine Funktion, die auf $\Omega \cap K_r(P_0)$ definiert ist. Wir bezeichnen dann $w(T^{-1}(z)) = \tilde{w}(z)$. Wenn $w(x)$ verallgemeinerte Ableitungen bis zur Ordnung k besitzt, so gilt

$$D^\beta w(x) = \sum_{|\gamma| \leq |\beta|} K_\gamma^\beta(x)\, D^\gamma \tilde{w}(T(x)), \qquad |\beta| \leqq k,$$

wobei die Koeffizienten $K_\gamma^\beta(x)$ von den Ableitungen der Funktionen $z_i(x)$, $i = 1, \ldots, n$, bis zur Ordnung k abhängen.

Speziell verwenden wir die Bezeichnung $\Delta w(T^{-1}(z)) = \tilde{\Delta}\tilde{w}(z)$, wobei $\tilde{\Delta}$ der auf die z-Koordinaten transformierte Laplaceoperator ist.

5.

Die Differenzenquotienten einer Funktion $w(x)$ schreiben wir in der Form

$$D^1_{\pm\Delta_i} w(x) = \frac{1}{\pm\Delta x_i}\left(\overset{\pm i}{w}(x) - w(x)\right) = \frac{1}{\pm\Delta x_i}\left(w\left(\overset{\pm i}{x}\right) - w(x)\right)$$

$$= \frac{1}{\pm\Delta x_i}\,[w(x_1, \ldots, x_{i-1}, x_i \pm \Delta x_i, x_{i+1}, \ldots, x_n) - w(x_1, \ldots, x_n)],$$

$$i = 1, \ldots, n, \qquad \Delta x_i > 0.$$

Wenn $v \in L_q(\Omega)$, $w \in L_{q'}(\Omega)$, $\dfrac{1}{q} + \dfrac{1}{q'} = 1$, und $\Omega' = \operatorname{supp} w$ echtes Teilgebiet von Ω ist, so gilt für alle $\Delta x_i < d(\Omega', \Gamma)$ die Beziehung

$$\int_\Omega v \cdot D^1_{\pm\Delta_i} w\, dx = -\int_\Omega w \cdot D^1_{\mp\Delta_i} v\, dx, \qquad i = 1, 2, \ldots, n. \tag{6}$$

Sei Ω' beliebiges echtes Teilgebiet von Ω. Wenn $w \in W_q^l(\Omega)$, $q > 1$, so gilt für alle $|\alpha| \leqq l - 1$ und $\Delta x_i < d(\Omega', \Gamma)$ die Abschätzung

$$\int_{\Omega'} |D^1_{\pm\Delta_i} D^\alpha w|^q \, dx \leqq \|w\|^q_{W_q^l(\Omega)}. \tag{7}$$

Den Nachweis der Existenz verallgemeinerter Ableitungen einer gegebenen Funktion führen wir mit Hilfe von

Lemma 1. *Seien $w \in L_q(\Omega)$, $q > 1$, $\Delta x_i < c_i$. Ferner sei Ω' ein Teilgebiet von Ω mit der Eigenschaft, daß für jedes $x \in \Omega'$ gilt $\overset{+i}{x} \in \Omega$. Wenn dann gleichmäßig für alle $\Delta x_i < c_i$ gilt*

$$\int_{\Omega'} |D^1_{\Delta_i} w|^q \, dx \leqq C^*,$$

so besitzt die Funktion w in Ω' eine verallgemeinerte Ableitung $D_i^1 w$, die zum Raum $L_q(\Omega')$ gehört. Außerdem gilt die Abschätzung

$$\|D_i^1 w\|^q_{L_q(\Omega')} \leqq C^*.$$

6.

Die folgenden Aussagen sind Hilfsmittel für den Nachweis von Regularitätseigenschaften der schwachen Lösungen u_i, $i = 1, 2, 3$, der Randwertprobleme (1_i), (2).

Lemma 2. *Wenn die Funktion $v \in L_1(\Omega)$ für jedes $\varphi \in \mathcal{D}(\Omega)$ die Abschätzung*

$$|(v, \Delta\varphi)| + |(v, \varphi)| \leqq \varkappa \|\varphi\|_{W_2^2(\Omega)} \tag{8}$$

mit einer von φ unabhängigen Konstanten $\varkappa$ erfüllt, so gilt $v \in L_2(\Omega)$.

Beweis. Aus der Abschätzung (8) folgt

$$\sup_{\varphi \in \mathcal{D}(\Omega)} \frac{|(v, \Delta\varphi)|}{\|\varphi\|_{W_2^2(\Omega)}}, \quad \sup_{\varphi \in \mathcal{D}(\Omega)} \frac{|(v, (\Delta - I)\,\varphi)|}{\|\varphi\|_{W_2^2(\Omega)}} \leqq \varkappa. \tag{9}$$

Die Operatoren Δ, $\Delta - I$ genügen aber für jedes $\varphi \in \mathcal{D}(\Omega)$ den Abschätzungen

$$\|\Delta\varphi\|_{L_2(\Omega)}, \; \|(\Delta - I)\,\varphi\|_{L_2(\Omega)} \geqq K \|\varphi\|_{W_2^2(\Omega)} \tag{10}$$

mit einer von φ unabhängigen Konstanten $K > 0$. Damit folgt aus (9):

$$\sup_{\varphi \in \mathcal{D}(\Omega)} \frac{|(v, \Delta\varphi)|}{\|\Delta\varphi\|_{L_2(\Omega)}}, \quad \sup_{\varphi \in \mathcal{D}(\Omega)} \frac{|(v, (\Delta - I)\,\varphi)|}{\|(\Delta - I)\,\varphi\|_{L_2(\Omega)}} \leqq \frac{\varkappa}{K}.$$

Man erhält dann die Behauptung, wenn man berücksichtigt, daß wenigstens eine der Mengen $\{\Delta\varphi\}$, $\{(\Delta - I)\,\varphi\}$, $\varphi \in \mathcal{D}(\Omega)$ dicht ist in $L_2(\Omega)$.

Bemerkung. Für die Funktion v aus Lemma 2 gilt die Abschätzung

$$\|v\|_{L_2(\Omega)} \leqq \frac{\varkappa}{K}. \tag{11}$$

Satz 2. *Sei Ω' ein beliebiges echtes Teilgebiet von Ω. Wenn die Funktion $v \in L_1(\Omega)$ für jedes $\varphi \in \mathscr{D}(\Omega)$ der Abschätzung*

$$|(v, \Delta\varphi)| + |(v, \varphi)| \leqq \varkappa\|\varphi\|_{W_2^1(\Omega)} \tag{12}$$

mit einer von φ unabhängigen Konstanten $\varkappa$ genügt, so gilt $v \in W_2^1(\Omega')$ und

$$\|v\|_{W_2^1(\Omega')} \leqq C \cdot \varkappa.$$

Beweis. Zunächst erhalten wir aus (11) mit Hilfe von Lemma 2 $v \in L_2(\Omega)$. Sei ψ eine beliebige Funktion aus $\mathscr{D}(\Omega')$. Wir setzen sie durch Null auf ganz Ω fort. Für $\Delta x_i < d(\Omega', \Gamma)$, $i = 1, \ldots, n$, gilt dann auch $D^1_{-\Delta_i}\psi \in \mathscr{D}(\Omega)$ und damit nach Voraussetzung (12):

$$|(v, \Delta D^1_{-\Delta_i}\psi)| + |(v, D^1_{-\Delta_i}\psi)| \leqq \varkappa\|D^1_{-\Delta_i}\psi\|_{W_2^1(\Omega)}$$

für jedes $\psi \in \mathscr{D}(\Omega')$. Hieraus erhalten wir unter Verwendung der Beziehung (6) und der Abschätzung (7):

$$|(D^1_{\Delta_i}v, \Delta\psi)| + |(D^1_{\Delta_i}v, \psi)| \leqq C \cdot \varkappa\|\psi\|_{W_2^1(\Omega')}$$

für jedes $\psi \in \mathscr{D}(\Omega')$, wobei die Konstante C von ψ und Δx_i nicht abhängt. Aus Lemma 2 und der Abschätzung (11) folgern wir dann, daß gleichmäßig für alle $\Delta x_i < d(\Omega', \Gamma)$ gilt

$$\|D^1_{\Delta_i}v\|_{L_2(\Omega')} \leqq C \cdot \varkappa.$$

Die Behauptung des Satzes folgt dann aus Lemma 1.

Satz 3. *Wenn die Funktion $v \in L_1(\Omega)$ für jedes $\varphi \in \mathscr{D}(\Omega)$ der Abschätzung* (12) *mit einer von φ unabhängigen Konstanten $\varkappa$ genügt, so gilt $v \in W_2^1(\Omega)$ und*

$$\|v\|_{W_2^1(\Omega)} \leqq C \cdot \varkappa.$$

Beweis. Unter Berücksichtigung von Satz 2 genügt es, die behauptete Eigenschaft lokal am Rande des Gebietes nachzuweisen.

Sei $P_0 \in \Gamma$ ein beliebiger, fixierter Punkt. Wir verwenden die Bezeichnungen von Punkt 4. Außerdem setzen wir

$$G = \left\{z \,\middle|\, \sum_{i=1}^{n-1} z_i^2 < \frac{\varrho^2}{4}, \quad 0 < z_n < \frac{\varrho}{2}\right\}.$$

Sei $\tilde{\psi} = \tilde{\psi}(z)$ eine beliebige Funktion aus $\mathscr{D}(Z)$. Die Funktion $\psi(x) = \tilde{\psi}(T(x))$ gehört dann zu $\mathscr{D}(T^{-1}(Z))$, wobei $T^{-1}(Z)$ das Urbild von Z bez. der Abbildung T ist. Setzen wir $\psi(x)$ durch Null außerhalb von $T^{-1}(Z)$ fort, so gilt auch $\psi \in \mathscr{D}(\Omega)$. Nach Voraussetzung (12) haben wir dann

$$|(v, \Delta\psi)_{L_2(T^{-1}(Z))}| + |(v, \psi)_{L_2(T^{-1}(Z))}| \leqq \varkappa\|\psi\|_{W_2^1(T^{-1}(Z))},$$

oder nach Übergang zu den z-Koordinaten

$$|(\tilde{v}, \tilde{\Delta}\tilde{\psi})_{L_2(Z)}| + |(\tilde{v}, \tilde{\psi})_{L_2(Z)}| \leqq C \cdot \varkappa \|\tilde{\psi}\|_{W_2^1(Z)} \tag{13}$$

für alle $\tilde{\psi} \in \mathscr{D}(Z)$.

Sei weiter $\tilde{\chi} = \tilde{\chi}(z)$ eine beliebige Funktion aus $\mathscr{D}(G)$. Wir setzen sie durch Null außerhalb von G fort. Für $\Delta z_i < \frac{\varrho}{2}$, $i = 1, \ldots, n$, gilt dann $D^1_{-\Delta_i}\tilde{\chi} \in \mathscr{D}(Z)$. Die Abschätzung (13) mit $\tilde{\psi} = D^1_{-\Delta_i}\tilde{\chi}$ liefert dann

$$|(\tilde{v}, \tilde{\Delta} D^1_{-\Delta_i}\tilde{\chi})| + |(\tilde{v}, D^1_{-\Delta_i}\tilde{\chi})| \leqq C \cdot \varkappa \|D^1_{-\Delta_i}\tilde{\chi}\|_{W_2^1(Z)}$$

oder nach Anwendung der Beziehung (6) und der Abschätzung (7)

$$|(D^1_{\Delta_i}\tilde{v}, \tilde{\Delta}\tilde{\chi})| + |(D^1_{\Delta_i}\tilde{v}, \tilde{\chi})| \leqq C \cdot \varkappa \|\tilde{\chi}\|_{W_2^2(G)}$$

für jedes $\tilde{\chi} \in \mathscr{D}(G)$, wobei die Konstante C weder von der Funktion $\tilde{\chi}$ noch von Δz_i, $i = 1, \ldots, n$, abhängt.

Hieraus folgern wir analog zum Beweis von Satz 2 $\tilde{v} \in W_2^1(G)$ und

$$\|\tilde{v}\|_{W_2^1(G)} \leqq C \cdot \varkappa.$$

Damit ist der Satz bewiesen.

7.

Wir formulieren nun die Regularitätsaussagen für die Lösungen u_i der Randwertaufgaben (1_i), (2). Die spezielle Gestalt der Operatoren M_i erlaubt es, dazu die Aussagen von Satz 3 heranzuziehen. Dabei nehmen wir generell an, daß die entsprechenden Voraussetzungen I, II.i., III.i., IV erfüllt sind.

Satz 4. *Sei $p > 2$ beliebig für $n = 2, 3, 4$ und $2 < p \leqq \frac{2n}{n-4}$ für $n > 4$. Dann hat die Lösung u_1 des Randwertproblems (1_1), (2) folgende Eigenschaften:*

a) *Wenn $f_\alpha \in L_2(\Omega)$, $|\alpha| \leqq 2$, so gilt $u_1 \in W^2_{2(p-1)}(\Omega)$ und*

$$\|u_1\|^{p-1}_{W^2_{2(p-1)}(\Omega)} \leqq C \cdot \sum_{|\alpha| \leqq 2} \|f_\alpha\|_{L_2(\Omega)}.$$

b) *Wenn $f_\alpha \in W_2^1(\Omega)$, $|\alpha| \leqq 2$, so gilt $|\Delta u_1|^{p-1} \in W_2^1(\Omega)$ und*

$$\| |\Delta u_1|^{p-1} \|_{W_2^1(\Omega)} \leqq C \cdot \sum_{|\alpha| \leqq 2} \|f_\alpha\|_{W_2^1(\Omega)}.$$

c) *Wenn $f_\alpha \in W_2^2(\Omega)$, $|\alpha| \leqq 2$, so gilt $B_1(x, \Delta u_1) \in W_2^2(\Omega)$ und*

$$\|B_1(x, \Delta u_1)\|_{W_2^2(\Omega)} \leqq C \cdot \sum_{|\alpha| \leqq 2} \|f_\alpha\|_{W_2^2(\Omega)}.$$

Satz 5. *p soll dieselben Voraussetzungen erfüllen wie in Satz 4. Dann hat die Lösung* u_2 *des Randwertproblems* (1_2), (2) *folgende Eigenschaften:*

a) *Wenn* $f_\alpha \in L_2(\Omega)$, $|\alpha| \leqq 2$, *so gilt* $u_2 \in W^2_{2(p-1)}(\Omega)$ *und*

$$\|u_2\|^{p-1}_{W_2^2(p-1)(\Omega)} \leqq C\Big(1 + \sum_{|\alpha|\leqq 2} \|f_\alpha\|_{L_2(\Omega)}\Big).$$

b) *Wenn* $f_\alpha \in W_2^1(\Omega)$, $|\alpha| \leqq 2$, *so gilt* $u_2 \in W_2^3(\Omega)$, $|\Delta u_2|^{p-1} \in W_2^1(\Omega)$ *und*

$$\int_\Omega (1 + |\Delta u_2|^{p-2})^2 \sum_{j=1}^n (D_j^1 \Delta u_2)^2\, dx \leqq C\left(1 + \sum_{|\alpha|\leqq 2} \|f_\alpha\|^2_{W_2^1(\Omega)}\right).$$

c) *Wenn* $f_\alpha \in W_2^2(\Omega)$, $|\alpha| \leqq 2$, *sowie* $n = 2, 3, 4$, $p \geqq 3$, *so gilt* $u_2 \in W_2^4(\Omega)$ *und*

$$\int_\Omega (1 + |\Delta u_2|^{p-2})^2 \sum_{j,k=1}^n (D_j^1 D_k^1 \Delta u_2)^2\, dx \leqq C\left(1 + \sum_{|\alpha|\leqq 2} \|f_\alpha\|^4_{W_2^2(\Omega)}\right).$$

Satz 6. *p soll dieselben Voraussetzungen erfüllen wie in Satz 4. Dann hat die Lösung* u_3 *des Randwertproblems* (1_3), (2) *folgende Eigenschaften:*

a) *Wenn* $f_\alpha \in L_2(\Omega)$, $|\alpha| \leqq 2$, *so gilt* $u_3 \in W^2_{2(p-1)}(\Omega)$ *und*

$$\|u_3\|^{p-1}_{W_2^2(p-1)(\Omega)} \leqq C\left(1 + \sum_{|\alpha|\leqq 2} \|f_\alpha\|_{L_2(\Omega)}\right).$$

b) *Wenn* $f_\alpha \in W_2^1(\Omega)$, $|\alpha| \leqq 2$, *so gilt* $u_3 \in W_2^3(\Omega)$, $|\Delta u_3|^{p-1} \in W_2^1(\Omega)$ *und*

$$\int_\Omega (1 + |\Delta u_3|)^{2(p-2)} \sum_{j=1}^n (D_j^1 \Delta u_3)^2\, dx \leqq C\left(1 + \sum_{|\alpha|\leqq 2} \|f_\alpha\|^2_{W_2^1(\Omega)}\right).$$

c) *Wenn* $f_\alpha \in W_2^2(\Omega)$, $|\alpha| \leqq 2$, *sowie* $n = 2, 3, 4$, *so gilt* $u_3 \in W_2^4(\Omega)$ *und*

$$\int_\Omega (1 + |\Delta u_3|)^{2(p-2)} \sum_{j,k=1}^n (D_j^1 D_k^1 \Delta u_3)^2\, dx \leqq C\left(1 + \sum_{|\alpha|\leqq 2} \|f_\alpha\|^4_{W_2^2(\Omega)}\right).$$

8.

Die in den Sätzen 4, 5, 6 formulierten Regularitätsaussagen lassen sich ohne weiteres übertragen auf solche Randwertaufgaben, in denen anstelle des Operators Δ ein beliebiger anderer elliptischer Differentialoperator zweiter Ordnung mit konstanten Koeffizienten steht, der die Abschätzung (10) erfüllt.

Die hier benutzte Methode läßt sich aber auch auf Gleichungen höherer Ordnung anwenden, und zwar auf solche, in denen anstelle des Operators Δ der Operator Δ^l oder ein anderer elliptischer Differentialoperator der Ordnung $2l$ mit konstanten Koeffizienten steht, für den eine Abschätzung von der Art (10) gilt.

Literatur

[1] Lions, J. L., Quelques méthodes de résolution des problèmes aux limites non linéaires. Paris 1969.

[2] Müller, M., J. Nečas, Über die Regularität der schwachen Lösungen von Randwertaufgaben für quasilineare elliptische Differentialgleichungen höherer Ordnung. Czechoslovak Math. J. **25**, 227–239 (1975).

[3] Müller, M., Über die Regularität der Lösungen quasilinearer elliptischer Differentialgleichungen höherer Ordnung am Rande des Gebietes. Math. Nachr. **68**, 141–160 (1975).

Anschrift des Verfassers:

Humboldt-Universität Berlin
Sektion Mathematik
108 Berlin
Unter den Linden
DDR

On local Hölder continuity and monotonicity of superposition operators

Wolfdietrich Müller and Gottfried Bruckner, Berlin

Abstract. The concepts of local Hölder continuity and locally uniform monotonicity are defined, and relations between local and global continuity and monotonicity properties are investigated for superposition mappings between Lebesgue spaces.

1. Introduction

One year ago, at the Nonlinear Operator Summer School of Stará Lesná, G. Bruckner [1] presented several results on the local Lipschitz continuity of superposition mappings between Lebesgue spaces. In the sequel, some more general results have been obtained making use of the concept of local Hölder continuity, and also monotonicity properties (which are in some sense dual to the continuity properties considered) have been investigated (cf. Bruckner [2]). In this paper, after recalling some definitions from [2], we shall prove that, for superposition operators, Hölder continuity on any arbitrarily small fixed ball implies local (and in some special situations even non-local) Hölder continuity on the whole space, and the Hölder function obtained during the proof will turn out to be the best possible one. A similar result will be derived for monotonicity. As a conclusion, some remarks on possible combinations of continuity and monotonicity properties will be given.

2. Definitions

Let X and W be normed linear spaces and A a mapping from X into W.

Definition 1. If there exist a non-negative real valued function $H(s, t)$ of the non-negative real arguments s, t with the property $H(s', t') \geqq H(s, t)$ for $s' \geqq s$, $t' \geqq t$ and a real number a, $0 < a \leqq 1$, such that the inequality

$$\|Ax - Ay\| \leqq H(\|x\|, \|y\|)\, \|x - y\|^a \tag{1}$$

holds for all $x, y \in X$ then the operator $A : X \to W$ is said to be *locally Hölder continuous with exponent a*. If (1) holds with $a = 1$ then A is said to be *locally Lipschitz continuous* (in correspondence with the definition given in [1]).

The function $H(s, t)$ will be called *Hölder* (*or Lipschitz*) *function*. If this function can be chosen to be a constant, $H(s, t) \equiv H_0 = \text{const}$, then A is Hölder continuous (Lipschitz continuous in case $a = 1$) on the whole space X.

For monotonicity considerations let be $W = X^*$, X^* the conjugate space of X, and let $(.,.)$ denote the pairing between elements of X^* and X.

Definition 2. If there exist a non-negative real valued function $c(s, t)$ of the non-negative real arguments s, t with the property $c(s', t') \leqq c(s, t)$ for $s' \geqq s$, $t' \geqq t$ and a real number $b \geqq 2$ such that the inequality

$$(Ax - Ay, x - y) \geqq c(\|x\|, \|y\|)\, \|x - y\|^b \tag{2}$$

holds for all $x, y \in X$ then the operator $A: X \to X^*$ is said to be *locally uniformly monotone with exponent b*. If (2) holds with $b = 2$ then A is said to be *locally strongly monotone*.

The function $c(s, t)$ will be called monotonicity function. If it can be chosen to be a constant, $c(s, t) \equiv c_0 = \text{const}$, then A is uniformly monotone (strongly monotone in case $b = 2$) on the whole space X.

In what follows X, W will be assumed to be Lebesgue spaces,

$$X = L_p(G, B_1), \qquad W = L_q(G, B_2),$$

where B_1, B_2 are Banach spaces, $G \subset R^n$ is a domain of finite measure, p, q are real numbers greater than 1. Norms in L_p, L_q will be denoted by $\|.\|$, norms in B_1, B_2 by $|.|$.

An operator $A: L_p \to L_q$ will be called a *superposition operator* (see e.g. [3]) if there exists a family of operators

$$A(t): B_1 \to B_2, \qquad t \in G$$

satisfying the Carathéodory condition (i.e. $A(t)\, u$ is continuous in u for every $t \in G$ and measurable in t for every $u \in B_1$) such that

$$(Ax)\,(t) = A(t)\, x(t) \qquad \text{a.e. in } G.$$

3. Local Hölder continuity

Local Hölder continuity as defined above implies Hölder continuity on every bounded subset of the space X. In particular, let $K_r(z) = \{x: \|x - z\| \leqq r\}$ denote the closed ball in X with center z and radius r, then from (1) we get an inequality

$$\|Ax - Ay\| \leqq H_0 \|x - y\|^a \qquad \text{for } x, y \in K_r(z) \tag{3}$$

(with $H_0 = H(\|z\| + r, \|z\| + r)$) for every z and r. We are going to prove that, for superposition operators, this implication is true also in the opposite direction: a Hölder inequality (3) on any fixed ball $K_r(z)$ implies local Hölder continuity with the same exponent a on the whole space X. Only the exponent a has to be suitably chosen in order to get non-trivial results.

Definition 3. A superposition operator $A: L_p \to L_q$ is called *non-trivial* if there exist a measurable subset $\Delta_0 \subset G$ with mes $\Delta_0 > 0$, a real number $\beta > 0$ and two elements $u, v \in B_1$ such that

$$|A(t)\,u - A(t)\,v| \geqq \beta \qquad \text{for} \quad t \in \Delta_0. \tag{4}$$

Theorem 1. *There exists no non-trivial superposition operator* $A: L_p \to L_q$ *satisfying an inequality* (3) *with an exponent* $a > \frac{p}{q}$.

Proof. Let A be a non-trivial superposition operator from L_p to L_q satisfying an inequality (3) in some ball $K_r(z)$. Let be $\Delta \subset \Delta_0$ and x, y defined by

$$x(t) = \begin{cases} u & \text{for } t \in \Delta \\ z(t) & \text{for } t \in G \setminus \Delta \end{cases}, \quad y(t) = \begin{cases} v & \text{for } t \in \Delta \\ z(t) & \text{for } t \in G \setminus \Delta \end{cases}.$$

Then by definition

$$\|x - y\|^a = \left[\int_G |x(t) - y(t)|^p\, dt\right]^{\frac{a}{p}} = (\text{mes}\,\Delta)^{\frac{a}{p}}\, |u - v|^a$$

and from (4)

$$\|Ax - Ay\| \geqq \left[\int_G |A(t)\,x(t) - A(t)\,y(t)|^q\, dt\right]^{\frac{1}{q}} = (\text{mes}\,\Delta)^{\frac{1}{q}}\, \beta,$$

therefore

$$\|Ax - Ay\| \geqq \left[\frac{\beta}{|u - v|^a}\,(\text{mes}\,\Delta)^{\frac{1}{q} - \frac{a}{p}}\right] \|x - y\|^a.$$

Now let mes $\Delta \to 0$; then $x \to z$, $y \to z$, and for $a > \frac{p}{q}$ the term in brackets becomes unbounded in contradiction to (3), q.e.d.

Theorem 2. *Let A be a superposition operator from L_p to L_q satisfying an inequality* (3) *with* $H_0 > 0$, $0 < a \leqq \min\left\{1, \frac{p}{q}\right\}$ *on some ball* $K_r(z)$. *Then*

$$\|Ax - Ay\| \leqq H_0 \max\left\{1, \left\|\frac{x - z}{r}\right\|^{\frac{p}{q} - a}, \left\|\frac{y - z}{r}\right\|^{\frac{p}{q} - a}\right\} \|x - y\|^a \tag{5}$$

for all $x, y \in L_p$.

Proof. For $x, y \in K_r(z)$ inequality (5) reduces to (3). Assume

$$\max\left\{\left\|\frac{x - z}{r}\right\|, \quad \left\|\frac{y - z}{r}\right\|\right\} = Q > 1. \tag{6}$$

For a first step of the proof, we suppose that the element z is piecewise constant with respect to t, and we show that then (5) is true for two fixed elements x, y that are also piecewise constant with respect to t and satisfy the additional condition that the pth

power of the quantity Q defined in (6) is a rational number:

$$Q^p = \frac{M}{N} \tag{7}$$

(M, N integers, $M > N > 0$). We suppose that there are measurable sets $G_i^{(1)}$ ($i = 1, \ldots, I$) such that $\bigcup_{i=1}^{I} G_i^{(1)} = G$, $G_i^{(1)} \cap G_k^{(1)} = \emptyset$ ($i \neq k$) and

$$x(t) = \text{const} \quad \text{for} \quad t \in G_i^{(1)} \quad (i = 1, \ldots, I),$$

and sets $G_j^{(2)}$, $G_k^{(3)}$ with analogous properties such that

$$y(t) = \text{const} \quad \text{for} \quad t \in G_j^{(2)} \quad (j = 1, \ldots, J),$$

$$z(t) = \text{const} \quad \text{for} \quad t \in G_k^{(3)} \quad (k = 1, \ldots, K).$$

Let every intersection $G_{ijk} = G_i^{(1)} \cap G_j^{(2)} \cap G_k^{(3)}$ be split up into M disjoint sets of equal measure. Let us denote these sets by G_{ijkm} ($m = 1, \ldots, M$). We define sets G_m by

$$G_m = \bigcup_{i=1}^{I} \bigcup_{j=1}^{J} \bigcup_{k=1}^{K} G_{ijkm}$$

and elements u_m, v_m by

$$u_m(t) = \begin{cases} x(t) - z(t) & \text{for} \quad t \in G_m \\ 0 & \text{for} \quad t \in G \setminus G_m \end{cases},$$

$$v_m(t) = \begin{cases} y(t) - z(t) & \text{for} \quad t \in G_m \\ 0 & \text{for} \quad t \in G \setminus G_m \end{cases}.$$

Then, because of $\operatorname{mes} G_m = \frac{1}{M} \operatorname{mes} G$, all elements u_m have the same norm,

$$\|u_m\|^p = \frac{1}{M} \|x - z\|^p \qquad (m = 1, \ldots, M),$$

and moreover, for sums of different elements the pth powers of the norms simply add: the relation

$$\|u_{m_1} + \cdots + u_{m_L}\|^p = \|u_{m_1}\|^p + \cdots + \|u_{m_L}\|^p = \frac{L}{M} \|x - z\|^p \tag{8}$$

is true for any combination $\{m_1, \ldots, m_L\}$ ($m_1 < \cdots < m_L$) of L numbers from $\{1, \ldots, M\}$, $1 \leqq L \leqq M$, and there are similar relations for the v_m and for the differences $u_m - v_m$,

$$\|v_{m_1} + \cdots + v_{m_L}\|^p = \frac{L}{M} \|y - z\|^p, \tag{9}$$

$$\|(u_{m_1} + \cdots + u_{m_L}) - (v_{m_1} + \cdots + v_{m_L})\|^p = \frac{L}{M} \|x - y\|^p. \tag{10}$$

Now take $L = N$ (N being defined by (7)), denote by Γ the set of all combinations of N numbers from $\{1, \ldots, M\}$, introduce the notations

$$\gamma = \{m_1, \ldots, m_N\} \in \Gamma,$$

$$u_\gamma = u_{m_1} + \cdots + u_{m_N}, \qquad v_\gamma = v_{m_1} + \cdots + v_{m_N}$$

and

$$\tilde{x}_\gamma = u_\gamma + z, \qquad \tilde{y}_\gamma = v_\gamma + z. \tag{11}$$

Then, because of (8), (6), (7),

$$\|\tilde{x}_\gamma - z\| = \|u_\gamma\| = \left(\frac{N}{M}\right)^{\frac{1}{p}} \|x - z\| \leqq \left(\frac{N}{M}\right)^{\frac{1}{p}} Qr = r,$$

and, because of (9), (6), (7),

$$\|\tilde{y}_\gamma - z\| \leqq r,$$

such that $\tilde{x}_\gamma, \tilde{y}_\gamma \in K_r(z)$ for all $\gamma \in \Gamma$. So (3) can be applied. Using (11) and (10) we get

$$\|A\tilde{x}_\gamma - A\tilde{y}_\gamma\| \leqq H_0\|\tilde{x}_\gamma - \tilde{y}_\gamma\|^a = H_0\|u_\gamma - v_\gamma\|^a = H_0 \left(\frac{N}{M}\right)^{\frac{a}{p}} \|x - y\|^a. \tag{12}$$

Now, by definition,

$$\|A\tilde{x}_\gamma - A\tilde{y}_\gamma\|^q = \int_G |A\tilde{x}_\gamma(t) - A\tilde{y}_\gamma(t)|^q \, dt = \sum_{m \in \gamma} \int_{G_m} |Ax(t) - Ay(t)|^q \, dt,$$

such that, because of (12),

$$\sum_{m \in \gamma} \int_{G_m} |Ax(t) - Ay(t)|^q \, dt \leqq \left[H_0 \left(\frac{N}{M}\right)^{\frac{a}{p}} \|x - y\|^a\right]^q \tag{13}$$

for all $\gamma \in \Gamma$. Note that the right-hand side of (13) is independent of γ. Summing up the inequalities (13) for all $\gamma \in \Gamma$ and taking into account that

$$\sum_{\gamma \in \Gamma} 1 = \binom{M}{N}, \qquad \sum_{\gamma \in \Gamma} \sum_{m \in \gamma} d_m = \frac{N}{M} \binom{M}{N} \sum_{m=1}^{M} d_m \tag{14}$$

and

$$\sum_{m=1}^{M} \int_{G_m} |Ax(t) - Ay(t)|^q \, dt = \|Ax - Ay\|^q$$

we obtain

$$\frac{N}{M} \binom{M}{N} \|Ax - Ay\|^q \leqq \binom{M}{N} \left[H_0 \left(\frac{N}{M}\right)^{\frac{a}{p}} \|x - y\|^a\right]^q$$

or, using (7),

$$\|Ax - Ay\| \leqq H_0 \left(\frac{N}{M}\right)^{\frac{a}{p}-\frac{1}{q}} \|x - y\|^a = H_0 Q^{\frac{p}{q}-a} \|x - y\|^a.$$

So, because of (6), inequality (5) is shown to be satisfied.

This proof works for every combination of piecewise constant elements x, y, z satisfying an inequality (7) with suitably chosen M, N. Since these elements are dense in L_p and the bound given by (5) is independent of I, J, K, M, N, inequality (5) is valid for all $x, y, z \in L_p$, q.e.d.

Corollary 1. *Under the assumptions of Theorem 2 the operator A is*

(i) *locally Hölder continuous if* $0 < a < \min\left\{1, \frac{p}{q}\right\}$,

(ii) *Hölder continuous if* $a = \frac{p}{q} < 1$,

(iii) *locally Lipschitz continuous if* $a = 1, p > q$,

(iv) *Lipschitz continuous if* $a = 1, p = q$.

Remark. In cases (i) and (iii), for $z = 0$ the Hölder (or Lipschitz) function

$$H(\|x\|, \|y\|) = H_0 \max\left\{1, \left\|\frac{x}{r}\right\|^{\frac{p}{q}-a}, \left\|\frac{y}{r}\right\|^{\frac{p}{q}-a}\right\}$$

is the best possible one. In cases (ii) and (iv) (and for every $z \in L_p$) (5) implies $H(\|x\|, \|y\|) \equiv H_0$.

4. Locally uniform monotonicity

Suppose $B_2 = B_1^*, \frac{1}{p} + \frac{1}{q} = 1$; then $L_p = L_p(G, B_1)$ and $L_p^* = L_q(G, B_2)$ are conjugate spaces. Denote by $[.,.]$ the pairing between elements of B_1^* and B_1 and by $(.,.)$ the pairing between elements of L_p^* and L_p. Assume a monotonicity inequality of the form

$$(Ax - Ay, x - y) \geqq c_0\|x - y\|^b \quad \text{for} \quad x, y, \in K_r(z) \tag{15}$$

($c_0 > 0, b \geqq 2$) to be fulfilled for some z, r.

Theorem 3. *There exists no superposition operator $A: L_p \to L_p^*$ satisfying an inequality* (15) *with an exponent $b < p$.*

Proof. See [2].

Theorem 4. *Let A be a superposition operator from L_p to L_p^* satisfying an inequality* (15) *with $c_0 > 0$, $b \geqq \max\{2, p\}$ on some ball $K_r(z)$. Then*

$$(Ax - Ay, x - y) \geqq c_0 \min\left\{1, \left\|\frac{x-z}{r}\right\|^{p-b}, \left\|\frac{y-z}{r}\right\|^{p-b}\right\} \|x - y\|^b \tag{16}$$

for all $x, y \in L_p$.

Proof. The technique of proof is the same as for Theorem 2. As before, the elements $\tilde{x}_\gamma, \tilde{y}_\gamma$ defined by (11) belong to $K_r(z)$, such that, because of (15),

$$(A\tilde{x}_\gamma - A\tilde{y}_\gamma, \tilde{x}_\gamma - \tilde{y}_\gamma) \geqq c_0\|\tilde{x}_\gamma - \tilde{y}_\gamma\|^b$$

for all $\gamma \in \Gamma$. By definition,

$$(A\tilde{x}_\gamma - A\tilde{y}_\gamma, \tilde{x}_\gamma - \tilde{y}_\gamma) = \sum_{m\in\gamma} \int_{G_m} [Ax(t) - Ay(t), x(t) - y(t)]\, dt,$$

so, using (10), we obtain

$$\sum_{m\in\gamma} \int_{G_m} [Ax(t) - Ay(t), x(t) - y(t)]\, dt \geqq c_0 \left(\frac{N}{M}\right)^{\frac{b}{p}} \|x - y\|^b.$$

Summing up these inequalities for all $\gamma \in \Gamma$ and making use of (14) and of

$$\sum_{m=1}^{M} \int_{G_m} [Ax(t) - Ay(t), x(t) - y(t)]\, dt = (Ax - Ay, x - y)$$

we get

$$\frac{N}{M}\binom{M}{N}(Ax - Ay, x - y) \geqq \binom{M}{N} c_0 \left(\frac{N}{M}\right)^{\frac{b}{p}} \|x - y\|^b$$

and, using (7),

$$(Ax - Ay, x - y) \geqq c_0 \left(\frac{N}{M}\right)^{\frac{b}{p}-1} \|x - y\|^b = c_0Q^{p-b}\|x - y\|^b.$$

All further arguments run along the pattern of the proof of Theorem 2.

Corollary 2. *Under the assumptions of Theorem 4 the operator A is*

(i) *locally uniformly monotone if* $b > \max\{2, p\}$,

(ii) *uniformly monotone if* $b = p > 2$,

(iii) *locally strongly monotone if* $b = 2, p > 2$,

(iv) *strongly monotone if* $b = p = 2$.

Remark. In cases (i) and (iii), for $z = 0$ the monotonicity function

$$c(\|x\|, \|y\|) = c_0 \min\left\{1, \left\|\frac{x}{r}\right\|^{p-b}, \quad \left\|\frac{y}{r}\right\|^{p-b}\right\}$$

is the best possible one. In cases (ii) and (iv) (and for every $z \in L_p$) (16) implies $c(\|x\|, \|y\|) \equiv c_0$.

5. Conclusions

In the theory of monotone operators in a Hilbert space, an operator is appreciated to be "good" if it is at the same time strongly monotone and Lipschitz continuous. So the question arises whether in more general spaces one can find operators that have these two "good" properties at least in some local sense. As the results obtained above show, for the class of superposition mappings from a Lebesgue space $L_p(G, B_1)$ into its conjugate space $L_{p/(p-1)}(G, B_1^*)$ the answer to this question is negative. Superposition operators that are simultaneously locally strongly monotone and locally Lipschitz continuous can exist only for $p = 2$ (this is the Hilbert space situation, and in this case the local properties defined in this paper are equivalent to the corresponding non-local properties). But there are "positive" results in the following sense: For $p < 2$ there exist operators that are locally strongly monotone and Hölder continuous (with exponent $a = p - 1$), and for $p > 2$ there exist operators that are locally Lipschitz continuous and uniformly monotone (with exponent $b = p$). Examples are given in [2].

References

[1] Bruckner, G., On local Lipschitz continuity of superposition operators. Comment. Math. Univ. Carolinae 17, 593—608 (1976).

[2] —, Über lokale Hölder-Stetigkeit und lokale gleichmäßige Monotonie von Superpositionsabbildungen zwischen Lebesgueschen Räumen. Math. Nachr. (to appear).

[3] Красносельский, М. А., П. П. Забрейко, Е. И. Пустыльник, П. Е. Соболевский, Интегральные операторы в пространствах суммируемых функций. ,,Наука", Москва 1966.

Address of the authors:

Akademie der Wissenschaften der DDR
Zentralinstitut für Mathematik und Mechanik
108 Berlin
Mohrenstr. 39
DDR

An iterated difference method for evolution inequalities

Reiner Nürnberg, Berlin

Abstract. In this paper we study the convergence of two approximation methods for the solution of evolution inequalities with nonlinear strongly monotone lipschitzian operators in Hilbert spaces.

At first we consider a difference method using a discretization of the space as well as of the time variable. Here we restrict ourselves to a Crank-Nicolson scheme.

Then we combine this method with an iteration algorithm to an iterated projection method. This approximation method is more constructive.

1.

Let V and H be real separable Hilbert spaces (with the norms $\|.\|$ and $|.|$, respectively), where H contains V as a subspace, the injection of V into H is continuous, and its range is dense in H. Identifying H with H' (the dual of H) and H' with a subspace of V' (the dual of V), we have:

$$V \subset H \subset V', \tag{H1}$$

where the injections are continuous, and each space is dense in the following one. We denote by $\langle .,. \rangle$ the scalar product in H as well as the duality between V and V'.

Then, we introduce the following Hilbert spaces:

$$\mathscr{V} = L^2(0, T; V), \qquad \mathscr{H} = L^2(0, T; H), \qquad \mathscr{V}' = L^2(0, T; V'),$$

where $(.,.) = \int_0^T \langle .,. \rangle \, dt$ is the scalar product in $\mathscr{H}$ and the duality between $\mathscr{V}$ and $\mathscr{V}'$.

Using the dense continuous embedding

$$\mathscr{V} \subset \mathscr{H} \subset \mathscr{V}',$$

we define the space

$$\mathscr{W} = \left\{ v \in \mathscr{V} : \frac{dv}{dt} \in \mathscr{V}' \right\},$$

where $\frac{d}{dt}$ means the derivative in the sense of distributions.

Moreover, let C be a closed convex subset of V with $0 \in C$ (for simplicity only). Then

$$\mathscr{C} = \{v \in \mathscr{V} : v(t) \in C \text{ a.e. on } [0, T]\} \tag{H2}$$

is a closed convex subset of $\mathscr{V}$.

Finally, let $A : \mathscr{V} \to \mathscr{V}'$ be an operator from $\mathscr{V}$ into $\mathscr{V}'$. Throughout all the paper we shall suppose that A satisfies the condition: for any $t \in [0, T]$ and $u, v \in \mathscr{V}$ the relation $u = v$ a.e. on $(0, t)$ implies $Au = Av$ a.e. on $(0, t)$.

We want to approximate the solution $u \in \mathscr{C}$ of the evolution inequality (cf. Brezis [2], Lions [11]):

$$\left.\begin{aligned} &\left(\frac{dv}{dt} + Au, v - u\right) \geqq (f, v - u) \qquad \forall v \in \mathscr{C} \cap \mathscr{W},\ v(0) = u_0, \\ &u(0) = u_0, \end{aligned}\right\} \tag{1}$$

where $f \in \mathscr{V}'$ and $u_0 \in \bar{C}^H$ (the closure of C in H).

Remark 1. If a $u \in \mathscr{C}$ satisfies the condition

$$\left(\frac{dv}{dt}, v - u\right) \geqq (g, v - u) \qquad \forall v \in \mathscr{C} \cap \mathscr{W}, \quad v(0) = u_0,$$

for any $g \in \mathscr{V}'$, then $u \in C(0, T; H)$ and $u(0) = u_0$; hence the initial value problem (1) has a sense.

Remark 2. For a strongly monotone continuous operator A the existence and uniqueness of the solution u of the evolution inequality (1) is assured (cf. Brezis [2]).

Remark 3. If the solution u of (1) is regular, i.e. $u \in \mathscr{W}$, then (1) is equivalent to

$$\left.\begin{aligned} &\left(\frac{du}{dt} + Au, v - u\right) \geqq (f, v - u) \qquad \forall v \in \mathscr{C}, \\ &u(0) = u_0. \end{aligned}\right\} \tag{2}$$

Remark 4. Putting $C = V$, we have that (1) is equivalent to the evolution equation

$$\left.\begin{aligned} &\frac{du}{dt} + Au = f, \\ &u(0) = u_0. \end{aligned}\right\} \tag{3}$$

In the following we are going to approximate the solution u of the evolution inequality (1) by a difference scheme.

Firstly, we suppose an approximation of the space V to be given; that means a family $\{V_h, p_h, r_h\}$ (where $h = (h_1, \ldots, h_n)$, $0 < h_i < h_i^0$, is a parameter), where

(i) V_h is a finite dimensional space;

(ii) $p_h : V_h \to V$ is a linear continuous injective operator, $\|\cdot\|_h = \|p_h \cdot\|$ is the norm in V_h;

(iii) $r_h : V \to V_h$ is a continuous operator (H3)

(cf. Aubin [1] and Temam [15]). The operators p_h and r_h are called *prolongation* and *restriction operators*, respectively.

By H_h we denote the space $V_h \equiv V_h'$ provided with the scalar product $\langle .,. \rangle_h = \langle p_h ., p_h . \rangle$ and the associated norm $|.|_h = |p_h .|$ induced by H.

By $C_h = p_h^{-1}(C)$ we define a closed convex subset of V_h.

Now, connecting this approximation of the space V with a discretization of the time variable, we introduce an approximation $\{\mathscr{V}_{hk}, p_{hk}, r_{hk}\}$ of $\mathscr{V}$ (where the second parameter appearing is $k = \frac{T}{m}$, $m \in N$). We define:

(i) $\mathscr{V}_{hk} = l_k^2 (1, \ldots, m; V_h)$ with the norm

$$\|.\|^2_{\mathscr{V}_{hk}} = k \cdot \sum_{i=1}^{m} \|.\|_h^2;$$

(ii) $p_{hk}: \mathscr{V}_{hk} \to \mathscr{V}$ with $p_{hk} v_{hk} = \sum_{i=1}^{m} p_h v_{hk}^i \omega_k^i$ for $v_{hk} \in \mathscr{V}_{hk}$, where ω_k^i is the characteristic function of the interval $[(i-1)\, k, ik[$.

(iii) $r_{hk}: \mathscr{V} \to \mathscr{V}_{hk}$ with $(r_{hk} v)^i = \frac{1}{k} \int_0^T r_h v \omega_k^i \, dt$,

$i = 1, \ldots, m, \quad$ for $\quad v \in \mathscr{V}$.

We note that the so defined prolongation operator is an injective one, and $\|.\|_{\mathscr{V}_{hk}} = \|p_{hk} .\|_{\mathscr{V}}$.

By $\mathscr{H}_{hk}$ we denote the space $\mathscr{V}_{hk} \equiv \mathscr{V}'_{hk}$ provided with the scalar product $(.,.)_{hk} = (p_{hk} ., p_{hk} .)$.

By $\mathscr{C}_{hk} = p_{hk}^{-1}(\mathscr{C})$ we define a closed convex subset of $\mathscr{V}_{hk}$.

Furthermore, to formulate a Crank-Nicolson scheme, we shall use the following notations:

$$\mathscr{W}_{hk} = H_h \times \mathscr{V}_{hk};$$

$$(\nabla_k v_{hk})^i = \frac{v_{hk}^i - v_{hk}^{i-1}}{k}, \qquad v_{hk} \in \mathscr{W}_{hk};$$

$$(\Delta v_{hk})^i = \frac{v_{hk}^i + v_{hk}^{i-1}}{2}, \qquad v_{hk} \in \mathscr{W}_{hk}.$$

By this definition ∇_k is a finite difference operator, and Δ is an embedding from $\mathscr{W}_{hk}$ into $\mathscr{V}_{hk}$. Then we introduce two prolongation operators from $\mathscr{W}_{hk}$ into $\mathscr{V}$ and $\mathscr{W}$, respectively, in the following manner:

$$p_{hk}^0 = p_{hk} \Delta, \qquad p_{hk}^1 v_{hk} = v_{hk}^0 + \int_0^t p_{hk} \nabla_k v_{hk} \, ds \quad \text{for} \quad v_{hk} \in \mathscr{W}_{hk};$$

we note that $\frac{d}{dt} p_{hk}^1 = p_{hk} \nabla_k$.

We associate now with the evolution inequality (1) the following family of approximating inequalities:

For fixed (h, k), to find a $u_{hk} \in \mathscr{W}_{hk}$ with $\Delta u_{hk} \in \mathscr{C}_{hk}$ such that

$$\left.\begin{aligned}&(\nabla_k u_{hk} + A_{hk}(\Delta u_{hk}), v_{hk} - \Delta u_{hk})_{hk}\\&\geqq (f_{hk}, v_{hk} - \Delta u_{hk})_{hk} \qquad \forall\, v_{hk} \in \mathscr{C}_{hk},\\&u_{hk}^0 = u_{0h},\end{aligned}\right\} \tag{4}$$

where $A_{hk}: \mathscr{V}_{hk} \to \mathscr{V}'_{hk}$ approximates $A: \mathscr{V} \to \mathscr{V}'$, $f_{hk} \in \mathscr{V}'_{hk}$ approximates $f \in \mathscr{V}'$ and $u_{0h} \in C_h$ approximates $u_0 \in \overline{C}^H$. More precisely, we shall assume the following consistency hypotheses to be fulfilled:

$$\left.\begin{aligned}&\text{(i) } A_{hk} = \hat{p}_{hk} A p_{hk} \text{ (where } \hat{p}_{hk} \text{ denotes the transpose of } p_{hk});\\&\text{(ii) } f_{hk} \text{ converges discretely to } f \text{ for } (h, k) \to 0, \text{ i.e.}\\&\qquad \|f_{hk} - \hat{p}_{hk} f\|_{\mathscr{V}'_{hk}} \to 0 \text{ for } (h, k) \to 0;\\&\text{(iii) } u_{0h} \text{ converges strongly to } u_0 \text{ in } H \text{ for } h \to 0, \text{ i.e.}\\&\qquad p_h u_{0h} \to u_0 \text{ in } H \text{ for } h \to 0.\end{aligned}\right\} \tag{H4}$$

Remark 5. For a strongly monotone continuous operator A_{hk} the existence and uniqueness of the solution u_{hk} of the difference inequality (4) is assured.

For the proof of a convergence theorem we shall need a lemma on the compatibility between $\dfrac{d}{dt}$ and $\mathscr{C}$:

Lemma 1 (cf. Brezis [2]). *Under the hypotheses* (H1) *and* (H2), *for all* $v \in \mathscr{C}$ *and* $v_0 \in \overline{C}^H$ *there exists a 'regularizing' sequence* $v_n \in \mathscr{C} \cap \mathscr{W}$ *with* $v_n(0) = v_0$, *satisfying the conditions:*

(i) $v_n \to v$ *in* $\mathscr{V}$;

(ii) $\displaystyle\varlimsup_{n\to\infty} \max_{t\in[0,T]} \int_0^t \left\langle \frac{dv_n}{dt}, v_n - v \right\rangle ds \leqq 0;$

moreover, if the additional assumption $v \in C(0, T; H)$ *is fulfilled, then there exists a 'regularizing' sequence* v_n, *satisfying also the condition:*

(iii) $v_n \to v$ *in* $C(0, T; H)$.

We shall suppose the approximation of C having the convergence property:

$$\left.\begin{aligned}&\text{for all } v \in C \text{ there exists a sequence } v_h \in C_h \text{ such that}\\&p_h v_h \to v \text{ in } V \text{ for } h \to 0.\end{aligned}\right\} \tag{H5}$$

Then we get the following assertion on the approximation of $\mathscr{C} \cap \mathscr{W}$:

Lemma 2. *Under the hypotheses* (H1) *to* (H3) *and* (H5), *for all* $v \in \mathscr{C} \cap \mathscr{W}$ *there exists an 'approximating' sequence* $v_{hk} \in C_h \times C_{hk}$ *satisfying the conditions:*

(i) $p_{hk}^0 v_{hk} \to v$ *in* $\mathscr{V}$ *for* $(h, k) \to 0$;

(ii) $\frac{d}{dt} p_{hk}^1 v_{hk} \to \frac{dv}{dt}$ *in* $\mathscr{V}'$ *for* $(h, k) \to 0$

(*hence*: $p_{hk}^1 v_{hk} \to v$ *in* $C(0, T; H)$ *for* $(h, k) \to 0$).

Using these lemmas we shall prove the convergence of u_{hk} to u:

Theorem 1. *Suppose* (H1) *to* (H5) *to be satisfied. If*

$$\left.\begin{array}{l} A: \mathscr{V} \to \mathscr{V}' \text{ is continuous and strongly monotone, i.e.} \\ (Au - Av, u - v) \geqq c\|u - v\|_{\mathscr{V}}^2 \ \forall u, v \in \mathscr{V}, c > 0, \end{array}\right\} \qquad \text{(H6)}$$

then the solution u_{hk} *of* (4) *converges strongly to the solution* u *of* (1) *in* $\mathscr{V} \cap C(0, T; H)$ *for* $(h, k) \to 0$, *i.e.* $p_{hk}^0 u_{hk} \to u$ *in* $\mathscr{V}$ *and* $p_{hk}^1 u_{hk} \to u$ *in* $C(0, T; H)$.

Proof. We note that under these assumptions (1) and (4) have unique solutions u and u_{hk}, respectively.

Putting $v_{hk} = 0$ into (4) and using the relation

$$2 \cdot (\nabla_k u_{hk}, \Delta u_{hk})_{hk} = |u_{hk}^m|_h^2 - |u_{hk}^0|_h^2,$$

we get

$$\big(A_{hk}(\Delta u_{hk}), \Delta u_{hk}\big)_{hk} \leqq (f_{hk}, \Delta u_{hk})_{hk} + \frac{1}{2} |u_{0h}|_h^2.$$

Hence, according to (H4) and (H6),

$$c\|p_{hk}^0 u_{hk}\|_{\mathscr{V}}^2 \leqq \|f_{hk} - \hat{r}_{hk} A(0)\|_{\mathscr{V}'_{hk}} \|p_{hk}^0 u_{hk}\|_{\mathscr{V}} + \frac{1}{2} |p_h u_{0h}|^2.$$

It follows that $\|p_{hk}^0 u_{hk}\|_{\mathscr{V}}$ is bounded, and hence there exists an $u_* \in \mathscr{V}$ and a sequence $(h', k') \to 0$ such that

$$p_{h'k'}^0 u_{h'k'} \rightharpoonup u_* \qquad \text{(converges weakly) in } \mathscr{V}.$$

Now, we put

$$\bar{v}_{hk}^i = \begin{cases} v_{hk}^i & \text{if } i \leqq j \\ u_{hk}^i & \text{otherwise} \end{cases}$$

into (4), where $v_{hk} \in C_h \times \mathscr{C}_{hk}$ and $j \in \{1, \ldots, m\}$ are arbitrary. Since

$$2 \cdot k \cdot \sum_{i=1}^{j} \langle (\nabla_k u_{hk} - \nabla_k v_{hk})^i, (\Delta u_{hk} - \Delta v_{hk})^i \rangle_h$$

$$= |u_{hk}^j - v_{hk}^j|_h^2 - |u_{hk}^0 - v_{hk}^0|_h^2$$

and A_{hk} is strongly monotone, we have

$$\frac{1}{2} |u_{hk}^j - v_{hk}^j|_h^2 + ck \sum_{i=1}^{j} \|(\Delta u_{hk} - \Delta v_{hk})^i\|_h^2$$

$$\leqq k \sum_{i=1}^{j} \langle \big(f_{hk} - \nabla_k v_{hk} - A_{hk}(\Delta v_{hk})\big)^i, (\Delta u_{hk} - \Delta v_{hk})^i \rangle_h + \frac{1}{2} |u_{0h} - v_{hk}^0|_h^2.$$

From this it follows

$$\max\left\{\frac{1}{2}\|p^1_{hk}u_{hk} - p^1_{hk}v_{hk}\|^2_{C(0,T;H)},\ c\|p^0_{hk}u_{hk} - p^0_{hk}v_{hk}\|^2_{\mathscr{V}}\right\}$$

$$\leq \max_{t\in[0,T]}\int_0^t\left\langle f - \frac{d}{dt}p^1_{hk}v_{hk} - Ap^0_{hk}v_{hk}, p^0_{hk}u_{hk} - p^0_{hk}v_{hk}\right\rangle ds$$

$$+ \|f_{hk} - \hat{r}_{hk}f\|_{\mathscr{V}'_{hk}}\|p^0_{hk}u_{hk} - p^0_{hk}v_{hk}\|_{\mathscr{V}} + \frac{1}{2}|u_{0h} - v^0_{hk}|_h^2.$$

Choosing an 'approximating' sequence v_{hk} for any $v \in \mathscr{C} \cap \mathscr{W}$, according to Lemma 2, we obtain

$$\overline{\lim_{(h',k')\to 0}}\ \max\left\{\frac{1}{2}\|p^1_{h'k'}u_{h'k'} - v\|^2_{C(0,T;H)},\ c\|p^0_{h'k'}u_{h'k'} - v\|^2_{\mathscr{V}}\right\}$$

$$\leq \max_{t\in[0,T]}\int_0^t\left\langle\frac{dv}{dt}, v - u_*\right\rangle ds + \|f - Av\|_{\mathscr{V}'}\|v - u_*\|_{\mathscr{V}} + \frac{1}{2}|v(0) - u_0|^2.$$

If we put a 'regularizing' sequence $v_n \in \mathscr{C} \cap \mathscr{W}$ with $v_n(0) = u_0$ for $u_* \in \mathscr{C}$ (the existence of which results from Lemma 1) into the relation just derived, we find

$$p^0_{h'k'}u_{h'k'} \to u_* \text{ in } \mathscr{V}, \text{ for } (h', k') \to 0.$$

Using this and putting an 'approximating' sequence for any $v \in \mathscr{C} \cap \mathscr{W}$ with $v(0) = u_0$ into the difference inequality (4), we get

$$\left(\frac{dv}{dt} + Au_*, v - u_*\right) \geq (f, v - u_*),$$

that is $u_* = u$; u_* is the solution of (1).

It can be shown in the same way, that from every subsequence of $p^0_{hk}u_{hk}$, one can extract a subsequence which converges to u in $\mathscr{V}$. This proves that $p^0_{hk}u_{hk} \to u$ in $\mathscr{V}$, for $(h, k) \to 0$.

Finally, since $u \in C(0, T; H)$ we can choose a 'regularizing' sequence v_n satisfying also $v_n \to u$ in $C(0, T; H)$. Putting this sequence into the relation derived above, we find

$$p^1_{hk}u_{hk} \to u \text{ in } C(0, T; H), \text{ for } (h, k) \to 0.$$

Thus the proof of the theorem is concluded.

Remark 6. In the case of an evolution equation ($C = V$), we have

$$\left\|\hat{r}_{hk}\frac{d}{dt}(p^1_{hk}u_{hk} - u)\right\|_{\mathscr{V}'_{hk}} \to 0, \text{ for } (h, k) \to 0;$$

furthermore, if the approximation of V is self-adjoint, i.e. r_h is the transpose of p_h, then

$$\frac{d}{dt}p^1_{hk}u_{hk} \to \frac{du}{dt} \text{ in } \mathscr{V}', \text{ for } (h, k) \to 0.$$

Remark 7. To prove Theorem 1 it is sufficient to suppose V to be a reflexive Banach space and $\mathcal{V} = L^p(0, T; V)$, where $p \geqq 2$, and $A: \mathcal{V} \to \mathcal{V}'$ to be a continuous strictly monotone coercive operator of the type (S_+).

Remark 8. Theorem 1 not only holds for a Crank-Nicolson scheme but also for the case in which $(\Delta v_{hk})^i = \Theta v_{hk}^i + (1 - \Theta) v_{hk}^{i-1}$ and $\Theta \in \left[\frac{1}{2}, 1\right]$, particularly for an implicit scheme.

2.

Now we are going to consider an iterative method for the solution of the approximating inequality (4), and after that an iterated difference method for the solution of the evolution inequality (1).

Firstly, we transform the inequality (4) into an equivalent fixed point problem. For this, let $S^r_{\mathcal{C}_{hk}, u_{0h}}: \mathcal{V}'_{hk} \to \mathcal{W}_{hk}$ be the operator defined by the equation (where $r > 0$ is a constant)

$$u_{hk} = S^r_{\mathcal{C}_{hk}, u_{0h}}(g_{hk}), \qquad g_{hk} \in \mathcal{V}'_{hk},$$

where $u_{hk} \in \mathcal{W}_{hk}$ with $\Delta u_{hk} \in \mathcal{C}_{hk}$ is the solution of the difference inequality

$$\left.\begin{array}{l} \left(r\nabla_k u_{hk} + J_h(\Delta u_{hk}), v_{hk} - \Delta u_{hk}\right)_{hk} \geqq (g_{hk}, v_{hk} - \Delta u_{hk})_{hk} \ \forall\ v_{hk} \in \mathcal{C}_{hk}, \\ u_{hk}^0 = u_{0h}. \end{array}\right\} \tag{5}$$

Here $J_h: V_h \to V_h'$ denotes the dual mapping. Since V_h is a Hilbert space, J_h is linear.

Using the operator of projection onto a closed convex subset, we obtain the following system of equations which is equivalent to (5):

$$\left.\begin{array}{l} u_{hk}^0 = u_{0h}, \\ u_{hk}^i = 2P_{C_h}^{2\frac{r}{k}} \left(2\,\frac{r}{k}\, I_h + J_h\right)^{-1} \left(2\,\frac{r}{k}\, u_{hk}^{i-1} + g_{hk}^i\right) - u_{hk}^{i-1}, \\ i = 1, \ldots, m; \end{array}\right\} \tag{6}$$

where $I_h: H_h \to H_h$ denotes the identity mapping, and $P_{C_h}^{\vartheta}: V_h \to V_h$ denotes the operator of projection onto C_h in the space V_h provided with the norm $\|\cdot\|_{h,\vartheta}^2 = \|\cdot\|_h^2 + \vartheta|\cdot|_h^2$.

On the other hand, by definition of $S^r_{\mathcal{C}_{hk}, u_{0h}}$ the inequality (4) is equivalent to the fixed point problem:

To find an $u_{hk} \in \mathcal{W}_{hk}$ such that

$$u_{hk} = S^r_{\mathcal{C}_{hk}, u_{0h}}\left(J_h(\Delta u_{hk}) - r(A_{hk}(\Delta u_{hk}) - f_{hk})\right) \tag{7}$$

for any $r > 0$.

Before we shall prove the convergence of the iterative method corresponding to the fixed point problem (7), we give two lemmas:

Lemma 3. *Under the assumptions* (H1) *to* (H3), *for any* $u_{0h} \in C_h$ *and* $r > 0$ *the operator* $S^r_{\mathscr{C}_{hk}, u_{0h}} : \mathscr{V}'_{hk} \to \mathscr{W}_{hk}$ *satisfies the Lipschitz condition*

$$\max\left\{\left(\frac{r}{2}\right)^{\frac{1}{2}} \|u_{hk} - v_{hk}\|_{C(\{0,\dots,m\};H_h)}, \|\Delta u_{hk} - \Delta v_{hk}\|_{\mathscr{V}_{hk}}\right\}$$
$$\leqq \|f_{hk} - g_{hk}\|_{\mathscr{V}'_{hk}} \ \forall\, f_{hk}, g_{hk} \in \mathscr{V}'_{hk},$$

where $u_{hk} = S^r_{\mathscr{C}_{hk}, u_{0h}}(f_{hk})$ *and* $v_{hk} = S^r_{\mathscr{C}_{hk}, u_{0h}}(g_{hk})$.

Proof. This assertion directly follows from the definition of the operator $S^r_{\mathscr{C}_{hk}, u_{0h}}$ by the inequality (5).

Lemma 4 (cf. Browder and Petryshyn [4]). *If*

$$\left.\begin{array}{l} A: \mathscr{V} \to \mathscr{V}' \text{ is a strongly monotone lipschitzian operator, i.e.} \\ (Au - Av, u - v) \geqq c\|u - v\|^2_{\mathscr{V}} \ \forall\, u, v \in \mathscr{V}, c > 0, \\ \|Au - Av\|_{\mathscr{V}'} \leqq L\|u - v\|_{\mathscr{V}} \ \forall\, u, v \in \mathscr{V}, L > 0, \end{array}\right\} \tag{H7}$$

then the operator $T^r: \mathscr{V} \to \mathscr{V}'$ *defined by* $T^r u = Ju - r(Au - f)\ \forall\, u \in \mathscr{V}, f \in \mathscr{V}'$, *is lipschitzian with the constant* $k_r = (1 - 2cr + L^2r^2)^{\frac{1}{2}}$.

Now, we can give the following assertion about the iterative method for the solution of (4):

Theorem 2. *Suppose* (H1) *to* (H4) *and* (H7) *to be satisfied. If* $0 < r < \frac{2c}{L^2}$, *then for fixed* (h, k) *the sequence* w^n_{hk} *defined by the algorithm*

$$w^n_{hk} = S^r_{\mathscr{C}_{hk}, u_{0h}}\left(J_h(\Delta w^{n-1}_{hk}) - r(A_{hk}(\Delta w^{n-1}_{hk}) - f_{hk})\right) \tag{8}$$

for any $w^0_{hk} \in \mathscr{W}_{hk}$ *converges strongly to the solution* u_{hk} *of* (4) *in* $\mathscr{W}_{hk}$; *more precisely, the following estimation holds*:

$$\max\left\{\left(\frac{r}{2}\right)^{\frac{1}{2}} \|w^n_{hk} - u_{hk}\|_{C(\{0,\dots,m\};H_h)}, \|\Delta w^n_{hk} - \Delta u_{hk}\|_{\mathscr{V}_{hk}}\right\}$$
$$\leqq \frac{(k_r)^n}{1 - k_r} \|\Delta w^1_{hk} - \Delta w^0_{hk}\|_{\mathscr{V}_{hk}}.$$

Proof. From Lemma 3 and Lemma 4 it follows that we can apply the principle of contractive mappings, since $k_r < 1$ for $0 < r < \frac{2c}{L^2}$.

Finally, we combine the iterative method with the difference method for the solution of the evolution inequality (1):

Theorem 3. *Assume* (H1) *to* (H5) *and* (H7) *to be fulfilled. Furthermore we suppose the restriction operator* $r_h: V \to V_h$ *to satisfy* $r_h p_h v_h = v_h, \ \forall\, v_h \in V_h$, *and the Lipschitz condition*

$$\|r_h u - r_h v\|_h \leqq \|u - v\| \qquad \forall\, u, v \in V.$$

If $0 < r < \frac{2c}{L^2}$, *then, for a given sequence* $(h_n, k_n) \to 0$, *the sequence* w_n *defined by the algorithm (write here* n *instead of* (h_n, k_n)*):*

$$w_n = S^r_{\mathscr{C}_n, u_{0n}}\left(J_n r_n p^0_{n-1} w_{n-1} - r(A_n r_n p^0_{n-1} w_{n-1} - f_n)\right) \tag{9}$$

for any $w_0 \in \mathscr{W}_0$ *converges strongly in* $\mathscr{V} \cap C(0, T; H)$ *to the solution* u *of* (1), *i.e.*

$$p_n{}^0 w_n \to u \text{ in } \mathscr{V} \text{ and } p_n{}^1 w_n \to u \text{ in } C(0, T; H);$$

more precisely, the following estimations hold:

$$\|p_n{}^0 w_n - u\|_{\mathscr{V}} \leqq (k_r)^n \|p_0{}^0 w_0 - u\|_{\mathscr{V}} + 2 \sum_{j=1}^{n} (k_r)^{n-j} \|p_j{}^0 u_j - u\|_{\mathscr{V}},$$

$$\left(\frac{r}{2}\right)^{\frac{1}{2}} \|p_n{}^1 w_n - u\|_{C(0,T;H)} \leqq (k_r)^n \|p_0{}^0 w_0 - u\|_{\mathscr{V}}$$

$$+ 2 \sum_{j=1}^{n-1} (k_r)^{n-j} \|p_j{}^0 u_j - u\|_{\mathscr{V}}$$

$$+ \|p_n{}^0 u_n - u\|_{\mathscr{V}} + \left(\frac{r}{2}\right)^{\frac{1}{2}} \|p_n{}^1 u_n - u\|_{C(0,T;H)}.$$

Proof. Using the prolongation operator $p^0_{hk}: \mathscr{W}_{hk} \to \mathscr{V}$ and the restriction operator $r_{hk}: \mathscr{V} \to \mathscr{V}_{hk}$, we transform the fixed point problem (7) in the space $\mathscr{W}_{hk}$ into an equivalent one in the space $\mathscr{V}$:

$$(p^0_{hk} u_{hk}) = p^0_{hk} S^r_{\mathscr{C}_{hk}, u_{0h}}\left(J_h r_{hk}(p^0_{hk} u_{hk}) - r(A_{hk} r_{hk}(p^0_{hk} u_{hk}) - f_{hk})\right).$$

So we have for the sequence $(h_n, k_n) \to 0$ a sequence of fixed point problems, the fixed points of which converge strongly to the solution u of (1), since (7) is equivalent to (4) and the assumptions of Theorem 1 are satisfied. From Lemma 3 and Lemma 4 it follows that the corresponding mappings are contractive with the constant $k_r < 1$ not depending on n. Hence we can apply general results concerning projection iteration methods (cf. Gajewski [5], Kluge [10]).

Remark 8. At every step of the iterative methods (8) and (9) according to (6) we have to calculate the solutions of a number of linear equations and their projections onto a closed convex set in a finite dimensional space. So we have the same situation as in the case of iterative methods for the solution of variational inequalities.

Concluding remarks

Projection methods and iterated projection methods for the solution of variational inequalities were considered in Kluge [8, 9], Mosco [12] and Sibony [14].

Gajewski and Gröger [6] introduced an iterative method for the solution of nonlinear evolution equations, where in every step a linear evolution equation has to be solved, and they combined this method with a projection method continuous in time to an iterated projection method for the solution of (3).

In this paper we combine an iterated method of this form with a difference method for the solution of evolution inequalities.

Difference methods for evolution equations with linear operators were given in Raviart [13] and Aubin [1].

References

[1] Aubin, J. P., Approximation des espaces de distributions et des opérateurs différentiels. Bull. Soc. Math. France Mém. **12**, 1–138 (1967).

[2] Brezis, H., Équations et inéquations non linéaires dans les espaces vectoriels en dualité. Ann. Inst. Fourier **18**, 115–175 (1968).

[3] Brezis, H., Opérateurs maximaux monotones et semi-groupes de contractions dans les espaces de Hilbert. Notas de Matemática, North-Holland Mathematics Studies **5** (1973).

[4] Browder, F. E., and W. V. Petryshyn, Construction of fixed points of nonlinear mappings in Hilbert space. J. Math. Anal. Appl. **20**, 197–228 (1967).

[5] Gajewski, H., Iterationsverfahren bei Folgen kontraktiver Operatoren. Mber. Dt. Akad. Wiss. **11**, 818–826 (1969).

[6] Gajewski, H., and K. Gröger, Ein Projektions-Iterationsverfahren für Evolutionsgleichungen. Math. Nachr. **72**, 119–136 (1976).

[7] Kluge, R., Ein Projektions-Iterations-Verfahren bei Fixpunktproblemen und Gleichungen mit monotonen Operatoren. Mber. Dt. Akad. Wiss. **11**, 599–609 (1969).

[8] Kluge, R., Dissertation B, Berlin 1970.

[9] Kluge, R., Näherungsverfahren zur approximativen Lösung nichtlinearer Variationsungleichungen. Math. Nachr. **51**, 343–356 (1971).

[10] Kluge, R., On a class of iteration methods. Math. Nachr. **73**, 7–18 (1976).

[11] Lions, J. L., Quelques méthodes de résolution des problèmes aux limites non linéaires. Dunod-Gauthier-Villars, Paris 1969.

[12] Mosco, U., Convergence of convex sets and solutions of variational inequalities. Advances in Math. **3**, 510–585 (1969).

[13] Raviart, P. A., Sur l'approximation de certaines équations d'évolution linéaires et non linéaires. J. Math. Pures Appl. **46**, 11–183 (1967).

[14] Sibony, M., Sur l'approximation d'équations et inéquations aux dérivées partielles non linéaires de type monotone. J. Math. Anal. Appl. **34**, 502–564 (1971).

[15] Temam, R., Numerical Analysis. Reidel Publishing Company, Holland 1973.

Address of the author:

Akademie der Wissenschaften der DDR
Zentralinstitut für Mathematik und Mechanik
108 Berlin
Mohrenstr. 39
DDR

Monotone Einschließung von Inversen positiv zerlegbarer Elemente durch ein quadratisch konvergentes Verfahren

Jochen W. Schmidt, Dresden

Abstract. In a partially ordered ring an iterative method is constructed in order to approximate the inverse of a given element. By the method two sequences are generated with the following properties: First the inverse is monotonously enclosed by the sequences and secondly both sequences converge to the inverse quadratically.

1. Einleitung

In der Mitteilung wird die Fragestellung behandelt, zu einem vorgegebenen Element a die als vorhanden vorausgesetzte Inverse a^{-1} durch zwei überlinear konvergente Folgen (x_n) und (y_n) monoton einzuschließen. Von W. Mönch [2] ist für den Fall $a^{-1} \geqq 0$ eine Konstruktionsvorschrift für derartige Folgen auf der Grundlage des Schulz-Verfahrens angegeben worden. Für $a \geqq 0$ ist die Aufgabe von J. W. Schmidt [3] durch die zugehörige bedingte Iteration

$$x_{n+1} = \sup (x_n, u_n) \quad \text{mit} \quad u_n = x_n + x_n(e - ax_n), \tag{1}$$

$$y_{n+1} = \inf (y_n, v_n) \quad \text{mit} \quad v_n = x_n + y_n(e - ax_n) \tag{2}$$

gelöst worden. In der vorliegenden Note wird allgemeiner angenommen, daß a positiv zerlegbar ist, daß es also zwei Elemente

$$b \geqq 0, \qquad c \geqq 0 \qquad \text{mit } a = b - c \tag{3}$$

gibt, und aufgezeigt, daß dann das Verfahren

$$x_{n+1} = \sup (x_n, u_n) \qquad \text{mit} \tag{4}$$

$$u_n = \frac{1}{2} \{x_n + y_n + x_n(e - bx_n + cy_n) + y_n(e - by_n + cx_n)\},$$

$$y_{n+1} = \inf (y_n, v_n) \qquad \text{mit} \tag{5}$$

$$v_n = \frac{1}{2} \{x_n + y_n + y_n(e - bx_n + cy_n) + x_n(e - by_n + cx_n)\}$$

zum Ziel führt: Unter geeigneten Voraussetzungen wird nachgewiesen, daß die Folgen (x_n), (y_n) die Inverse a^{-1} bei quadratischer Konvergenz monoton eingrenzen.

Für $c = 0$ und folglich $a \geqq 0$ stimmen die Verfahren (1), (2) und (4), (5) nicht überein; man wird in diesem Fall (1), (2) vorziehen, da im Vergleich zu (4), (5) bedeutend weniger Multiplikationen erforderlich sind.[1] Die in [3] für den wichtigen Sonderfall quadratischer Matrizen beschriebene Möglichkeit zur Gewinnung von Startmatrizen x_0, y_0 kann für (4), (5) übernommen werden. Die getroffene Annahme, daß mit z_0 eine hinreichend gute Näherung für a^{-1} zur Verfügung steht, geht von der Vorstellung aus, daß eine sinnvolle Anwendung des Verfahrens (4), (5) ebenso wie beim Verfahren (1), (2) vor allem darin besteht, für eine z. B. mit einem direkten Verfahren gewonnene Näherung z_0 eine Fehlereingrenzung der Form $z_0 - w \leqq a^{-1} \leqq z_0 + w$ und nötigenfalls eine iterative Nachverbesserung mit monotoner Einschließung vorzunehmen.

Für Literatur kann auf die in [2] und [3] genannten Arbeiten sowie auf G. Alefeld und J. Herzberger [1], Abschn. 18, für vergleichbare Verfahren auf Grundlage der Intervallrechnung verwiesen werden.

2. Hilfsmittel

Die Untersuchungen werden in einem Ring R mit dem Nullelement 0 und einem Einselement e durchgeführt. Durch einen Kegel $K \subset R$ mit den Eigenschaften

$$K + K \subset K, KK \subset K, K \cap (-K) = \{0\}$$

werde in R durch

$$x \leqq y \text{ für } x, y \in R \text{ genau dann, wenn } y - x \in K$$

eine Halbordnung erklärt. Darüber hinaus sei R ein Verband, d. h., zu $x, y \in R$ existieren stets $\sup(x, y) \in R$, $\inf(x, y) \in R$. Jedes Element $a \in R$ ist dann gemäß (3) positiv zerlegbar; man braucht nur

$$b = a^+ = \sup(a, 0), \qquad c = a^- = \sup(-a, 0)$$

zu setzen. Mit $|a| = a^+ + a^- \in R$ werde der Betrag von $a \in R$ bezeichnet.

Weiterhin sei in R eine Norm mit den Eigenschaften

$$\|x\| = 0 \quad \text{genau für} \quad x = 0, \qquad \|-x\| = \|x\|,$$

$$\|x + y\| \leqq \|x\| + \|y\|, \qquad \|xy\| \leqq \|x\| \, \|y\|$$

[1] Die Anzahl der Multiplikationen in (4), (5) läßt sich, worauf mich W. Nauber aufmerksam machte, noch herabsetzen, wenn man zur Berechnung von u_n und v_n zunächst

$$\frac{1}{2}(v_n + u_n) = m_n + m_n(e - am_n), \qquad \frac{1}{2}(v_n - u_n) = d_n(b + c)\, d_n$$

mit Hilfe von

$$m_n = \frac{1}{2}(y_n + x_n), \qquad d_n = \frac{1}{2}(y_n - x_n)$$

errechnet.

vorhanden, bezüglich welcher der Kegel K normal sei, d. h.,

$$0 \leqq x \leqq y \quad \text{impliziert} \quad \|x\| \leqq \|y\|.$$[2]

Die Invertiermenge von R werde mit I bezeichnet,

$$I = \{x \in R : x^{-1} \in R \quad \text{mit} \quad xx^{-1} = x^{-1}x = e \quad \text{vorhanden}\}.$$

Sie sei nicht leer.

Ein wesentliches Beispiel für einen Ring mit den genannten Eigenschaften stellt die Menge der (m, m)-Matrizen $x = (x_{ik})$ mit üblicher Addition und Multiplikation, mit der natürlichen Halbordnung

$$x \leqq y, \quad \text{wenn} \quad x_{ik} \leqq y_{ik} \quad \text{für} \quad i, k = 1, \ldots, m$$

und mit der Zeilensummennorm

$$\|x\| = \max_{i=1,\ldots,m} \sum_{k=1}^{m} |x_{ik}|$$

dar. Die Matrizen $\sup(x, y)$ und $\inf(x, y)$ werden durch elementweise Maximum- bzw. Minimum-Bildung berechnet,

$$\big(\sup(x, y)\big)_{ik} = \max(x_{ik}, y_{ik})$$

$$\big(\inf(x, y)\big)_{ik} = \min(x_{ik}, y_{ik}), \qquad i, k = 1, \ldots, m.$$

3. Monotone Einschließung

Der Ring R sei von der beschriebenen Art; jedoch wird zunächst von den Eigenschaften, welche mit der Norm zusammenhängen, kein Gebrauch gemacht. Von $a \in R$ wird angenommen, daß sogar $a \in I$, während $a = b - c$, z. B. $b = a^+$ und $c = a^-$, eine positive Zerlegung von a darstelle.

Satz 1. *Es sei $a \in I$, für die Startelemente $x_0, y_0 \in R$ gelte*

$$x_0 \leqq a^{-1} \leqq y_0, \tag{6}$$

und es bezeichne $a = b - c$, $b \geqq 0$, $c \geqq 0$, eine positive Zerlegung von a. Unter diesen Annahmen bilden die Folgen (x_n), (y_n) aus den Verfahren (4), (5) *eine monotone Einschließung für a^{-1}, d. h., es ist für $n = 0, 1, \ldots$*

$$x_0 \leqq \cdots \leqq x_{n-1} \leqq x_n \leqq a^{-1} \leqq y_n \leqq y_{n-1} \leqq \cdots \leqq y_0. \tag{7}$$

[2] Außerdem sei in R eine Multiplikation mit $\frac{1}{2}$ möglich, d. h. zu jedem $x \in R$ existiere genau ein $y \left(= \frac{1}{2} x\right) \in R$ mit $y + y = x$, und es mögen $\frac{1}{2} K \subset K$ und $\left\|\frac{1}{2} x\right\| = \frac{1}{2} \|x\|$ gelten.

Beweis durch Induktion: Für $n = 0$ ist (7) mit (6) vorausgesetzt worden. Für den Schluß von n auf $n + 1$ folgt aus

$$x_n \leqq a^{-1} \leqq y_n$$

wegen $b \geqq 0$, $c \geqq 0$ zunächst

$$bx_n \leqq ba^{-1} \leqq by_n, \qquad -cy_n \leqq -ca^{-1} \leqq -cx_n$$

und durch Addition

$$bx_n - cy_n \leqq (b - c)\, a^{-1} = e \leqq by_n - cx_n.$$

Daher werden

$$u_n \leqq \frac{1}{2} \{x_n + y_n + a^{-1}(e - bx_n + cy_n) + a^{-1}(e - by_n + cx_n)\} = a^{-1},$$

$$v_n \geqq \frac{1}{2} \{x_n + y_n + a^{-1}(e - bx_n + cy_n) + a^{-1}(e - by_n + cx_n)\} = a^{-1}.$$

Aus $u_n \leqq a^{-1}$, $x_n \leqq a^{-1}$ erhält man unmittelbar auf Grund der Definition des Supremums

$$x_n \leqq x_{n+1} = \sup(x_n, u_n) \leqq a^{-1},$$

ebenso gewinnt man aus $v_n \geqq a^{-1}$, $y_n \geqq a^{-1}$ über die Definition des Infimums

$$y_n \geqq y_{n+1} = \inf(y_n, v_n) \geqq a^{-1}.$$

Das war für den Satz zu zeigen.

Aus [3] werde als Ergänzung das analoge Ergebnis über das Verfahren (1), (2) zitiert.

Satz 2. *Es sei $a \in I$, $a \geqq 0$, und für die Startelemente bestehe die Ungleichung* (6). *Dann ist die monotone Einschließung* (7) *ebenfalls für das Verfahren* (1), (2) *gültig.*

Beweis. Für den Schluß von n auf $n + 1$ folgen aus $x_n \leqq a^{-1} \leqq y_n$ über $ax_n \leqq e$ für die dortigen u_n, v_n die Abschätzungen

$$u_n \leqq x_n + a^{-1}(e - ax_n) = a^{-1},$$

$$v_n \geqq x_n + a^{-1}(e - ax_n) = a^{-1}$$

und somit die gewünschten Ungleichungen

$$x_n \leqq x_{n+1} \leqq a^{-1} \leqq y_{n+1} \leqq y_n.$$

4. Konvergenzgeschwindigkeit

Der vorangehende Satz 1 kann, wenn jetzt alle in 2. beschriebenen Eigenschaften von R eingesetzt werden, durch die folgende Konvergenzaussage vervollständigt werden.

Satz 3. *Es sei* $a = b - c \in I$ *mit* $b \geqq 0$, $c \geqq 0$, *und die Startelemente* x_0, y_0 *mögen außer* (6) *noch*

$$\|e - bx_0 + cy_0\| < 1, \qquad \|e - by_0 + cx_0\| < 1 \tag{8}$$

erfüllen. Zusätzlich zur monotonen Einschließung (7) *erhält man dann die quadratische Konvergenz der Folgen* (x_n), (y_n) *aus* (4), (5) *gegen* a^{-1}.

Beweis. Man bestätigt unter Verwendung der Abkürzungen

$$z_n = e - bx_n + cy_n, \qquad w_n = e - by_n + cx_n$$

durch einfache Abschätzung bzw. Rechnung

$$0 \leqq z_{n+1} \leqq e - bu_n + cv_n = \frac{1}{2}\{w_n^2 + z_n^2\},$$

$$0 \geqq w_{n+1} \geqq e - bv_n + cu_n = \frac{1}{2}\{w_n z_n + z_n w_n\}.$$

Mit der Abkürzung

$$r_n = \max(\|z_n\|, \|w_n\|)$$

erhält man daher über die Normalität des Kegels

$$\|z_{n+1}\| \leqq r_n^2, \qquad \|w_{n+1}\| \leqq r_n^2,$$

also insgesamt

$$r_{n+1} \leqq r_n^2. \tag{9}$$

Die Voraussetzung (8) beinhaltet $r_0 < 1$, womit sich aus (9) ergibt, daß (r_n) und folglich auch (z_n) und (w_n) quadratisch konvergente Nullfolgen sind. Wegen

$$0 \leqq y_{n+1} - x_{n+1} \leqq v_n - u_n = \frac{1}{2}(y_n - x_n)(b + c)(y_n - x_n)$$

konvergiert die Folge $(y_n - x_n)$ ebenfalls quadratisch gegen 0.

Beachtet man schließlich

$$a^{-1} - x_n = a^{-1}z_n + a^{-1}c(x_n - y_n),$$

$$a^{-1} - y_n = a^{-1}w_n + a^{-1}c(y_n - x_n),$$

erhält man die behauptete quadratische Konvergenz der Folgen (x_n), (y_n) gegen a^{-1}.

Für das Verfahren (1), (2) lautet die analoge Konvergenzaussage aus [3] folgendermaßen.

Satz 4. *Es sei* $a \in I$, $a \geqq 0$, *und neben* (6) *gelte*

$$\|e - ax_0\| < 1. \tag{10}$$

Dann schließen die Folgen (x_n), (y_n) *aus* (1), (2) *die Inverse* a^{-1} *monoton ein und konvergieren quadratisch gegen diesen Wert.*

Beweis. Mit $z_n = e - ax_n$, $w_n = e - ay_n$ erhält man

$$0 \leqq z_{n+1} \leqq e - au_n = z_n^2, \qquad 0 \geqq w_{n+1} \geqq e - av_n = w_n z_n$$

und über die Normalität des Kegels

$$\|z_{n+1}\| \leqq \|z_n\|^2, \qquad \|w_{n+1}\| \leqq \|w_n\| \, \|z_n\|,$$

so daß sich (z_n) und (w_n) wegen $\|z_0\| < 1$ als Nullfolgen erweisen. Setzt man wieder $r_n = \max(\|z_n\|, \|w_n\|)$, so ist auch (r_n) eine Nullfolge. Darüber hinaus besteht die Ungleichung $r_{n+1} \leqq r_n^2$, welche die Konvergenzgeschwindigkeit 2 für (r_n) und somit auch für (z_n) und (w_n) impliziert. Aus

$$a^{-1} - x_n = a^{-1} z_n, \qquad a^{-1} - y_n = a^{-1} w_n$$

folgt daher die Behauptung des Satzes.

5. Startmatrizen

Jetzt sei R speziell der Ring der (m, m)-Matrizen mit der natürlichen Halbordnung und der Zeilensummennorm. Die Matrix

$$h = (h_{ik}) \quad \text{mit} \quad h_{ik} = 1 \quad \text{für} \quad i, k = 1, \ldots, m$$

erlaubt für jede Matrix $x \in R$ die Eingrenzung

$$-\|x\| \, h \leqq x \leqq \|x\| \, h.$$

Die Konstruktion von Startmatrizen x_0, y_0 für die Iteration (4), (5) kann von [3] übernommen werden, wenn wie dort angenommen wird, daß mit z_0 eine hinreichend gute Näherung für a^{-1} verfügbar ist.

Satz 5. *Bezeichnet man mit*

$$d = e - az_0 \tag{11}$$

den Defekt von z_0 *und gilt*

$$\|d\| < 1, \tag{12}$$

wird die Forderung (6) *an die Startmatrizen von*

$$x_0 = z_0 - w, \qquad y_0 = z_0 + w \quad \text{mit} \quad w = \beta h |d|, \qquad \beta = \frac{\|z_0\|}{1 - \|d\|}$$

erfüllt. Ist zusätzlich

$$(1 + \beta \|b + c\| \, \|h\|) \, \|d\| < 1,$$

trifft auch die Forderung (8) (*bzw.* (10)) *zu.*

Beweis. Wegen (12) ist $(e - d)^{-1}$ vorhanden und gilt

$$\|(e - d)^{-1}\| \leqq \frac{1}{1 - \|d\|};$$

aus (11) ergibt sich

$$\begin{aligned} a^{-1} &= z_0(e - d)^{-1} = z_0 + z_0(e - d)^{-1} d \\ &= z_0 + z_0(e - d)^{-1} d^+ - z_0(e - d)^{-1} d^- . \end{aligned}$$

Beachtet man $d^+ \geqq 0$, $d^- \geqq 0$ und

$$-\beta h \leqq z_0(e - d)^{-1} \leqq \beta h,$$

folgt unmittelbar die erste Behauptung:

$$\begin{aligned} -w = -\beta h d^+ - \beta h d^- &\leqq z_0(e - d)^{-1} d^+ - z_0(e - d)^{-1} d^- \\ &= a^{-1} - z_0 \leqq \beta h d^+ + \beta h d^- = w . \end{aligned}$$

Für den zweiten Teil errechnet man

$$0 \leqq e - bx_0 + cy_0 = d + \beta(b + c)\, h|d|,$$

$$0 \geqq e - by_0 + cx_0 = d - \beta(b + c)\, h|d|,$$

also ist in Bestätigung von (8)

$$\left.\begin{array}{l} \|e - bx_0 + cy_0\| \\ \|e - by_0 + cx_0\| \end{array}\right\} \leqq (1 + \beta\|b + c\|\,\|h\|)\,\|d\| < 1 .$$

Eine Weiterentwicklung zu dem Verfahren (4), (5) ist inzwischen in der Mitteilung [4] vorgestellt worden.

Literatur

[1] Alefeld, G., und J. Herzberger, Einführung in die Intervallrechnung. Bibliographisches Institut, Mannheim—Wien—Zürich 1974.

[2] Mönch, W., Monotone Einschließung von positiven Inversen. Z. Angew. Math. Mech. **53**, 207—208 (1973).

[3] Schmidt, J. W., Monotone Einschließung von Inversen positiver Elemente durch Verfahren vom Schulz-Typ. Computing **16**, 211—219 (1976).

[4] Schmidt, J. W., und W. Nauber, Über Verfahren zur zweiseitigen Approximation inverser Elemente. Computing **17**, 59—67 (1976).

Anschrift des Verfassers:

Technische Universität Dresden
Sektion Mathematik
8027 Dresden
Mommsenstr. 13
DDR

On a class of nonlinear integral equations of Volterra-Hammerstein type

Klaus R. Schneider, Berlin

1. The aim of this paper is to describe a class of nonlinear integral equations of Volterra-Hammerstein type whose solutions may be obtained via the solutions of a nonlinear equation in a finite dimensional space. The general idea consists in the extension of the principle of degenerated kernel [1]. The considerations were stimulated by investigations on existence and numerical approximation of forced oscillations [2].

The following notations will be used throughout the paper. We denote by $|.|$ the Euclidean norm in the space R_n, let S be the interval $[a, b] \subset R_1$, $a < b$, and $L_2(S, R_n)$ be the space of square integrable functions mapping S into R_n.

2. Firstly we shall consider in $L_2(S, R_n)$ the linear Fredholm integral equation

$$\varphi(t) = \eta(t) + \int_a^t G(t, s)\,\varphi(s)\,ds + K(t)\int_a^b M(s)\,\varphi(s)\,ds \tag{2.1}$$

assuming

A1. *$G(t, s)$ is a $n \times n$ matrix whose elements are square integrable on $S \times S$, $M(t)$ is a $n \times n$ matrix whose elements belong to $L_2(S, R_1)$.*

A2. *$K(t)$ is a $n \times n$ matrix whose elements belong to $L_2(S, R_1)$, η is an element of $L_2(S, R_n)$.*

The equation (2.1) is equivalent to the system

$$\varphi(t) - \int_a^t G(t, s)\,\varphi(s)\,ds = \psi(t), \tag{2.2}$$

$$\psi(t) = \eta(t) + K(t)\int_a^b M(s)\,\varphi(s)\,ds. \tag{2.3}$$

Under the assumption A1 the equation (2.2) has for each $\psi \in L_2(S, R_n)$ in $L_2(S, R_n)$ a unique solution

$$\varphi(t) = (L\psi)(t) \equiv \psi(t) + \sum_{m=1}^{\infty} (G^m\psi)(t), \tag{2.4}$$

where G^m denotes the m-th iterated kernel of $G(t, s)$. Obviously, L is a linear operator.

If we substitute (2.4) into (2.3), we get

$$\psi(t) = \eta(t) + K(t) \int_a^b M(s) (L\psi) (s)\, ds. \tag{2.5}$$

Using the notations

$$\mu = \int_a^b M(s) (L\psi) (s)\, ds, \qquad \nu = \int_a^b M(s) (L\eta) (s)\, ds,$$

$$N = \int_a^b M(s) (LK) (s)\, ds$$

we obtain from (2.5) the system of linear equations

$$(I - N) \mu = \nu \tag{2.6}$$

for the determination of μ. Consequently, we have:

Lemma 2.1. *Suppose that* A1 *and* A2 *hold. Then the equations* (2.1) *and* (2.6) *are equivalent.*

Remark 2.1. If the linear operator L is explicitely unknown, we may approximate L by the operator L_n defined by

$$(L_n\psi) (t) = \psi(t) + \sum_{m=1}^{n} (G^m\psi) (t).$$

If there exists an integer n_0 such that $\det (I - N_n) \neq 0 \; \forall n \geqq n_0$, where $N_n = \int_a^b M(s) \times (L_nK) (s)\, ds$, then the equation (2.1) has a unique solution in $L_2(S, R_n)$.

3. Now we consider the nonlinear integral equation of Volterra-Hammerstein type in the space $L_2(S, R_n)$

$$\varphi(t) = \eta(t) + \int_a^t g(t, s, \varphi(s))\, ds + K(t) \int_a^b h(s, \varphi(s))\, ds. \tag{3.1}$$

Analogously to (2.2), (2.3), we write (3.1) in the equivalent form

$$\varphi(t) = \psi(t) + \int_a^t g(t, s, \varphi(s))\, ds \equiv (V\varphi) (t), \tag{3.2}$$

$$\psi(t) = \eta(t) + K(t) \int_a^b h(s, \varphi(s))\, ds. \tag{3.3}$$

In the first step we look for conditions under which the nonlinear integral equation of Volterra type (3.2) has a unique solution in $L_2(S, R_n)$ for each $\psi \in L_2(S, R_n)$. By this way an operator W is defined, mapping ψ into the solution φ of (3.2), i.e.

$$\varphi = W\psi. \tag{3.4}$$

In the second step we substitute (3.4) into (3.3) and reduce (3.3) to a nonlinear equation in R_n.

To prove that (3.2) has a unique solution for $\psi \in L_2(S, R_n)$ we use the contraction principle.

A3. *$g: S \times S \times R_n \to R_n$ is measurable for each $u \in L_2(S, R_n)$ on $S \times S$ and satisfies the following conditions:*

i. *There exists a measurable function $m(t, s)$ on $S \times S$ such that*

$$\int_a^b \left(\int_a^t m(t, s)\, ds \right)^2 dt < +\infty, \tag{3.5}$$

$$|g(t, s, u(s))| \leqq m(t, s) \tag{3.6}$$

$\forall u \in L_2(S, R_n)$ and almost everywhere in $S \times S$.

ii. *There exists a measurable function $l(t, s)$ on $S \times S$ such that*

$$\left\{ \int_a^b \int_a^t l^4(t, s)\, ds\, dt \right\}^{1/4} = c_1 < +\infty, \tag{3.7}$$

$$|g(t, s, u_1(s)) - g(t, s, u_2(s))| \leqq l(t, s)\, |u_1(s) - u_2(s)| \tag{3.8}$$

$\forall u_1, u_2 \in L_2(S, R_n)$ and almost everywhere in $S \times S$.

Theorem 3.1. *Suppose that* A3 *is satisfied. Then the nonlinear Volterra equation* (3.2) *has for each $\psi \in L_2(S, R_n)$ a unique solution in $L_2(S, R_n)$.*

Proof. Firstly we shall prove that the operator V defined in (3.2) maps $L_2(S, R_n)$ into $L_2(S, R_n)$. To this end we have to show for $\varphi \in L_2(S, R_n)$

$$\int_a^b \left| \int_a^t g(t, s, \varphi(s))\, ds \right|^2 dt < +\infty.$$

By (3.6) and (3.5) we have

$$\int_a^b \left| \int_a^t g(t, s, \varphi(s))\, ds \right|^2 dt \leqq \int_a^b \left(\int_a^t m(t, s)\, ds \right)^2 dt < +\infty.$$

Now we will prove that V is a strict contraction on $L_2(S, R_n)$. We introduce in $L_2(S, R_n)$ by

$$\|\varphi\|_{L_2,k} = \left(\int_a^b |\varphi(t)|^2\, e^{-2kt}\, dt \right)^{1/2}, \qquad k > 0$$

the norm $\|\cdot\|_{L_2,k}$, which is equivalent to the norm $\|\cdot\|_{L_2}$,

$$\|\varphi\|_{L_2} = \left(\int_a^b |\varphi(t)|^2\, dt \right)^{1/2}.$$

Let φ_1 and φ_2 be elements of $L_2(S, R_n)$. By (3.2), (3.8) and (3.7) we get

$$|(V\varphi_1)(t) - (V\varphi_2)(t)| \leqq \int_a^t \left|g\big(t, s, \varphi_1(s)\big) - g\big(t, s, \varphi_2(s)\big)\right| ds$$

$$\leqq \int_a^t l(t, s)\,|\varphi_1(s) - \varphi_2(s)|\,ds \leqq \left(\int_a^t l^2(t, s)\,e^{2ks}\,ds\right)^{1/2} \left(\int_a^b e^{-2ks}\,|\varphi_1(s) - \varphi_2(s)|^2\,ds\right)^{1/2}$$

$$\leqq \left(\int_a^t l^4(t, s)\,ds\right)^{1/4} \left(\frac{e^{4kt} - e^{4ka}}{4k}\right)^{1/4} \|\varphi_1 - \varphi_2\|_{L_2,k}.$$

Thus, using Schwarz inequality,

$$\|V\varphi_1 - V\varphi_2\|_{L_2,k} \leqq \left(\int_a^b \int_a^t l^4(t, s)\,ds\,dt\right)^{1/4} \left(\int_a^b \frac{1 - e^{4k(a-t)}}{4k}\,dt\right)^{1/4} \|\varphi_1 - \varphi_2\|_{L_2,k}$$

$$= c_1 \left(\frac{b - a}{4k} + \frac{1 - e^{4k(a-b)}}{16k^2}\right)^{1/4} \|\varphi_1 - \varphi_2\|_{L_2,k}.$$

Therefore, there is a number $k_0 > 1$ such that for $k \geqq k_0$

$$c_1 \left(\frac{b - a}{4k} + \frac{1 - e^{4k(a-b)}}{16k^2}\right)^{1/4} \leqq c < 1.$$

Consequently, V is strictly contractive in $L_2(S, R_n)$ with respect to the norm $\|\cdot\|_{L_2,k_0}$. Using the equivalence of norms $\|\cdot\|_{L_2}$ and $\|\cdot\|_{L_2,k}$ and Banach's fixed point theorem we obtain that (3.2) possesses for $\psi \in L_2(S, R_n)$ a unique solution in $L_2(S, R_n)$, if the assumption A3 holds.

Corollary 3.1. *Under the assumption* A3 *there exists an operator* $W: L_2(S, R_n) \to L_2(S, R_n)$, *which maps* ψ *into the solution* φ *of* (3.2) *and satisfies*

$$\|W\psi_1 - W\psi_2\|_{L_2,k_0} \leqq \frac{1}{1 - c}\,\|\psi_1 - \psi_2\|_{L_2,k_0}. \tag{3.9}$$

In the following we suppose

A4. $h\big(s, u(s)\big)$ *is measurable on* S *for each* $u \in L_2(S, R_n)$.
Now we substitute $\varphi = W\psi$ into (3.3) and obtain

$$\psi(t) = \eta(t) + K(t) \int_a^b h\big(s, (W\psi)(s)\big)\,ds. \tag{3.10}$$

Using

$$\beta = \int_a^b h\big(s, (W\psi)(s)\big)\,ds \tag{3.11}$$

we get from (3.10)

$$(W\psi)(t) = W(\eta + K\beta)(t). \tag{3.12}$$

Thus, we obtain from (3.11) and (3.12) the following determining equation for β

$$\beta = \int_a^b h(s, W(\eta + K\beta)(s))\, ds \equiv F(\beta). \tag{3.13}$$

Consequently, we have:

Theorem 3.2. *Suppose that* A2—A4 *hold. Then each solution of the nonlinear equation* (3.13) *in* R_n *defines uniquely a solution of the nonlinear integral equation* (3.1).

In general, the operator F is explicitely unknown. This complicates the application of Theorem 3.2 for establishing existence results of equation (3.1). In the next section we give two ways to obtain existence results for solutions of (3.1) using Theorem 3.2.

4. The first way to obtain results about the existence of a solution of the equation (3.1) consists in applying the property of Lipschitz continuity of the operator F.

Theorem 4.1. *Assume that* A2—A4 *hold. Furthermore we suppose:*

i. $K(s)$ *is a* $n \times n$ *matrix such that*

$$\int_a^b \|K(s)\|^4\, ds < +\infty,$$

where $\|M\|$ *denotes the norm of a* $n \times n$ *matrix* (m_{ij}) *defined by* $\|M\| = \left\{ \sum_{1 \leq i,k \leq n} m_{ik}^2 \right\}^{1/2}$.

ii. *There exists a measurable function* $r(s)$ *on* S *such that*

$$\frac{1}{1-c} \left\{ \int_a^b r^4(s)\, ds \int_a^b \|K(s)\|^4\, ds\, \frac{(e^{4k_0(b-a)} + e^{4k_0(a-b)} - 2)}{16k_0^2} \right\}^{1/4} \leq d < 1, \tag{4.1}$$

$$|h(s, u_1(s)) - h(s, u_2(s))| \leq r(s)\, |u_1(s) - u_2(s)| \tag{4.2}$$

$\forall u_1, u_2 \in L_2(S, R_n)$ *and almost everywhere in* S.

Then the nonlinear integral equation (3.1) *has for each* $\psi \in L_2(S, R_n)$ *a unique solution* $\varphi \in L_2(S, R_n)$.

Proof. It is easy to verify that the operator F defined in (3.13) maps R_n into R_n. For $\beta_1, \beta_2 \in R_n$ we have by (3.13), (4.2), (3.9) and (4.1)

$$\begin{aligned}
|F(\beta_1) - F(\beta_2)| &\leq \int_a^b |h(s, W(\eta + K\beta_1)(s)) - h(s, W(\eta + K\beta_2)(s))|\, ds \\
&\leq \int_a^b r(s)\, |W(\eta + K\beta_1)(s) - W(\eta + K\beta_2)(s)|\, ds \\
&\leq \left\{ \int_a^b r^2(s)\, e^{2k_0 s}\, ds \right\}^{1/2} \|W(\eta + K\beta_1) - W(\eta + K\beta_2)\|_{L_2,k_0} \\
&\leq \left\{ \int_a^b r^4(s)\, ds \int_a^b e^{4k_0 s}\, ds \right\}^{1/4} \frac{1}{1-c} \|K\beta_1 - K\beta_2\|_{L_2,k_0} \\
&\leq \frac{1}{1-c} \left\{ \int_a^b r^4(s)\, ds \int_a^b e^{4k_0 s}\, ds \int_a^b e^{-4k_0 s}\, ds \int_a^b \|K(s)\|^4\, ds \right\}^{1/4} |\beta_1 - \beta_2| \\
&\leq d|\beta_1 - \beta_2|.
\end{aligned} \tag{4.3}$$

By Banach's fixed point theorem F has a unique fixed point β_* in R_n. Consequently, equation (3.1) has by Theorem 3.2 a unique solution in $L_2(S, R_n)$, q.e.d.

Remark 4.1. Since W is in general explicitely unknown, we can not use F for a direct numerical approximation of β_*. But in case F is strictly contractive there is an approximation W_n of W such that F_n, $F_n(\beta) \equiv \int_a^b h(s, W_n(\eta + K\beta)(s))\, ds$, is also strictly contractive.

The second way for establishing existence results consists in applying the powerful tool of Brouwer degree.

Let G be a bounded open region in R_n containing $\beta = 0$. Let $f: G \to R_n$ be a continuous function such that $f(x) \neq 0$ $\forall x \in \partial G$, where ∂G denotes the boundary of G. Then we denote by $d(f, G, 0)$ the mapping degree of f with respect to G at 0. From the definition of $d(f, G, 0)$ we obtain immediately the general theorem:

Theorem 4.2. *Suppose that* A2—A4 *hold. Furthermore let G be a bounded open region in R_n containing $\beta = 0$ such that $d(I - F, G, 0) \neq 0$. Then there exists at least one solution of* (3.1).

The following theorems give sufficient conditions for the assumptions of Theorem 4.2 to be satisfied.

Theorem 4.3. *Assume that* A2—A4 *hold. Furthermore we suppose*

A5. *There exists a function $\delta_1(s) \in L_1(S, R_1)$ such that*

$$\int_a^b \delta_1(s)\, ds \leqq c_3, \tag{4.4}$$

$$|h(s, u(s))| \leqq \delta_1(s) \qquad \forall u \in L_2(S, R_n) \tag{4.5}$$

and almost everywhere in S.

A6. $$\|h(s, u_1(s)) - h(s, u_2(s))\|_{L_2} \to 0 \tag{4.6}$$

as $\|u_1 - u_2\|_{L_2} \to 0$.

Then the equation (3.1) *has at least one solution.*

Proof. We consider in R_n the ball $B_{c_3} = \{\beta : |\beta| \leqq c_3\}$. For $\beta \in B_{c_3}$ we have by (3.13) and (4.4)

$$|F(\beta)| \leqq \int_a^b |h(s, W(\eta + K\beta)(s))|\, ds \leqq \int_a^b \delta_1(s)\, ds \leqq c_3,$$

i.e. $FB_{c_3} \subset B_{c_3}$.

For $\beta_1, \beta_2 \in B_{c_3}$ we obtain from (3.9)

$$\|W(\eta + K\beta_1) - W(\eta + K\beta_2)\|_{L_2,k_0}$$

$$\leqq \frac{1}{1 - c} \left\{ \int_a^b \|K(s)\|^2 \, e^{-2k_0 s}\, ds \right\}^{1/2} |\beta_1 - \beta_2|. \tag{4.7}$$

It follows from (4.3), (4.6) and (4.7) that F is continuous. Then, by Brouwer's fixed point theorem F has at least one fixed point in B_{c_3}. Consequently, the equation (3.1) has at least one solution.

Now we consider the equation

$$\varphi(t) = \int_a^t G(t, s)\, \varphi(s)\, ds + M(t) \int_a^b h\big(s, \varphi(s)\big)\, ds \tag{4.8}$$

under the assumptions A1, A4 and A5. Then the operator W is linear.

Theorem 4.4. *Suppose that* A1, A4 *and* A6 *are satisfied. Additionally we assume*

i. $h(t, x) = h_1(t, x) + h_2(t, x)$, *where*

$$h_1(t, -x) = -h_1(t, x) \qquad \forall (t, x) \in R_1 \times R_n.$$

ii. *There is a number* $r > 0$ *such that*

$$\left| \int_a^b \left[h_1\big(s, W(M\beta)(s)\big) + \mu h_2\big(s, W(M\beta)(s)\big)\right] ds \right| \neq r$$

for $|\beta| = r$ *and* $\mu \in [0, 1]$.

Then the equation (4.8) *has at least one solution.*

Proof. We consider in R_n the sphere $\Sigma_r = \{\beta : |\beta| = r\}$. Under the above assumption the operator

$$F_\mu(\beta) \equiv \int_a^b \left[h_1\big(s, W(M\beta)(s)\big) + \mu h_2\big(s, W(M\beta)(s)\big)\right] ds$$

is continuous in μ and β for $\mu \subset [0, 1]$ and satisfies

$$F_\mu(\beta) \neq \beta \quad \text{for} \quad \beta \in \Sigma_r \quad \text{and} \quad \mu \in [0, 1].$$

Because

$$F_0(-\beta) = \int_a^b h_1\big(s, W(M(-\beta))(s)\big)\, ds = \int_a^b h_1\big(s, -(WM\beta)(s)\big)\, ds = -F_0(\beta)$$

we have $d(I - F_0, B_r, 0) \neq 0$. Since $F_0(\beta)$ and $F_1(\beta) = F(\beta)$ are homotopic, we get

$$d(I - F, B, 0) \neq 0.$$

Consequently, (4.8) has at least one solution.

5. Equations of the form (3.1) appear for example in handling of boundary value problems of ordinary differential equations using Green's function. Consider, e.g., the periodic boundary value problem

$$\frac{dx}{dt} = A(t)\, x + f(t, x), \qquad x(0) = x(T). \tag{5.1}$$

If we suppose

i. $f(t, x)$ is a continuous mapping of $R_1 \times R_n$ into R_n satisfying $f(t + T, x) = f(t, x)$ $\forall (t, x) \in R_1 \times R_n$.

ii. The linear system

$$\frac{dy}{dt} = A(t)\, y \tag{5.2}$$

is non-critical, i.e. there exists the matrix $\big(I - X(T)\big)^{-1}$, where $X(t)$ is the fundamental matrix of (5.2) satisfying $X(0) = I$

then the boundary value problem (5.1) is equivalent to the Volterra-Hammerstein integral equation

$$\varphi(t) = X(t) \int_0^t X^{-1}(s)\, f\big(s, \varphi(s)\big)\, ds + X(t)\, \big(I - X(T)\big)^{-1} X(T) \int_0^T X^{-1}(s)\, f\big(s, \varphi(s)\big)\, ds\,. \tag{5.3}$$

A detailed investigation of (5.3) using the method described above can be found in [2, 3].

References

[1] Michlin, S. G., Vorlesungen über lineare Integralgleichungen. Berlin 1962.

[2] Schneider, K. R., Über eine Methode zur Untersuchung erzwungener Schwingungen. Math. Nachr. **76**, 109–122 (1977).

[3] Schneider, K. R., Investigations of nonlinear integral equation in the theory of nonlinear oscillations. *In:* Proceedings of the VII International Conference on Nonlinear Oscillations. These Abhandl. 1977, Nr. 4N, 221–236.

Address of the author:

Akademie der Wissenschaften der DDR
Zentralinstitut für Mathematik und Mechanik
108 Berlin
Mohrenstr. 39
DDR

Universal approximation in the finite element method

Karel Segeth, Praha

Abstract. The properties of the universal approximation by hill functions and by systems of hill functions are surveyed in the paper. These ideas are employed in solving a model one-dimensional second-order boundary-value problem by the finite element method. The numerical results illustrate the advantages of certain appropriate choices of the systems of hill functions.

In the finite element method, the principal question is an appropriate choice of the trial functions with the desired approximation properties. An approach to this problem called the approximation by hill functions is studied e.g. by Babuška [1], and Fix and Strang [3]. We briefly survey the properties of this approximation in the one-dimensional case and some its generalizations given in [4, 5].

In conclusion, we present a numerical example illustrating the behavior of the universal hill functions and their systems in practical computation.

Universal approximation

Choosing a function φ with a compact support and calling it a hill function, we construct in the one-dimensional Euclidean space R the approximation of the form

$$\sum_{k=-\infty}^{\infty} c_k \varphi\left(\frac{x}{h} - k\right)$$

where c_k are coefficients and $0 < h < 1$ is a parameter. Considering the approximation of this kind, we can show that it is necessary for the Fourier transform of the hill function φ to have zeros at the points $2\pi j$ for all non-zero integers j. The multiplicity t of these zeros determines the highest order of approximation attainable. For any function $f \in W_2^{\beta}(R)$ and $\beta \geqq \alpha \geqq 0$ there exist a constant C independent of h and coefficients c_k such that

$$\left\| f(x) - \sum_{k=-\infty}^{\infty} c_k \varphi\left(\frac{x}{h} - k\right)\right\|_{\alpha} \leqq C h^{\mu} \|f\|_{\beta}$$

where $\mu = \min(\beta - \alpha, t - \alpha)$ (see [1]).

Considering the approximation in a more general form,

$$\sum_{k=-\infty}^{\infty} c_k \omega\left(\left(\frac{x}{h} - k\right)\eta(h)\right)$$

where $\eta(h)$ is a certain increasing function and $\eta(0) = 0$, we can obtain the universal approximation. Namely, let $\omega(x)$ be an infinitely smooth rapidly decreasing function, $\Lambda(t)$ its Fourier transform, and $\Lambda(0) \neq 0$. Then for any $f \in W_2^\beta(R)$, $\varepsilon > 0$ and $\beta \geqq \alpha \geqq 0$, there exist a constant C independent of h and coefficients c_k such that

$$\left\| f(x) - \sum_{k=-\infty}^{\infty} c_k \omega\left(\left(\frac{x}{h} - k\right) \eta(h)\right) \right\|_\alpha \leqq C h^{\beta-\alpha-\varepsilon} \|f\|_\beta$$

(see [4]).

Therefore the approximation of this type is universal, i.e., for any approximated function f, we obtain the best possible order of approximation limited only by the smoothness of f. Analogously to the hill function φ, the function ω is called the universal hill function.

The Fourier transform Λ of the universal hill function ω has — in general — no zeros at all. The quality of approximation is achieved only by the employment of the so-called Λ-admissible function η.

In a further generalization, we may use the system $\{\omega_y\}$ of universal hill functions with some positive parameter y instead of a single function ω and study the approximation of the form

$$\sum_{k=-\infty}^{\infty} c_k \omega_{\eta(h)}\left(\left(\frac{x}{h} - k\right) \eta(h)\right).$$

Then this approximation is also universal and the above estimate holds for the system $\{\omega_y\}$, too (see [5]).

Moreover, the system $\{\omega_y\}$ can be chosen so that the Fourier transform Λ_y of ω_y has zeros at the points $\pm 2\pi j/y$; $j = 1, \ldots, J$ for some positive J (with some of these zeros being possibly multiple). This property of ω_y is a certain analogy to the corresponding property of the hill function φ. As a consequence, the error of the universal approximation by the system $\{\omega_y\}$ decreases (see [5]).

This effect is of great practical importance since it shows for relatively large h. The computation for small h is numerically rather unstable and we are forced to employ methods giving good results for large h like the method presented.

Approximating a function f the support of which is compact, we expect the support of the constructed approximation to be also compact. However, this is possible only in the case when the universal hill function ω_y has also a compact support. In practice, this $\omega_y(x)$ need not have a compact support but may decrease (at ∞) so rapidly that (from a numerical point of view) its values are negligible for $|x|$ greater than some $Y = Y(h) > 0$. Putting

$$\begin{aligned} \omega_{y,Y}(x) &= \omega_y(x), & |x| &< Y, \\ &= 0, & |x| &\geqq Y \end{aligned}$$

and constructing the approximation of the form

$$\sum_{k=-\infty}^{\infty} c_k \omega_{\eta(h),Y(h)}\left(\left(\frac{x}{h} - k\right) \eta(h)\right),$$

we can show that the universality is preserved (see [5]).

A numerical example

Numerical experiments fully confirm the above statements.

We solved the same one-dimensional second-order boundary-value problem as in [2, 4, 5], i.e., the ordinary equation

$$-u''(x) + cu(x) = g(x), \qquad x \in (0, \pi), \quad c > 0$$

with the boundary conditions

$$u'(0) = u'(\pi) = 0$$

and the right-hand part

$$g(x) = -\sin\left(d\left(x - \frac{1}{2}\pi\right)\right), \qquad d > 0.$$

The exact solution of this problem is

$$u(x) = -\frac{1}{d^2 + c}\sin\left(d\left(x - \frac{1}{2}\pi\right)\right)$$

as long as d is an odd integer. (There is a misprint in the formula for the general exact solution in [5]. The first term on the right-hand part should be preceded by a minus sign.)

The problem was solved by the finite element method with the trial functions $\omega_{\eta(h),Y(h)}\big((x/h - k)\,\eta(h)\big)$ and with five different systems $\{\omega_y\}$ of universal hill functions. Some of the results have been published by the author in [5].

The computation has been carried out in single precision on IBM 370/135 and MINSK 22 computers with various choices of the functions $\eta(h)$, $Y(h)$ and various values of the parameters c, d of the problem. The system of linear algebraic equations obtained was solved by the Gaussian elimination.

A typical result is shown in Fig. 1 where the scale of the variable $N = \pi/h$ (horizontal) is linear while the scale of the error (vertical) is logarithmic. The actual values of the parameters used in computation of the solution in Fig. 1 are

$$\eta(h) = \frac{3.8}{0.5 + \log^{1.0001} h^{-1}},$$

$$Y(h) = \frac{11}{6} + \frac{2}{3}\log^{1.0001} h^{-1},$$

$$c = 0.25, \qquad d = 3.$$

The error is measured by the quantity

$$\left(\frac{1}{N+1}\sum_{n=0}^{N} |U(nh) - u(nh)|^2\right)^{\frac{1}{2}}$$

where U is the approximate solution.

The individual lines in Fig. 1 correspond to the five various choices of the systems $\{\omega_y\}$. In general, the graph shows that the error decreases (as $N \to \infty$) more rapidly than any polynomial of a finite degree in all the five cases. For large N, the error increases due to the round-off.

Putting

$$\omega_y(x) = P(xy)\,\omega(x)$$

where P is a polynomial and

$$\omega(x) = e^{-x^2},$$

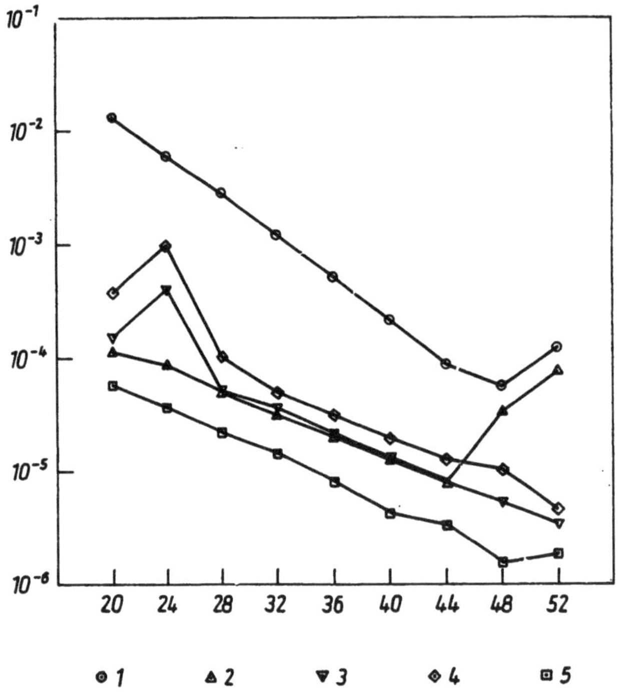

Fig. 1

we can obtain Λ_y with the desired zeros by the appropriate choice of P. The four lines in the graph correspond to the following arrangement of the zeros of Λ_y:

1 — no zero,

2 — a simple zero at $\pm 2\pi/y$,

3 — simple zeros at $\pm 2\pi/y$, $\pm 4\pi/y$,

4 — a double zero at $\pm 2\pi/y$.

The results confirm that the requirement $\Lambda_y(\pm 2\pi/y) = 0$ influences the error very strongly. On the other hand, the lines 3 and 4 show that further zeros of Λ_y or their higher multiplicity cannot improve the error of the approximation any more.

The fifth system $\{\omega_y\}$ is chosen in such a way that

$$\Lambda_y(t) = \frac{\sin\left(\frac{1}{2}\,yt\right)}{\frac{1}{2}\,yt}\,\Lambda(t)$$

where Λ is the Fourier transform of the above function $\omega(x) = \mathrm{e}^{-x^2}$. Then Λ_y has simple zeros at all the points $2\pi j/y$ where j is a non-zero integer. The corresponding system is

$$\omega_y(x) = \frac{1}{y}\,\frac{\sqrt{\pi}}{2}\left(\Phi\left(x + \frac{1}{2}\,y\right) - \Phi\left(x - \frac{1}{2}\,y\right)\right)$$

where Φ is the so-called error function, i.e.,

$$\frac{\sqrt{\pi}}{2}\,\Phi'(x) = \mathrm{e}^{-x^2}.$$

Line 5 in the graph represents the error obtained in approximation by this system of universal hill functions. It is apparent that the requirement $\Lambda_y(2\pi j/y) = 0$ for all the non-zero integers j makes the error less than in all the preceding cases. On the other hand, it is obvious that the complexity of the computation with this fifth system of universal hill functions is considerable.

References

[1] Babuška, I., Approximation by hill functions. Comment. Math. Univ. Carolinae **11**, 787–811 (1970).

[2] Babuška, I., J. Segethová, K. Segeth, Numerical experiments with the finite element method I. Tech. Note BN-669, Institute for Fluid Dynamics and Applied Mathematics, University of Maryland, College Park, August 1970.

[3] Fix, G., G. Strang, Fourier analysis of the finite element method in Ritz-Galerkin theory. Studies in Appl. Math. **48**, 265–273 (1969).

[4] Segeth, K., Universal approximation by hill functions. Czechoslovak Math. J. **22**, 612–640 (1972).

[5] Segeth, K., Universal approximation by systems of hill functions. Apl. Mat. **19**, 403–436 (1974).

[6] Strang, G., G. J. Fix, An analysis of the finite element method. Prentice-Hall, Englewood Cliffs, N. J. 1973.

Address of the author:

Mathematical Institute of the Czechoslovak Academy of Sciences
Žitná 25
11567 Praha 1
Czechoslovakia

To the maximum principle for one-dimensional heat transfer equation control problems

Friedmar Unger, Freiberg

Abstract. For a class of boundary control problems at the one-dimensional heat transfer equation in the cylinder case a maximum principle of Pontryagin-type is presented. Further is refered to some other classes of heat control problems also with distributed control, which are suited to formulate maximum principles on a similar way.

Using a new method of L. Bittner in [1] L. v. Wolfersdorf presented a maximum principle for processes being described by a Hammerstein system of integral equations with weakly singular kernels. In [2] he obtained a maximum principle for a class of heat transfer equation control problems, reducing these to problems of the structure considered in [1]. The application of this technique to further classes of one-dimensional heat transfer equation control problems, describing especially cooling or drying processes at plate, ball, cylinder or coaxial cylinders, leads to the following statements:

1. Let $u(r, t)$ be a generalized (in the sense exposed in [5]) solution of the problem

$$u_t = u_{rr} + \frac{1}{r} u_r + \Phi(r, t) \qquad (0 < r < 1, \quad 0 < t \leqq T), \tag{1}$$

$$\left.\begin{aligned} &u(r, 0) = 0 \quad (0 < r < 1), \qquad u(0, t) \text{ finite } \quad (0 \leqq t \leqq T), \\ &u_r(1, t) = \chi\big(t, u(1, t), v(t)\big) \qquad (0 < t \leqq T) \end{aligned}\right\} \tag{2}$$

with the following assumptions:

a.)

$$f(r, t) = \int_0^1 \int_0^t G(r, \xi, t - \tau)\, \Phi(\xi, \tau)\, d\tau\, d\xi \tag{3}$$

$$\Big(G(r, \xi, t) = \sum_{n=0}^{\infty} \frac{2}{J_0^2(\varrho_n)}\, \xi J_0(\varrho_n \xi)\, J_0(\varrho_n r)\, e^{-\varrho_n^2 t},$$

ϱ_n $(n = 1, 2, \ldots)$ the non-negative zeros of $J_1(\varrho)$ arranged in ascending order of magnitude$\Big)$ is given, continuous on $[0, 1] \times [0, T]$ and χ is given, continuous on $[0, T] \times R^1 \times R^1$ and has there a continuous partial derivative χ_u;

b.) The admissible control functions $v(t)$ are bounded and measurable (or piecewise continuous) on $[0, T]$ and satisfy the restriction

$$v(t) \in V \quad \forall t \in [0, T] \quad \text{with} \quad V = V(t) = \{v \in R^1 \mid c(t) \leqq v \leqq d(t)\} \tag{4}$$

$(c(t), d(t)$ given on $[0, T]$, piecewise continuous, $c(t) \leqq d(t))$.

We look for an admissible control $v(t)$ so, that

$$I = \sum_{i=0}^{2} \alpha_i I_i = \min \tag{5}$$

where

$$I_0 = \int_0^T \varrho_0(t) \, [u(1, t) - q_0(t)]^2 \, dt,$$

$$I_1 = \int_0^1 \varrho_1(r) \, [u(r, T) - q_1(r)]^2 \, dr, \qquad I_2 = \int_0^T \varrho_2(t) \, v^2(t) \, dt$$

(ϱ_0, ϱ_2 given on $[0, T]$, continuous, nonnegative; ϱ_1 given so, that $\varrho_1(r) \equiv 0$ on $[0, \varepsilon]$ $\big(\varepsilon \in (0, 1]\big)$ and continuous and non-negative on $[\varepsilon, 1]$; q_0, q_1 given on $[0, T]$ and $[0, 1]$, respectively, continuous; $\alpha_i \geqq 0$, $\sum_{i=0}^{2} \alpha_i > 0$, given).

Generalized solution of (1), (2) is each

$$u(r, t) = \int_0^t \int_0^1 G(r, \xi, t - \tau) \, \Phi(\xi, \tau) \, d\xi \, d\tau + \int_0^t G(r, 1, t - \tau) \, \chi\big(\tau, w(\tau), v(\tau)\big) \, d\tau \tag{6}$$

formed with a continuous solution of the integral equation

$$w(t) = f(1, t) + \int_0^t G(1, 1, t - \tau) \, \chi\big(\tau, w(\tau), v(\tau)\big) \, d\tau. \tag{7}$$

Therefore the problem, equivalently to the above, consists in the determination of an admissible control $v(t)$ and a corresponding (determined by (7)) state $w(t)$ $\big(\equiv u(1, t)\big)$ so, that $I = \min$. This problem is one of the type considered in the appendix in [2].

The following maximum principle is valid:

If $v^0(t)$ is an optimal control for the problem formulated above and $u^0(r, t)$ a corresponding optimal state, then

$$\max_{v \in V} H(t, v) = H\big(t, v^0(t)\big) \quad \textit{for almost every} \quad t \in [0, T] \tag{8}$$

where

$$H(t, v) = \alpha_2 \varrho_2(t) \, v^2 - \chi\big(t, u^0(1, t), v\big) \cdot z(1, t) \tag{9}$$

with

$$z(r, t) = 2\alpha_2 \int_0^1 \varrho_1(\xi) \, \big(u^0(\xi, T) - q_1(\xi)\big) \, G(\xi, r, T - t) \, d\xi$$
$$+ \int_t^T G(1, r, \tau - t) \, \zeta(\tau) \, d\tau. \tag{10}$$

$\zeta(t)$ is a bounded and measurable solution of the adjoint integral equation

$$\zeta(t) = \chi_u\big(t, u^0(1, t), v^0(t)\big) \, z(1, t) + 2\alpha_0 \varrho_0(t) \, \big(u^0(1, t) - q_0(t)\big). \tag{11}$$

The "adjoint state function" $z(r, t)$ can be interpreted as generalized (in the sense indicated above) solution of the following adjoint boundary value problem:

$$z_t = -z_{rr} + \frac{1}{r} z_r - \frac{1}{r^2} z \tag{12}$$

$$\left.\begin{aligned} z(r, T) &= 2\alpha_1 \varrho_1(r) \big(u^0(r, T) - q_1(r)\big), \quad z(0, t) \text{ finite}, \\ z_r(1, t) &= 2\alpha_0 \varrho_0(t) \big(u^0(1, t) - q_0(t)\big) \\ &\quad + \Big(1 + \chi_u\big(t, u^0(1, t), v^0(t)\big)\Big) \cdot z(1, t). \end{aligned}\right\} \tag{13}$$

2. A similar structured assertion under modified assumptions can be obtained for certain problems with boundary and distributed control in the case of the plate [5, 4]. Using the results obtained there and the substitution $u(r, t) = \frac{1}{r} U(r, t)$ one can get an analogous result for an adequate problem at the ball [4].

3. The called method, to produce maximum principles, further is applicable to the following control problem at coaxial cylinders [6]:

$$w_t = a^2 \left[w_{rr} + \frac{1}{r} w_r\right] + \omega(r, t) \qquad (0 < r < R_1, \quad 0 < t \leqq T), \tag{14}$$

$$\left.\begin{aligned} &w(r, 0) = 0 \quad (0 < r < R_1), \quad w(0, t) \text{ finite} \quad (0 \leqq t \leqq T), \\ &w_r(R_1, t) = \varphi\big(t, w(R_1, t), W(R_1, t), u(t)\big) \qquad (0 < t \leqq T); \end{aligned}\right\} \tag{15}$$

$$W_t = A^2 \left[W_{rr} + \frac{1}{r} W_r\right] + \Omega(r, t) \qquad (R_1 < r < R_2, \quad 0 < t \leqq T), \tag{16}$$

$$\left.\begin{aligned} &W(r, 0) = 0 \quad (R_1 < r < R_2), \quad W_r(R_2, t) = \psi\big(t, W(R_2, t), U(t)\big), \\ &W_r(R_1, t) = \Phi\big(t, w(R_1, t), W(R_1, t), u(t)\big) \qquad (0 < t \leqq T); \end{aligned}\right\} \tag{17}$$

$$\begin{aligned} I - \sum_{i=1}^{2} \alpha_i \int_0^T \varrho_i(t) \, [W(R_i, t) - q_i(t)]^2 \, dt \\ + \alpha_3 \int_0^T \varrho_3(t) \, [w(R_1, t) - q_3(t)]^2 \, dt \\ + \alpha_4 \int_0^{R_1} \varrho_4(r) \, [w(r, T) - q_4(r)]^2 \, dr \\ + \alpha_5 \int_{R_1}^{R_2} \varrho_5(r) \, [W(r, T) - q_5(r)]^2 \, dr \\ + \alpha_6 \int_0^T \varrho_6(t) \, u(t)^2 \, dt + \alpha_7 \int_0^T \varrho_7(t) \, U(t)^2 \, dt = \min \end{aligned} \tag{18}$$

with certain assumptions concerning the given a, A, ω, Ω, R_1, R_2, φ, Φ, ψ, T, h, α_j, ϱ_j $(j = 1, \ldots, 7)$, $q_1, \ldots, q_5$ and the control functions u and U.

Especially a control problem at the hollow cylinder can be treated.
The case, that the third conditions in (15) and (17) are replaced by

$$\lambda_1 w(R_1, t) = \lambda_2 W(R_1, t) + g_1(t) \qquad (0 \leqq t \leqq T) \tag{19}$$

$$\mu_1 w_r(R_1, t) = \mu_2 W_r(R_1, t) + g_2(t)$$

($\lambda_1, \lambda_2, \mu_1, \mu_2, g_1, g_2$ given) and the functional (18) without the second and sixth term of the sum is to be minimized, is in treatment.

References

[1] v. Wolfersdorf, L., Optimale Steuerung bei Hammersteinschen Integralgleichungen mit schwach singulären Kernen. Math. Operationsf. Statist. **6**, 609–626 (1975).

[2] v. Wolfersdorf, L., Optimale Steuerung einer Klasse nichtlinearer Aufheizungsprozesse. Z. Angew. Math. Mech. **55**, 353–362 (1975).

[3] Unger, F., Zum Maximumprinzip bei Randsteuerproblemen der eindimensionalen Wärmeleitungsgleichung. Z. Angew. Math. Mech. **56**, 25–32 (1976).

[4] Unger, F., Zum Maximumprinzip bei der eindimensionalen Wärmeleitungsgleichung mit Rand- und verteilter Steuerung. Z. Angew. Math. Mech. **56**, 123–124 (1976).

[5] v. Wolfersdorf, L., Maximum principle for control processes governed by mildly nonlinear parabolic equations. In this volume, 273–291.

[6] Unger, F., Ein Maximumprinzip bei Steuerproblemen bei der eindimensionalen Wärmeleitungsgleichung — Randsteuerung bei achsenparallelen Kreiszylindern. Z. Angew. Math. Mech. **57** (1977).

Address of the author:

Bergakademie Freiberg
Sektion Mathematik
92 Freiberg
Bernhard-v.-Cotta-Str. 2
DDR

Über die Regularität für eine nichtlineare Variationsungleichung höherer Ordnung mit einseitiger Beschränkung auf dem Rand

Michael Wulst, Berlin

Abstract. In this paper is studied the regularity of a non-linear variational inequality for a higher order differential operator on a bounded domain, the solution of which is required to lie above zero on the boundary.

1. Einleitung

Extremalaufgaben mit Beschränkungen auf dem Rande führen zur Behandlung von Variationsungleichungen. Über deren Lösbarkeit gibt es bereits eine Vielzahl von Ergebnissen, vgl. z. B. [6, 1]. Regularitätsaussagen sind vor allem für Probleme mit Differentialoperatoren 2. Ordnung bewiesen worden, [1, 4, 7], für Aufgaben mit nichtlinearen Differentialoperatoren höherer Ordnung und Beschränkungen im Gebiet sind solche im allgemeinen nur für das Innere des zugrunde gelegten Gebietes bekannt [2, 3]. In der vorliegenden Arbeit werden nichtlineare Variationsungleichungen mit Differentialoperatoren $2m$-ter Ordnung und einseitiger Beschränkung auf dem Rand untersucht. Es wird gezeigt, daß hierfür Regularitätsaussagen im ganzen Gebiet gelten.

Für ein beschränktes Gebiet $\Omega \subset E_N$ mit hinreichend glattem Rand Γ betrachten wir die folgende Aufgabe

$$\int_\Omega \sum_{|i|\leq m}' A_i(x, Du)\, D^i(u-v)\, dx \leq \int_\Omega \sum_{|i|\leq m}' f_i D^i(u-v)\, dx \tag{1.1}$$

für alle v aus K, wobei

$$K = \{v \subset H^m(\Omega) : v \geqq 0 \text{ f. ü. auf } \Gamma\}, \quad m \geqq 1 \text{ ganzz.}$$

Sei $\xi = \{\xi_\beta\}_{|\beta|\leq m}$ ein Element des s-dimensionalen Euklidischen Raumes E_s, wobei s die Anzahl aller Ableitungen der Ordnung $|\beta| \leqq m$ angibt.

Dann sollen die A_i und f_i die folgenden Voraussetzungen erfüllen:

$$A_i \in C^1(\bar{\Omega} \times R_s), \tag{1.2}$$

$$\sum_{|i|\leq m} |A_i(x, \xi)| + \sum_{j=1}^{N} \left| \frac{\partial}{\partial x_j} A_i(x, \xi) \right| \leqq C \left(1 + \sum_{|l|\leq m} |\xi_l| \right), \tag{1.3}$$

$$\sum_{|i|,|j|\leqq m} \left| \frac{\partial}{\partial \xi_j} A_i(x, \xi) \right| \equiv \sum_{|i|,|j|\leqq m} |A_{ij}(x, \xi)| \leqq C, \tag{1.4}$$

$$\sum_{|i|,|j|\leqq m} A_{ij}(x, \xi)\, t_i t_j \geqq C_0 \sum_{|i|\leqq m} t_i^2, \tag{1.5}$$

wobei $c_0 > 0$ und $t_i \in R$,

$$f_i \in L_2(\Omega). \tag{1.6}$$

Die Existenz und Eindeutigkeit der Lösung $u \in K$ der Aufgabe (1.1) sind durch die Bedingungen (1.2)—(1.6) gesichert.

Das Ziel der Arbeit besteht im Nachweis der Zugehörigkeit der Lösung u des Problems (1.1) zu $H^{m+1}(\Omega)$, sofern nur die f_i zu $H^1(\Omega)$ gehören. Von entscheidender Bedeutung dabei ist, daß $\mathcal{D}(\Omega)$ in K enthalten ist.

2. Definitionen

Das Gebiet Ω soll einen hinreichend oft stetig differenzierbaren Rand Γ besitzen [10].

Mit $E(\overline{\Omega})$ bezeichnen wir die Menge aller reellen, unendlich oft differenzierbaren und in $\overline{\Omega}$ zusammen mit allen Ableitungen stetigen Funktionen.

$\mathcal{D}(\Omega)$ ist die Teilmenge von $E(\Omega)$ aller Funktionen mit kompaktem Träger. Mit $H^m(\Omega)$ bezeichnen wir den Raum aller reellen Funktionen u, die zusammen mit allen Ableitungen $D^i u$, $|i| \leqq m$, (im Sinne von Distributionen) quadratisch integrierbar sind. Die Norm ist gegeben durch

$$\|u\|^2_{H^m(\Omega)} = \int_\Omega \sum_{|i|\leqq m} |D^i u|^2\, dx.$$

Alle positiven Konstanten in Sätzen und Beweisen bezeichnen wir mit c oder c_i.

Die Differenzenquotienten schreiben wir in folgender Form:

$$D^1_{\pm\Delta_i} u(x) = \frac{1}{\pm\Delta x_i} \left(\overset{\pm i}{u}(x) - u(x) \right) = \frac{1}{\pm\Delta x_i} \left(u\left(\overset{\pm i}{x} \right) - u(x) \right)$$
$$= \frac{1}{\pm\Delta x_i} \left(u(x_1, \ldots, x_{i-1}, x_i \pm \Delta x_i, x_{i+1}, \ldots, x_N) - u(x_1, \ldots, x_N) \right).$$

3. Abschätzungen

Satz 3.1. *Für die Lösung u von* (1.1) *gilt unter den Voraussetzungen* (1.2)—(1.5)

$$u \in H^{m+1}_{\mathrm{loc}}(\Omega),$$

falls die f_i aus $H^1_{\mathrm{loc}}(\Omega)$ sind.

Den Beweis dieses Satzes führt man auf die Behandlung einer Variationsgleichung zurück, was möglich ist, weil $\mathcal{D}(\Omega)$ in K enthalten ist.

Um die Aussage von Satz 3.1 für das gesamte Gebiet Ω zeigen zu können, beschränken wir uns in unserer Betrachtung auf ein Randstreifenstück des Gebietes.

Seien $\bar{x} \in \Gamma$ beliebig und $(y_1, \ldots, y_N)$ ein lokales Koordinatensystem mit dem Ursprung in $\bar{x}$, dessen y_N-Achse parallel zur inneren Normale an Γ in $\bar{x}$ verläuft. Da die Voraussetzungen (1.2)—(1.6) invariant bezüglich Drehung und Translation des Koordinatensystems sind, können wir annehmen, daß die Ungleichung (1.1) und die Bedingungen (1.2)—(1.6) bereits in dem lokalen Koordinatensystem mit dem Ursprung in $\bar{x}$ notiert sind.

Da wir den Rand Γ als hinreichend glatt vorausgesetzt haben, genügt der Durchschnitt $K_r(\bar{x}) \cap \Gamma$ der Gleichung

$$y_N = w(y'), \qquad y' = (y_1, \ldots, y_{N-1}),$$

wobei w eine glatte Funktion ist, die in dem durch die Projektion von $\Gamma \cap K_r(\bar{x})$ auf die Hyperebene $y_N = 0$ entstehenden Gebiet definiert ist. $K_r(\bar{x})$ bezeichnet eine Kugel mit dem Mittelpunkt in $\bar{x}$ und r ihren Radius.

In $K_r(\bar{x})$ können wir nun neue Koordinaten $(z_1, \ldots, z_N)$ wie folgt einführen:

$$z_i = T_i(y') = y_i, \qquad i = 1, \ldots, N-1,$$

$$z_N = T_N(y) = y_N - w(y').$$

Die Abbildungen T_j und T_j^{-1}, $j = 1, \ldots, N$, sind hinreichend oft stetig differenzierbar (entsprechend der Glattheit von w bzw. Γ). Für die Jacobi-Determinante haben wir $D(z)/D(y) = 1$. Wenn wir $\varrho > 0$ hinreichend klein wählen, dann gehört der Zylinder

$$\bar{G}_1 = \left\{(z_1, \ldots, z_N) : |z'| < \frac{\varrho}{2}, \qquad 0 \leqq z_N \leqq \frac{\varrho}{2}\right\}$$

zum Bild von $\overline{K_r(\bar{x})}$ bez. der Abbildung T. Das trifft dann auch für

$$\bar{G}_2 = \left\{(z_1, \ldots, z_N) : |z'| \leqq \frac{\varrho}{4}, \qquad 0 \leqq z_N \leqq \frac{\varrho}{4}\right\}$$

und

$$\bar{G}_3 = \left\{(z_1, \ldots, z_N) : |z'| < \frac{\varrho}{8}, \qquad 0 \leqq z_N \leqq \frac{\varrho}{8}\right\}$$

zu.

Wir führen folgende neue Bezeichnungen ein:

$$u(T^{-1}(z)) = \tilde{u}(z),$$

$$v(T^{-1}(z)) = \tilde{v}(z),$$

$$f_i(T^{-1}(z)) = \tilde{f}_i(z),$$

$$D_y{}^i u(y) = D_y{}^i \tilde{u}(T(y)) = \sum_{|\alpha| \leqq |i|} M_\alpha{}^i(y)\, D_z{}^\alpha \tilde{u}(T(y)), \qquad |i| \leqq m,$$

wobei die Koeffizienten $M_\alpha^{i}(y)$ von den Ableitungen der Funktionen T_j, $j = 1, \ldots, N$, bis zur $|i|$-ten Ordnung abhängen;

$$A_i\left(T^{-1}(z), \sum_{|\alpha| \leq |\beta|} M_\alpha^{\beta}(T^{-1}(z))\, D^\alpha \tilde{u}(z)\right) = \tilde{A}_i(z, D\tilde{u}(z)).$$

Man kann nachprüfen, daß die Eigenschaften (1.2)—(1.4), (1.6) für die $\tilde{A}_i$ und $\tilde{f}_i$ entsprechend auch gelten.

Weiterhin ergibt sich, daß die zur Matrix

$$\mathfrak{N} = \mathfrak{M}^t \mathfrak{A} \mathfrak{M} = \{N_{\gamma\delta}\}_{|\gamma|,|\delta| \leq m}$$

mit

$$\mathfrak{M} = \{M_\alpha^{\beta}\}_{|\beta|,|\alpha| \leq m} \quad \text{und} \quad \mathfrak{A} = \{A_{ij}\}_{|i|,|j| \leq m}$$

gehörende quadratische Form wegen (1.5) und der Tatsache, daß $\mathfrak{M}$ nicht entartet, positiv definit ist.

Satz 3.2. *Der Rand Γ von Ω sei $(m + 1)$-mal stetig differenzierbar, die f_i sollen aus $H^1(\Omega)$ sein.*

Unter den Voraussetzungen (1.2)—(1.5) *gilt*

$$D^\alpha \frac{\partial \tilde{u}}{\partial z_i} \in L_2(G_3), \qquad i = 1, \ldots, N - 1, \quad |\alpha| = m, \tag{3.1}$$

wobei $u(y) = \tilde{u}(T(y))$ Lösung der Variationsungleichung (1.1) *ist.*

Beweis. Wir wenden die Methode der Verschiebungen an, die z. B. in [6] von Lions benutzt wurde. Wir beschränken uns darauf, (3.1) für $i = 1$ zu zeigen. Dazu führen wir die folgenden Bezeichnungen ein:

$$(\tilde{A}_i)_{\pm h} = \tilde{A}_i(z_1 \pm h, z_2, \ldots, z_N,\ D\tilde{u}(z_1 \pm h, z_2, \ldots, z_N)),$$

$$v_{\pm h}(z) = v(z_1 \pm h, z_2, \ldots, z_N),$$

$$z_\theta = (z_1 + \theta h, z_2, \ldots, z_N),$$

$$\tilde{u}_\theta = \tilde{u} + \theta(\tilde{u}_h - \tilde{u}),$$

$$D_y^{\beta} u_\theta(T^{-1}(z)) = \sum_{|\gamma| \leq |\beta|} M_\gamma^{\beta}(z^{-1}(T_\theta))\, D_z^{\gamma} \tilde{u}_\theta(z);$$

$\mathfrak{Z} = \mathfrak{Z}(z)$ sei eine Funktion mit den Eigenschaften:

i) $\mathfrak{Z} \in \mathcal{D}(E_N)$,

ii) $\mathfrak{Z} \equiv 1$ in $\bar{G}_3$,

iii) $0 \leq \mathfrak{Z}(z) \leq 1$,

iv) $\mathfrak{Z}(z) = 0$ für alle z, die nicht in $\bar{G}_2$ liegen.

Die Funktion $\mathfrak{Z}(T(y))$ setzen wir außerhalb von $T^{-1}(G_2)$ durch Null auf ganz Ω fort. Dann gilt:

$$u(y) \pm u(y)\, \mathfrak{Z}^{2m}(T(y)) \in K,$$

denn es ist

$$u(y) \cdot \left(1 \pm \mathfrak{Z}^{2m}\left(T(y)\right)\right) \geqq 0 \quad \text{auf } \Gamma,$$

wegen der Beschaffenheit von $u(y)$ und $\mathfrak{Z}\left(T(y)\right)$.

Deshalb sind auch die Funktionen

$$u\left(\overset{+h}{y}\right) \mathfrak{Z}^{2m}\left(T(y)\right), \qquad u\left(\overset{-h}{y}\right) \mathfrak{Z}^{2m}\left(T\left(\overset{-h}{y}\right)\right)$$

und

$$u(y) \pm u(y) \cdot \mathfrak{Z}^{2m}\left(T\left(\overset{-h}{y}\right)\right)$$

aus K, wobei

$$\overset{\pm h}{y} = T^{-1}\left(z \pm \vec{h}\right)$$

mit $\vec{h} = (h, 0, \ldots, 0)$.

Folglich können wir in (1.1)

$$v(y) = u(y) \pm u(y) \cdot \mathfrak{Z}^{2m}\left(T\left(\overset{-h}{y}\right)\right)$$

setzen und bekommen nach Transformation auf die $(z_1, \ldots, z_N)$-Koordinaten wegen $\operatorname{supp} \mathfrak{Z} \subset\subset G_2$ und wenn $h < \frac{\varrho}{4}$ ist

$$\int_{G_1} \sum_{|i| \leqq m} \sum_{|\alpha| \leqq |i|} (\tilde{A}_i M_\alpha^i - \tilde{f}_i M_\alpha^i)_h \, D^\alpha(\tilde{u}_h \cdot \mathfrak{Z}^{2m}) \, dz = 0. \tag{3.2}$$

Aus (1.1) folgt andererseits allgemein

$$\int_{\Omega} \sum_{|i| \leqq m} \left(A_i(y, Du) - f_i(y)\right) D^i v \, dy \geqq 0 \tag{3.3}$$

für alle v aus K, woraus mit

$$v(y) = u\left(\overset{-h}{y}\right) \cdot \mathfrak{Z}^{2m}\left(T\left(\overset{-h}{y}\right)\right)$$

$$\int_{G_1} \sum_{|i| \leqq m} \sum_{|\alpha| \leqq |i|} (\tilde{A}_i M_\alpha^i - \tilde{f}_i M_\alpha^i)_h \, D^\alpha(\tilde{u} \cdot \mathfrak{Z}^{2m}) \, dz \geqq 0 \tag{3.4}$$

folgt. Weiter bekommt man aus (1.1) mit

$$v(y) = u(y) \pm u(y) \, \mathfrak{Z}^{2m}\left(T(y)\right)$$

bzw. aus (3.3) mit

$$v(y) = u\left(\overset{+h}{y}\right) \cdot \mathfrak{Z}^{2m}\left(T(y)\right)$$

$$\int_{G_1} \sum_{|i| \leqq m} \sum_{|\alpha| \leqq |i|} (\tilde{A}_i M_\alpha^i - \tilde{f}_i M_\alpha^i) \, D^\alpha(\tilde{u} \cdot \mathfrak{Z}^{2m}) \, dz = 0 \tag{3.5}$$

bzw.

$$\int_{G_1} \sum_{|i| \leqq m} \sum_{|\alpha| \leqq |i|} (\tilde{A}_i M_\alpha^i - \tilde{f}_i M_\alpha^i) \, D^\alpha(\tilde{u}_n \mathfrak{Z}^{2m}) \, dz \geqq 0. \tag{3.6}$$

(3.2), (3.4)—(3.6) liefern dann gemeinsam

$$\int_{G_1} \sum_{|i|\leq m} \sum_{|\alpha|\leq|i|} \left((\tilde{A}_i M_\alpha^i)_h - \tilde{A}_i M_\alpha^i\right) D^\alpha(\tilde{u}_n \mathfrak{Z}^{2m} - \tilde{u}\mathfrak{Z}^{2m})\, dz$$
$$\leq \int_{G_1} \sum_{|i|\leq m} \sum_{|\alpha|\leq|i|} \left((\tilde{f}_i M_\alpha^i)_h - \tilde{f}_i M_\alpha^i\right) D^\alpha(\tilde{u}_n \mathfrak{Z}^{2m} - \tilde{u}\mathfrak{Z}^{2m})\, dz. \tag{3.7}$$

Die linke Seite von (3.7) können wir unter Verwendung des Mittelwertsatzes, der Produktregel, (1.5) und der Eigenschaft von $\mathfrak{R}$ nach unten abschätzen und erhalten dann:

$$\int_{G_1} \sum_{|j|\leq m} \left(\frac{1}{h} D^j(\tilde{u}_n - \tilde{u})\, \mathfrak{Z}^m\right)^2 dz$$
$$\leq C \int_{G_1} \sum_{|i|\leq m} \sum_{|\alpha|\leq|i|} \left|\frac{1}{h} \frac{\partial}{\partial z_1} \left(\tilde{A}_i(z_\theta, D\tilde{u}_\theta)\, M_\alpha^i\left(T^{-1}(z_\theta)\right)\right) D^\alpha(\tilde{u}_n - \tilde{u}) \cdot \mathfrak{Z}^{2m}\right| dz$$
$$+ C \int_{G_1} \sum_{|i|\leq m} \sum_{|\alpha|\leq|i|} \sum_{\gamma<\alpha} \left|\frac{1}{h^2} \left((\tilde{A}_i M_\alpha^i)_h - \tilde{A}_i M_\alpha^i\right) D^\gamma(\tilde{u}_n - \tilde{u})\, D^{\alpha-\gamma}\mathfrak{Z}^{2m}\right| dz$$
$$+ C \int_{G_1} \sum_{|i|\leq m} \sum_{|\alpha|\leq|i|} \left|\frac{1}{h^2} \left((\tilde{f}_i M_\alpha^i)_h - \tilde{f}_i M_\alpha^i\right) D^\alpha(\tilde{u}_n - \tilde{u})\, \mathfrak{Z}^{2m}\right| dz$$
$$+ C \int_{G_1} \sum_{|i|\leq m} \sum_{|\alpha|\leq|i|} \sum_{\gamma<\alpha} \left|\frac{1}{h^2} \left((\tilde{f}_i M_\alpha^i)_h - \tilde{f}_i M_\alpha^i\right) D^\gamma(\tilde{u}_n - \tilde{u})\, D^{\alpha-\gamma}\mathfrak{Z}^{2m}\right| dz. \tag{3.8}$$

Mit Hilfe von (1.2)—(1.4) und der Tatsache, daß $f_i \in H^1(\Omega)$, kann man die rechte Seite von (3.8) nach oben abschätzen, und man bekommt schließlich

$$\int_{G_1} \sum_{|j|\leq m} \left(\frac{1}{h} D^j(\tilde{u}_n - \tilde{u})\, \mathfrak{Z}^m\right)^2 dz \leq c(\tilde{u}; \tilde{f}_i),$$

wobei c eine von $\tilde{u}$ und $\tilde{f}_i$ abhängende Konstante ist. Daraus ist ersichtlich, daß

$$\frac{1}{h} (\tilde{u}_h - \tilde{u})$$

beschränkt ist in $H^m(G_3)$, woraus sich sofort die Behauptung ergibt.

Satz 3.3. *Unter den Voraussetzungen von Satz* 3.2 *gilt*

$$D^\alpha \frac{\partial \tilde{u}}{\partial z_N} \in L_2(G_3), \qquad |\alpha| = m.$$

Beweis. Sei $\varphi \in \mathcal{D}(G_3)$ beliebig und außerhalb von G_3 weiter durch Null fortgesetzt. Wenn $\Delta z_N < \frac{\varrho}{8}$, so ist $D^1_{-\Delta_N}\varphi \in \mathcal{D}(G_2)$ und damit

$$v(y) = u(y) \pm D^1_{-\Delta_N}\varphi\left(T(y)\right) \in K.$$

Mit diesem speziellen v in (1.1) erhalten wir

$$\int_{G_2} \sum_{|i|\le m} \sum_{|\alpha|\le|i|} D^1_{\Delta_N}\big(\tilde{A}_i(z, D\tilde{u})\, M_\alpha{}^i\big)\, D^\alpha\varphi\, dz$$
$$= -\int_{G_2} \sum_{|i|\le m} \sum_{|\alpha|\le|i|} \tilde{f}_i(z)\, M_\alpha{}^i\big(T^{-1}(z)\big)\, D^\alpha(D^1_{-\Delta_N}\varphi)\, dz. \tag{3.9}$$

Auf die linke Seite von (3.9) wenden wir wieder den Mittelwertsatz an und spalten dann die Summanden mit $\mu = (0, \ldots, 0, m)$ nacheinander ab. Wir erhalten:

$$\int_{G_2} N_{\mu\mu}(z_\theta, Du_\theta)\, D^1_{\Delta_N} D^\mu \tilde{u} D^\mu\varphi\, dz$$
$$= -\int_{G_2} \sum_{|i|,|j|\le m} \sum_{\substack{|\gamma|\le|i|\\ |\alpha|\le|i|\\ \mu\ne\alpha}} A_{i\nu}\big(T^{-1}(z_\theta), Du_\theta\big)\, M^\nu_{j\theta} M^i_{\mu\theta} D^1_{\Delta_N} D^j\tilde{u} D^\alpha\varphi\, dz$$
$$- \int_{G_2} \sum_{|i|\le m} \sum_{|\alpha|\le|i|} \frac{\partial}{\partial z_N}\, \tilde{A}_i(z_\theta, D\tilde{u}_\theta)\, M^i_{\alpha\theta} D^\alpha\varphi\, dz$$
$$- \int_{G_2} \sum_{|i|\le m} \sum_{|\alpha|\le|i|} \sum_{K=1}^{N} \frac{\partial}{\partial y_k}\, M_\alpha{}^i\big(T^{-1}(z_\theta)\big)\, \frac{\partial}{\partial z_N}\, T_k{}^{-1}(z_\theta)\, \tilde{A}_i(z_\theta, D\tilde{u}_\theta)\, D^\alpha\varphi\, dz$$
$$+ \int_{G_2} \sum_{|i|\le m} \sum_{|\alpha|\le|i|} \tilde{f}_i(z)\, D^\alpha(D^1_{-\Delta_N}\varphi)\, dz$$
$$- \int_{G_2} \sum_{|i|,|\nu|\le m} \sum_{\substack{|j|\le m\\ j\ne\mu}}{}' A_{i\nu}\big(T^{-1}(z_\theta), Du_\theta\big)\, M^\nu_{j\theta} M^i_{\mu\theta} D^1_{\Delta_N} D^j\tilde{u} D^\mu\varphi\, dz, \tag{3.10}$$

wobei $M^i_{\alpha\theta} = M_\alpha{}^i\big(T^{-1}(z_\theta)\big)$ ist.

Die rechte Seite von (3.10) können wir unter Verwendung der Bedingungen an die $\tilde{A}_i$ und ihre Ableitungen, die (1.2)—(1.4) entsprechen, sowie (3.1) nach oben abschätzen, so daß man erhält

$$\sum_{|\alpha|\le m} \left| \int_{G_2} N_{\mu\mu} D^1_{\Delta_N} D^\mu\tilde{u} D^\alpha\varphi\, dz \right| \le C(\tilde{u}; \tilde{f}_i)\, \|\varphi\|_{H^m(G_2)}.$$

Wenn wir die Abschätzung (3.6) aus (8)

$$\|w\|_{L_2(\Omega)} \le C \sup_{\varphi\in \dot{C}^\infty(\Omega)} \frac{\sum\limits_{|\alpha|\le m} \left| \int_\Omega w D^{2\alpha}\varphi\, dx \right|}{\|\varphi\|_{H^{2m}(\Omega)}}$$

für $w = D^1_{\Delta_N} D^\mu \tilde{u}$ benutzen, folgt unter Berücksichtigung von $N_{\mu\mu} \ge c_1 > 0$

$$\int_{G_2} (D^1_{\Delta_N} D^\mu\tilde{u})^2\, dz \le C(\tilde{u}, \tilde{f}_i)$$

gleichmäßig für alle $\Delta Z_N < \frac{\varrho}{8}$, woraus sich sofort das Gewünschte ergibt.

Satz 3.4. *Unter den Voraussetzungen von Satz 3.2 gilt für die Lösung u des Problems* (1.1)

$$u \in H^{m+1}(\Omega).$$

Diese Behauptung folgt dann aus den Sätzen (3.1)—(3.3).

Literatur

[1] Brezis, H., G. Stampacchia, Sur la régularité de la solution d'inéquations elliptiques. Bull. Soc. math. France **96**, 153—180 (1968).
[2] Frehse, J., Zum Differenzierbarkeitsproblem bei Variationsungleichungen höherer Ordnung. Abh. Math. Seminar Univ. Hamburg **36**, 140—149 (1971).
[3] —, On the regularity of the solution of the biharmonic variational inequality. Manuscr. math. **9**, 91—103 (1973).
[4] Gerhardt, C., Regularity of solutions of nonlinear variational inequalities. Arch. rat. Mech. Anal. **52**, 389—393 (1973).
[5] Ладыженская, О. А., Н. Н. Уральцева, Линейные и квазилинейные уравнения эллиптического типа. Москва 1964.
[6] Lions, J. L., Quelques méthodes de resolution des problèmes aux limites non linéaires. Paris 1969.
[7] Lewy, H., G. Stampacchia, On the regularity of the solution of a variational inequality. Comm. pure appl. Math. **22**, 153—188 (1969).
[8] Müller, M., Über die Regularität der Lösungen quasilinearer elliptischer Differentialgleichungen höherer Ordnung am Rande des Gebietes. Math. Nachr. **68**, 141—160 (1975).
[9] —, J. Nečas, Über die Regularität der schwachen Lösungen von Randwertaufgaben für quasilineare elliptische Differentialgleichungen höherer Ordnung. Czechoslovak. math. J. **25**, 227—239 (1975).
[10] Nečas, J., Les méthodes directes en théorie des équations elliptiques. Prag 1967.
[11] Müller, M., Das Differenzenverfahren für gewisse quasilineare elliptische Differentialgleichungen II. Math. Nachr. **49**, 215—236 (1971).

Anschrift des Verfassers:

Humboldt-Universität Berlin
Sektion Mathematik
108 Berlin
Unter den Linden 6
DDR

On some nonlinear problems in critical reactor control

Rostislav Zezula, Praha

Abstract. In this paper, a construction of the relation $G = SF$ is described between the optimization function F and control regime function G used in two frequently occurring mathematical models of critical reactor control.

1. Introductory remarks

Reactor criticality is a typical linear problem of determining the simple positive dominant eigenvalue λ_1 and the corresponding nonnegative eigenfunction x_1 of the generalized eigenvalue problem $Ax = \lambda Bx$ in a suitable inner product space X, where A, B are linear operators in X with domains $D(A) \subset D(B) = X$, $\overline{D(A)} = X$, such that $\exists B^{-1}$ and $B^{-1}A$ is of Perron-Frobenius type [4].

Despite of it, some reactor criticality problems must be formulated mathematically as nonlinear problems, e.g., critical reactor control problems, in which one of the physical and geometrical parameters of the reactor core is to be chosen as unknown control vector which together with the state vector of the neutron flux is to be determined from the neutron balance equation endowed by the criticality and optimality conditions.

2. A class of nonlinear critical reactor control problems

Let us consider a concrete nuclear reactor assembly in $\Omega \subset E_3$, consisting of the reactor core $\Omega_C \subset \Omega$, filled with a medium formed by neutron gas, fuel and moderator, and of the reflector $\Omega_R = \Omega \setminus \Omega_C$ formed by a moderator.

The state of the neutron gas is described by the neutron gas density $\Phi = \Phi(x, v, t)$: $E_i \to E_1$, $i = 3, \ldots, 7$ where x denotes the radius vector, v the velocity vector of the neutron and $t \in E_1$ stands for the time variable. The shape and the material composition of the reactor assembly is described by some geometrical and material parameters from which we choose the fuel density distribution $M(x, t)$ as the control vector.

Reactor criticality means the stationary case $\Phi = \Phi(x, v)$, $M = M(x)$. Let us consider the following simplest mathematical model of a critical reactor (the so called one group diffusion model for desk geometry, with $a > 0$ as critical core halfthickness and $b > a$

as reflector thickness [1]):

$$\frac{d^2\Phi}{dx^2} + \omega^2\big(M, k(M)\big)\,\Phi = 0 \qquad \text{for } x \in (0, b) \setminus \{a\}, \tag{2.1}$$

ω^2 and k given functions

$$\Phi(x) > 0 \qquad \text{for } x \in (0, b) \text{ (criticality condition)} \tag{2.2}$$

$$M(x) = 0,\ \omega^2\big(0, k(0)\big) = \text{const} < 0 \qquad \text{for } x \in (a, b) \equiv \Omega_R \text{ (reflector conditions)} \tag{2.2a}$$

$$\left.\frac{d\Phi}{dx}\right|_{x=0} = 0 \tag{2.3}$$

$$\Phi(b) = 0 \tag{2.3a}$$

(boundary conditions)

$$\Phi|_{a-} = \Phi|_{a+} \tag{2.4}$$

$$\left.\frac{d\Phi}{dx}\right|_{a-} = \varkappa \left.\frac{d\Phi}{dx}\right|_{a+}, \qquad \varkappa = \text{const} > 0 \tag{2.4a}$$

(transient conditions)

For a given $M = M(x)$ fulfilling (2.2a) and fixed outer reactor boundary b, (2.1)—(2.4a) is a linear eigenparameter problem whose solution is the couple $\{a, \Phi_a(x)\}$ where the functional $a = a(M)$ is the eigenparameter (called the critical core halfthickness) and $\Phi_a = \Phi_a(x)$ is the corresponding eigenvector. In [3], an algorithm for constructive solution of the eigenparameter problems like (2.1)—(2.4a) is described, consisting in solving, for fixed $a \in (0, b)$ and for $\Phi(0) = \Phi_0$ the initial value problem (2.1)—(2.3) in $(0, a)$, together with the boundary value problem (2.1)—(2.2a), (2.3a), (2.4) in (a, b) and testing the validity of the remaining transient condition (2.4a) for each $a \in (0, b)$.

In critical reactor control problems, both the neutron gas state vector Φ and control vector M are considered as unknown. Denoting by U the *class of admissible control vectors* $M(x)$, for which the criticality problem (2.1)—(2.4a) has the unique solution $\{a, \Phi_a(x)\}$, the following two simple mathematical models of critical reactor control are used frequently [3]:

1) *Optimal control model*

Criticality problem (2.5a)

(2.1)—(2.4a) for $M \in U$.

Moreover, the optimality condition

$$\exists M_0 \in U, \quad I_1(M_0) = \inf_{M \in U} I_1(M), \tag{2.6a}$$

Technical control model

Criticality problem (2.5b)

(2.1)—(2.4a) for $M \in U$.

Moreover, the control regime condition

$$G\big(\Phi_a(x), M(x)\big) = C_1 = \text{const} \neq 0 \tag{2.6b}$$

for $x \in \big(0, a(M)\big)$, $M \in U$

and $\{a(M), \Phi_a\}$ solving (2.1)—(2.4a),

where $I_1(M)$ in (2.6a) is the *optimization functional* defined by

$$I_1(M) = \int_0^{a(M)} F\big(\Phi_a(x), M(x)\big)\,dx \quad \text{with} \quad M \in U, \{a(M), \Phi_a(x)\} \text{ solving (2.1)—(2.4a)} \tag{2.7a}$$

($F = F(\Phi, M) \in C(E_2)$ given *optimization function*) and $G = G(\Phi, M) \in C(E_2)$ in (2.6b) is a given *control regime function*.

The solution of the optimal control problem (2.5a), (2.6a) is the triple $\{a_0^F, M_0^F(x) \in U, \Phi_{a_0}^F(x)\}$ depending on $F \in C(E_2)$ and in general not unique, for which (2.5a) and (2.6a) hold. The solution of the technical control model (2.5b) and (2.6b) is the triple $\{a_1^G, M_1^G(x) \in U, \Phi_{a_1}^G(x)\}$ depending on $G \in C(E_2)$ and in general not unique, for which (2.5b) and (2.6b) are valid.

We are interested mostly in the cases when both the above control models describe the same physical situation:

Definition 2.1. We say that both the optimal control problem (2.5a) and (2.6a) and the technical control problem (2.5b), (2.6b) are *equivalent* iff we have for some solutions $\{a_0^F, M_0^F(x), \Phi_{a_0}^F(x)\}$, $\{a_1^G, M_1^G(x), \Phi_{a_1}^G(x)\}$ resp. of (2.5a), (2.6a) or (2.5b), (2.6b) resp.

$$a_0^F = a_1^G, \quad \text{and} \quad M_0^F(x) = M_1^G(x), \Phi_{a_0}^F(x) = \Phi_{a_1}^G(x) \quad \text{for} \quad \forall x \in \langle 0, b \rangle. \tag{2.8}$$

Now, let us consider the following critical reactor control problem:

Problem 2.1 [3—9]. For the case of equivalent optimal control problem (2.5a), (2.6a) and technical control problem (2.5b), (2.6b), to construct the relation S between the optimal control function F and the control regime function G, i.e.,

$$G = SF. \tag{2.9}$$

For the sake of simplicity, let us assume the validity of the implicite function theorem for the control regime function $G = G(\Phi, M)$, i.e., let us make

Assumption 2.1.

$$\Phi = C_1 H(M), \qquad H(M) \in C_2(E), \qquad G(C_1 H(M), M) \equiv C_1 > 0 \tag{2.10}$$
$$\text{for } \forall x \in (0, a(M)), \qquad M = M(x) \in U \qquad \text{and} \qquad \forall C_1 > 0$$

and moreover

$$H(M) > 0, \qquad \frac{dH}{dM} \neq 0 \qquad \text{for } \forall x \in (0, a(M)) \text{ and } M \in U. \tag{2.10a}$$

Let us denote by $\mathcal{G}$ the class of those $G \in C(E_2)$ for which (2.10), (2.10a), i.e., Assumption 2.1, holds.

Inserting for $G \in \mathcal{G}$ the relation (2.10) in the criticality problem (2.1)—(2.4a) and using (2.10a), we get the following *nonlinear eigenparameter problem* for $M = M(x)$, with the eigenparameter $a = a_2^G$ and $M = M_C^G(x)$ as the corresponding eigenvector:

$$\frac{d^2M}{dx^2} + \frac{d^2H}{dM^2}\left[\frac{dH}{dM}\right]^{-1}\left(\frac{dM}{dx}\right)^2 + \left[\frac{dH}{dM}\right]^{-1} \omega^2(M, k(M))\, H(M) = 0$$
$$\text{for} \quad x \in (0, a(M)), \tag{2.11}$$

$$M(x) > 0 \qquad \text{for } x \in (0, a(M)), \tag{2.12}$$

$$M(x) = 0, \qquad \omega^2(0, k(0)) = \text{const} < 0 \quad \text{for } x \in (a(M), b), \tag{2.12a}$$

$$\left.\begin{aligned} &\left.\frac{dM}{dx}\right|_{x=0} = 0 \\ &M(0) \equiv M_0 = C > 0 \quad \text{arbitrary} \end{aligned}\right\} \begin{aligned} &\text{initial conditions} && (2.13) \\ &\text{for } M(x) = M_C{}^G(x), && (2.13\text{a}) \end{aligned}$$

$$\left.\begin{aligned} &C_1 H(M)|_{a-} = \Phi_R|_{a+} \\ &C_1 \left(\frac{dH}{dM}\frac{dM}{dx}\right)\bigg|_{a-} = \varkappa \left.\frac{d\Phi_R}{dx}\right|_{a+} \end{aligned}\right\} \begin{aligned} &\text{transient conditions} && (2.14) \\ &\text{for } M(x) = M_C{}^G(x), && (2.14\text{a}) \end{aligned}$$

where $\Phi_R = \Phi_R{}^G(x)$, $x \in (a(M), b\rangle$ is the solution of the nonhomogeneous boundary value problem (2.1)—(2.2a), (2.3a), (2.14) in the reflector, i.e., for $M \equiv 0$.

The solution $\{a_2{}^G, M_C{}^G(x)\}$ of the nonlinear eigenparameter problem (2.11)—(2.14a) can be obtained by the same *constructive algorithm* as for (2.1)—(2.4a), i.e., by solving, for fixed $a \in (0, b)$, the initial value problem (2.11)—(2.12), (2.13), (2.13a) for $M(x)$ in $(0, a)$, together with the boundary value problem (2.1)—(2.2), (2.12a), (2.3a), (2.14) for $\Phi = \Phi_R{}^G(x)$ in $(a, b) \equiv \Omega_R$ and testing the validity of the remaining transient condition (2.14a), which gives us the eigenparameter $a = a_2{}^G$ as the least value of $a \in (0, b)$ for which the test of (2.14a) is fulfilled.

Denoting by $M_C{}^G(x)$ the corresponding solution of the initial value problem (2.11) to (2.12), (2.13), (2.13a) in $(0, a_2{}^G)$, by $\Phi_R = \Phi_R{}^G(x)$ the solution of the boundary value problem (2.1)—(2.2), (2.12a), (2.3a), (2.14) in $(a_2{}^G, b)$ and setting

$$M_2{}^G(x) = \begin{cases} M_C{}^G(x) & \text{for} \quad x \in \langle 0, a_2{}^G\rangle \\ 0 & \text{for} \quad x \in (a_2{}^G, b\rangle \end{cases} \tag{2.15}$$

$$\Phi^G_{a_2}(x) = \begin{cases} C_1 H\big(M_C{}^G(x)\big) & \text{for} \quad x \in \langle 0, a_2{}^G\rangle \\ \Phi_R{}^G(x) & \text{for} \quad x \in (a_2{}^G, b\rangle \end{cases} \tag{2.16}$$

we see that, the couple $\{a_2{}^G, M_C{}^G(x)\}$ being the solution of the nonlinear eigenparameter problem (2.11)—(2.14a), the so constructed triple $\{a_2{}^G, M_2{}^G(x), \Phi^G_{a_2}(x)\}$ is the solution of the technical control problem (2.5b), (2.6b).

Now, let us define the following sets $\neq \emptyset$ by assumption:

Definition 2.2.

$$W_F \stackrel{\text{def}}{=} \{F \in C(E_2) \mid \exists \, \{a_0{}^F, M_0{}^F(x) \in U, \Phi^F_{a_0}(x)\} \text{ solving (2.5a), (2.6a)}\}, \tag{2.17}$$

$$W_G \stackrel{\text{def}}{=} \left\{ G \in \mathscr{G} \mid \exists \, \{a_2{}^G, M_C{}^G(x)\} \quad \begin{aligned} &\text{solving (2.11)—(2.14a)} \\ &\text{and such that } M_2{}^G(x) \in U. \end{aligned} \right\} \tag{2.17a}$$

Definition 2.3.

$$G \in W_G \Rightarrow U_G \stackrel{\text{def}}{=} \{M_2{}^G(x)\}, \qquad F \in W_F \Rightarrow U_F \stackrel{\text{def}}{=} \{M_0{}^F(x)\}, \tag{2.18}$$

$$U_1 = \bigcup_{G \in W_G} U_G, \qquad U_0 = \bigcup_{F \in W_F} U_F. \tag{2.18a}$$

Further, let us make the following

Assumption 2.2. Let us have

$$G_1, G_2 \in W_G \Rightarrow U_{G_1} \cap U_{G_2} = \emptyset \qquad \text{for } G_1 \neq G_2. \tag{2.19}$$

Moreover, let us assume that also

$$F_1, F_2 \in W_F \Rightarrow U_{F_1} \cap U_{F_2} = \emptyset \qquad \text{for } F_1 \neq F_2. \tag{2.19a}$$

Thus, the following definition makes sense:

Definition 2.4. Let R, T resp. be the mappings with domain U_0, U_1 resp. and range W_F, W_G resp. defined in the following way:

$$M \in U_1 \Rightarrow T(M) = G \in W_G \qquad \text{such that } M \in U_G, \tag{2.20}$$

$$M \in U_0 \Rightarrow R(M) = F \in W_F \qquad \text{such that } M \in U_F. \tag{2.20a}$$

We shall show the validity of the following assertions [4]:

Lemma 2.1. *Let $M \in U_1$. Let $\{a_2^{TM}, M_C^{TM}(x)\}$, where T is given by* (2.20), *be the corresponding solution of* (2.11)—(2.14a).
Then, for $M_2^{TM}(x)$ given by (2.15) *we have*

$$T(M_2^{TM}) = TM. \tag{2.21}$$

Further, $\{a_2^{TM}, \Phi_{a_2}^{TM}(x)\}$ (where $\Phi_{a_2}^{TM}(x)$ is given by (2.16)) *is the unique solution of the criticality problem* (2.1)—(2.4a) *for $M = M_2^{TM}$, and $\{a_2^{TM}, M_2^{TM}(x), \Phi_{a_2}^{TM}(x)\}$ is the unique solution of the technical control problem* (2.5b), (2.6b), *i.e., we have*

$$a_2^{TM} = a_1^{TM}, \quad M_2^{TM}(x) \equiv M_1^{TM}(x), \quad \Phi_{a_2}^{TM}(x) \equiv \Phi_{a_1}^{TM}(x). \tag{2.21a}$$

Proof. By (2.20), $M \in U_1 \Rightarrow \exists!\, T(M) \in W_G$ such that $M \in U_{TM}$. By (2.18), we have $M_2^{TM} \in U_{TM}$ where $M_2^{TM} = M_2^{TM}(x)$ is given by (2.15). Therefore, by (2.20), we have $T(M_2^{TM}) = TM$, so that also $a_2^{T(M_2^{TM})} = a_2^{TM}$, $\Phi_{a_2}^{T(M_2^{TM})} = \Phi_{a_2}^{TM}$. Because $M_2^G(x) \in U$ for $\forall\, G \in W_G$ by (2.17a), we have $U_1 \subset U$, so that (2.1)—(2.4a) has for $M = M_2^{TM}$ a unique solution, which, by construction, for $\Phi_{a_2}^{TM}(x)$ given by (2.16) and (2.20), must be the couple $\{a_{a_2}^{TM}, \Phi_{a_2}^{TM}(x)\}$, $TM = T(M_2^{TM})$. In virtue of (2.10), (2.10a), to each solution $\{a_2^G, M_C^G(x)\}$, $G = TM \in W_G$ of (2.11)—(2.14a) there corresponds by (2.15), (2.16) a solution $\{a_2^G, M_2^G(x), \Phi_{a_2}^G(x)\}$ of (2.5b), (2.6b). Q.E.D.

Clearly, conditions ensuring that the triple $\{a_2^{TM}, M_2^{TM}(x), \Phi_{a_2}^{TM}(x)\}$ is also a solution of (2.5a), (2.6a) will be of interest, i.e., we have to study the equivalence problem for both the control problems (2.5a), (2.6a) and (2.5b), (2.6b):

Corollary 2.1. *Let us have*

$$M \in U_1 \cap U_0, \qquad M_0^{RM}(x) \equiv M_1^{TM}(x), \tag{2.22}$$

this implies that both the reactor control problems (2.5a), (2.6a) *with $F = RM$ and* (2.5b), (2.6b) *with $G = TM$ are equivalent, i.e., also*

$$a_0^{RM} = a_1^{TM}, \qquad \Phi_{a_0}^{RM}(x) = \Phi_{a_1}^{TM}(x). \tag{2.22a}$$

Proof. By (2.20a), $M \in U_1 \cap U_0 \Rightarrow \exists!\, R(M) \in W_F$ such that $\{a_0^{RM}, M_0^{RM}(x), \Phi_{a_0}^{RM}(x)\}$ solves (2.5a), (2.6a). Moreover, by (2.20), $M \in U_1 \cap U_0 \Rightarrow \exists!\, TM \in W_G$ such that, in virtue of Lemma 2.1, we have $M_2^{TM} \equiv M_1^{TM}$, $T(M) = T(M_2^{TM}) = T(M_1^{TM})$, the couple $\{a_1^{TM} = a_2^{TM}, \Phi_{a_1}^{TM}(x) \equiv \Phi_{a_2}^{TM}(x)\}$ with $\Phi_{a_1}^{TM}(x) \equiv \Phi_{a_2}^{TM}(x)$ given by (2.16) is the unique solution of the criticality problem (2.1)—(2.4a) for $M = M_2^{TM} \equiv M_1^{TM}$, which appears in both the control problems (2.5a), (2.6a) and (2.5b), (2.6b). Because $\{a_0^{RM}, \Phi_{a_0}^{RM}\}$ also solves uniquely (2.1)—(2.4a) for $M = M_0^{RM}$, and $M_0^{RM} \equiv M_1^{TM}$ by assumption, we must have $a_0^{RM} = a_1^{TM}$, $\Phi_{a_0}^{RM}(x) \equiv \Phi_{a_1}^{TM}(x)$. Q.E.D.

For the needs of the construction of the mapping S between F and G, let us define the following *link functional*:

Definition 2.5.

$$M \in U_1 \Rightarrow I_2^F(M) \overset{\text{def}}{=} \int_0^{a_2^{TM}} F\big(\Phi_{a_2}^{TM}(x), M_2^{TM}(x)\big)\, dx. \tag{2.23}$$

Further, let us make the following main assumption:

Assumption 2.3.

$$\emptyset \neq W_F^* \overset{\text{def}}{=} \{F \in W_F \mid \exists!\, M^* = M^*(F) \in U_1 \text{ such that } I_2^F(M^*) = \inf_{M \in U_1} I_2^F(M)\}. \tag{2.24}$$

Using Assumption 2.3 and Definitions 2.4, 2.5 we shall define constructively the mapping $S : W_F^* \to W_G$ in (2.9) as follows:

Definition 2.6. For $F \in W_F^* \neq \emptyset$ let us have

$$G = SF \Leftrightarrow G = TM^*(F) \in W_G. \tag{2.25}$$

By (2.24) and (2.25), we have transformed Problem 2.1, i.e., the problem of constructing the mapping S between the optimization function F and the control regime function $G = SF$ to the solution of the following optimization problem for the link functional $I_2^F(M)$:

Problem 2.2. To characterize the class $W_F^* \subset W_F$ of the optimization functions F for which Assumption 2.3 is valid, and to find effective algorithms for constructing $M^*(F)$ and $T: U_1 \to W_G$.

Lemma 2.2. *Let $G \in W_G$, $F \in W_F^*$. Let W_F, W_G fulfil Assumption 2.2.*

Then to each optimization function $F \in W_F^* \subset W_F$ there corresponds a unique control regime function $G = SF \in W_G$ given by (2.25), i.e.,

$$SF = TM^*(F), \tag{2.26}$$

where the mapping $F \to M^*(F) : W_F^* \to U_1$ is given by (2.24).

If we assume, moreover, that for $M^* = M^*(F)$ given by (2.24) we have also $M^* \in U_0$, $M_0^{RM^*} \in U_F$, and that the functional $I_1(M)$ given by (2.7a) has at most one minimum $I_1(M_0^F)$ on $U_0 \cap U_1$, then

$$F = RM^*(F) \quad \text{and} \quad G \equiv TM^*(F) = SRM^*(F). \tag{2.27}$$

Proof. We shall need the following

Lemma 2.3. *Let* $M \in U_1$, $F \in W_F$. *Let*

$$\{M_2^{TM} \mid M \in U_1, M_2^{TM} \quad \text{given by (2.15)}\} = U_1 \tag{2.28}$$

Then we have $I_2^F(M) = I_1(M_2^{TM})$, *i.e.,*

$$\mathscr{R}(I_1(M)|_{U_1}) = \mathscr{R}(I_2^F(M)) \tag{2.29}$$

so that the inequality

$$\inf_{M \in U} I_1(M) \leqq \inf_{M \in U_1} I_2^F(M) = \inf_{M \in U_1} I_1(M)$$

holds.

Proof. Follows from Definition 2.23 and Lemma 2.1.

Proving Lemma 2.2, we see that, by assumptions, $\emptyset \neq W_F^*$ and both the definitions 2.4 of T, R and 2.6 of S are meaningful, so that we have $SF = TM^*(F) \in W_G$ for each $F \in W_F^*$. If, moreover, $M^* = M^*(F) \in U_F \subset U_0$, then $RM^* \in W_F$ is well defined by (2.20a), and by (2.17), $\{a_0^{RM^*}, M_0^{RM^*} \in U, \Phi_{a_0}^{RM^*}\}$ solves (2.5a), (2.6a) so that, by (2.18), (2.18a) we have $M_0^{RM^*} = M_0^F$ in virtue of the additional assumptions in Lemma 2.2. Thus, by definition (2.20a) we have $RM^* = R(M_0^{RM^*}) = R(M_0^F) = F$, so that also $SF = TM^*(F) = SRM^*(F)$. Q.E.D.

Lemma 2.2 opens the question, under what conditions the optimization problems (2.24) for $I_2^F(M)$ and (2.6a) for $I_1(M)$ are the same:

Theorem 2.1. *Let* $F \in W_F^*$. *Let for* $\{a_0^F, M_0^F(x), \Phi_{a_0}^F(x)\}$ *solving* (2.5a), (2.6a) *we have* $M_0^F \in U_1$. *Let* $M_0^F = M_2^{TM_0^F}$ *where* $M_2^{TM_0^F}$ *is given by* (2.16) *for* $G = TM_0^F$. *Let the optimization functional* $I_1(M)$ *in* (2.6a) *have for* $F \in W_F^*$ *at most one minimum value on* $U_1 \subset U$.

Then both the reactor control problems (2.5a), (2.6a) *for* F *and* (2.5b), (2.6b) *for* $G = TM_0^F$ *are equivalent, i.e.,*

$$a_0^F = a_2^{TM_0^F} \quad \text{and} \quad M_0^F(x) = M_2^{TM_0^F}(x), \Phi_{a_0}^F(x) = \Phi_{a_2}^{TM_0^F}(x) \quad \text{for} \quad \forall x \in \langle 0, b \rangle. \tag{2.30}$$

Further, we have

$$G \equiv SF = TM_0^F, \qquad F = RM_0^F. \tag{2.31}$$

If we suppose, moreover, that both the operators T, R *resp. given by* (2.20), (2.20a) *resp. are injective, then*

$$F = RT^{-1} \cdot TM_0^F, \quad SF = TR^{-1} F \quad \text{so that} \quad S = TR^{-1} \quad \text{and} \quad \exists\, S^{-1} = RT^{-1}, \tag{2.32}$$

i.e., in this case $S = TR^{-1}$ *is also injective.*

Proof. Because $M_0^F \in U_1$ by assumption, we have by (2.20) and Lemma 2.1: $TM_0^F \in W_G$ and $\exists\, M_2^{TM_0^F} \in U_1$ such that $\{a_2^{TM_0^F}, \Phi_{a_2}^{TM_0^F}(x)$ given by (2.16)$\}$ is the unique solution of (2.1)—(2.4a) for $M = M_2^{TM_0^F}$. But, by $F \in W_F^*$ and (2.7a), $\{a_0^F, \Phi_{a_0}^F(x)\}$ also solves uniquely (2.1)—(2.4a) for $M = M_0^F$, and $M_0^F = M_2^{TM_0^F}$ by assumption, so that we have also $a_2^{TM_0^F} = a_0^F$, $\Phi_{a_2}^{TM_0^F}(x) \equiv \Phi_{a_0}^F(x)$. By (2.17)—(2.18a) and (2.20a) we have $M_0^F \in U_0$ and $F = R(M_0^F)$. By Lemma 2.2, to $\forall F \in W_F^*$ there corresponds the unique $G = SF \in W_G$ given by $SF = TM^*(F)$, where $M^*(F)$ is given by (2.24). Clearly, by assumption, we have $I_1(M_0^F) = I_1(M_2^{TM_0^F}) = I_2^F(M_0^F) \geqq I_2^F(M^*(F))$. But, by Lemma 2.3, we have $I_1(M_0^F) \leqq I_2^F(M^*(F))$ so that $I_1(M_0^F) = I_2^F(M^*(F))$ and $M_0^F \in U_1$, $M^*(F) \in U_1$. Therefore, by the unicity assumptions for $I_1(M)$ and $I_2(M)$ on U_1, we must have $M_0^F = M^*(F) \in U_1$, so that, by (2.20), we obtain $SF \equiv TM^*(F) = TM_0^F$. If we assume further that both T and R are injective with T^{-1}, R^{-1} as the corresponding inverses, then we have $F = RM_0^F = RT^{-1} \cdot TM_0^F = RT^{-1}SF$ and $SF = TM_0^F = TR^{-1} \cdot RM_0^F = TR^{-1}F$ so that $S = TR^{-1}$ and $\exists\, S^{-1} = RT^{-1}$. Q.E.D.

Thus, we see that the construction of the mapping S, $G = SF$, depends heavily on the construction of the mapping T defined by (2.20) and on the constructive solving of the minimizing problem for $I_2^F(M)$ in (2.24). The last of these two tasks being sufficiently known [10], we shall give some information about the first one:

Lemma 2.4. *Let $H(M) \in C_2(M(0), M(a))$ fulfil (2.10a). Let be $H(M) = H_1(M) \cdot H_2(M)$, where H_1 is injective and H_1^{-1} homogeneous.*

Then if we set

$$G(\Phi, M) = \frac{H_1^{-1}(\Phi)}{H_2(M)} \tag{2.33}$$

we have for $\forall\, C_1 > 0$:

$$\Phi = C_1 H(M) \Rightarrow G(\Phi, M) \equiv C_1 > 0 \qquad \text{so that } G(\Phi, M) \in \mathscr{G}. \tag{2.33a}$$

Proof. Substituting $\Phi = C_1H(M)$ into $G(\Phi, M) = \dfrac{H_1^{-1}(\Phi)}{H_2(M)}$ we obtain

$$\frac{H_1^{-1}(\Phi)}{H_2(M)} = \frac{H_1^{-1}(C_1H(M))}{H_2(M)} = \frac{C_1H_1^{-1} \cdot H_1H_2(M)}{H_2(M)} \equiv C_1 > 0.$$

Thus $G = G(\Phi, M)$ given by (2.33) fulfils (2.10), so that $G \in \mathscr{G}$. Q.E.D.

Theorem 2.2. *Let $M = M(x) \in U_1$ be injective, so that we have*

$$\exists\, x = x(M) \quad \text{such that} \quad x(M(x)) \equiv x, \quad M(x(M)) \equiv M. \tag{2.34}$$

Let

$$H(M) = \frac{1}{C_1} \Phi_{a(M)}(x(M)) \in C_2(M(0), M(a(M))) \tag{2.35}$$

(where $\{a(M), \Phi_{a(M)}(x)\}$ is the solution of (2.5a) for $M = M(x)$) fulfil the assumptions of Lemma 2.4. Let $G = G(\Phi, M)$ given by (2.33) fulfil the assumption

$$G(\Phi, M) \in W_G. \tag{2.36}$$

Let be

$$M_2^G = M_2^{TM} \tag{2.37}$$

(where M_2^G and M_2^{TM} are given by (2.15) and T is given by (2.20)).

Then we have

$$TM = \frac{H_1^{-1}(\Phi)}{H_2(M)} \qquad \textit{where} \qquad C_1 H_1(M)\, H_2(M) = \Phi_{a(M)}\big(x(M)\big). \tag{2.38}$$

Proof. By (2.17a)—(2.18a) $U_1 \subset U$, so that $M \in U_1 \subset U \Rightarrow \exists!\ \{a(M), \Phi_{a(M)}(x)\}$ solving (2.5a) for $M \in U_1$ considered. By the assumptions (2.34), (2.36) and Lemma 2.4, to $H(M)$ given by (2.35) there corresponds $G(\Phi, M) \in W_G \subset \mathscr{G}$ given by (2.33). But, by (2.17a), $G \in W_G \Rightarrow \{a_2{}^G, M_2{}^G, \Phi^G_{a_2}\}$ (where $M_2{}^G$ is given by (2.15) and $\Phi^G_{a_2}$ is given by (2.16)) solves (2.5b), (2.6b) and $\{a_2{}^G, \Phi^G_{a_2}\}$ solves uniquely (2.5a) for $M = M_2{}^G$. Because $M_2{}^{TM} = M_2{}^G$ by assumption and $T(M) = T(M_2{}^{TM})$ by (2.20), $T(M_2{}^G) = G$ by (2.20), we have $TM = T(M_2{}^{TM}) = T(M_2{}^G) = G$. Q.E.D.

Remark 2.2. A generalisation of the above results to more complicated reactor criticality models and control problems in other branches of applied physics is also possible [11].

Remark 2.3. There remains a lot of open problems of existence type which we have simply assumed (as physically acceptable), because our main goal was the construction of the correspondence mapping S between the optimization function F in (2.6a) and the control regime function $G = SF$ in (2.6b) [4].

References

[1] Weinberg, A. M., Wigner, E. P., The physical theory of neutron chain reactors. The Univ. of Chicago Press, 2nd ed. 1959.
[2] Zezula, R., Reactor criticality as a free boundary type problem. Report ÚVT ČVUT (to appear).
[3] Zezula, R., On a class of nonlinear eigenvalue problems arising in reactor physics. Report ÚVT ČVUT I/73. Theory of Nonlinear Operators — Proceedings of a Summer School, Hiddensee, GDR, 1972, 313—324. Akademie-Verlag, Berlin 1974.
[4] Zezula, R., On two nonlinear problems in critical reactor calculations. ÚVT ČVUT-Collection of papers 1975 (Theory of Programming, Numerical Analysis, Data Processing).
[5] Goertzel, G., Minimum critical mass and flat flux. J. Nuclear Energy 2, 193—201 (1956).
[6] Wilkins, J. E., Nucl. Sci. Eng. 1, 68 (1956).
[7] Bartošek, V., Zezula, R., Flat flux in a slab reactor with natural uranium. J. Nuclear Energy, Parts A/B, 20, 129—134 (1966).
[8] Zezula, R., A sufficient condition for flattening of the thermal neutron flux and some related problems (in onedimensional geometries). Apl. Mat. 14, 134—145 (1969).
[9] Zezula, R., Onedimensional homogenized reactor with natural Uranium and with flattened specific output. Apl. Mat. 14, 323—336 (1969).
[10] Fučík, S., Nečas, J., Souček, V., Introduction to calculus of variations (in Czech). Lecture Notes UK, SPN, Prague 1972.
[11] Zezula, R., On a correspondence principle in control theory (to appear).

Address of the author:

Mathematical Institute of Charles University
Sokolovská 83
18600 Praha 8
Czechoslovakia

Bei Fragen zur Produktsicherheit wenden Sie sich bitte an:
If you have any questions regarding product safety,
please contact:

Walter de Gruyter GmbH
Genthiner Straße 13
10785 Berlin
productsafety@degruyterbrill.com